前 言
——— PREFACE

随着全球化的不断加强,教育也在快速全球化。我国与不同国家的高校进行合作,建立中外合作办学机构,不但提升了我国教育的国际竞争力,也培养了大量高层次国际化人才。2017年全国教育工作会议工作报告中指出:"中外合作办学是实现不出国留学的重要手段,也是借鉴世界经验、引进优质资源的试验田。要下力气建设高水平示范性中外合作办学机构和项目,全面发挥中外合作办学辐射作用,深化对国内教育教学改革推动作用。"

大学物理(双语)是工程技术类各专业的一门重要的必修基础课;同时,它也是培养科学素质的一门重要课程。本课程的基本概念、基本理论和基本方法是构成科学素养的重要组成部分,对激发探索和创新精神、培养逻辑思维能力、提升整体科学素养、形成科学价值观起着其他课程无法代替的作用。

为了适应当前教育改革的需要,响应教育部"推进新工科建设与发展,开展新工科研究和实践"的号召,适应面向未来新技术和新产业发展的需要,在"长安大学教材建设项目"支持推动下,我们编写了本书。

本书的主要特色和创新点如下:

(1) 将基础物理知识与新技术应用有机融合。每项高新技术的产生和发展都与物理学的发展密不可分,本书努力将基础的物理知识与现代高新工程技术相结合。全书每章均选取一个合适的关键物理知识点,阐明它们在高新工程技术中的应用,实现理论和实际的有机融合。例如,在第1章质点运动学中,介绍完求解质点运动的两类常见问题这一知识点后,又分析了此计算方法在高速铁路建设所使用的北斗惯导小车中的应用;在第3章动量、能量和动量矩中,介绍完动量定理这一知识点后,又分析了其在火箭飞行控制中的应用。通过将物理知识与实际工程技术融合,将理论与实际应用紧密连接起来,解决学生关心的"学物理有什么用"的问题,激发学生的学习兴趣。

(2) 将基础物理知识向当今科学前沿延伸。在教育部高等学校非物理类专业物理基础课程教学指导分委员会制定的"非物理类理工学科大学物理课程教学基本要求"中提到,大学物理课程的基本内容是几十甚至几百年前就建立起来的理论体系,其中有些内容不免与现实脱节。为此,本书在每章主要内容的后面增加了延伸阅读,努力将本章的基础物理知识向当今科学前沿延伸。例如,第4章增加了"从猫下落翻身到运动生物学"阅读材料,第5章增加了"非线性振动"阅读材料。这样可以使学生尽早接触前沿科学技术的发展脉搏和现代物理的前沿课题,具有鲜明的时代特色。

(3) 将课程思政元素融入物理知识的学习中。为了响应教育部"推动课程思政全程融入课堂教学"的号召,本书精心选择和组织教材内容,将课程思政元素融入物理知识的学习中。本书在每章主要内容的后面都增加了"科学家简介"一栏,介绍与本章物理知识密切相关的国内外知名物理学家,介绍他们的学习环境、成长经历和学术成就。例如,在第1章中,通过介绍伽利略的生平,让学生感受物理学家追求真理、实事求是的科学态度,以及不向宗教迷信势力妥协的可贵品质；在第 3、4、6 章中，通过介绍钱学森、钱伟长、邓稼先等人的生平，让学生感受科学家的爱国主义情怀。同时，在每章中引入我国物理学理论的发展和技术应用成果，如我国自主研制的"北斗导航系统""光学干涉绝对重力仪"和"墨子号量子通信卫星"等，以增强学生的民族自豪感，激发学生的爱国情怀。

（4）将物理理论处理的问题实际化。本书在每章的课后习题中，除了常规的思考题和练习题外，还增加了一些接近实际情形、难度稍大的提升题，这类采用物理知识处理的问题与生活中的实际情况更接近，使物理理论指导工科实践的意义得到彰显，也有助于激发学生学习物理理论的兴趣。

本书由长安大学应用物理系教材编写组共同编写，王真、侯兆阳担任主编，沈浩、令狐佳珺、王欢、邹鹏飞、王学智担任副主编。王真编写了第 5、6、8、10 章，侯兆阳编写了第 11 章，沈浩编写了第 1、2、9 章，令狐佳珺编写了第 4、7、14 章，王欢编写了第 12 章，邹鹏飞编写了第 8 章，王学智编写了第 3、13 章。王真和侯兆阳对全书进行了校对和审定。长安大学应用物理系在线课程建设组提供了讲课视频资料。在本书的编写过程中，西安电子科技大学出版社刘小莉等编辑给予了大力协助，在此表示诚挚的谢意。

由于编者水平有限，书中难免存在不足之处，敬请广大读者批评指正。

编　者
2024 年 8 月

目 录
CONTENTS

Part 1 Mechanics

Chapter 1 Kinematics ········· 2
  1.1 Particle, reference frame and coordinate system ········· 2
    1.1.1 Particle and particle system ········· 2
    1.1.2 Reference frame and coordinate system ········· 3
  1.2 Physical quantities describing particle motion ········· 5
    1.2.1 Position vector ········· 5
    1.2.2 Displacement ········· 6
    1.2.3 Velocity ········· 7
    1.2.4 Acceleration ········· 8
    1.2.5 Two types of fundamental problems in kinematics ········· 9
    1.2.6 Application of particle kinematics in engineering technology ········· 11
  1.3 Planar curve motion in the natural coordinate system ········· 12
    1.3.1 Natural coordinate system ········· 12
    1.3.2 Planar curve motion in the natural coordinate system ········· 12
  1.4 Angular description of circular motion and the relationship between angular quantity and linear quantity ········· 14
    1.4.1 Angular displacement ········· 15
    1.4.2 Angular velocity and angular acceleration ········· 15
    1.4.3 Relationship between angular quantity and linear quantity in circular motion ········· 15
  1.5 Relative motion ········· 17
Scientist Profile ········· 19
Extended Reading ········· 19
Discussion Problems ········· 21
Problems ········· 21
Challenging Problems ········· 23

Chapter 2 Dynamics ········· 24
  2.1 Newton's laws ········· 24
    2.1.1 Newton's first law ········· 24
    2.1.2 Newton's second law ········· 25
    2.1.3 Newton's third law ········· 26

2.2　Several common forces in mechanics ……………………………………………………… 27
　2.2.1　Gravitational force and gravity ……………………………………………………… 27
　2.2.2　Elastic force …………………………………………………………………………… 28
　2.2.3　Friction ………………………………………………………………………………… 29
2.3　Application of Newton's laws ……………………………………………………………… 31
2.4　Inertial frame and non-inertial frame ……………………………………………………… 35
　2.4.1　Inertial frame and non-inertial frame ……………………………………………… 35
　2.4.2　Inertial force in a non-inertial frame ……………………………………………… 36
Scientist Profile ……………………………………………………………………………………… 37
Extended Reading …………………………………………………………………………………… 38
Discussion Problems ………………………………………………………………………………… 40
Problems ……………………………………………………………………………………………… 41
Challenging Problems ……………………………………………………………………………… 42

Chapter 3　Momentum, Energy and Angular Momentum …………………………… 44

3.1　Law of momentum for a particle and particle system …………………………………… 44
　3.1.1　Law of momentum for a particle …………………………………………………… 44
　3.1.2　Law of momentum for a particle system ………………………………………… 47
　3.1.3　Application of the law of momentum in engineering technology ……………… 49
3.2　Conservation of momentum ……………………………………………………………… 51
3.3　Work, kinetic energy and kinetic energy theorem ……………………………………… 53
　3.3.1　Definition of work …………………………………………………………………… 53
　3.3.2　Kinetic energy theorem …………………………………………………………… 56
3.4　Conservative force, work and potential energy ………………………………………… 58
　3.4.1　Characteristics of work done by several common forces ………………………… 58
　3.4.2　Potential energy ……………………………………………………………………… 61
　3.4.3　Potential energy curve ……………………………………………………………… 63
3.5　Work-energy theorem and conservation of mechanical energy ………………………… 65
　3.5.1　Work-energy theorem for a particle system ……………………………………… 65
　3.5.2　Work-energy theorem ……………………………………………………………… 66
　3.5.3　Conservation of mechanical energy ……………………………………………… 67
3.6　Conservation of energy …………………………………………………………………… 69
3.7　Angular momentum and conservation of angular momentum ………………………… 69
　3.7.1　Angular momentum of a particle relative to a certain point …………………… 70
　3.7.2　Angular momentum theorem for a particle relative to a fixed point …………… 71
Scientist Profile ……………………………………………………………………………………… 75
Extended Reading …………………………………………………………………………………… 76
Discussion Problems ………………………………………………………………………………… 77
Problems ……………………………………………………………………………………………… 78
Challenging Problems ……………………………………………………………………………… 79

Chapter 4　Fundamentals of Rigid Body Mechanics …………………………………… 81

4.1　Kinematics of rotation about a fixed axis ………………………………………………… 81

    4.1.1 Definition of a rigid body ·· 81
    4.1.2 Basic motion of a rigid body ·· 82
    4.1.3 Description of rotation about a fixed axis ·· 82
  4.2 Fundamentals of dynamics of rigid body rotation about a fixed axis ················ 84
    4.2.1 Torque ·· 84
    4.2.2 Law of rotation (Newton's second law for rotation) ··························· 86
  4.3 Calculation of rotational inertia ·· 87
  4.4 Application of the law of rotation ·· 91
  4.5 Kinetic energy and work in rotational motion ··· 93
    4.5.1 Work done by torque ·· 93
    4.5.2 Kinetic energy of rotation ·· 94
    4.5.3 Kinetic energy theorem of rigid body rotation about a fixed axis ········ 94
  4.6 Angular momentum of a rigid body and conservation of angular momentum ····· 97
    4.6.1 Angular momentum of a rigid body ·· 97
    4.6.2 Angular momentum theorem ··· 97
    4.6.3 Conservation of angular momentum ·· 98
    4.6.4 Application of conservation of angular momentum in engineering technology ········· 98
  Scientist Profile ··· 101
  Extended Reading ··· 102
  Discussion Problems ·· 104
  Problems ··· 105
  Challenging Problems ·· 107

Chapter 5 Mechanical Oscillation ·· 109

  5.1 Simple harmonic motion ·· 109
    5.1.1 Definition of simple harmonic motion ··· 109
    5.1.2 Simple harmonic motion and uniform circular motion ····················· 115
  5.2 Energy in simple harmonic motion ·· 117
  5.3 Combination of simple harmonic motions ··· 119
    5.3.1 Combination of two SHMs on the same line and with the same frequency ············ 119
    5.3.2 Combination of two SHMs on the same line and with different frequencies ············ 122
    5.3.3 Combination of two SHMs with the same frequency and perpendicular to each other ······ 123
    5.3.4 Combination of two SHMs with different frequencies and perpendicular to each other ··· 125
  5.4 Damped and forced oscillation ·· 125
    5.4.1 Damped oscillation ·· 125
    5.4.2 Forced oscillation ·· 128
    5.4.3 Resonance ·· 128
    5.4.4 Application of the resonance phenomenon in engineering technology ········ 129
  Scientist Profile ··· 131
  Extended Reading ··· 132
  Discussion Problems ·· 134
  Problems ··· 135
  Challenging Problems ·· 137

Chapter 6 Mechanical Wave ... 138

6.1 Generation and basic characteristics of a mechanical wave ... 138
- 6.1.1 Formation of a mechanical wave ... 138
- 6.1.2 Transverse and longitudinal waves ... 139
- 6.1.3 Wave line and wave surface ... 140
- 6.1.4 Characteristic physical quantities describing a wave ... 141

6.2 Wave function of a plane simple harmonic wave ... 143
- 6.2.1 Wave function of a plane simple harmonic wave ... 143
- 6.2.2 Physical meaning of the wave function ... 144

6.3 Wave energy and energy flux density ... 147
- 6.3.1 Wave energy ... 147
- 6.3.2 Average energy flux density vector ... 150
- 6.3.3 Amplitudes of plane and spherical waves ... 151

6.4 Huygens' principle ... 152

6.5 Wave interference ... 154
- 6.5.1 Principle of superposition of waves ... 154
- 6.5.2 Interference of waves ... 154

6.6 Standing wave ... 158
- 6.6.1 Standing wave experiment ... 158
- 6.6.2 Standing wave equation ... 159

6.7 The Doppler effect ... 163
- 6.7.1 The Doppler effect ... 163
- 6.7.2 Application of the Doppler effect in engineering technology ... 166
- 6.7.3 Shockwave ... 167

Scientist Profile ... 168
Extended Reading ... 169
Discussion Problems ... 171
Problems ... 172
Challenging Problems ... 174

Part 2 Thermology

Chapter 7 Kinetic Theory of Gas ... 176

7.1 Microscopic characteristics of thermal motion of a gas system ... 177
- 7.1.1 Microscopic characteristics of a gas system ... 177
- 7.1.2 Statistical law of thermal motion of gas molecules ... 178

7.2 Pressure of ideal gas ... 179
- 7.2.1 Microscopic model of ideal gas ... 179
- 7.2.2 Pressure formula of ideal gas ... 179
- 7.2.3 Statistical significance and microscopic nature of the pressure formula ... 181

7.3 Microscopic interpretation of temperature ... 182

7.4　Energy equipartition theorem ……………………………………………………………… 185
　　7.4.1　Energy equipartition theorem ………………………………………………………… 185
　　7.4.2　Internal energy of ideal gas …………………………………………………………… 187
7.5　Law of the Maxwell speed distribution ………………………………………………… 188
　　7.5.1　Function of speed distribution ………………………………………………………… 189
　　7.5.2　Law of the Maxwell speed distribution ……………………………………………… 190
　　7.5.3　Three statistical speeds of speed distribution ……………………………………… 190
7.6　Law of the Boltzmann distribution ……………………………………………………… 193
　　7.6.1　Law of the Maxwell velocity distribution …………………………………………… 193
　　7.6.2　Law of the Boltzmann distribution …………………………………………………… 193
　　7.6.3　Application of the law of the Boltzmann distribution in engineering technology ………… 195
7.7　Mean free path of a gas molecule ………………………………………………………… 196
Scientist Profile ……………………………………………………………………………………… 199
Extended Reading …………………………………………………………………………………… 200
Discussion Problems ………………………………………………………………………………… 202
Problems ……………………………………………………………………………………………… 202
Challenging Problems ……………………………………………………………………………… 203

Chapter 8　Fundamentals of Thermodynamics ……………………………………………… 205

8.1　Basic concepts of thermodynamics ……………………………………………………… 206
　　8.1.1　Equilibrium state parameters …………………………………………………………… 206
　　8.1.2　State equation of ideal gas ……………………………………………………………… 207
8.2　The first law of thermodynamics ………………………………………………………… 207
　　8.2.1　Quasi-static process ……………………………………………………………………… 207
　　8.2.2　Work and heat in a quasi-static process ……………………………………………… 208
　　8.2.3　Internal energy in a quasi-static process ……………………………………………… 209
　　8.2.4　The first law of thermodynamics ……………………………………………………… 210
8.3　Heat capacity ………………………………………………………………………………… 211
　　8.3.1　Definition of heat capacity ……………………………………………………………… 211
　　8.3.2　Molar heat capacity at constant volume ……………………………………………… 211
　　8.3.3　Molar heat capacity at constant pressure …………………………………………… 212
　　8.3.4　Relationship between molar heat capacity at constant pressure and
　　　　　　molar heat capacity at constant volume ……………………………………………… 212
8.4　Application of the first law of thermodynamics ……………………………………… 214
　　8.4.1　Isovolumetric process …………………………………………………………………… 214
　　8.4.2　Isobaric process …………………………………………………………………………… 215
　　8.4.3　Isothermal process ………………………………………………………………………… 216
　　8.4.4　Adiabatic process ………………………………………………………………………… 217
8.5　Cycle process ………………………………………………………………………………… 221
　　8.5.1　Cycle process ……………………………………………………………………………… 221
　　8.5.2　Efficiency of a heat engine ……………………………………………………………… 222
　　8.5.3　Coefficient of performance ……………………………………………………………… 222
　　8.5.4　The Carnot cycle …………………………………………………………………………… 224

8.5.5　Application of the thermodynamic cycle process in engineering technology ……… 227
8.6　The second law of thermodynamics ……………………………………………………… 229
　　8.6.1　Formulation of the second law of thermodynamics …………………………… 229
　　8.6.2　Carnot's theorem …………………………………………………………………… 231
8.7　Statistical significance of the second law of thermodynamics and
　　 the principle of entropy increase ……………………………………………………… 233
　　8.7.1　Statistical significance of the second law of thermodynamics ……………… 233
　　8.7.2　Principle of entropy increase …………………………………………………… 235
Scientist Profile ……………………………………………………………………………………… 236
Extended Reading …………………………………………………………………………………… 237
Discussion Problems ………………………………………………………………………………… 239
Problems ……………………………………………………………………………………………… 239
Challenging Problems ……………………………………………………………………………… 242

Part 1 Mechanics

Mechanics is one of the oldest and most developed disciplines in physics. It is generally believed that it originated from the statement that force produced motion by the ancient Greek scholar Aristotle in the 4th century BC. The principle of a lever was also discussed in the ancient Chinese book *Mojing*. Mechanics developed into a systematic and independent discipline when Newton published *Mathematical Principles of Natural Philosophy* in 1687. Based on the research of scientists such as Galileo, Descartes, and Huygens, Newton proposed the famous three laws of motion through analysis and summary, laying the foundation for classical mechanics. Therefore, classical mechanics is also called Newtonian mechanics.

The study of mechanics is about the mechanical motion of objects. In its long-term development, a rigorous complete theoretical and research method system has formed. Based on the observation and analysis of a large number of physical phenomena and experimental facts, physical models have been established and then inferences and predictions made through rigorous mathematical deduction and logical reasoning have been tested, modified and improved in practice. Therefore, mechanics is praised as the most perfect and universal theory. It was not until the beginning of the 20th century that mechanics was discovered to have certain limitations in the fields of microscopic and high-speed motion that were later eliminated by quantum mechanics and relativity respectively. However, in general technical fields (such as civil engineering, mechanical manufacturing and water conservancy facilities), mechanics is still an indispensable basic theory.

This part mainly introduces particle mechanics, rotation of a rigid body about a fixed axis, mechanical vibration and mechanical waves.

Chapter 1

Kinematics

The French scientist Descartes once said, "Give me matter and motion, and I will construct the world." Matter and motion are two inseparable concepts. Matter is the carrier of motion, and motion is one of the forms of matter's existence. Motion is absolute; stillness is relative.

This chapter adopts the concepts, operations, and methods of vector and calculus, mainly studies the physical quantities that describe the mechanical motion of particles and the relationship between them but does not explore the reasons for the changes in the motion quantities.

1.1 Particle, reference frame and coordinate system

1.1.1 Particle and particle system

1. Particle

Any object has a certain size and shape, and even small molecules, atoms, and other microscopic particles are no exception. Generally speaking, the change of the size and shape of the object will have an influence on the movement of the object. However, if the size and shape of the object do not play a role in the problem we are studying, or if the effect is insignificant and negligible, we can approximately regard the object as a particle which is a point with mass but without size and shape and is an ideal physical model.

Whether an object can be regarded as a particle is discretionary. For example, when the earth revolves around the sun, the average distance from the earth to the sun is about 10^4 times the radius of the earth. Therefore, the movement of all points on the earth relative to the sun can be regarded as the same. At this time, the size and shape of the earth can be ignored, and the earth can be regarded as a particle. But when studying the rotation of the earth, it will be impractical to regard it as a particle.

2. Particle system

When an object cannot be regarded as a particle, the whole object can be regarded as a system composed of many particles. We can consider a mechanical system containing two or more particles as a system of particles. Each particle in the particle system can not only be affected by the force of external objects—the external force, but also be affected by the interaction force between the particles in the particle system—the internal force. The distinction between external and internal forces depends on the selection of the particle system. For the earth-moon system composed of the earth and the moon, the gravitational force of the sun on the earth and the moon is an external force, and the gravitational force between the earth and the moon is an internal force. Solids, gases, and liquids can also be regarded as particle systems when studying some of their physical properties. By knowing the motion of each particle in the particle system, we can find out the motion of the whole object. Therefore, the study of the motion of the particle is the basis on which to study the motion of the object.

 ## 1.1.2 Reference frame and coordinate system

1. Reference frame

All matter in nature is moving all the time, and motion is an inherent property of matter and beyond human consciousness, which indicates the absoluteness of motion. Although motion is absolute, the description of the motion of an object is relative. The same object is in different motion conditions relative to different observers. For example, when a train passes a platform, the person standing on the platform sees the train moving forward, but from the perspective of the passenger sitting quietly in the carriage, the train keeps still for him or her while the platform is moving backward. Therefore, when describing the position and position changes of an object, it is always necessary to select other objects as references and then examine how the object in question moves relative to the reference object. With different reference objects, the descriptions of the motion of the object are also different. This is the relativity of motion descriptions.

The object or object system selected as the reference for describing the motion is called the reference object, and the three-dimensional space fixed with the reference object is called the reference space. A change in the position of an object is always accompanied by a change in time. Therefore, we need a clock to study the movement. The reference space and the clock connected to it constitute a reference frame. However it is customary to refer to the reference object as the reference frame, and it is not necessary to specify the reference space and clock connected to it. The motion of the same object is different with respect to different reference frames. For example, a free-falling object near the ground, with the earth as the reference frame, moves in a straight line; with a train traveling at a constant speed as the reference frame, it moves in a curve. Generally speaking, in studying the motion of an object in kinematics (not in dynamics), the reference frame should be chosen with the nature of the problem and convenience in mind. After the reference frame is selected, to quantitatively express the position of the object relative to the reference frame, an appropriate coordinate system must also be established on the reference frame.

2. Coordinate system

To quantitatively determine the position and motion state of an object relative to the reference frame, a certain coordinate system must be established on the reference frame, so that the position and motion state of the object at a certain moment can be represented by a set of coordinates. The most commonly used coordinate systems are as follows.

(a) **Cartesian coordinate system**

Select any point on a reference object as the origin, and then select three axes X, Y and Z; then the position of the mass point is determined. The unit vectors along the X, Y and Z axes are **i**, **j**, and **k** respectively, as shown in Figure 1.1.1(a).

(b) **Polar coordinate system**

The cartesian coordinate system is the most commonly used, but for some motions, such as the movement of particles under the action of centripetal force, it is desirable to use a plane polar coordinate system instead of a rectangular one. Take a point O on the plane as the pole and a ray starting from the pole as the polar axis, and a polar coordinate system results. In the system, the position of a particle is represented by two coordinates r and θ. r is the distance from the particle to the pole, called the polar diameter, and θ is the angle between the line connecting the particle with the pole and the polar axis, called the polar angle. In the polar coordinate system, the unit vectors in the radial and transverse directions are e_r and e_θ respectively, with e_r representing the direction of increasing r and e_θ the direction of increasing θ. e_r and e_θ are perpendicular to each other, but it should be noted that they are not constant vectors because they vary with positions of the particle, as shown in Figure 1.1.1(b).

(c) **Natural coordinate system**

Another coordinate system is called the natural coordinate system, which is often used when the motion trajectory of an object is known. Establish a curved "coordinate axis" along the trajectory of a particle, select any point O' on the trajectory as the "origin", and use the length S of the arc from the origin O' to the particle as the position coordinate of the particle. The direction of coordinate increase is artificially specified, and the arc length S is called natural coordinate. The two-dimensional coordinate system established here with the tangential unit vector e_τ and the normal unit vector e_n is called the natural coordinate system. For example, if a train A runs along the track and a certain point O' (such as a station) is taken as the starting point of timing on the track, then the position can be determined by the length of the orbital curve $O'A$, as shown in Figure 1.1.1(c).

In addition to the coordinate systems introduced above, also commonly used are the spherical coordinate system, cylindrical coordinate system, and so on. The motion state of an object is completely determined by the reference system and has nothing to do with the selection of the coordinate system. Different coordinate systems use different variables to describe the motion but the motion state of the corresponding object does not change.

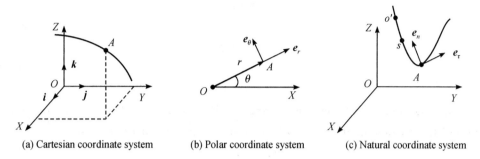

(a) Cartesian coordinate system   (b) Polar coordinate system   (c) Natural coordinate system

Figure 1.1.1   Coordinate systems

1.2   Physical quantities describing particle motion

1.2.1   Position vector

When we study the motion of a particle, we must first determine the position of the particle relative to the reference frame. As shown in Figure 1.2.1, when the coordinate system is determined, the position of the particle is determined by the vector from the coordinate origin O to the particle P, i.e. OP. This vector is called the position vector and is often represented by r.

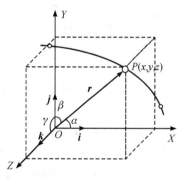

Figure 1.2.1   Position vector

In the Cartesian coordinate system, the position vector r can be expressed as

$$r = xi + yj + zk \qquad (1.2.1)$$

where i, j and k are the unit vectors along the three coordinate axes; x, y and z are the three components of the position vector r, and are scalars. From the three components of the position vector, its magnitude and direction can be obtained:

Magnitude of the position vector:

$$|r| = r = \sqrt{x^2 + y^2 + z^2}$$

Direction of the position vector:

$$\cos\alpha = \frac{x}{r}, \quad \cos\beta = \frac{y}{r}, \quad \cos\gamma = \frac{z}{r}$$

When a particle moves, its position vector r changes with time, that is to say, the position vector r is a function of time t, which means that the components x, y and z of the position vector are also functions of time. The functional formula representing the motion process is called the motion equation, which can be expressed as

$$r = r(t) \tag{1.2.2}$$

or

$$\begin{cases} x = x(t) \\ y = y(t) \\ z = z(t) \end{cases} \tag{1.2.3}$$

If time t is eliminated from the parametric Equation (1.2.3), the motion trajectory equation of the particle is obtained, which is also called the orbit equation.

$$f(x, y, z) = 0 \tag{1.2.4}$$

The equations of motion can also be expressed with other coordinates. If polar coordinates are selected, we have

$$\begin{cases} r = r(t) \\ \theta = \theta(t) \end{cases} \tag{1.2.5}$$

When the natural coordinate system is selected, we have

$$s = s(t) \tag{1.2.6}$$

1.2.2 Displacement

As shown in Figure 1.2.2, suppose a particle moves along a curved track, at time t it is at P_1, at time $t + \Delta t$ it moves to P_2, and the position vectors of $\boldsymbol{OP}_1$ and $\boldsymbol{OP}_2$ are r_1 and r_2 respectively, then the increment of position vector in the time interval Δt can be expressed as

$$\Delta r = r_2 - r_1 \tag{1.2.7}$$

We call it the displacement from position P_1 to P_2, which is a physical quantity that describes the magnitude and direction of the object's position increment.

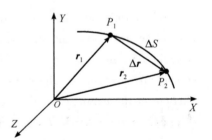

Figure 1.2.2 Displacement

In the Cartesian coordinate system, the expression for displacement is

$$\begin{aligned} \Delta r &= r_2 - r_1 = (x_2 - x_1)i + (y_2 - y_1)j + (z_2 - z_1)k \\ &= \Delta x i + \Delta y j + \Delta z k \end{aligned} \tag{1.2.8}$$

The magnitude of displacement can be calculated by

$$|\Delta r| = \sqrt{(x_2 - x_1)^2 + (y_2 - y_1)^2 + (z_2 - z_1)^2} = \sqrt{\Delta x^2 + \Delta y^2 + \Delta z^2} \tag{1.2.9}$$

The magnitude of displacement can only be denoted as $|\Delta r|$, not as Δr. Usually, Δr represents the length increment of two position vectors, i.e., $\Delta r = |r_2| - |r_1|$, while $|\Delta r|$ represents the magnitude of displacement.

It should be noted that displacement represents a change in the position of an object, not the distance traveled by the particle. As shown in Figure 1.2.2, the displacement is the directed segment, and its magnitude is the length of the secant line $|\Delta \boldsymbol{r}|$. The distance is the length ΔS of the curve $\overparen{P_1 P_2}$ which is a scalar quantity. In general, $|\Delta \boldsymbol{r}| \neq \Delta S$. Only when Δt approaches 0, do we have $|d\boldsymbol{r}| \approx dS$.

1.2.3 Velocity

To study the motion of a particle, it is necessary to know its displacement as well as how long it takes to cover this displacement, that is, its velocity.

As shown in Figure 1.2.3, within the time interval Δt, the change of the particle position causes the displacement $\Delta \boldsymbol{r}$, and then the ratio of $\Delta \boldsymbol{r}$ to Δt is called the average velocity of the particle in the time interval Δt:

$$\bar{\boldsymbol{v}} = \frac{\Delta \boldsymbol{r}}{\Delta t} \quad (1.2.10)$$

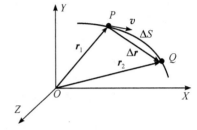

Figure 1.2.3 The velocity vector at point P is tangent to the particle's trajectory

The direction of the average velocity is the same as that of the displacement $\Delta \boldsymbol{r}$.

The average velocity is not accurate enough to describe the speed of motion. Since the motion of a particle can be fast or slow in the time interval Δt and the direction can also change continuously, the average velocity cannot reflect the real details of the motion of the particle. If we want to know the exact motion of the particle at a certain moment or a certain position, Δt should be reduced as much as possible, that is, $\Delta t \to 0$, hence the limit value of the average velocity, which is called **instantaneous velocity** or **velocity** for short. The mathematical expression is

$$\boldsymbol{v} = \lim_{\Delta t \to 0} \bar{\boldsymbol{v}} = \lim_{\Delta t \to 0} \frac{\Delta \boldsymbol{r}}{\Delta t} = \frac{d\boldsymbol{r}}{dt} \quad (1.2.11)$$

The direction of the velocity is that of the limit of the displacement $\Delta \boldsymbol{r}$ when $\Delta t \to 0$, that is, in the tangent direction of the orbit where the particle is located and toward the side where the particle moves.

In the Cartesian coordinate system, Equation (1.2.11) can be expressed as

$$\boldsymbol{v} = \frac{d\boldsymbol{r}}{dt} = \frac{dx}{dt}\boldsymbol{i} + \frac{dy}{dt}\boldsymbol{j} + \frac{dz}{dt}\boldsymbol{k} \quad (1.2.12)$$

or its component expressions

$$v_x = \frac{dx}{dt}, \quad v_y = \frac{dy}{dt}, \quad v_z = \frac{dz}{dt} \quad (1.2.13)$$

The magnitude of the velocity $|\boldsymbol{v}| = \left|\frac{d\boldsymbol{r}}{dt}\right|$ is called speed, often represented as v. In the Cartesian coordinate system, the speed v can be expressed as

$$v = |\boldsymbol{v}| = \left[\left(\frac{dx}{dt}\right)^2 + \left(\frac{dy}{dt}\right)^2 + \left(\frac{dz}{dt}\right)^2\right]^{\frac{1}{2}} \quad (1.2.14)$$

The SI units of speed and velocity are m·s^{-1}.

1.2.4 Acceleration

When a particle moves, its speed and direction may change with time, and acceleration is a physical quantity that describes how fast the speed changes.

As shown in Figure 1.2.4, within the time interval Δt, the particle moves from position P_1 to P_2, with a velocity increment of $\Delta v = v_2 - v_1$. We define the average acceleration $\bar{a}$ as

$$\bar{a} = \frac{\Delta v}{\Delta t} \tag{1.2.15}$$

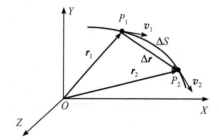
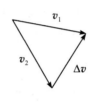

(a) Velocity vectors at point P_1 and point P_2　　　(b) Velocity increment from point P_1 to point P_2

Figure 1.2.4　Velocity increment

To accurately describe the change of particle velocity, **instantaneous acceleration**, namely **acceleration**, must be introduced.

The acceleration of a particle at a certain moment or a certain position is equal to the limit of the average acceleration when Δt tends to be zero near the moment, and its mathematical expression is

$$a = \lim_{\Delta t \to 0} \bar{a} = \lim_{\Delta t \to 0} \frac{\Delta v}{\Delta t} = \frac{dv}{dt} = \frac{d^2 r}{dt^2} \tag{1.2.16}$$

Acceleration is a vector whose direction is the same as that of the limit of Δv. The SI unit of acceleration is m·s^{-2}.

In the Cartesian coordinate system, the acceleration can be expressed as

$$a = \frac{dv}{dt} = \frac{dv_x}{dt}i + \frac{dv_y}{dt}j + \frac{dv_z}{dt}k = \frac{d^2 x}{dt^2}i + \frac{d^2 y}{dt^2}j + \frac{d^2 z}{dt^2}k \tag{1.2.17}$$

or its component expressions

$$a_x = \frac{dv_x}{dt} = \frac{d^2 x}{dt^2}, \quad a_y = \frac{dv_y}{dt} = \frac{d^2 y}{dt^2}, \quad a_z = \frac{dv_z}{dt} = \frac{d^2 z}{dt^2} \tag{1.2.18}$$

Its magnitude is

$$a = |a| = \left[\left(\frac{dv_x}{dt}\right)^2 + \left(\frac{dv_y}{dt}\right)^2 + \left(\frac{dv_z}{dt}\right)^2\right]^{\frac{1}{2}} = \left[\left(\frac{d x^2}{dt^2}\right)^2 + \left(\frac{d y^2}{dt^2}\right)^2 + \left(\frac{d z^2}{dt^2}\right)^2\right]^{\frac{1}{2}} \tag{1.2.19}$$

1.2.5 Two types of fundamental problems in kinematics

In kinematics, there are two kinds of common problems in solving particle motion.

The first type of problem: with the equation of motion $r = r(t)$ given, find the displacement, velocity, and acceleration of the particle. Such problems can be solved by derivation according to the definition.

The second type of problem: with the acceleration $a = a(t)$ or velocity $v = v(t)$ and the initial conditions given, find the position and displacement of the particle, etc. Such problems can be solved using the integral method.

Example 1.2.1 A particle moves on the XOY plane; the equations of motion are $x = 3t + 5$ and $y = \frac{1}{2}t^2 + 3t - 4$, where the unit of time is the second (s) and the units of x and y are the meter (m).

(1) Using time t as a variable, write the expression of the particle's position vector;

(2) Find the position vectors of the particle at $t = 1$ s and $t = 2$ s, and calculate the displacement of the particle within this 1 s;

(3) Write the vector expression of the particle's velocity, and calculate the velocity of the particle at $t = 4$ s;

(4) Write the vector expression of the particle's acceleration, and calculate the acceleration of the particle at $t = 4$ s.

Solution (1) The expression of the particle's position vector is

$$r = (3t+5)\boldsymbol{i} + \left(\frac{1}{2}t^2 + 3t - 4\right)\boldsymbol{j} \text{ m}$$

(2) Substitute $t = 1$ and $t = 2$ into the above equation, and we have

$$r_1 = 8\boldsymbol{i} - 0.5\boldsymbol{j} \text{ m}$$
$$r_2 = 11\boldsymbol{i} + 4\boldsymbol{j} \text{ m}$$

The displacement of the particle within this 1 s is

$$\Delta r = r_2 - r_1 = 3\boldsymbol{i} + 4.5\boldsymbol{j} \text{ m}$$

(3) According to the definition of velocity, we have

$$\boldsymbol{v} = \frac{d\boldsymbol{r}}{dt} = 3\boldsymbol{i} + (t+3)\boldsymbol{j} \text{ m} \cdot \text{s}^{-1}$$

The velocity of the particle at $t = 4$ s is

$$\boldsymbol{v}_4 = 3\boldsymbol{i} + 7\boldsymbol{j} \text{ m} \cdot \text{s}^{-1}$$

(4) According to the definition of acceleration, we have

$$\boldsymbol{a} = \frac{d\boldsymbol{v}}{dt} = \boldsymbol{j} \text{ m} \cdot \text{s}^{-2}$$

Thus, the acceleration of the particle is constant, and the acceleration of the particle at $t = 4$ s is still $\boldsymbol{j}$ m·s^{-2}.

Example 1.2.2 A running motorboat moves in a straight line after the engine is turned off. Its acceleration direction is opposite to the velocity direction, and its magnitude is proportional to the square of the speed, that is, $a = \dfrac{dv}{dt} = -kv^2$, where k is a constant. The velocity of the motorboat is v_0 when the engine is turned off. Find

(1) the relationship between the velocity of the motorboat and time;

(2) the relationship between the driving distance of the motorboat and time.

Solution For a one-dimensional problem, since the motion is along a straight line, the straight line is regarded as the X axis. The displacement, velocity and acceleration of the particle can be regarded as scalars. After we have determined the positive direction of the X axis, the positive and negative signs are sufficient to indicate the direction of the relevant quantity. We assume that the initial velocity direction is the positive direction of the X axis.

(1) Rewrite the acceleration $a = \dfrac{dv}{dt} = -kv^2$ as

$$dv = -kv^2 dt$$

Separating variables and taking the integral at both sides of the above equation, we have

$$\int_{v_0}^{v} \dfrac{dv}{v^2} = \int_{0}^{t} -k\, dt$$

Therefore

$$v = \dfrac{v_0}{1+kv_0 t}$$

(2) Rewrite the velocity $v = \dfrac{dx}{dt} = \dfrac{v_0}{1+kv_0 t}$ as

$$dx = \dfrac{v_0}{1+kv_0 t} dt$$

Taking the integral at both sides, we have

$$\int_{0}^{x} dx = \int_{0}^{t} \dfrac{v_0}{1+kv_0 t} dt$$

Thus

$$x = \dfrac{1}{k} \ln(1 + kv_0 t)$$

Example 1.2.3 When an object is suspended on a spring and moves in the vertical direction, its acceleration $a = -ky$, where k is a constant and y is the coordinate measured at any time with the equilibrium position as the origin. Assuming that the coordinate of the vibrating object is y_0 when the velocity is v_0, find the function relationship between velocity and y coordinate.

Solution Take the equilibrium position of the object as the origin of the Y axis, and take the vertical downward direction as the positive direction of the axis.

Rewrite the acceleration $a = \dfrac{dv}{dt} = -ky$ as

$$a = \frac{\mathrm{d}v}{\mathrm{d}t} = \frac{\mathrm{d}v}{\mathrm{d}t}\frac{\mathrm{d}y}{\mathrm{d}y} = v\,\frac{\mathrm{d}v}{\mathrm{d}y} = -ky$$

Separating variables and taking the integrals at both sides, we have

$$\int_{v_0}^{v} v\,\mathrm{d}v = \int_{y_0}^{y} -ky\,\mathrm{d}y$$

Then

$$v = \sqrt{v_0^2 + k(y_0^2 - y^2)}$$

1.2.6 Application of particle kinematics in engineering technology

Particle kinematics is the basis of the study of mechanical problems in physics. It is closely related to many other movements, such as the movement of atoms and other microscopic particles, molecular thermal motion and electromagnetic motion. Therefore, the establishment of particle kinematics laid the foundation for other major discoveries in physics later on. At the same time, the theory of particle kinematics has many important applications in engineering technology, such as the inertial navigation system.

The inertial navigation system is a navigation parameter solution system that uses a gyroscope and accelerometer as sensitive devices. It is an autonomous navigation system that does not rely on external information and does not radiate energy to the outside. The gyroscope is used to form a navigation coordinate system, stabilize the measurement axis of the accelerometer in this coordinate system and give the heading and attitude angle. The accelerometer is used to measure the acceleration of a moving body, then velocity is obtained by integrating it once over time and displacement is obtained by integrating the velocity once over time.

The inertial navigation system has the ability to work in all weather conditions and at all times and spaces and has high accuracy short-term navigation parameters. It is suitable for precise navigation and control of moving carriers in various environments such as in the sea, on land, in the air, underwater, and in space, and is of great military significance. With the development of electronic technology and the exploration of commercial value, the application of inertial navigation technology has expanded to civil fields such as vehicle navigation, rail transit, tunnels, fire positioning and indoor positioning. It is even widely used in drones, autonomous driving, and portable positioning terminals.

In order to maximize the smoothness of the track during high-speed railway construction, workers need to accurately obtain the three-dimensional position coordinates of the track, track spacing, etc., so as to achieve high-precision measurement of various geometric parameters such as track direction, height, gauge, and level. The traditional method of obtaining these parameters is to use a track gauge or a total station to perform semi-automatic measurement. Recently, the Chinese scientific and technological personnel have independently developed the "Beidou inertial navigation vehicle", as shown in Figure

1.2.5. This device integrates a domestic satellite navigation receiver and inertial navigation system that supports Beidou-3. By using the Beidou satellite positioning technology, it can effectively suppress the accumulation of errors in the integration process of the inertial navigation system, allowing the entire system to maintain a high-precision level for a long time. While ensuring measurement accuracy, the Beidou

Figure 1.2.5 Beidou inertial navigation vehicle

inertial navigation vehicle has increased operating efficiency by more than 20 times, greatly reducing measurement costs and field complexity.

1.3 Planar curve motion in the natural coordinate system

In this section, we will study the general curvilinear motion of a particle on a plane, for which the natural coordinate system is relatively simple, and the circular motion is only a special case of it.

1.3.1 Natural coordinate system

As shown in Figure 1.3.1, when the trajectory of a particle is known, take any point O on the trajectory as the coordinate origin, and the orbit length S of the particle from the origin can be used to determine the position P of the particle at any time.

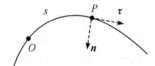

Figure 1.3.1 Natural coordinate system

The unit vectors (e_τ, e_n) in the tangential direction τ and the normal direction n of the trajectory are used as its independent coordinate direction. Such a coordinate system is called a natural coordinate system, and S is called a natural coordinate. It is more convenient to describe general curvilinear motion with natural coordinates.

1.3.2 Planar curve motion in the natural coordinate system

1. Velocity

The magnitude of the velocity is speed and its direction is in the tangent direction of the trajectory, the same as τ, and the unit vector is e_τ. Thus, the velocity can be expressed as

$$v = v e_\tau = \frac{ds}{dt} e_\tau \qquad (1.3.1)$$

2. Acceleration

Suppose that a particle moves along a curved track, as shown in Figure 1.3.2 (a). At time t, it is located at P_1 and at time $t+\Delta t$, it arrives at P_2. The velocities of the particle at P_1 and P_2 are $\boldsymbol{v}_1$ and $\boldsymbol{v}_2$, respectively. The increment of velocity in the time interval is $\Delta \boldsymbol{v}$.

Figure 1.3.2(b) shows the relationship between $\boldsymbol{v}_1$, $\boldsymbol{v}_2$ and $\Delta \boldsymbol{v}$, where $\Delta \boldsymbol{v}$ is the vector $\boldsymbol{BC}$. Intercept $|\boldsymbol{AD}|=|\boldsymbol{AB}|=|\boldsymbol{v}_1|$ from the vector $\boldsymbol{AC}$, and then the remaining part can be expressed as $|\boldsymbol{DC}|=|\boldsymbol{AC}|-|\boldsymbol{AB}|=|\boldsymbol{v}_2|-|\boldsymbol{v}_1|$, marked as $\Delta \boldsymbol{v}_\tau$. Connect BD and mark its vector as $\Delta \boldsymbol{v}_n$. Thus, the velocity increment $\Delta \boldsymbol{v}$ is divided into two parts, namely $\Delta \boldsymbol{v} = \Delta \boldsymbol{v}_n + \Delta \boldsymbol{v}_\tau$. According to the definition of acceleration, we have

$$\boldsymbol{a} = \lim_{\Delta t \to 0} \frac{\Delta \boldsymbol{v}}{\Delta t} = \lim_{\Delta t \to 0} \frac{\Delta \boldsymbol{v}_n}{\Delta t} + \lim_{\Delta t \to 0} \frac{\Delta \boldsymbol{v}_\tau}{\Delta t}$$

Let $\lim\limits_{\Delta t \to 0} \dfrac{\Delta \boldsymbol{v}_n}{\Delta t} = \boldsymbol{a}_n$ and $\lim\limits_{\Delta t \to 0} \dfrac{\Delta \boldsymbol{v}_\tau}{\Delta t} = \boldsymbol{a}_\tau$, and we have

$$\boldsymbol{a} = \boldsymbol{a}_n + \boldsymbol{a}_\tau \tag{1.3.2}$$

The physical meanings of $\boldsymbol{a}_n$ and $\boldsymbol{a}_\tau$ will be discussed below.

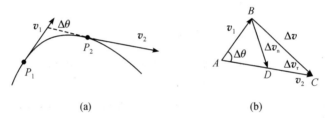

Figure 1.3.2　Tangential and normal acceleration

The direction of $\boldsymbol{a}_n$ is consistent with that of the limit of $\Delta \boldsymbol{v}_n$ when $\Delta t \to 0$. As shown in Figure 1.3.2 (b), when $\Delta t \to 0$, $\Delta \theta \to 0$; the direction of the limit of $\Delta \boldsymbol{v}_n$ is perpendicular to $\boldsymbol{v}_1$. Therefore, when the particle is located at point B, the direction of $\boldsymbol{a}_n$ is along the normal of the trajectory curve at this point. Thus, $\boldsymbol{a}_n$ is usually called the normal acceleration. The magnitude of $\boldsymbol{a}_n$ is

$$a_n = |\boldsymbol{a}_n| = \left|\lim_{\Delta t \to 0} \frac{\Delta \boldsymbol{v}_n}{\Delta t}\right|$$

From Figure 1.3.2 (b), $|\Delta \boldsymbol{v}_n| = v_1 \Delta \theta$ when $\Delta t \to 0$. Note that point B can be any point on the circumference. Remove the subscript of v_1, and we get

$$a_n = v \lim_{\Delta t \to 0} \frac{\Delta \theta}{\Delta t} = v \frac{d\theta}{dt} \tag{1.3.3}$$

Since $\dfrac{d\theta}{dt} = \dfrac{d\theta}{ds} \dfrac{ds}{dt} = v \dfrac{1}{\rho}$, where $\rho = \dfrac{ds}{d\theta}$ is the radius of curvature of the curvature circle passing through point B, Equation (1.3.3) can be written as

$$a_n = \frac{v^2}{\rho}$$

The direction of $\boldsymbol{a}_\tau$ is consistent with that of the limit of $\Delta \boldsymbol{v}_\tau$ when $\Delta t \to 0$. As shown in

Figure 1.3.2(b), when $\Delta t \to 0$, $\Delta \theta \to 0$; the limit of $\Delta \boldsymbol{v}_\tau$ will be in the tangent direction at point B. So $\boldsymbol{a}_\tau$ is called the tangential acceleration. The magnitude of $\boldsymbol{a}_\tau$ is

$$a_\tau = |\boldsymbol{a}_\tau| = \left| \lim_{\Delta t \to 0} \frac{\Delta \boldsymbol{v}_\tau}{\Delta t} \right| = \frac{dv}{dt}$$

In summary, the total acceleration of a particle moving in a curve is

$$\boldsymbol{a} = a_n \boldsymbol{e}_n + a_\tau \boldsymbol{e}_\tau = \frac{v^2}{\rho} \boldsymbol{e}_n + \frac{dv}{dt} \boldsymbol{e}_\tau \tag{1.3.4}$$

That is, the acceleration of a particle in the curve motion is equal to the vector sum of the normal acceleration and the tangential acceleration. The magnitude of acceleration is

$$a = |\boldsymbol{a}| = \sqrt{a_n^2 + a_\tau^2} = \left[\left(\frac{v^2}{\rho} \right)^2 + \left(\frac{dv}{dt} \right)^2 \right]^{\frac{1}{2}} \tag{1.3.5}$$

The direction of acceleration can be determined by the following formula

$$\tan\theta = \frac{a_n}{a_\tau}$$

For a particle in a uniform circular motion with a radius R, since the velocity only changes in direction but not in magnitude, the tangential acceleration of the particle at any time is zero and

$$\rho = R, \quad a_n = \frac{v^2}{R}$$

Therefore, the resultant acceleration of uniform circular motion is

$$\boldsymbol{a} = \boldsymbol{a}_n = a_n \boldsymbol{e}_n$$

When a particle makes a nonuniform circular motion of radius R,

$$a_\tau = \frac{dv}{dt}, \quad a_n = \frac{v^2}{R}$$

$$\boldsymbol{a} = a_n \boldsymbol{e}_n + a_\tau \boldsymbol{e}_\tau = \frac{v^2}{R} \boldsymbol{e}_n + \frac{dv}{dt} \boldsymbol{e}_\tau$$

It can be seen that the normal acceleration reflects the change of the velocity direction while the tangential acceleration reflects the change of the velocity magnitude. When a particle moves in a straight line with variable speed, due to $\rho \to \infty$, the normal acceleration of the particle at any time is zero. Hence $\boldsymbol{a} = \boldsymbol{a}_\tau = a_\tau \boldsymbol{e}_\tau$.

1.4 Angular description of circular motion and the relationship between angular quantity and linear quantity

For a particle in a circular motion, since the radius of curvature of its orbit is equal everywhere and the direction of velocity is always along the tangent of the circle, it is more convenient to use the plane polar coordinate system to describe the circular motion.

1.4.1 Angular displacement

As shown in Figure 1.4.1, a particle moves in a circle with a radius R around the center O. The center of the circle is selected as the origin of the polar coordinates, and OO' is the polar axis. When the particle moves along the circle, the polar radius is a constant R. The angle θ formed by the polar radius and the polar axis is called the angular position. It is usually specified that the angle θ obtained counterclockwise from the polar axis is positive; otherwise it is negative, so the angular position θ is a scalar. The position of the particle at any time t can be completely determined by the angular position θ. At this time, θ is a function of t, which can be expressed as

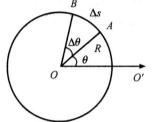

Figure 1.4.1 Angular representation of circular motion

$$\theta = \theta(t) \qquad (1.4.1)$$

That is the kinematic equation expressed by the angular position of the particle in circular motion.

As shown in Figure 1.4.1, the particle is located at point A at time t, and its angular position is θ. After a time interval Δt, the particle moves to point B, and the angular position θ increases by $\Delta\theta$. $\Delta\theta$ can uniquely describe the position change experienced by the particle and is called angular displacement. For the plane circular motion of the particle, there are only two possible directions for its angular displacement. Generally, it is stipulated that the counterclockwise angular displacement is a positive value; the clockwise angular displacement is a negative value. In the International System of Units, the unit of angular displacement is the radian(rad).

1.4.2 Angular velocity and angular acceleration

As with introducing velocity and acceleration, we can also introduce angular velocity and angular acceleration, namely

$$\omega = \lim_{\Delta t \to 0} \frac{\Delta\theta}{\Delta t} = \frac{d\theta}{dt} \qquad (1.4.2)$$

$$\beta = \lim_{\Delta t \to 0} \frac{\Delta\omega}{\Delta t} = \frac{d\omega}{dt} \qquad (1.4.3)$$

In the International System of Units, the units of angular velocity and angular acceleration are radian per second ($rad \cdot s^{-1}$) and radian per second squared ($rad \cdot s^{-2}$), respectively.

1.4.3 Relationship between angular quantity and linear quantity in circular motion

When a particle moves in a circle, it can be described by either linear or angular quantities. There is a certain relationship between them.

As shown in Figure 1.4.1, there is a relationship between arc length Δs and angle $\Delta\theta$: $\Delta s = R\Delta\theta$ and when $\Delta\theta \to 0$,

$$ds = R\, d\theta \tag{1.4.4}$$

According to this expression, it is not difficult to prove that in a circular motion, there is the following relationship between linear and angular quantities

$$\begin{cases} v = \dfrac{ds}{dt} = R\dfrac{d\theta}{dt} = R\omega \\ a_n = \dfrac{v^2}{R} = R\omega^2 \\ a_\tau = \dfrac{dv}{dt} = R\dfrac{d\omega}{dt} = R\beta \end{cases} \tag{1.4.5}$$

Similar to a particle's linear motion with a constant acceleration, such formulae exist in the uniformly accelerated circular motion with the angular acceleration β constant

$$\begin{cases} \omega = \omega_0 + \beta t \\ \theta = \theta_0 + \omega_0 t + \dfrac{1}{2}\beta t^2 \\ \omega^2 - \omega_0^2 = 2\beta(\theta - \theta_0) = 2\beta\Delta\theta \end{cases} \tag{1.4.6}$$

where θ_0 and ω_0 are the initial angular position and initial angular velocity, respectively.

Example 1.4.1 Suppose a particle moves in a circle with a radius of 1 m, and its angular position (expressed in radians) can be expressed by the equation $\theta = 2 + t^3$.

(1) When $t = 2$ s, what are its normal acceleration and tangential acceleration?

(2) At what moment do the tangential and normal accelerations have the same value?

Solution From the kinematic equation $\theta = 2 + t^3$, the angular velocity of the particle can be obtained as

$$\omega = \frac{d\theta}{dt} = 3t^2$$

The angular acceleration is

$$\beta = \frac{d\omega}{dt} = 6t$$

(1) According to the relationship between the angle quantity and the line quantity, we can get

$$a_\tau = R\beta = 6t, \quad a_n = R\omega^2 = 9t^4$$

When $t = 2$ s, we can get

$$a_\tau = 12 \text{ m}\cdot\text{s}^{-2}, \quad a_n = 144 \text{ m}\cdot\text{s}^{-2}$$

(2) Let $a_\tau = a_n$ at time t. Since

$$a_\tau = R\beta = 6t, \quad a_n = R\omega^2 = 9t^4$$

then

$$t = \sqrt[3]{\frac{2}{3}} \text{ s}$$

1.5 Relative motion

The description of motion is relative. When we choose different reference frames, the description of the motion of the same object will be different. For example, in studying the free-falling motion of a small ball in a train carriage in uniform motion, if the ground is selected as the reference system, the small ball will perform a parabolic motion; if the train itself is selected as the reference system, the small ball will move in a straight line. The description of the motion of the ball is different because the train, as the reference frame, is moving relative to the ground, as the reference frame. When studying the motion of objects near the ground, we usually regard the ground as a "stationary reference frame", and other reference frames that move relative to the ground as a "moving reference frame".

For the same object, its motion relative to a stationary reference frame is called absolute motion, and its motion relative to a moving reference frame is called relative motion. The motion of a moving reference frame relative to a stationary reference frame is called convected motion. It is rather complicated to discuss the relationship between absolute motion and relative motion of the same object. Here we only describe the situation when the moving reference frame is in translational motion relative to the stationary reference frame.

When an object moves relative to another, if a straight line arbitrarily drawn in the moving object always remains parallel to itself, we call this motion translation. All points in a translational object move at the same speed and acceleration. As shown in Figure 1.5.1, there is a stationary reference system S, on which the coordinate system $OXYZ$ is fixed; there is also a moving reference system S', on which the coordinate system $O'X'Y'Z'$ is fixed, and S' system translates with velocity v_0 relative to S system. When a particle moves in the space, its position vectors in the S system and S' system are respectively r (called absolute position vector) and r' (called relative position vector), and the position vector of the origin O' relative to O is r_0 (called the convected position vector). According to the triangle rule of vector addition, there results the following relationship:

$$r = r' + r_0 \qquad (1.5.1)$$

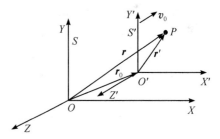

Figure 1.5.1 Relative motion

That is, the absolute position vector is equal to the vector sum of the relative position vector and the convected position vector. Taking the time derivative at both sides of Equation (1.5.1), we get

$$v = v' + v_0 \tag{1.5.2}$$

where $v = \dfrac{dr}{dt}$ is the velocity of the particle relative to the S system, called the absolute velocity, $v' = \dfrac{dr'}{dt}$ is the velocity of the particle relative to the S' system, called the relative velocity, and $v_0 = \dfrac{dr_0}{dt}$ is the velocity of the S' system relative to the S system, called the convected velocity. Differentiating both sides of Equation (1.5.2) with respect to time, we get

$$a = a' + a_0 \tag{1.5.3}$$

where $a = \dfrac{dv}{dt}$ is the acceleration of the particle relative to the S system, called the absolute acceleration, $a' = \dfrac{dv'}{dt}$ is the acceleration of the particle relative to the S' system, called the relative acceleration and $a_0 = \dfrac{dv_0}{dt}$ is the acceleration of the S' system relative to the S system, called the convected acceleration. r, v and a describe the absolute motion of the particle, r', v' and a' describe the relative motion of the particle, and r_0, v_0 and a_0 describe the convected motion of the moving reference frame relative to the stationary reference frame.

It should be pointed out that Equations (1.5.1), (1.5.2) and (1.5.3) are only applicable to the case where the moving speed of the object is much smaller than the speed of light. When the moving speed of the object can be compared with the speed of light, the above equations do not hold and must be modified considering the relativistic effect.

Example 1.5.1 A person is riding a bicycle at speed v westward, and the north wind is blowing at speed v (to the ground). What are the speed and direction of the wind encountered by the cyclist?

Solution Let the ground be the stationary reference frame E, the moving motion reference frame M be the person, and the wind be the moving object P. Then absolute velocity: $v_{PE} = v$, the direction is south; convected velocity: $v_{ME} = v$, the direction is west. The problem is transformed into how to determine the value and direction of the relative velocity v_{PM}.

$$v_{PE} = v_{PM} + v_{ME}$$

According to Figure 1.5.2,

$$|v_{ME}| = |v_{PE}| = v$$

Then

$$\angle \alpha = 45°$$

$$v_{PM} = \sqrt{v_{MP}^2 + v_{PE}^2} = \sqrt{2}\, v$$

The direction of v_{PM} is 45° south by east.

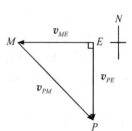

Figure 1.5.2 Figure of Example 1.5.1

Scientist Profile

Galileo Galilei (1564 – 1642) was an Italian mathematician, physicist, astronomer, and a pioneer of scientific revolution. As the inventor of the pendulum needle and the thermometer, he has made great contributions to human beings in science and is one of the founders of modern experimental science.

In history, he first integrated mathematics, physics and astronomy on the basis of scientific experiments, which expanded, deepened and changed human understanding of the movement of matter and the universe. Galileo summarized the law of free fall, the law of inertia and the principle of relativity. Thus he overthrew Aristotle's many assumptions of physics, laid the foundation of classical mechanics, refuted the Ptolemaic geocentric system, and strongly supported Copernicus' heliocentric theory.

Through systematic experiments and observations, he overthrew the traditional speculative view of nature and created a modern science based on experimental facts and with a strict logical system. Therefore, he is known as "Father of Modern Mechanics" and "Father of Modern Science".

Extended Reading

Development history of "meter" and "second"

The movement of matter occurs in space and time. To quantitatively describe the motion of matter in the reference frame, it is necessary to measure the space interval and the time interval. Therefore, the study of the motion of matter must involve the two concepts of space and time. Space and time are also the objects of physical research. Space reflects the extensiveness of matter, which is associated with the change of the size and position of objects; time reflects the continuity and persistence of physical events.

The basic unit of spatial length is the "meter", the most widely used length unit in the world whose formation has undergone nearly 300 years. In 1791, with painstaking efforts of Lagrange, chairman of the French Committee of Weights and Measures, the unit of length, which affected the world, appeared. The French authorities stipulated that the earth meridian passing through Paris, which is one 40 millionth of the meridian length, was defined as 1 meter. France promulgated the "meter system" in 1812, and in 1837, started compulsory implementation in the country, so that the meter system first took root in France. In 1872, the decision to produce the international original ruler was made at the

World Length Conference held in Paris. In 1875, the International Committee of Weights and Measures held a meeting in Paris. Representatives of 17 governments including France, Germany, the United States and Russia, jointly signed the Metric Convention and established the International Bureau of Weights and Measures. It was recognized that the "meter system" was one of the greatest scientific undertakings born in the French Revolution. "Meter" was determined as the standard international length unit, which has been used ever since.

At the end of the 19th century, experiment found that the red spectral line of natural cadmium (Cd) had very good clarity and reproducibility. In 1927, the international agreement decided to use this line as the length standard of spectroscopy, and for the first time, the non-physical standard had been used to define meter. After the 1960s, the emergence of laser led to the discovery of a more superior light source, which could make the length measurement more accurate—as long as a certain time interval was determined, the product of the speed of light and this time interval could define the length unit. In 1960, the 11th International Conference on Measurement adopted the definition of light wave meter: 1 meter was 1,650,763. 73 times the wavelength of the radiation caused by the energy level transition between 2P10 and 5d5 of a krypton-86 atom in a vacuum. In October 1983, the 17th International Metrology Conference adopted a new definition of the meter: The meter was the length of the distance traveled by light in a vacuum in a time interval of 1/299,792,458 seconds. The new definition of the meter had great scientific significance, and from then on the speed of light c became an exact number. By unifying the units of length and time, highly accurate time measurement can be used to greatly improve the accuracy of length measurement.

The basic unit of time is the "second", originally defined as 1/86,400 of the average solar day. The precise definition of the "mean solar day" was established by astronomers, but measurements showed that mean solar days could not guarantee necessary accuracy. In order to define the unit of time more precisely, in 1960 the 11th International Conference on Measurement approved a definition based on the tropical year: The second was 1/31, 556,925. 974,7 of the tropical year at 12 o'clock on January 0, 1900. However, the precision of this definition still could not meet the requirements of precision metrology. At the 13th International Metrology Conference in 1967, by the atomic energy level transition measurement technology at that time, the definition of second was changed to: 1 second is the duration of 9,192,631,770 cycles of radiation corresponding to the transition of two hyfine energy levels of the ground state of a cesium-133 atom in a zero magnetic field. This timing standard makes the accuracy of time measurement reach $10^{-12} \sim 10^{-13}$.

In physics, the upper limit of time involved is about 10^{38} s, which is the lower limit for the lifetime of a proton. The time scale involved in Newtonian mechanics is approximately $10^{-3} \sim 10^{15}$ s, which is from the period of sound vibrations to the period of the sun's rotation around the center of the galaxy. Time scales in particle physics are very small, and there is a "long-lived" elementary particle called muon, whose life is only 10^{-6} s. The shortest lived particles are some resonant particles, for example Z^0 and $W^{\pm}$, whose lifetime is only about 10^{-24} s. The smallest time currently involved in physics is 5.4×10^{-44} s called Planck time, and for a time smaller than it, the concept of time may no longer apply.

Discussion Problems

1.1 What is the kinematic equation of a particle? How many forms of particle motion equation have you learned?

1.2 Answer the following questions and give examples that match your answer:

(1) Can an object have a constant speed and a changing velocity at the same time?

(2) When the velocity is zero, is the acceleration necessarily zero? When the acceleration is zero, is the velocity necessarily zero?

(3) Is it possible that the acceleration of an object keeps decreasing while the velocity keeps increasing?

(4) When an object has an acceleration with constant magnitude and direction, can its velocity direction change?

1.3 The kinematic equations of a particle are $x=x(t)$, $y=y(t)$. When calculating the particle velocity and acceleration, someone used the following way: first according to $r=\sqrt{x^2+y^2}$, he got $r(t)$. And then according to $|v|=\left|\dfrac{dr}{dt}\right|$ and $|a|=\left|\dfrac{d^2 r}{dt^2}\right|$, he calculated the magnitude of the velocity and acceleration of the particle. Do you think this approach is right? If not, what is wrong?

1.4 A particle moves in a circle, and the velocity increases uniformly with time. Do you think the magnitudes of a_n, a_τ and a all change with time? How does the angle between the total acceleration a and the velocity v change over time?

Problems

1.5 A particle moves on the XOY plane and its position (in meters) as a function of time (in seconds) is $x=t^2+1$, $y=3t+5$.

(1) Write the expression for the position vector of the particle in terms of time t;

(2) Calculate the position vector when $t=1$ s and $t=2$ s, and calculate the displacement of the particle within this 1 second;

(3) Calculate the average velocity from $t=0$ s to $t=4$ s;

(4) Find the vector expression of the particle's velocity, and calculate the velocity of the particle at $t=4$ s;

(5) Calculate the average acceleration of the particle from $t=0$ s to $t=4$ s;

(6) Write the vector expression of the particle's acceleration, and calculate the acceleration of the particle at $t=4$ s (Please express the position vector, displacement, average velocity, instantaneous velocity, average acceleration, and instantaneous acceleration as vectors in the Cartesian coordinate system).

1.6 As shown in Figure T1-1, there is a small boat on the lake and someone uses a light rope to pull the boat through a fixed pulley at a height h on the shore. Suppose the person draws the rope at a uniform speed v_0, the rope does not stretch, and the lake is still. Find the magnitude of velocity and acceleration of the boat when the distance from the shore is s.

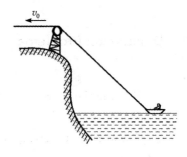

Figure T1-1 Figure of Problem 1.6

1.7 A particle's motion equations are $x=t^2$, $y=(t-1)^2$ in SI units. Try to find

(1) the trajectory equation of the particle;

(2) the displacement of the particle from $t=1$ s to $t=2$ s;

(3) the velocity and acceleration of the particle when $t=2$ s.

1.8 It is known that the kinematic equation of a particle is $r=2t\boldsymbol{i}+(2-t^2)\boldsymbol{j}$. Try to find

(1) the trajectory equation of the particle;

(2) the displacement of the particle from $t=1$ s to $t=2$ s;

(3) the velocity and acceleration of the particle when $t=2$ s.

1.9 A particle moves along a straight line with an acceleration of $a=-2x$ in SI units. Try to find the relationship between the velocity v of the particle and the position coordinate x. Let $v_0=4$ m·s^{-1} when $x=0$.

1.10 It is known that a particle moves in a straight line and its acceleration is $a=4+3t$ m·s^{-2}. At the moment it starts to move, $x=5$ m, $v=0$. Find the velocity and position of the particle at $t=10$ s.

1.11 The radius of a flywheel is 0.4 m and it is moving with an angular acceleration $\beta=0.2$ rad·s^{-2}. Find the velocity, normal acceleration, tangential acceleration, and resultant acceleration of each point on the edge at $t=2$ s.

1.12 A particle moves along a circle with a radius of 1 m and the equation of motion is $\theta=2+3t^3$, where the unit of θ is the radian and the unit of t is the second.

(1) When $t=2$ s, what are the tangential and normal accelerations of the particle?

(2) When the direction of the acceleration forms an angle of 45° with the radius, what is its angular displacement?

1.13 A particle moves along the circle with radius R according to the law of $S=v_0 t - \frac{1}{2}bt^2$, where S is the arc length from the mass point to a certain point on the circumference and v_0 and b are constant. Find

(1) the acceleration of the particle at time t;

(2) the value of t when the acceleration is numerically equal to b.

1.14 A person walks up an immobile 15m-long escalator in 90 s. When standing on the same escalator, now moving, the person is carried the same distance in 60 s. How much time would it take that person to walk up the moving escalator? Does the answer depend on the length of the escalator?

Challenging Problems

1.15 In the polar coordinate system, the equations of motion for a particle are $r = r_0 + v_0 t$, $\theta = \omega t$, where r_0, v_0 and ω are constants.

(1) What is the trajectory of the mass point?

(2) How does the velocity of the particle change with time?

1.16 A small ball is thrown from a height h with a horizontal velocity v_x, and continues to jump forward after colliding with the ground. Assuming that the ball has no friction in the horizontal direction and the ratio of the speed after the collision in the vertical direction to the speed before the collision is k ($k < 1$), which is called the rebound coefficient.

(1) What are the time and horizontal distance before the ball comes to rest?

(2) The ratio of the horizontal speed of the ball to its speed at the first rebound is a constant. For different rebound coefficients, what are the characteristics of the trajectory of the ball?

1.17 Assume that the opponent's missile strikes our side at an angle θ with the horizontal plane and with the initial velocity v_0. It is in a projectile motion (Air resistance is not considered). The speed of our missile is always v_2 and the direction is always pointing to the opponent's missile. Find

(1) the flight trajectory of our missile;

(2) the time required for our missile to hit the opponent's missile.

Chapter 2

Dynamics

Based on Newton's law, this chapter analyzes the interaction between objects to reveal the reasons for the change in particle motion state and the laws to follow. Newton's three laws of motion are the core of classical mechanics and the basic laws in particle dynamics; his law of universal gravitation unifies the motion of the universe and the motion of the earth for the first time in human history, which provides powerful support for the heliocentric theory. His theory of natural science finally breaks the shackles of religion.

2.1 Newton's laws

2.1.1 Newton's first law

Newton's first law: Any object will remain at rest or in a state of uniform linear motion until an external force makes it change this state.

Notes about Newton's first law:

(1) The first law introduces two important concepts of force and inertia.

The first law states that every object has the property of "maintaining its state of rest or moving at a uniform speed along a straight line" when it is not subjected to external force. This property is inertia. Therefore, the first law is often called the law of inertia. When the motion of an object changes, it is the result of the action of other objects, and the physical representation of this action is "force".

(2) The first law is a generalization and summary of a large number of observations and experimental facts.

The first law cannot be directly proved experimentally, because there are no "isolated" objects in the world that are completely immune to the effects of other objects. We are convinced that the first law is correct because the results derived from it are consistent with the experimental facts.

(3) The first law defines the inertial frame.

According to the first law, we can always find a reference frame in which objects (particles) remain at rest or move in a straight line with uniform velocity under the condition that there are no forces acting on them. We call such a frame of reference an inertial frame. In this sense, the first law defines the inertial frame.

To sum up, the first law is rich in content. It not only puts forward the concepts of force and inertia but also defines the inertial frame.

2.1.2 Newton's second law

Suppose a particle has a mass m and its velocity at a certain moment is v, the product mv is called the momentum of the particle.

Newton's second law can be expressed as: The net force F on a particle is proportional to the rate of change of momentum with time t, and its mathematical expression is

$$F = \frac{d(mv)}{dt} \tag{2.1.1}$$

This is the general form of Newton's second law. If m is a constant, it can be expressed as

$$F = m\frac{dv}{dt} = ma \tag{2.1.2}$$

In the case of objects moving at a much lower speed than the speed of light or in general engineering problems, m is regarded as a constant.

Notes about Newton's second law:

(1) The second law gives the quantitative relationship between the three physical quantities of force, mass, and acceleration. When the mass of the object is constant, the magnitude of the acceleration a is proportional to the magnitude of the net external force, and the direction of the acceleration is consistent with the direction of the net external force. Under the action of the same external force, objects with different mass will obtain different accelerations, and the magnitude of the acceleration is inversely proportional to the mass of the object: The greater the mass, the smaller the acceleration, and vice versa. Thus, mass is a measure of the inertia of an object.

(2) Newton's second law reflects the instantaneous change of force and momentum with time.

If the momentum of a particle changes at a certain moment, the particle must be subjected to a force at that moment. If m is constant and F is the instantaneous external force at a certain moment, a is the instantaneous acceleration at the corresponding moment. Once an external force acts on an object, the corresponding acceleration will be generated immediately. If the external force is changed, the acceleration will change accordingly. Once the external force is removed, the acceleration will disappear immediately.

(3) Newton's second law states the vector relationship between force and acceleration.

If a particle is subjected to a resultant force in a certain direction, an acceleration must be generated in this direction, which can be expressed in the following form in the Cartesian coordinate system $OXYZ$.

$$\begin{cases} F_x = ma_x = m\dfrac{dv_x}{dt} = m\dfrac{d^2x}{dt^2} \\ F_y = ma_y = m\dfrac{dv_y}{dt} = m\dfrac{d^2y}{dt^2} \\ F_z = ma_z = m\dfrac{dv_z}{dt} = m\dfrac{d^2z}{dt^2} \end{cases} \qquad (2.1.3)$$

When a particle moves in a curve on a plane, in the natural coordinate system, Newton's second law can be written as

$$\boldsymbol{F} = m\boldsymbol{a} = m(\boldsymbol{a}_\tau + \boldsymbol{a}_n) = m\dfrac{dv}{dt}\boldsymbol{e}_\tau + m\dfrac{v^2}{\rho}\boldsymbol{e}_n \qquad (2.1.4)$$

Its component expressions are

$$\boldsymbol{F}_\tau = m\boldsymbol{a}_\tau = m\dfrac{dv}{dt}\boldsymbol{e}_\tau$$

$$\boldsymbol{F}_n = m\boldsymbol{a}_n = m\dfrac{v^2}{\rho}\boldsymbol{e}_n$$

In the equations, $\boldsymbol{F}_\tau$ is called the tangential force, $\boldsymbol{F}_n$ is called the normal force (or centripetal force), and ρ is the radius of curvature.

2.1.3 Newton's third law

Neither Newton's first law nor second law explains what is the nature of force and what properties it has. Newton's third law answers these questions.

Newton's third law: Force is the action of an object on another object. If an object exerts a force $\boldsymbol{F}$ on another object, the latter simultaneously gives the former a reaction force $\boldsymbol{F}'$; $\boldsymbol{F}$ and $\boldsymbol{F}'$ are equal in magnitude and opposite in direction, acting on the same straight line. Its mathematical expression is

$$\boldsymbol{F} = -\boldsymbol{F}' \qquad (2.1.5)$$

Notes about Newton's third law:

(1) Although the action force and the reaction force are equal in magnitude and opposite in direction and act on the same straight line, their acting points are different. They are acting on two different objects respectively. Depending on the acting point of the force, the resulting effect may be different.

(2) Action force and reaction force always appear in pairs and they always arise and disappear at the same time.

(3) Action force and reaction force are of the same nature.

Chapter 2 Dynamics 27

2.2 Several common forces in mechanics

The task of particle dynamics is to study the motion of particles under the action of force. As far as we know, there are four types of interactions between objects in nature: gravitational interaction, electromagnetic interaction, strong interaction, and weak interaction. Gravitational interaction exists between all objects, and it only manifests obvious effects between massive objects (such as celestial bodies and the earth). The manifestation of this interaction is universal gravitation. Gravity is a manifestation of universal gravitation. Electromagnetic interaction generally exists in all charged bodies, and the electrostatic force and magnetic field force between charged particles are the manifestations of this interaction. The electromagnetic force is much stronger than gravity, and the electrostatic force between electrons and protons alone is 10^{39} times stronger than gravity. The strong and weak interactions act in the nucleus and manifest themselves as short-range forces. For example, the strong interaction exists between hadrons such as protons, neutrons, and muons. The force range of the strong interaction is about 10^{-15} m and its magnitude is about 10 times that of the electrostatic force. Meanwhile, the force range of the weak interaction is about 10^{-17} m, and its magnitude is about 10^{-2} that of the electrostatic force. Here several common forces will be introduced such as gravitational force, gravity, elastic force and friction.

2.2.1 Gravitational force and gravity

Newton believed that there is gravitation not only between celestial bodies, but also between other objects, which is called **gravitational force**.

There are two particles with mass m_1 and m_2 respectively. The distance between them is r. The magnitude of the universal gravitational force F between them is proportional to the mass product $m_1 m_2$, and is inversely proportional to the square of their distance r^2. The direction of F is along the line connecting them, as shown in Figure 2.2.1.

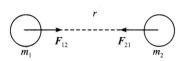

Figure 2.2.1 Schematic diagram of gravitational force

The magnitude of gravitational force can be expressed as

$$F = G \frac{m_1 m_2}{r^2} \qquad (2.2.1)$$

The gravitational force of the particle m_1 on the particle m_2 is expressed as

$$\boldsymbol{F}_{21} = -G \frac{m_1 m_2}{r^2} \boldsymbol{r}_0 \qquad (2.2.2)$$

where $G = 6.67 \times 10^{-11}$ N · m² · kg⁻² is called the gravitational constant, the positive direction of r_0 is defined as from the force-applying particle to the force-receiving particle, that is, from m_1 to m_2, and the negative sign indicates that the direction of the gravitational force F_{21} is opposite to the direction of r_0, pointing to the force-applying particle m_1. When calculating the gravitational force of m_2 acting on m_1, the positive direction of r_0 should be from m_2 to m_1.

Note: When applying the law of universal gravitation, pay attention to the definition of the positive direction of r_0. In addition, Equation (2.2.2) is only applicable to gravity between two particles.

The force exerted by the earth on an object near the ground is called the gravity on the object, expressed by G, and the magnitude of the gravity is called the weight. If the influence of the earth's rotation is ignored, the gravity of an object is approximately equal to the universal gravitational force of the earth on it, and its direction is straight downward, pointing to the center of the earth.

$$G = mg \tag{2.2.3}$$

If the object is located at a height h near the ground, then from the above two equations, we have

$$mg = G \frac{mm_e}{(r_e + h)^2}$$

In the equation, m_e is the mass of the earth and r_e is the radius of the earth. Since the object is near the ground, $r_e + h \approx r_e$ and

$$mg = G \frac{mm_e}{r_e^2}$$

Therefore

$$g = G \frac{m_e}{r_e^2}$$

Substituting the mass of the earth $m_e = 5.977 \times 10^{24}$ kg and the radius of the earth $r_e = 6370$ km into the above formula, the acceleration of gravity $g = 9.82$ m·s^{-2} can be obtained. Usually, we approximate the gravitational acceleration of objects near the ground as 9.8 m·s^{-2} in calculations.

2.2.2 Elastic force

The force produced by an object to restore its original shape due to deformation is called elastic force, and its direction depends on the deformation of the object. Several common elastic forces are described below.

1. Spring force

A spring will be deformed (elongated or compressed) under the action of external force. At the same time, the spring resists the deformation and exerts on the force-applying object a force, which is the elastic force of the spring, as shown in Figure 2.2.2.

Fix one end of a spring, and connect the other end to an object placed on a horizontal plane. Take the position of the object when the spring is not stretched or compressed as the coordinate origin O to establish a coordinate system OX, and the point O is called the equilibrium position of the object. Experiments have shown that within the elastic limit, the elastic force can be expressed as

$$F = -kx \qquad (2.2.4)$$

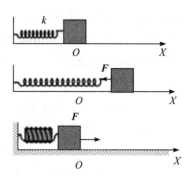

Figure 2.2.2 Spring force

In the formula, x is the displacement of the object relative to the equilibrium position (origin), and its magnitude is the elongation (or compression) amount of the spring. The proportional coefficient k is called the stiffness coefficient of the spring, which characterizes the mechanical properties of the spring. The unit is $N \cdot m^{-1}$. The above formula shows that the magnitude of the elastic force is proportional to the elongation (or compression) amount of the spring, and the direction of the elastic force is opposite to the displacement direction. This law is also known as Hooke's law.

2. Normal force

The normal force is caused by the deformation of objects that are pressed against each other, and generally the amount of deformation is small. For example, when a heavy object is placed on the table, the table is deformed by the weight, which produces an upward elastic force F_N, that is, the supporting force of the table on the heavy object. As shown in Figure 2.2.3, the object will also be deformed when it is pressed by the table top, thus generating a downward elastic force F'_N, i.e., the pressure of the heavy object on the table top. The normal force is always perpendicular to the contact surface between objects or the common tangent plane of the contact point.

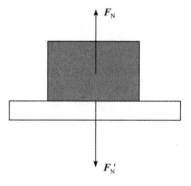

Figure 2.2.3 Normal force

3. Tension

When a soft rope is stretched and deformed by an external force, it will generate elastic force. At the same time, there is also a mutual elastic force between the inner sections of the rope. This elastic force is called tension.

2.2.3 Friction

For two objects that are in contact with each other and squeeze each other, when they have relative motion or relative motion tendency, a force that hinders relative motion will be generated on the contact surface of the two, and this force is called friction force. The frictional force is generated between objects in direct contact, and its direction is in the tangent direction of the contact surface of the two objects, and is opposite to the direction of their relative motion or relative motion tendency.

1. Static friction

Suppose an object is placed on the support surface (the ground, an inclined plane, etc.) and the external force F acts on it, so that the object forms a sliding tendency relative to the support surface but does not move, as shown in Figure 2.2.4.

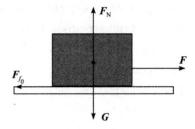

Figure 2.2.4 Static friction

At this time, there will be a friction force between the object and the supporting surface, which will balance with the external force F, so that the object remains still relative to the supporting surface. This friction force is called static friction and is denoted as F_{f_0}. The magnitude of the static friction force F_{f_0} is related to the external force F on the object. When the external force increases to a certain extent, the object will start to slide. The static friction force at this time is called the maximum static friction force, denoted as $F_{f_{max}}$. Experiments show that the maximum static friction force is proportional to the magnitude of the normal support force F_N (also known as normal pressure) between the contact surfaces, that is

$$F_{f_{max}} = \mu_0 F_N \tag{2.2.5}$$

In the formula, μ_0 is called the coefficient of static friction, which is related to the material properties, roughness, dry and wet condition and other factors of the contact surfaces of two objects, and is usually determined by experiment.

Obviously, the magnitude of the static friction is between zero and the maximum static friction, that is,

$$0 < F_{f_0} \leqslant F_{f_{max}} \tag{2.2.6}$$

2. Kinetic friction

When the external force F acting on the above objects exceeds the maximum static friction, then relative motion occurs and the friction between the two contact surfaces is called sliding friction. The direction of the sliding friction force is opposite to the direction of the relative sliding between the two objects, and the magnitude of the sliding friction force F_f is also proportional to the magnitude of the normal support force F_N, that is

$$F_f = \mu F_N \tag{2.2.7}$$

In the formula, μ is called the coefficient of sliding friction, which is usually slightly smaller than the coefficient of static friction. Sometimes, it can be approximately considered as $\mu = \mu_0$ in calculations.

3. Viscous force

When an object moves in a fluid, the fluid resists the movement of the object, and the resistance is called wet friction or viscous resistance. The reasons for and laws of fluid resistance are relatively complicated. When the velocity of the object relative to the fluid is not very large, the fluid resistance is mainly viscous resistance, which is proportional to the magnitude of the velocity, that is

$$F_f = -cv \tag{2.2.8}$$

where c is called the viscous damping coefficient.

For a spherical object, when the speed is not too large, the viscous force is

$$F_f = -6\pi r \eta v \tag{2.2.9}$$

The "$-$" sign indicates that the direction of the force is opposite to the direction of the velocity of the object. In the formula, r is the radius of the spherical object, v is the velocity of the object, and η is the viscous coefficient of the fluid. The magnitude of the fluid resistance is related to the size, shape, velocity of the object, and the properties of the object and the fluid. As the object moves faster in the fluid, the relationship between the viscous resistance and the velocity is nonlinear and complicated.

2.3 Application of Newton's laws

Newton's laws are applied in two ways: ① to determine the acceleration, velocity, and position of a particle as a function of time when given all the forces acting on the particle; ② to determine the forces acting on a particle when given the acceleration, velocity or position of the particle as a function of time. The general methods and steps of applying Newton's laws to solve problems are as follows:

(1) Select the research object. It is necessary to "isolate" the research object from other objects related to it;

(2) According to the characteristics of the problem, select an appropriate coordinate system to simplify the calculation;

(3) Check the movement situation, analyze the force, and draw the force diagram;

(4) Establish the corresponding kinetic equation;

(5) Solve the equation, and analyze and discuss the result.

Example 2.3.1 As shown in Figure 2.3.1, a light rope of length l is tied to a small ball of mass m at one end, and is tied to a fixed point O at the other end. At the beginning, the ball is at the lowest position. If the ball obtains the initial velocity v_0 as shown in the figure, the ball will make a circular motion in the vertical plane. Find the speed of the ball at any position and the tension of the string.

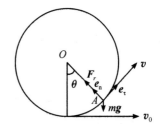

Figure 2.3.1 Figure of Example 2.3.1

Solution At $t=0$, the ball is at the lowest point, and the velocity is v_0. Assume that

at time t, the ball is at point A, the light rope forms an angle θ with the plumb line, and the velocity is v, as shown in Figure 2.3.1. At this time, the ball is subjected to gravity $m\boldsymbol{g}$ and the pulling force of the rope $\boldsymbol{F}_r$. Since the mass of the rope is ignored, the tension in the rope is equal to the pulling force of the rope on the ball. According to Newton's second law, the equation of motion of the ball is

$$\boldsymbol{F}_r + m\boldsymbol{g} = m\boldsymbol{a} \tag{1}$$

In order to list the components of the motion equation, we set up the natural coordinate system as shown in the figure. At point A, select the direction of the velocity v as the tangent direction, and its unit vector is $\boldsymbol{e}_\tau$; the direction of point A pointing to the center O is taken as the normal direction, and its unit vector is $\boldsymbol{e}_n$. Then the component expressions of Equation (1) in the tangential direction and the normal direction are respectively

$$F_r - mg\cos\theta = ma_n$$
$$-mg\sin\theta = ma_\tau$$

In non-uniform circular motion, normal acceleration $a_n = v^2/l$ and tangential acceleration $a_\tau = dv/dt$, so the above two equations can be rewritten as

$$F_r - mg\cos\theta = m\frac{v^2}{l} \tag{2}$$

$$-mg\sin\theta = m\frac{dv}{dt} \tag{3}$$

In Equation (3), we have $\dfrac{dv}{dt} = \dfrac{dv}{d\theta}\dfrac{d\theta}{dt}$. According to the definition of angular velocity ($\omega = d\theta/dt$) and the relationship between angular velocity and linear velocity ($v = l\omega$), the above equation can be further rewritten as

$$\frac{dv}{dt} = \frac{v}{l}\frac{dv}{d\theta}$$

Substituting this relation into Equation (3), we have

$$v\,dv = -gl\sin\theta\,d\theta$$

Integrating the above equation and adding initial conditions, we have

$$\int_{v_0}^{v} v\,dv = -gl\int_{0}^{\theta}\sin\theta\,d\theta$$

Therefore

$$v = \sqrt{v_0^2 + 2lg(\cos\theta - 1)} \tag{4}$$

Substituting this relation into Equation (2), we have

$$F_r = m\left(\frac{v_0^2}{l} - 2g + 3g\cos\theta\right) \tag{5}$$

It can be seen from Equation (4) that the velocity of the ball is related to the position, that is, $v = v(\theta)$. As the angle θ increases from 0 to π, the velocity of the ball decreases while as the angle θ increases from π to 2π, the velocity of the ball increases. The ball moves in a circular motion with variable speed.

It can also be seen from Equation (5) that during the process of the ball rising from the lowest point to the highest point, as the angle θ increases, the tension F_T of the rope on the ball gradually decreases, reaching the minimum at the highest point. As the ball descends from the highest point to the bottom, the tension F_T increases gradually and reaches the maximum value at the bottom.

Example 2.3.2 As shown in Figure 2.3.2, a thin rope spans the fixed pulley, and objects with mass m_1 and m_2 are suspended on one side of the rope respectively ($m_1 > m_2$). Assume that the mass of the pulley and the mass of the string are negligible. There is no slip between the pulley and the string and the friction of the axle is negligible.

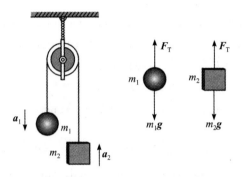

Figure 2.3.2 Figure (a) of Example 2.3.2

(1) Find the accelerations of the objects and the tension of the string after they are released;

(2) If the above device is fixed on the top of the elevator as shown in Figure 2.3.3, try to find the accelerations of the two objects relative to the elevator and the tension of the string when the elevator moves vertically upwards with acceleration a relative to the ground.

Solution (1) Select the ground as the inertial reference frame, and make a force diagram as shown in Figure 2.3.2.

Considering that the mass of the string and the pulley is negligible, the forces F_{T1} and F_{T2} on the two objects by the string should be equal to the tension F_T of the rope, that is, $F_{T1} = F_{T2} = F_T$. At the same time, the magnitude of the acceleration should be the same, that is, $a_1 = a_2 = a$, and their directions are shown in the figure. According to Newton's second law, we have

$$m_1 g - F_T = m_1 a$$
$$F_T - m_2 g = m_2 a$$

Simultaneously solving the above two equations, the magnitude of the acceleration of the two objects and the tension of the rope can be obtained as

$$a = \frac{m_1 - m_2}{m_1 + m_2} g, \quad F_T = \frac{2 m_1 m_2}{m_1 + m_2} g$$

(2) Select the ground as the inertial reference frame, and the acceleration of the elevator relative to the ground is a, as shown in Figure 2.3.3.

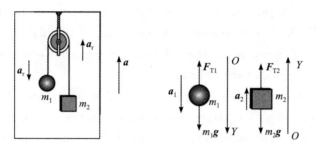

Figure 2.3.3 Figure (b) of Example 2.3.2

Let a_r be the acceleration of the object m_1 relative to the elevator; then its acceleration relative to the ground is $a_1 = a_r + a$. Considering that the mass of the string and the pulley is negligible, we have $F_{T1} = F_{T2} = F_T$. According to Newton's second law, we have

$$m_1 g + F_{T1} = m_1 a_1$$

Considering that the object m_1 moves along the Y axis and $a_1 = a_r - a$, the above equation is rewritten as

$$m_1 g - F_{T1} = m_1 g - F_T = m_1 a_1 = m_1 (a_r - a) \tag{1}$$

Since the length of the rope remains unchanged, the acceleration of the object m_2 relative to the elevator is also a_r. The acceleration of the object m_2 relative to the ground should be $a_2 = a_r + a$. The motion equation of the object m_2 is

$$F_T - m_2 g = m_2 a_2 = m_2 (a_r + a) \tag{2}$$

According to Equation (1) and Equation (2), the magnitude of the acceleration of objects m_1 and m_2 relative to the elevator can be obtained as

$$a_r = \frac{m_1 - m_2}{m_1 + m_2}(g + a) \tag{3}$$

Substituting Equation (3) into Equation (1), the tension of the light rope can be obtained as

$$F_T = \frac{2 m_1 m_2}{m_1 + m_2}(g + a)$$

Example 2.3.3 A metal ball with mass m and a radius r is released from the water surface and sinks to the bottom of the water. The buoyancy of the ball is ignored during the falling process, and the viscous resistance is $F_r = 6\pi r \eta v$. The direction of viscous resistance is opposite to the direction of motion of the object. In the formula, r is the radius of the spherical object, v is the speed of motion, and η is the viscous coefficient of the fluid. If the ball sinks vertically, try to find the sinking velocity of the ball as a function of time.

Solution At time t, when the falling velocity of the ball is v, the magnitude of the net force is $mg - F_r$, and the direction is vertically downward. Assuming that the acceleration of the ball at this time is a, according to Newton's second law, the dynamic equation of the small ball can be obtained as

$$mg - F_r = ma$$

that is

$$mg - 6\pi\eta r v = m\frac{dv}{dt}$$

Let $b = 6\pi r\eta$, and the above equation can be rewritten as

$$mg - bv = m\frac{dv}{dt} \qquad (1)$$

Therefore

$$\frac{dv}{dt} = -\frac{b}{m}\left(v - \frac{mg}{b}\right) \qquad (2)$$

The small ball is released from rest, that is, when $t = 0$, $v_0 = 0$. Separating the variables in the above equation and taking the integral, we have

$$\int_0^v \frac{dv}{v - \left(\frac{mg}{b}\right)} = -\frac{b}{m}\int_0^t dt$$

Then, we get

$$v = \frac{mg}{b}\left[1 - e^{-(b/m)t}\right] \qquad (3)$$

According to Equation (3), the falling speed of the ball increases with time; when $t \to \infty$, $e^{-(b/m)t} \to 0$, and the sinking speed reaches the maximum value $v = mg/b$. Actually, it does not take an infinite time for the sinking speed to reach the maximum value. When the net force acting on the ball is zero, that is, $mg = F_r$, the ball starts to move in a straight line at a uniform speed in the vertical direction, and this speed is the maximum speed it can reach.

2.4 Inertial frame and non-inertial frame

2.4.1 Inertial frame and non-inertial frame

In kinematics, we can choose a reference frame to study the motion of an object, as long as the selected reference frame brings convenience and simplification to the research of object motion. In dynamics, when using Newton's laws of motion to study the motion of objects, can the reference system still be chosen arbitrarily? We will discuss it with the following example.

A small ball rests on a smooth table inside a train car. When the car moves in a straight line at a constant speed relative to the ground, observers in the car see the ball at rest relative to the table, while people on the side of the road see the ball moving in a straight line at a constant speed with the car. In this case, Newton's laws of motion are applicable no matter whether the car or the ground is used as the reference system. Because the ball is not subjected to external force in the horizontal direction, it remains at rest or in a state of uniform linear motion. When the car suddenly moves with a forward acceleration a relative to the ground, for the observer in the car, the ball moves relative to the table (car) with an acceleration of $-a$. But for the observer on the ground, the ball still maintains the original state of motion against the ground, that is, the acceleration is zero. If Newton's law is applicable with the earth as the reference system, then it can be concluded that the net force on the particle is zero, that is, $F=0$. If Newton's law is applied when the car is selected as the reference frame, then it can be concluded that the net force on the particle is not zero $F=-ma$. These two conclusions are obviously contradictory. Newton's law cannot be applied to the above two reference frames at the same time, that is, the reference frame cannot be chosen arbitrarily. Therefore, we call the frame of reference to which Newton's law of motion applies an inertial frame of reference, or inertial frame for short; otherwise, it is called a non-inertial frame.

To determine whether a frame of reference is an inertial frame, we can rely only on observation and experiment. Usually, we take the earth as an inertial frame even though the earth is not a strict inertial system. The closest star to the earth is the sun, and the distance between the two is about 1.5×10^8 km. Due to the existence of the sun, the center of the earth has an acceleration of 5.9×10^{-3} m·s^{-2} relative to the sun, which is the revolution acceleration. As for the acceleration caused by the rotation of the earth, it is even greater, reaching 3.4×10^{-2} m·s^{-2}. For most experiments that do not require high precision, the acceleration effect of this rotation can still be ignored.

2.4.2 Inertial force in a non-inertial frame

Newton's laws apply only to the inertial frame, not to the reference frame accelerated relative to the inertial frame.

In practical problems, it is often necessary to deal with the motion of objects in a non-inertial system. At this time, we need to introduce the concept of inertial force. Inertial force is a virtual force, which is the force from the acceleration effect of the reference system itself in the non-inertial system. Unlike real force, inertial force cannot find the corresponding force-applying object. Its magnitude is equal to the product of the mass m of the object and the acceleration of the non-inertial system a_0, but the direction is opposite to a_0. Inertial force can be expressed as

$$F_{\text{inertial}} = -ma_0$$

In the non-inertial system, if the real force on the object is F and the inertial force is F_{inertial}, the acceleration of the object with respect to this non-inertial system a_{relative} can be the same as Newton's law in form, and its relationship can be obtained as follows:

$$F + F_{\text{inertial}} = ma_{\text{relative}}$$

The following example can be explained by introducing inertial forces. As shown in

Figure 2.4.1, the train moves with an acceleration a_0 in the positive direction of the OX axis relative to the ground reference system. For the observer in the train, the ball on the smooth table moves with an acceleration of $-a_0$ in the negative direction of the OX axis. If we assume that there is an imaginary inertial force acting on the ball with mass m, i.e., $F_{inertial} = -ma_0$, then Newton's second law can also be applied to the non-inertial reference frame of the train. That is to say, for the observer in the train with an acceleration of a_0, there is an inertial force of ma_0 in the opposite direction of a_0 acting on the ball.

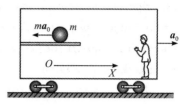

Figure 2.4.1 Inertial force

In general, if the force acting on an object contains inertial force $F_{inertial}$, the mathematical expression of Newton's second law can be expressed in a non-inertial system as

$$F + F_{inertial} = ma_{relative}$$
$$F - ma_0 = ma_{relative}$$

In the equation, a_0 is the acceleration of the non-inertial system relative to the inertial system, $a_{relative}$ is the acceleration of the object relative to the non-inertial system, and F is the net external force other than the inertial force acting on the object.

Inertial force has a wide range of applications in technology. For example, the accelerometers installed in the inertial navigation systems of missiles and ships determine the acceleration of the system by calculating the magnitude of the inertial force acting on them.

Newton's laws are the foundation of particle mechanics and the whole of classical mechanics, which provide theoretical guidance for human daily life, engineering science, space exploration and other activities. However, it should be noted that: ① Newton's laws are only valid for inertial systems; ② Newton's laws are only applicable to low-speed motion systems, that is, the speed of the object is much less than the speed of light in a vacuum; ③ The laws are only applicable to macroscopic objects, not microscopic particles, whose motion follows the laws of quantum mechanics.

 Scientist Profile

Isaac Newton(1642 – 1727) was a British physicist, mathematician, astronomer and natural philosopher, recognized as one of the greatest and most influential scientists in human history.

In 1687, Newton published a representative work *Mathematical Principles of Natural Philosophy*. The book, based on the basic concepts of mechanics (mass, momentum, inertia and force) and the basic laws (gravitation and Newton's three laws of motion) and using calculus he invented, established the complete

and rigorous system of classical mechanics, and combined the celestial mechanics and ground object mechanics for the first time in the history of physics; it provided strong theoretical support for heliocentrism and made the study of natural science finally free from the shackles of religion.

Newton also contributed to optics. He first discovered the dispersion of sunlight, and further determined the refractive index of different colors of light. On this basis he invented the reflecting telescope, laying the foundation for modern large optical astronomical telescopes. In mathematics, Newton and Leibniz each independently invented calculus, proved the generalized binomial theorem, proposed the "Newton method" to approach the zero point of a function, and contributed to the study of power series.

Extended Reading

From Newton, three-body to chaos: how does scientific cognition go from simple to complex

Since the 17th century, the classical mechanics system established on the basis of Newton's laws of motion has achieved great success in both natural science and engineering technology fields. Be it the star on the sky or the car and ship on the ground, and be it the large celestial body or the small particle, Newtonian mechanics can be widely applied. The return of Halley's comet at a predetermined time in 1757 and the discovery of Neptune at a predetermined orientation in 1846 testified to the brilliant achievements of Newtonian mechanics in astronomy. So much so that people believe that as long as the force and initial state of an object is understood, Newton's laws of motion can be used to determine the "past" and "future" of the object. If its initial conditions change slightly, the trajectory of the object will change only slightly. Newtonian mechanics is therefore known as "deterministic theory". The French mathematician Laplace put this deterministic thinking to its zenith in a famous statement: Suppose that a wise man knows at every instant all the forces that motivate nature and the relative positions of all the bodies that make it up. If this wise man is so profound that he can analyze such a multitude of data and condense into one formula the motions of the largest objects and the tiniest atoms in the universe, nothing is uncertain to him and the future unfolds before him as the past.

Since the 1960s, more and more research results have shown that in a "deterministic system" without external interference, there are also uncertain factors. Another French mathematician, Poincare, in the study of the "limited three-body problem", found that in the cross section of relative space, the motion of relatively small objects with relatively small mass is actually interwoven into an intricate spiderweb shape. This complex motion is highly unstable, and any small disturbance will cause the orbit of the object to deviate significantly after some time. Such a motion is therefore unpredictable after a period of time, because any small deviation in the initial conditions or in the calculation process can cause the calculated results to deviate significantly from the actual trajectory of the motion. Poincare's discovery shows that simple physical models can produce very complex motions, and deterministic equations can lead to unpredictable results.

The nonlinear characteristic of the system is the fundamental reason that the deterministic equations may have unpredictable results. In the 1960s, Lorentz, a famous meteorologist at the Massachusetts Institute of Technology, established a set of nonlinear differential equations when studying the atmosphere. This equation could only be solved numerically, iterating over and over again after the initial value was given. He had calculated a series of climate evolution data with a certain initial value, and when he turned on the computer again to examine the longer-term evolution of this series, in order to save himself from starting from scratch, he entered an intermediate data of the series as an initial value and then calculated it according to the same procedure. He had hoped for the same result as in the second half of the last series. Unexpectedly, however, after a short repetition, the new calculation quickly deviated from the original result, as shown in Figure ER 1. He soon realized that the computer wasn't malfunctioning, but that the problem was with the data he had entered as initial values. The computer originally stored the six-digit decimal 0.506,127, but he printed out a three-digit decimal 0.506 that he had entered this time. After further study, he found that for certain ordinary differential equations, even if the initial conditions differ by a very small value, the final results of the calculation may be very different. This phenomenon is the sensitive dependence of nonlinear dynamic systems on the initial conditions. He then likened the phenomenon to the butterfly effect: A butterfly flapping its wings in Brazil can cause a tornado in Texas. Later, people realized that this is actually a chaotic phenomenon, and the butterfly effect is a vivid description of the chaotic movement, and Lorentz himself is known as "Father of Chaos".

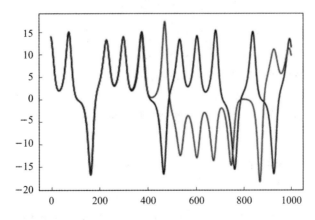

Figure ER 1　Lorentz's climate curve

Extreme sensitivity to initial values is a general and fundamental characteristic of chaotic motion. For two chaotic motions with just different initial values, their differences become larger and larger with time, and even increase exponentially with time. This rapid expansion of difference has serious consequences for the chaotic motion, because in principle, the initial value cannot be given completely accurately. As a result, under any actual given initial conditions, our prediction of the evolution of the chaotic motion will be exponentially reduced to zero. That is to say, we cannot predict the chaotic motion after a little longer time. In this way, the link between determinism and predictability is severed. Chaotic motion is deterministic, but it is unpredictable at the same time. Chaos is the chaos of determinism.

However, chaos does not simply represent a chaotic random motion, and there are some common rules for all kinds of chaotic motions. For example, all kinds of chaotic states have local instability and global stability, that is, the chaotic motion never repeats its orbit in a finite region, but it does not diverge indefinitely or tend to rest. They all have the same Feigenbaum constant, positive Lyyaplov index, positive measure entropy, fractional strange attractors, and continuous power spectrum. Chaotic attractors all have self-nested fractal structures and so on, as shown in Figure ER 2. It can be said that behind Newtonian mechanics there is a strange chaos, and in the depths of chaos there is a stranger "order".

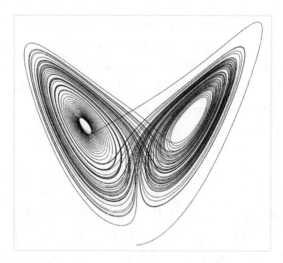

Figure ER 2 Strange attractors of the Lorentz chaotic system

Discussion Problems

2.1 Determine the features of the resultant force acting on a particle under the following conditions:

(1) The particle is in uniform linear motion;

(2) The particle is in uniform deceleration linear motion;

(3) The particle is in uniform circular motion;

(4) The particle is in uniformly accelerated circular motion.

2.2 Does friction necessarily hinder the motion of an object? Try to explain why a bicycle moves forward when a person rides it.

2.3 A metal ball is tied at one end of a rope and the other end is held by hand so that the ball is in a circular motion.

(1) When the ball is moving at the same angular speed, which one will break more easily, the long rope or the short? Why?

(2) When the ball is moving at the same linear speed, which one will break more easily, the long rope or the short? Why?

2.4 In Hooke's law $F_x = -kx$, what is the physical meaning of the minus sign?

2.5 What is the scope of application of Newton's laws?

Problems

2.6 A plate of mass 1 kg stands on the table, and another object of mass 2 kg is placed on the plate. The interface between the object and the plate and the interface between the plate and the table show the same coefficient of sliding friction $\mu_k = 0.25$ and the coefficient of static friction $\mu_s = 0.30$.

(1) If the horizontal force F is applied to pull the plate and both the object and the plate consequently move together with the acceleration of $a = 1 \text{ m} \cdot \text{s}^{-2}$, what is the interaction force between them? What is the interaction force between the plate and the table?

(2) What is the minimum force required to pull the plate out from under the object?

2.7 In Figure T2-1, the stiffness coefficients of the two springs are k_1 and k_2 respectively. Prove:

(1) when they are connected in series, the total stiffness coefficient k with k_1 and k_2 satisfies the equation $\dfrac{1}{k} = \dfrac{1}{k_1} + \dfrac{1}{k_2}$;

(2) when they are connected in parallel, the total stiffness coefficient $k = k_1 + k_2$.

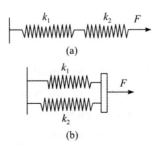

Figure T2-1 Figure of Problem 2.7

2.8 A small boat of mass m is sailing through a lake at the speed of v_0. When the sail falls, only the water resistance force $F = -kv^2$ is exerted on the boat. Find the relationship between the speed of the boat moving through the water and the time.

2.9 A small ball of mass 10 kg is tied to one end of a light rope with length l of 5 m to form a pendulum, as shown in Figure T2-2. The maximum angle from the vertical line is 60 degrees.

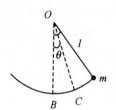

Figure T2-2 Figure of Problem 2.9

(1) What is the velocity of the ball when it passes through the vertical position? What is the tension of the rope at this point?

(2) At any position while $\theta < 60°$, what is the relation between the ball velocity v and θ? During this process, what are the acceleration of the ball and the tension in the rope?

(3) Assuming $\theta = 60°$, what are the acceleration of the ball and the tension in the rope?

2.10 The mass of the parachuter and the equipment totals m. On jumping from the tower, he opens the parachute immediately, when his speed can be roughly considered as zero. Then the air resistance is given by $R = kv^2$, which is proportional to the square of the speed. Find the parachuter's speed as a function of time t and ultimate speed v_T.

2.11 A particle of mass 16 kg moves in the XOY plane. A constant force is exerted on the particle with the force components $F_x = 6$ N and $F_y = -7$ N. When $t = 0$, $x = y = 0$, $v_x = -2$ m·s^{-1}, and $v_y = 0$. Find the position vector and velocity of the particle at $t = 2$ s.

2.12 A strip ring track of a certain height is fixed horizontally on a smooth table top with radius R, as shown in Figure T2-3. An object is sliding against the inner side of the orbit. The sliding friction coefficient between the object and the orbital plane is μ. Assuming the object passes point A with velocity v_0. Find

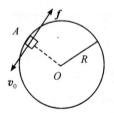

Figure T2-3 Figure of Problem 2.12

(1) the velocity of the object when sliding through distance S on the orbit after this moment;

(2) the time required for the object velocity to change from v_0 to v.

Challenging Problems

2.13 When a body moves in air, the magnitude of the resistance acting on the body can be expressed as $F = C\rho A v^2/2$, where ρ is the density of air, A is the effective cross-sectional area of the object, and C is the resistance coefficient. The limiting velocity of a parachute and a person is $v_T = 5$ m·s^{-1} and it starts to fall from static state.

(1) What are its speed and the falling height with respect to time?

(2) If the parachute does not open until the speed reaches 10 m·s^{-1}, what is its

motion pattern?

2.14 A steamboat is traveling in a straight line with velocity v_0. After the engine is off, the resistance f acting on the steamboat is in the opposite direction to the velocity v. The magnitude of the resistance is proportional to v^n with the proportionality coefficient of k_n. When n is chosen as positive numbers, find the speed of the steamboat and its distance as a function of time.

2.15 A chain of length l is placed onto a smooth table with part of the chain of length b hanging down, as shown in Figure T2 - 4. The chain starts to move from static state.

(1) What is the motion pattern of the chain?

(2) What are the time taken for the chain to slide out of the desktop and the speed at this time?

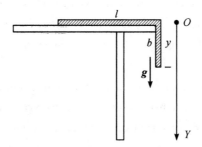

Figure T2 - 4 Figure of Problem 2.15

Chapter 3

Momentum, Energy and Angular Momentum

In mechanics, it is necessary to study not only the instantaneous effect of force, but also the cumulative effect of force on a particle or a particle system. At the same time, the mechanical characteristics of a particle's motion around a fixed point are also to be explored.

The main topics discussed in this chapter include: momentum of a particle and a particle system, the law of momentum, and the conservation of momentum; work, the work-energy relationship, and conservation of mechanical energy; angular momentum of a particle moving around a fixed point, the law of angular momentum, and the principle of conservation of angular momentum.

3.1 Law of momentum for a particle and particle system

3.1.1 Law of momentum for a particle

1. Momentum

The product of a particle's mass m and its velocity v is called the momentum of the particle, represented by the symbol $\boldsymbol{p}$.

$$\boldsymbol{p} = m\boldsymbol{v} \qquad (3.1.1)$$

Momentum $\boldsymbol{p}$ is a vector, whose direction is the same as that of the particle's velocity. Momentum is a state quantity, which means that when the motion state of the particle is determined, the momentum is also determined.

2. Impulse

From the mathematical expression of Newton's second law

$$\boldsymbol{F} = \frac{\mathrm{d}\boldsymbol{p}}{\mathrm{d}t} = \frac{\mathrm{d}}{\mathrm{d}t}(m\boldsymbol{v})$$

we have
$$\boldsymbol{F}\,\mathrm{d}t = \mathrm{d}(m\boldsymbol{v}) \tag{3.1.2}$$

This equation shows that the change in an object's momentum, $\mathrm{d}\boldsymbol{p}$, is determined by the product of the net force $\boldsymbol{F}$ acting on the object and the time $\mathrm{d}t$ during which it acts. In order to describe this cumulative effect of force over time, the product of force and the duration of its action is defined as impulse, usually represented by the symbol $\boldsymbol{I}$. Impulse is a vector, whose direction is along the change in momentum. Impulse is a process quantity, which is related not only to force, but also to the duration of its action. $\boldsymbol{F}\,\mathrm{d}t$ is called the elemental impulse of the net force $\boldsymbol{F}$ during the time $\mathrm{d}t$, represented by $\mathrm{d}\boldsymbol{I}$.

$$\mathrm{d}\boldsymbol{I} = \boldsymbol{F}\,\mathrm{d}t \tag{3.1.3}$$

If a varying force $\boldsymbol{F}$ acting on an object persists from time t_1 to time t_2, integrating the equation yields the cumulative effect of force $\boldsymbol{F}$ on the object over the time period, which can be represented as

$$\boldsymbol{I} = \int_{t_1}^{t_2} \boldsymbol{F}\,\mathrm{d}t = \int_{p_1}^{p_2} \mathrm{d}\boldsymbol{p} = \boldsymbol{p}_2 - \boldsymbol{p}_1 \tag{3.1.4}$$

where $\boldsymbol{p}_1$ is the momentum of the object at the initial time t_1, $\boldsymbol{p}_2$ is the momentum of the object at the final time t_2, and $\boldsymbol{p}_2 - \boldsymbol{p}_1$ is the increment of the object's momentum. The integral of force $\boldsymbol{F}$ with respect to time from t_1 to t_2, denoted as $\int_{t_1}^{t_2} \boldsymbol{F}\,\mathrm{d}t$, is defined as the impulse of the net force $\boldsymbol{F}$ over the time interval $\Delta t = t_2 - t_1$, represented by $\boldsymbol{I}$. Therefore, impulse is the accumulation of force over time.

If the change in momentum of a particle over a time interval $\Delta t = t_2 - t_1$ is $\Delta \boldsymbol{p}$ or the impulse is $\boldsymbol{I}$, then the average impulsive force is given by

$$\overline{\boldsymbol{F}} = \frac{\Delta \boldsymbol{p}}{\Delta t} = \frac{\boldsymbol{I}}{\Delta t} \tag{3.1.5}$$

3. Law of momentum for a particle

If the varying force $\boldsymbol{F}$ acts over a time interval $t_2 - t_1$, and $\boldsymbol{v}_1$ and $\boldsymbol{v}_2$ are the velocities of the object at time t_1 and t_2, respectively, then the impulse is

$$\int_{t_1}^{t_2} \boldsymbol{F}\,\mathrm{d}t = \int_{v_1}^{v_2} \mathrm{d}(m\boldsymbol{v}) = m\boldsymbol{v}_2 - m\boldsymbol{v}_1 \tag{3.1.6}$$

The integral on the left side $\int_{t_1}^{t_2} \boldsymbol{F}\,\mathrm{d}t$ is the impulse of the variable force $\boldsymbol{F}$ in the time interval $t_2 - t_1$, which is represented by $I = \int_{t_1}^{t_2} \boldsymbol{F}\,\mathrm{d}t$. The right side of the above equation represents the increment of the momentum of the object, which is $m\boldsymbol{v}_2 - m\boldsymbol{v}_1 = \boldsymbol{p}_2 - \boldsymbol{p}_1$. Therefore, Equation (3.1.6) can be further expressed as

$$\boldsymbol{I} = \boldsymbol{p}_2 - \boldsymbol{p}_1 \tag{3.1.7}$$

The equation above shows that the impulse of the total force acting on an object is equal to the change in the object's momentum. The relationship is known as the law of momentum for a particle and Equation (3.1.6) or (3.1.7) is the general expression of the law, also known as the integral form of the law.

The law of momentum shows that the accumulation effect of a force acting for a period of time is reflected in the change of the object's momentum during that time interval.

The law of momentum is a vector expression, and its component expressions in a Cartesian coordinate system are as follows:

$$\begin{cases} I_x = \int_{t_1}^{t_2} F_x \, dt = mv_{2x} - mv_{1x} \\ I_y = \int_{t_1}^{t_2} F_y \, dt = mv_{2y} - mv_{1y} \\ I_z = \int_{t_1}^{t_2} F_z \, dt = mv_{2z} - mv_{1z} \end{cases} \qquad (3.1.8)$$

The equations show that the impulse of a force acting on a particle during a certain time interval, when projected onto a coordinate axis, is equal to the change in momentum of the particle projected onto that axis during the same time interval. In other words, the impulse of the net force acting on a particle is equal to the change in momentum, which is called the component description of the law of momentum.

Example 3.1.1 The force F acting on a particle varies with time according to the relationship $F = 2t\boldsymbol{i} + 3t^2 \boldsymbol{j}$, where time is in seconds and force is in Newtons. Calculate the impulse of the force from 1 s to 3 s.

Solution According to the definition of impulse of variable force, we have

$$I = \int_{t_0}^{t} F \, dt = \int_{1}^{3} (2t\boldsymbol{i} + 3t^2 \boldsymbol{j}) \, dt = 8\boldsymbol{i} + 26\boldsymbol{j} \text{ N} \cdot \text{s}$$

It can be seen from the above equation that the magnitude of the impulse is $I = \sqrt{8^2 + 26^2} = \sqrt{740}$ N·s, and the angle between the direction of the impulse and the OX positive axis is $\theta = \arctan \frac{13}{4}$. This problem can also be solved by using the law of momentum.

Example 3.1.2 As shown in Figure 3.1.1, a small ball with mass $m = 0.15$ kg shoots at a smooth surface with an initial speed $v_0 = 10$ m·s^{-1}, at an incidence angle $\theta_1 = 30°$, and then bounces at an angle $\theta_2 = 60°$. Assuming the collision time $\Delta t = 0.01$ s, calculate the average impact force of the ball on the ground.

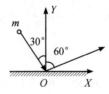

Figure 3.1.1 Figure of Example 3.1.2

Solution Taking the ball as the object of study, assuming the average impact force of the ground on the ball is $\overline{F}$, and the velocity of the ball after the collision is v, according to the law of momentum, we have

$$I_x = 0 = mv\sin\theta_2 - mv_0 \sin\theta_1$$

$$I_y = (\overline{F} - mg)\Delta t = mv\cos\theta_2 - (-mv_0 \cos\theta_1)$$

Then

$$v = v_0 \frac{\sin\theta_1}{\sin\theta_2}$$

$$\overline{F} = \frac{mv_0 \sin(\theta_1 + \theta_2)}{\Delta t \sin\theta_2} + mg$$

So

$$\overline{F} = \frac{0.15 \times 10}{0.01 \times \sqrt{3}/2} + 0.15 \times 9.8 \approx 175 \text{ N}$$

The average impact force of the ball on the ground is the reaction force of $\overline{F}$. In this problem, the effect of gravity is considered, but in fact, the gravity $mg = 0.15 \times 9.8 = 1.47$ N, which is less than 1% of F, so it can be ignored.

3.1.2 Law of momentum for a particle system

The conservation of momentum for a single particle has been discussed above. In practical problems, we often encounter systems consisting of several or many interactive particles, such as the earth and the moon, and a group of charged particles. Such a system composed of interactive particles is called a particle system. The forces of interaction between particles in the system are called the internal force of the system. The forces exerted by particles outside the system on any particle inside the system are called external force. Both the internal and external forces are relative to the system. It can be proved that the law of momentum still holds for the particle system.

First, let's discuss a system composed of two particles, as shown in Figure 3.1.2. Let the mass of the two particles be m_1 and m_2, and the external forces acting on each particle be F_1 and F_2, respectively. The internal forces between the two particles are f_{12} and f_{21}. Under the action of external and internal forces, the motion states of the two particles will change. The

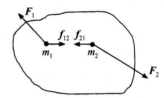

Figure 3.1.2 Internal and external forces of a particle system

acting time of force is $\Delta t = t_2 - t_1$, and the velocity of the particle with mass m_1 changes from v_{10} to v_1, while the velocity of the particle with mass m_2 changes from v_{20} to v_2.

Applying the law of momentum to the particle m_1, we have

$$\int_{t_1}^{t_2} (F_1 + f_{12}) dt = m_1 v_1 - m_1 v_{10}$$

Applying the law of momentum to the particle m_2, we have

$$\int_{t_1}^{t_2} (F_2 + f_{21}) dt = m_2 v_2 - m_2 v_{20}$$

Adding the above two equations, we get

$$\int_{t_1}^{t_2} (F_1 + F_2 + f_{12} + f_{21}) dt = (m_1 v_1 + m_2 v_2) - (m_1 v_{10} + m_2 v_{20}) \quad (3.1.9)$$

Since f_{12} and f_{21} are a pair of action force and reaction force, according to Newton's third law

$$f_{12} = -f_{21}$$

So

$$f_{12} + f_{21} = 0 \tag{3.1.10}$$

Substituting Equation (3.1.10) into Equation (3.1.9), we get

$$\int_{t_1}^{t_2} (F_1 + F_2) dt = (m_1 v_1 + m_2 v_2) - (m_1 v_{10} + m_2 v_{20}) \tag{3.1.11}$$

The left side of Equation (3.1.11) represents the impulse of the total external force acting on the system. The first term on the right side is the vector sum of the final momentum of the two particles in the system, called the final momentum of the system. The second term on the right side is the vector sum of the initial momentum of the two particles in the system, called the initial momentum of the system. The difference between these two terms represents the change in momentum of the system. Therefore, this equation states that the total impulse of the external forces acting on the system is equal to the change in momentum of the system.

This conclusion also applies to any system composed of multiple particles. Suppose there is a system composed of n particles. Since the internal forces in the system always appear in pairs, according to Newton's third law, the total of the impulses of the internal forces is always zero. Therefore, we have

$$\int_{t_1}^{t_2} \sum_{i=1}^{n} F_i \, dt = \sum_{i=1}^{n} m_i v_i - \sum_{i=1}^{n} m_i v_{i0} \tag{3.1.12}$$

The equation represents the law of momentum for a particle system, which can be stated as follows: The impulse of the external forces acting on a system is equal to the change in the total momentum of the system. This theorem shows that the impulse of external forces as a process quantity can be described by the difference between two momentums as state quantities. From the above analysis, only external forces can change the momentum of a system, while internal forces cannot.

Equation (3.1.12) is the vector expression of the law of momentum, and in the Cartesian coordinate system, its component forms are

$$\int_{t_1}^{t_2} \sum_{i=1}^{n} F_{ix} \, dt = \sum_{i=1}^{n} m_i v_{ix} - \sum_{i=1}^{n} m_i v_{i0x}$$

$$\int_{t_1}^{t_2} \sum_{i=1}^{n} F_{iy} \, dt = \sum_{i=1}^{n} m_i v_{iy} - \sum_{i=1}^{n} m_i v_{i0y} \tag{3.1.13}$$

$$\int_{t_1}^{t_2} \sum_{i=1}^{n} F_{iz} \, dt = \sum_{i=1}^{n} m_i v_{iz} - \sum_{i=1}^{n} m_i v_{i0z}$$

The above equations are the component forms of the law of momentum for a particle system, which state that the impulse of the system's total external force in a certain direction is equal to the increment of the total momentum of the system in that direction.

Chapter 3 Momentum, Energy and Angular Momentum   49

3.1.3 Application of the law of momentum in engineering technology

The law of momentum, one of the most fundamental laws of motion in physics, is the very principle behind rocket flight. In the rocket combustion chamber, the high-temperature and high-pressure gas generated by the combustion of fuel is continuously ejected from the rocket cavity to the rear, causing a forward thrust, which is the driving force that accelerates the rocket body. As the fuel continues to burn, the mass of the rocket body decreases, making the flying rocket a variable-mass object.

Suppose a rocket with mass M and velocity v is moving upward at time t. Then at time $t+\mathrm{d}t$, gas with mass $\mathrm{d}m$ is ejected at a relative velocity of u (downward) with respect to the rocket, as shown in Figure 3.1.3. The external force acting on the rocket system is $\boldsymbol{F}$. According to the law of momentum,

$$F\mathrm{d}t = (M-\mathrm{d}m)(v+\mathrm{d}v) + \mathrm{d}m(v+\mathrm{d}v-u) - Mv \tag{3.1.14}$$

Figure 3.1.3 Law of momentum in rocket launch

Assuming the rocket's motion is one-dimensional, with upward velocity as the positive direction, Equation (3.1.14) can be simplified as

$$F\mathrm{d}t = M\mathrm{d}v - u\mathrm{d}m \tag{3.1.15}$$

The second order small quantity $\mathrm{d}m\mathrm{d}v$ is omitted in Equation (3.1.15). Moving the second term on the right side of Equation (3.1.15) and dividing both sides by $\mathrm{d}t$,

$$F + u\frac{\mathrm{d}m}{\mathrm{d}t} = M\frac{\mathrm{d}v}{\mathrm{d}t} \tag{3.1.16}$$

Since $\mathrm{d}m = -\mathrm{d}M$ (where $\mathrm{d}m$ is the decrease in mass of the rocket), the equation of motion of the rocket is

$$F = M\frac{\mathrm{d}v}{\mathrm{d}t} + u\frac{\mathrm{d}M}{\mathrm{d}t} \tag{3.1.17}$$

When considering only gravity

$$F = -Mg = M\frac{\mathrm{d}v}{\mathrm{d}t} + u\frac{\mathrm{d}M}{\mathrm{d}t} \tag{3.1.18}$$

Assuming $t=0$, $v=v_0$ and $M=M_0$; integrating both sides of Equation (3.1.16)

$$\int_0^t -g\,\mathrm{d}t = \int_{v_0}^v \mathrm{d}v + u\int_{M_0}^M \frac{\mathrm{d}M}{M} \tag{3.1.19}$$

$$v = v_0 + u\ln\frac{M_0}{M} - gt \tag{3.1.20}$$

Equation (3.1.20), also known as the Tsiolkovsky rocket equation, is the motion equation for a rocket. It indicates that to increase the final flight speed of a rocket, two measures can be taken: ① increasing the value of u, which means raising the ejection velocity of the fuel; ② increasing the ratio of the total mass of the propellant and the rocket body to the mass of the rocket. In practice, there are three methods:

(1) Using high-energy propellant, i. e. , propellant with high specific impulse. But the increase in specific impulse is limited by the level of science and technology. Today, the commonly used high specific impulse chemical propellants are liquid oxygen and liquid hydrogen.

(2) Using high-strength structural materials to reduce the weight of the rocket structure. But this method is also limited by the current level of material science and technology.

(3) Increasing the mass of the rocket propellant. But simply increasing the mass of the propellant is not feasible. When the mass of the propellant increases, the volume of the storage tank also needs to increase, and the weight of the structure increases accordingly. Therefore, to achieve high speed, multi-stage rockets are generally used. Several single-stage rockets are connected together, and the first-stage rocket ignites first. When the first-stage rocket's fuel is depleted, it falls off by itself, and then the second-stage rocket starts to work, and so on. This can make the rocket achieve a high flight speed.

As early as the Song Dynasty, people tied tubes filled with gunpowder to arrow shafts. During flight, the reaction force produced by burning gunpowder made the arrow fly a longer distance. This type of reaction force-assisted arrow can be called the primitive solid rocket. In the 1880s, Swedish engineer Laval invented the Laval nozzle, which made the design of rocket engines more perfect. In 1903, the Russian Konstantin Tsiolkovsky proposed the idea and design principles of manufacturing large liquid rockets. In 1926, American rocket expert Goddard tested the first controllable liquid rocket. In 1944, Germany first used the controllable V-2 missile, which was powered by a liquid rocket engine, in war. After World War II , countries such as the Soviet Union and the United States successively developed various rocket weapons, including intercontinental ballistic missiles.

China began to develop new rockets in the 1950s. On April 24, 1970, China successfully launched its first artificial satellite, using the "Long March 1" three-stage carrier rocket. As the birthplace of rockets, China has already owned world advanced modern rocket technology and has steadily entered the international launch service market. On November 3, 2016, China's largest rocket in service, the Long March 5, was launched successfully. The total length of the rocket is 56. 97 meters, and its take off weight is about 869 tons. It has a carrying capacity of 25 tons for near-earth orbit and 14 tons for geostationary transfer orbit, which means that China has the ability to explore deeper space and has laid an important foundation for national major science and technology projects and major engineering projects, such as the third phase of lunar exploration and the first Mars exploration mission in the future.

3.2 Conservation of momentum

According to the mathematical expression of Newton's second law $\sum_{i=1}^{n} \boldsymbol{F}_i = \frac{\mathrm{d}\boldsymbol{p}}{\mathrm{d}t}$, when a system is not subjected to external force or the vector sum of external forces is zero, $\sum_{i=1}^{n} \boldsymbol{F}_i = 0$, we can get

$$\sum_{i=1}^{n} m_i \boldsymbol{v}_i = \sum_{i=1}^{n} m_i \boldsymbol{v}_{i0} = \text{constant vector} \tag{3.2.1}$$

The above equation shows that when the resultant external force on a system is always zero, the total momentum of the system remains unchanged. This is known as the law of conservation of momentum of the system.

In practical applications, the component expressions of the law of conservation of momentum in the Cartesian coordinate system are often used as follows:

$$\sum_{i=1}^{n} F_{ix} = 0, \quad \sum_{i=1}^{n} m_i v_{ix} = \sum_{i=1}^{n} m_i v_{i0x} = \text{constant}$$

$$\sum_{i=1}^{n} F_{iy} = 0, \quad \sum_{i=1}^{n} m_i v_{iy} = \sum_{i=1}^{n} m_i v_{i0y} = \text{constant}$$

$$\sum_{i=1}^{n} F_{iz} = 0, \quad \sum_{i=1}^{n} m_i v_{iz} = \sum_{i=1}^{n} m_i v_{i0z} = \text{constant}$$

The above equations show that when the component of the resultant external force on a system in a certain direction is constant, the component of the system's total momentum in that direction is conserved.

When applying the law of conservation of momentum, the following points should be noted:

(1) The condition for the conservation of momentum of the system is that the system is not subject to an external force or the resultant external force is zero. Sometimes the resultant external force on a system is not zero, but compared with the internal force of the system, the external force is much smaller. In such cases, the effect of the external force on the system can be ignored, and the momentum of the system is considered to be conserved. Problems such as explosions, collisions, and strikes can often be handled in this way.

(2) In many practical problems of mechanics, the resultant external force on a system is not equal to zero, but its component in a certain direction may be equal to zero. In such cases, although the total momentum of the system is not conserved, its component in that direction is conserved.

(3) The law of momentum and the law of conservation of momentum only hold in the

inertial frame. Since we derive the law of momentum and the law of conservation of momentum from Newton's law, they are only valid in an inertial frame. Therefore, when using them to solve a problem, an inertial frame must be selected as the reference frame, and the momentum of each object must be relative to the same inertial frame.

(4) The law of conservation of momentum is a universal law of nature. Although it is derived from Newton's law, the law should not be regarded as an inference of Newton's law. Modern scientific experiments and theoretical analysis have shown that in nature, both the interactions between celestial bodies and the interactions between microscopic particles such as protons, neutrons, and electrons obey the law of conservation of momentum. But, in the microscopic field, Newton's laws are not applicable.

The following examples illustrate the application of the law of conservation of momentum.

Example 3.2.1 As shown in Figure 3.2.1, a ball with mass m slides from rest in a quarter-circular chute with mass M. Assuming the radius of the chute is R and all friction can be neglected, find the distance the chute moves horizontally when the ball reaches the bottom of the chute.

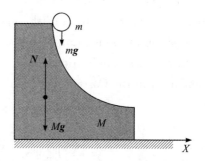

Figure 3.2.1 Figure of Example 3.2.1

Solution Take the system composed of m and M as the research object, and the external forces on the system are gravity $(M+m)\mathbf{g}$ and the supporting force of the ground on the chute $\mathbf{N}$. The directions of these external forces are in the vertical direction, and the component of the external forces on the system in the horizontal direction is zero, so momentum is conserved in the horizontal direction.

Assuming that in the sliding process, the sliding velocity of m relative to M is $\mathbf{u}$, the velocity of M relative to the ground is $\mathbf{v}$, and the horizontal right is the positive axis, then in the X direction

$$m(u_x - v) - Mv = 0$$

where u_x is the component of $\mathbf{u}$ in the X direction. Then,

$$u_x = \frac{m+M}{m} v$$

Assuming that the time m moves in the arc chute is t, and the distance it moves relative to M in the horizontal direction is R, there is

$$R = \int_0^t u_x \, dt = \frac{M+m}{m} \int_0^t v \, dt$$

So the distance the chute moves on the horizontal plane is

$$s = \int_0^t v \, dt = \frac{m}{M+m} R$$

Example 3.2.2 A homogeneous and flexible rope with mass m and length L is hung vertically so that its lower end just touches the ground, as shown in Figure 3.2.2. If the upper end of the rope is released from a stationary state, allowing it to freely fall to the ground, find the force exerted by the ground on the rope when it has fallen a length l.

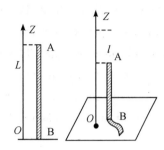

Figure 3.2.2 Figure of Example 3.2.2

Solution Take the ground as the origin of the coordinates, and the Z axis to be the vertical direction.

When the rope falls for a length of l, the ungrounded part of the rope is regarded as a particle system. The length of the rope that has not landed is z. The speed of the ungrounded part is

$$v = -\sqrt{2g(L-z)}\,\boldsymbol{k} = -\sqrt{2gl}\,\boldsymbol{k}$$

The mass is $\dfrac{M}{L}z$, and its momentum is

$$\boldsymbol{p} = \frac{M}{L}z\boldsymbol{v}$$

which is the total momentum of the system.

The external forces on the system are the gravity $-Mg\boldsymbol{k}$ and the force of the ground against the rope $\boldsymbol{F}$. Then the resultant force is $\boldsymbol{F} - Mg\boldsymbol{k}$; according to the law of momentum,

$$\boldsymbol{F} - Mg\boldsymbol{k} = \frac{\mathrm{d}\boldsymbol{p}}{\mathrm{d}t}$$

From the above equation, the force of the ground against the rope can be obtained as

$$\boldsymbol{F} = \frac{\mathrm{d}\boldsymbol{p}}{\mathrm{d}t} + Mg\boldsymbol{k} = \frac{\mathrm{d}}{\mathrm{d}t}\left(-\frac{M}{L}z\sqrt{2g(L-z)}\,\boldsymbol{k}\right) + Mg\boldsymbol{k}$$

Simplifying the above equatioin, we can get

$$\boldsymbol{F} = 3Mg\left(1 - \frac{z}{L}\right)\boldsymbol{k} = 3Mg\,\frac{l}{L}\boldsymbol{k}$$

In this example we analyze the motion of the entire rope and solve it by calculating the rate of change of the total momentum.

3.3 Work, kinetic energy and kinetic energy theorem

3.3.1 Definition of work

1. Work done by a constant force

The most basic definition of work in mechanics is the work done by a constant force.

As shown in Figure 3.3.1, an object undergoes linear motion and experiences a constant force $\boldsymbol{F}$, causing it to undergo displacement $\Delta \boldsymbol{r}$. The angle between $\boldsymbol{F}$ and $\Delta \boldsymbol{r}$ is θ. The work done by the constant force $\boldsymbol{F}$ is defined as the product of the component of the force in the direction of

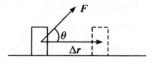

Figure 3.3.1 Work done by a constant force

displacement and the magnitude of the displacement. If W represents work, then we have

$$W = F\cos\theta \, |\Delta \boldsymbol{r}|$$

According to the definition of vector scalar product, the above equation can be written as

$$W = \boldsymbol{F} \cdot \Delta \boldsymbol{r} \tag{3.3.1}$$

That is, **the work done by a constant force is equal to the dot product of the force and the displacement of the particle.**

Work is a scalar quantity, and the unit of work is the Joule(J) in the International System of Units.

2. Work done by a variable force

If a particle moves along a path from point a to point b, as shown in Figure 3.3.2, and the magnitude and direction of the force acting on the particle are changing during the process, the

Figure 3.3.2 Work done by a variable force

work done by the variable force on the curved path can be calculated as follows. Firstly, the path is divided into many small segments, and any small segment of displacement can be denoted by $\Delta \boldsymbol{r}_i$. The force $\boldsymbol{F}_i$ acting on the particle during this segment of displacement can be regarded as a constant force, and the angle between $\boldsymbol{F}_i$ and the displacement $\Delta \boldsymbol{r}_i$ is θ_i; then the work done by the force over the displacement is given by

$$\Delta W_i = F_i \, |\Delta \boldsymbol{r}_i| \cos\theta_i = \boldsymbol{F}_i \cdot \Delta \boldsymbol{r}_i$$

When $\Delta \boldsymbol{r}_i$ is infinitely small $\Delta \boldsymbol{r}_i \to d\boldsymbol{r}$, the above equation can be expressed as

$$dW = \boldsymbol{F} \cdot d\boldsymbol{r} \tag{3.3.2}$$

dW is the element work done by the force $\boldsymbol{F}$ over the displacement element $d\boldsymbol{r}$.

Then add up all the elemental work along the whole path to get the work done by the force, and it can be expressed as

$$W_{ab} = \int_a^b \boldsymbol{F} \cdot d\boldsymbol{r} = \int_a^b F\cos\theta \, |d\boldsymbol{r}| \tag{3.3.3}$$

The distance element ds and the displacement element $d\boldsymbol{r}$ are equal in magnitude during the time period dt, that is, $|d\boldsymbol{r}| = ds$. Therefore, the above equation can be expressed as

$$W_{ab} = \int_a^b \boldsymbol{F} \cdot d\boldsymbol{r} = \int_a^b F\cos\theta \, ds \tag{3.3.4}$$

According to Equation (3.3.4), work is a process quantity, which is the accumulation of force in space. Generally speaking, the value of the integral is related to the integral path.

In Cartesian coordinates, **F** and d**r** can be written as

$$F = F_x\mathbf{i} + F_y\mathbf{j} + F_z\mathbf{k}$$
$$d\mathbf{r} = dx\mathbf{i} + dy\mathbf{j} + dz\mathbf{k}$$

Therefore, there is

$$dW = F_x dx + F_y dy + F_z dz$$
$$W = \int_a^b (F_x dx + F_y dy + F_z dz) \tag{3.3.5}$$

Equation (3.3.5) is the expression of the work done by the force on the object in the Cartesian coordinate system. The integrals in Equations (3.3.3), (3.3.4) and (3.3.5) are linear integrals along the curvilinear path ab.

If the object is subjected to several variable forces $F_1, F_2, \cdots, F_n$ at the same time, the net force **F** is

$$F = F_1 + F_2 + \cdots + F_n$$

The work done by the net force on an object is

$$W = \int_a^b F \cdot d\mathbf{r} = \int_a^b (F_1 + F_2 + \cdots + F_n) \cdot d\mathbf{r}$$
$$= \int_a^b F_1 \cdot d\mathbf{r} + \int_a^b F_2 \cdot d\mathbf{r} + \cdots + \int_a^b F_n \cdot d\mathbf{r} \tag{3.3.6}$$
$$= W_1 + W_2 + \cdots + W_n$$

The above equation shows that the work done by the net force on an object is equal to the algebraic sum of the work done by each component force. Obviously, the above result is based on the principle of the superposition of forces.

Example 3.3.1 A particle with a force of $F = 2y\mathbf{i} + 4j$ N acting on it, moves from point A at $x = -2$ to point B at $x = 3$. Calculate the work done by the force in the following cases shown in Figure 3.3.3.

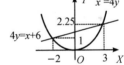

Figure 3.3.3 Figure of Example 3.3.1

(1) The particle moves along the parabolic path $x^2 = 4y$;

(2) The particle moves along the straight line $4y = x + 6$.

Solution Work done by the variable force

$$W = \int_A^B F \cdot d\mathbf{r} = \int_{x_A}^{x_B} F_x dx + \int_{y_A}^{y_B} F_y dy + \int_{z_A}^{z_B} F_z dz$$

(1) When the orbit of the particle is a parabola, the work done by the force is

$$W_1 = \int_{(x_1, y_1)}^{(x_2, y_2)} (F_x dx + F_y dy) = \int_{x_1}^{x_2} 2y dx + \int_{y_1}^{y_2} 4 dy$$
$$= \int_{-2}^{3} \frac{x^2}{2} dx + \int_{1}^{\frac{9}{4}} 4 dy \approx 10.8 \text{ J}$$

(2) When the orbit of the particle is a straight line, the work done by the force is

$$W_2 = \int_{(x_1, y_1)}^{(x_2, y_2)} (F_x dx + F_y dy) = \int_{x_1}^{x_2} 2y dx + \int_{y_1}^{y_2} 4 dy$$
$$= \int_{-2}^{3} \frac{1}{2}(x+6) dx + \int_{1}^{\frac{9}{4}} 4 dy = 21.25 \text{ J}$$

It can be seen that the work done by the force is related to the path.

3. Power

In order to reflect the speed at which a force does work on an object, in many practical situations, not only the total work completed but also the time required to complete the total work needs to be considered. In physics, power is introduced as a physical quantity that expresses how fast work is done on an object. **Power** is defined as the work done in a unit time, and if the work is ΔW in the time interval Δt, the **average power** during Δt is

$$\overline{P} = \frac{\Delta W}{\Delta t} \qquad (3.3.7)$$

If $\Delta t \to 0$, then the **instantaneous power at a certain time** is

$$P = \lim_{\Delta t \to 0} \frac{\Delta W}{\Delta t} = \frac{dW}{dt}$$

Because

$$\Delta W = F \Delta s \cos\theta$$

therefore

$$P = \lim_{\Delta t \to 0} F \cos\theta \frac{\Delta s}{\Delta t} = Fv\cos\theta = \boldsymbol{F} \cdot \boldsymbol{v} \qquad (3.3.8)$$

The above formula shows that the instantaneous power of the force $\boldsymbol{F}$ acting on an object is equal to the dot product of the force $\boldsymbol{F}$ acting on the object and the instantaneous velocity $\boldsymbol{v}$ of the object, or the instantaneous power is equal to the product of the component of the force in the direction of velocity and the magnitude of the velocity.

In the International System of Units, the unit of power is the Watt (W), $1\ \text{W} = 1\ \text{J} \cdot \text{s}^{-1}$.

3.3.2 Kinetic energy theorem

Experiments show that when a force does work on a particle, the kinetic energy of the particle changes.

A particle with mass m is under the action of the net force $\boldsymbol{F}$, it moves from point a to point b along the curve $\overset{\frown}{ab}$, and its velocities at the two points are respectively $\boldsymbol{v}_1$ and $\boldsymbol{v}_2$, as shown in Figure 3.3.4. The angle between $d\boldsymbol{r}$ and net force $\boldsymbol{F}$ is θ; then the elemental work done by the net force $\boldsymbol{F}$ on the particle is

$$dW = \boldsymbol{F} \cdot d\boldsymbol{r} = F\cos\theta |d\boldsymbol{r}| = F\cos\theta\, ds$$

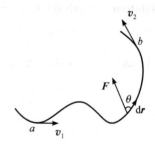

Figure 3.3.4 Kinetic energy theorem

According to Newton's second law, the equation of motion in the tangential direction at any time can be obtained as

$$F_\tau = ma_\tau = m\frac{dv}{dt}$$

F_τ is the projection of the net force $\boldsymbol{F}$ in the tangential direction, $F_\tau = F\cos\theta$, and $v = \dfrac{\mathrm{d}s}{\mathrm{d}t}$ can be obtained from $\mathrm{d}s = v\mathrm{d}t$, so the elemental work is

$$\mathrm{d}W = F\cos\theta\, \mathrm{d}s = m\,\frac{\mathrm{d}v}{\mathrm{d}t}\mathrm{d}s = m\,\frac{\mathrm{d}v}{\mathrm{d}s}\frac{\mathrm{d}s}{\mathrm{d}t}\mathrm{d}s = mv\,\mathrm{d}v$$

The work done by the resultant external force from a to b is

$$W = \int_a^b F\cos\theta\, \mathrm{d}s = \int_{v_1}^{v_2} mv\, \mathrm{d}v = \frac{1}{2}mv_2^2 - \frac{1}{2}mv_1^2 \tag{3.3.9}$$

$E_k = \dfrac{1}{2}mv^2$ is the kinetic energy of the particle when the velocity is v. In this way, $E_{k1} = \dfrac{1}{2}mv_1^2$ and $E_{k2} = \dfrac{1}{2}mv_2^2$ are the initial kinetic energy and the final kinetic energy of the object, respectively. Equation (3.3.9) can be written as

$$W = E_{k2} - E_{k1} \tag{3.3.10}$$

The above equation shows that the work done by the net force on a particle is equal to the increment of the particle's kinetic energy. This conclusion is known as the kinetic energy theorem of the particle.

Notes about the kinetic energy theorem:

(1) The kinetic energy theorem of the particle describes the relationship between work and the change in motion of the particle, specifically the change in kinetic energy. It states that any change in the kinetic energy of the particle is due to the net force acting on the particle, and the work done by the net force in a process is equal to the change of kinetic energy of the particle in the same process. Thus, work as a process quantity can be described by the change in a state quantity, and work is the measure of the change in kinetic energy.

(2) The kinetic energy theorem of the particle is derived from Newton's second law and, like Newton's second law, applies only in an inertial reference frame. Since both displacement and velocity are dependent on the selected frame of reference, when applying the kinetic energy theorem of the particle, both work and kinetic energy must be relative to the same inertial reference frame.

(3) Work reflects the spatial accumulation effect of force and is a process quantity, while kinetic energy describes the motion state of an object and is a single-valued function of the state. Once the motion state is determined, the object has a specific kinetic energy corresponding to it, making kinetic energy a state quantity. According to the kinetic energy theorem, the work done by the net force on the object is equal to the difference between the kinetic energy at the final state and that at the initial state, without involving the intermediate state. Therefore, applying the kinetic energy theorem to solve related mechanical problems is sometimes easier than applying Newton's second law.

Example 3.3.2 As shown in Figure 3.3.5, the conveyor sends a uniform object with length L and mass m to the horizontal platform on the right through the slideway at the initial speed v_0, and the front end of the object slides on the platform for a distance S before it stops. Friction on the slideway can be ignored, the friction coefficient between the object and the table is μ, and $S>L$. Calculate the initial velocity v_0 of the object.

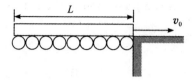

Figure 3.3.5 Figure of Example 3.3.2

Solution Since the object is uniform, before the whole object slides on the platform, its normal pressure on the platform can be considered to be proportional to the mass of its part on the platform, so the frictional force f_r on the platform is a variable force. The friction force is expressed as

$$0<x<L, \quad f_r = \mu \frac{m}{L} gx$$

$$x \geqslant L, \quad f_r = \mu mg$$

When the front end of the object stops at S, the work done by the friction force is

$$A = \int F\,dx = -\int f_r\,dx = -\int_0^L \mu \frac{m}{L} gx\,dx - \int_L^S \mu mg\,dx = -\mu mg\left(\frac{L}{2} + S - L\right)$$

$$= -\mu mg\left(S - \frac{L}{2}\right)$$

Then from the kinetic energy theorem we get

$$-\mu mg\left(S - \frac{L}{2}\right) = 0 - \frac{1}{2} mv_0^2$$

So

$$v_0 = \sqrt{2\mu g\left(S - \frac{L}{2}\right)}$$

3.4 Conservative force, work and potential energy

3.4.1 Characteristics of work done by several common forces

In general, an object moves from a certain position to another one through different paths, and the magnitude of the work done by the force on the object varies, and is related to the path taken. However, some forces are exceptions and the value of work done by these forces is independent of the path taken. Examples of such forces include gravity, gravitational force, and elastic force.

1. Work done by gravity

When an object moves near the ground, gravity will do work on the object. Consider an object with mass m moving along a curved path from the initial position to the final one.

The heights of the initial and final positions are h_a and h_b, respectively. The ground is used as the reference system, and a Cartesian coordinate system $OXYZ$ is established as shown in Figure 3.4.1 to calculate the work done by gravity on this curved path.

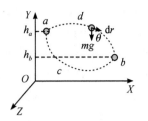

Figure 3.4.1 Work done by gravity

Select a displacement element $d\mathbf{r}$ near any point on the curve, and the work done by gravity on the object is given by

$$dW = \mathbf{F} \cdot d\mathbf{r} = mg\cos\theta |d\mathbf{r}|$$

As can be seen from the figure, $dy = -|d\mathbf{r}|\cos\theta$; substituting it into the above equation

$$dW = -mg\,dy$$

where dy is the projection of $d\mathbf{r}$ on the Y axis.

The total work done by gravity on the object is

$$W = \int_{y_1}^{y_2} -mg\,dy = -(mgy_2 - mgy_1) = mgh_a - mgh_b \qquad (3.4.1)$$

The above result shows that the work done by gravity on an object is only related to the initial position ($y_1 = h_a$) and end position ($y_2 = h_b$) of the object, but has nothing to do with the shape of the path the object travels. If the object moves along the $\widehat{adb}$ curve or $\widehat{acb}$ from point a to b, the work done by gravity is the same.

$$W_{\widehat{adb}} = W_{\widehat{acb}} = mgh_a - mgh_b$$

Further, it can be concluded that the work done by gravity on an object along a closed curve $\widehat{acbda}$ is

$$W_{\widehat{acbda}} = W_{\widehat{acb}} + W_{\widehat{bda}}$$

The direction of gravity is vertically downward, and the direction of displacement element along the curve $\widehat{bda}$ is opposite to the direction of the corresponding displacement element along the curve $\widehat{adb}$. The element work done on the curve $\widehat{bda}$ and on the curve $\widehat{adb}$ is equal in magnitude and opposite in sign, and the total work is

$$W_{\widehat{adb}} = -W_{\widehat{bda}}$$

Then

$$W_{\widehat{acbda}} = W_{\widehat{acb}} - W_{\widehat{adb}} = 0 \qquad (3.4.2)$$

Therefore, the characteristics of work done by gravity can also be expressed as follows: Under the action of gravity, when an object moves along a closed path, the work done by gravity is zero.

2. Work done by gravitational force

The artificial satellite is subject to the gravitational force of the earth when it moves, and the planets of the solar system are subject to the gravitational force of the sun when they move. This kind of problem can be attributed to the gravitational action on a moving particle from another fixed one. Now let's calculate the work done by the gravitational force on the moving particle.

The mass of a fixed particle is M. A moving particle with mass m reaches position b from a through a path $\overset{\frown}{acb}$, and the distances from the fixed particle to a and b are respectively r_a and r_b, as shown in Figure 3.4.2. During the movement, the magnitude and direction of the gravitational force acting on the particle m are changing, so this is a variable force problem.

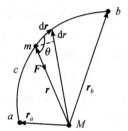

Figure 3.4.2 Work done by gravitation

When the particle m moves to one point with a distance r from the fixed particle, the displacement element $d\boldsymbol{r}$ is taken near this point. The gravitational force acting on the particle m can be considered a constant force with a magnitude of

$$F = G\frac{Mm}{r^2}$$

Then, the elemental work done by gravitational force $\boldsymbol{F}$ over the displacement element $d\boldsymbol{r}$ is

$$dW = F\cos\theta \,|d\boldsymbol{r}| \quad (3.4.3)$$

where θ is the angle between the gravitational force $\boldsymbol{F}$ and $d\boldsymbol{r}$, as shown in Figure 3.4.2.

$$|d\boldsymbol{r}|\cos(\pi-\theta) = dr$$
$$|d\boldsymbol{r}|\cos\theta = -dr \quad (3.4.4)$$

Substituting Equation (3.4.4) into Equation (3.4.3), we can get

$$dW = -F\,dr = -G\frac{Mm}{r^2}dr \quad (3.4.5)$$

Therefore, in the process of the particle m moving from point a to b along the curve $\overset{\frown}{acb}$, the work done by gravitational force is

$$W = \int_a^b dW = \int_{r_a}^{r_b} -G\frac{Mm}{r^2}dr = GMm\left(\frac{1}{r_b} - \frac{1}{r_a}\right) \quad (3.4.6)$$

The above equation shows that when the masses M and m are given, the work done by gravitational force is only related to the initial and final position of the particle m, and has nothing to do with the path it has taken.

3. Work done by elastic force

As shown in Figure 3.4.3, there is a light spring with a stiffness coefficient k on the horizontal table, its left end is fixed, and the right end is tied to an object of mass m. The

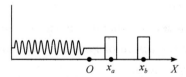

Figure 3.4.3 Work done by elastic force

spring's extension direction is in the positive direction of the X axis, and the coordinate origin O is at the position of the object when the spring is at its original length. When the coordinate of the object is x, the elastic force of the spring acting on the object is

$$F = -kx\bm{i}$$

While the object is moving from position a (the coordinate is x_a) to b (the coordinate is x_b), the elastic force is variable. But when the object is in the displacement element $\mathrm{d}x\bm{i}$, the elastic force can be regarded as a constant force, and the element work it does is

$$\mathrm{d}W = \bm{F} \cdot \mathrm{d}\bm{r} = -kx\bm{i} \cdot \mathrm{d}x\bm{i} = -kx\,\mathrm{d}x \tag{3.4.7}$$

In the process of the object moving from position a to b, the work done by the elastic force is

$$W = \int_{x_a}^{x_b} -kx\,\mathrm{d}x = \frac{1}{2}kx_a^2 - \frac{1}{2}kx_b^2 \tag{3.4.8}$$

It can be seen that within the elastic limit of the spring, the work done by the elastic force is only related to the initial and final positions of the spring, but has nothing to do with the elastic deformation process.

Similarly, if an object moves from position a to b and back to a, the elastic force does zero work during the whole process.

By calculating the work done by gravity, gravitational force, and elastic force, we can observe that they share a common characteristic. The work done is independent of the path taken by the object, and only depends on its initial and final positions. If the object makes a full circle along any closed path, the work done by the force is zero. Then the force is called a conservative force. Its mathematical expression can be written as

$$W = \oint_l \bm{F} \cdot \mathrm{d}\bm{r} = 0 \tag{3.4.9}$$

The force that satisfies Equation (3.4.9) is a conservative force, and the force that does not satisfy the equation is called a non-conservative force. From the above analysis, we can see that gravity, universal gravitation, and elastic force of a spring are all conservative forces. Friction, viscous force, etc. are non-conservative forces, and the work done by non-conservative forces is closely related to the path an object takes.

3.4.2 Potential energy

If a conservative force acts on an object in a system and the position of the object changes, the conservative force does work. Then the energy related to the position of the object will change. This energy determined by the position of the object is called the **potential energy** of the system and is expressed by E_p.

1. Gravity potential energy

In Equation (3.4.1), if $h_a = h$ and $h_b = 0$, then the work done by gravity is

$$W = mgh$$

Thus, the magnitude represents the ability of gravity to do work when an object is at height h (compared with $h = 0$). mgh is usually called the gravity potential energy of the system composed of the object and the earth, which is

$$E_p = mgh \tag{3.4.10}$$

2. Gravitational potential energy

From the relationship between the work done by the conservative force and potential energy and Equation (3.4.6), we can get

$$E_{pa} - E_{pb} = -G\frac{Mm}{r_a} + G\frac{Mm}{r_b}$$

Usually, we take infinity as the zero point of the potential energy. If r_b is the zero point of potential energy, that is $r_b \to \infty$, then the potential energy at point a is

$$E_{pa} = -G\frac{Mm}{r_a}$$

Thus, the **gravitational potential energy** of any point is

$$E_p = -G\frac{Mm}{r} \qquad (3.4.11)$$

where r is the distance of the particle from the center of gravitational force. The value of gravitational potential energy is always negative, which means that the potential energy of a particle at a finite distance from the center of gravitational force is always smaller than the potential energy at infinity.

3. Elastic potential energy

In the same way, from the relationship between the work done by the elastic force and potential energy and Equation (3.4.8) we can get

$$E_{pa} - E_{pb} = \frac{1}{2}kx_a^2 - \frac{1}{2}kx_b^2$$

If the position of a spring when it is at the natural original length is regarded as the zero point of elastic potential energy, that is, the elastic potential energy is zero at $x_b = 0$, then

$$E_{pa} = \frac{1}{2}kx_a^2$$

That is, the **elastic potential energy of the spring** is

$$E_p = \frac{1}{2}kx^2 \qquad (3.4.12)$$

E_{pa} represents the potential energy of the object in the system at the initial position a, and E_{pb} represents the potential energy of the object in the system at the final position b. Then when the object in the system moves from position a to position b, the work done by the conservative force on the object is equal to the negative value of the increment of the potential energy of the object

$$W_{ab} = -(E_{pb} - E_{pa}) = -\Delta E_p \qquad (3.4.13)$$

The work done by the conservative force on the object is determined by the potential energy change from the initial position to the final position. If the conservative force does positive work, $W_{ab} > 0$, then $E_{pb} < E_{pa}$ and the potential energy of the system decreases; if the conservative force does negative work, that is, the other force resists the conservative force and does positive work $W_{ab}<0$, then $E_{pb}>E_{pa}$ and the potential energy of the system increases.

Notes about potential energy:

(1) The concept of potential energy is introduced based on the characteristics of conservative force. Therefore, the concept of potential energy can only be introduced when there is a conservative force in the system.

(2) Potential energy belongs to the system. The gravitational potential energy is due to the earth's gravitational force acting on the object. The elastic potential energy is due to elastic force. It can be seen that the existence of potential energy depends on the interaction between objects and potential energy does not belong to a specific object but to an interacting system. We often refer to the "potential energy of an object", which is not entirely accurate.

(3) The magnitude of potential energy is a relative quantity. The value of potential energy is related to the selected reference point. When calculating the potential energy in a certain system, the chosen zero potential energy reference point must be specified. For example, if point b is the zero point of potential energy, that is $E_{pb}=0$, then using the above formula, we can calculate the potential energy of any other point relative to point b

$$W_{ab}=E_{pa}-0$$

$$E_{pa}=W_{ab}=\int_a^b \boldsymbol{F} \cdot \mathrm{d}\boldsymbol{r} \qquad (3.4.14)$$

The above equation shows that the potential energy of an object at a certain point is equal to the work done by the conservative force in the process of moving the object from this point to the zero point of potential energy.

(4) The difference in potential energy is absolute. The difference of potential energy between any two given points is always constant, irrespective of the choice of zero potential energy point.

(5) Potential energy is a state quantity and a single-valued function of the state. The magnitude of the potential energy changes with the relative position of the system.

(6) The unit of potential energy is the same as that of kinetic energy. In the International System of Units, the unit of potential energy is also the Joule (J), the same as that of work.

3.4.3 Potential energy curve

If there is a conservative force, the potential energy can be calculated. Once the coordinate system and the zero potential energy point are determined, the potential energy of an object is only a function of the position coordinate, which is $E_p=E_p(x, y, z)$. The curve that shows the variation of potential energy with the coordinate is **called the potential energy curve**. Figure 3.4.4 shows the curves of gravity potential energy, elastic potential energy and gravitational potential energy, respectively. The potential energy curve has very important applications in atomic physics, nuclear physics, molecular physics, solid state physics and other fields.

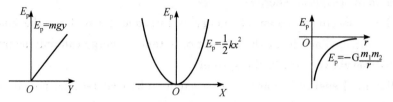

(a) Gravitational potential energy   (b) Elastic potential energy   (c) Gravitational potential energy

Figure 3.4.4 Potential energy function curves

Using the differential relationship between the conservative force and the potential energy function when the potential energy function is known, the magnitude and direction of the conservative force on each point in the conservative force field can be obtained

$$F = -\left(\frac{\partial E_p}{\partial x}i + \frac{\partial E_p}{\partial y}j + \frac{\partial E_p}{\partial z}k\right) \tag{3.4.15}$$

Example 3.4.1 The radius of the earth is R, its mass is M, and an object of mass m is at a height $2R$ above the ground as shown in Figure 3.4.5. Take the earth and the object as a system.

(1) If infinity is taken as the zero point of gravitational potential energy, what is the gravitational potential energy of the system?

(2) If the ground is taken as the zero point of gravitational potential energy, what is the gravitational potential energy of the system?

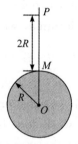

Figure 3.4.5 Figure of Example 3.4.1

Solution (1) As shown in the figure, suppose the object is at point P, and take infinity as the zero point of gravitational potential energy. When the object is $2R$ above the ground, the potential energy of the system is the work done by the gravitational force when the object moves from $2R$ to infinity

$$E_P = \int_{3R}^{\infty} \left(-\frac{GMm}{r^2}\right) dr = -\frac{GMm}{3R}$$

(2) Take the surface of the earth as the zero point of potential energy, and the potential energy of the object at a height $2R$ above the ground is the work done by the gravitational force when the object moves from there to the zero point of potential energy

$$E_P = \int_{3R}^{R} \left(-\frac{GMm}{r^2}\right) dr = \frac{2GMm}{3R}$$

It can be seen that the magnitude of potential energy is related to the selection of the zero potential energy point. Therefore, when analyzing the potential energy problem, it is necessary to indicate where the zero potential energy point is.

Chapter 3 Momentum, Energy and Angular Momentum 65

3.5 Work-energy theorem and conservation of mechanical energy

3.5.1 Work-energy theorem for a particle system

In practical problems, a particle system composed of multiple particles is often involved. In the following, we study the work-energy relationship of a particle system, i.e., the kinetic energy theorem of a particle system.

As shown in Figure 3.5.1, a particle system consists of n particles, and the forces acting on each particle are distinguished as the external forces and internal forces. Suppose that in a motion process, for an arbitrary particle (such as the ith particle), whose mass is m_i, the force acting on it is divided into external forces and internal forces. The work by the external forces is $W_{i,\,\text{ext}}$, and the work by the internal forces is $W_{i,\,\text{int}}$. The initial kinetic energy is $\frac{1}{2}m_i v_{i0}^2$, and the final kinetic energy is $\frac{1}{2}m_i v_i^2$. Applying the kinetic-energy theorem to the particle,

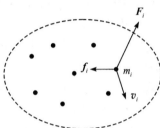

Figure 3.5.1 Particle system

$$W_{i,\,\text{ext}} + W_{i,\,\text{int}} = \frac{1}{2}m_i v_i^2 - \frac{1}{2}m_i v_{i0}^2 \tag{3.5.1}$$

Thus, for each particle in the system,

$$W_{1,\,\text{ext}} + W_{1,\,\text{int}} = \frac{1}{2}m_1 v_1^2 - \frac{1}{2}m_1 v_{10}^2$$

$$W_{2,\,\text{ext}} + W_{2,\,\text{int}} = \frac{1}{2}m_2 v_2^2 - \frac{1}{2}m_2 v_{20}^2$$

$$\cdots$$

$$W_{n,\,\text{ext}} + W_{n,\,\text{int}} = \frac{1}{2}m_n v_n^2 - \frac{1}{2}m_n v_{n0}^2$$

The sum of the above is

$$\sum_{i=1}^{n} W_{i,\,\text{ext}} + \sum_{i=1}^{n} W_{i,\,\text{int}} = \sum_{i=1}^{n} \frac{1}{2}m_i v_i^2 - \sum_{i=1}^{n} \frac{1}{2}m_i v_{i0}^2 \tag{3.5.2}$$

Let $W_{\text{ext}} = \sum_{i=1}^{n} W_{i,\,\text{ext}}$ be the total work done by the external forces acting on the particle system, $W_{\text{int}} = \sum_{i=1}^{n} W_{i,\,\text{int}}$ be the total work done by the internal forces acting on the particle system, $E_{k0} = \sum_{i=1}^{n} \frac{1}{2}m_i v_{i0}^2$ be the total initial kinetic energy of the particle system, and

Let $E_k = \sum_{i=1}^{n} \frac{1}{2} m_i v_i^2$ be the total final kinetic energy of the particle system.

Then
$$W_{ext} + W_{int} = E_k - E_{k0} \tag{3.5.3}$$

The equation above shows that the total work done by external forces acting on a particle system and the total work done by internal forces acting on each particle in the system are equal to the change in the total kinetic energy of the particle system. This relationship is known as the kinetic energy theorem of a particle system.

Comparing the kinetic energy theorem of a particle system with the momentum theorem of a particle system, we can see that the change in the momentum of the particle system depends only on external forces acting on it. However, the change in the kinetic energy of the particle system depends not only on external forces, but also on internal forces. For example, during the launch of a projectile, the explosion of gunpowder generates a force that propels the projectile forward and the gun barrel backward. This explosive force is an internal force of the system, and it does positive work on both the gun and the projectile. Thus, even though internal forces do not change the total momentum of the system, their work can change the total kinetic energy of the system.

3.5.2 Work-energy theorem

Internal forces in a system of particles can be divided into conservative internal forces and non-conservative internal forces. Correspondingly, the work done by internal forces on a particle system can be divided into two parts: the work done by conservative internal forces and the work done by non-conservative internal forces.

$$W_{int} = W_{C, int} + W_{NC, int}$$

Substituting the definition formula of potential energy Equation (3.4.13) and Equation (3.5.3) into the above expression, we can get

$$W_{ext} - (E_p - E_{p0}) + W_{NC, int} = E_k - E_{k0}$$
$$W_{ext} + W_{NC, int} = (E_k + E_p) - (E_{k0} + E_{p0})$$

Let $E_0 = E_{k0} + E_{p0}$, and $E = E_k + E_p$, which are the initial state mechanical energy and the final state mechanical energy of the system, respectively. Then the above equation can be simplified and rewritten as

$$W_{ext} + W_{NC, int} = E - E_0 \tag{3.5.4}$$

The sum of the work done by the external forces and the non-conservative internal forces is equal to the increment of the mechanical energy of the system, which is the work-energy theorem of the particle system. The work-energy theorem is obtained by introducing potential energy of the particle system, so it is only applicable to the inertial system.

Work done by both external and non-conservative internal forces in a system can cause changes in the mechanical energy of the system. The work done by external forces leads to the transformation between the energy of external objects and the mechanical energy of the system. When the external forces do positive work, the mechanical energy of the system increases. When the external forces do negative work, the mechanical energy of the system decreases. When the non-conservative internal forces do work, the transformation of mechanical energy and other forms of energy (such as chemical energy and thermal energy) occurs inside the system. When the non-conservative internal forces do positive work, the mechanical energy of the system increases, which means that other forms of energy are converted into mechanical energy. When the non-conservative internal forces do negative work, the mechanical energy of the system decreases, which indicates that the mechanical energy is converted into other forms of energy.

Note that the kinetic energy theorem and the work-energy theorem of a particle system both reveal the relationship between the change in energy of the system and work. The former indicates the relationship between the changes in kinetic energy and work, including the work of all forces, while the latter indicates the relationship between the changes in mechanical energy and work. Because the change in potential energy has already reflected the work done by conservative internal forces, it is only necessary to calculate the work done by forces except the conservative internal forces.

3.5.3 Conservation of mechanical energy

From the work-energy theorem Equation (3.5.4), the condition for the conservation of mechanical energy of the particle system is that in the changing process of the particle system from the initial state to the final state, the external forces and each pair of non-conservative internal forces do not do work. Then,

$$E_k + E_p = E_{k0} + E_{p0}$$
$$E = E_0 = C \qquad (3.5.5)$$

The equation shows that, when only conservative forces do work in the system, the kinetic energy and potential energy inside the particle system are transformed into each other, but the total mechanical energy is conserved. This is the conservation of mechanical energy.

Notes about the conservation of mechanical energy:

(1) The conservation of mechanical energy is a universal law in nature.

(2) When applying the conservation of mechanical energy, a certain inertial frame of reference must be selected. The mechanical energy of the system is conserved for the selected inertial frame of reference, but may not be conserved for another inertial frame of reference.

(3) Conservation of mechanical energy means that the mechanical energy of the system remains constant at any moment during a whole motion process, rather than only at the beginning and end of the process.

Example 3.5.1 As shown in Figure 3.5.2, a vertical spring with a stiffness coefficient k is fixed on the table at the lower end, and the upper end is connected to a wooden block with mass M, and it is in a static state. If a ball with mass m falls freely from a height of h, a completely inelastic collision with the wooden block occurs. Find

(1) the maximum compression amount ΔX of the spring;

(2) the mechanical energy loss ΔE of the entire system from the start to the maximum compression position.

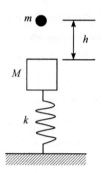

Figure 3.5.2   Figure of Example 3.5.1

Solution (1) In the initial stage, the ball falls freely. According to the conservation of mechanical energy, the speed of the ball before it collides with the wooden block can be obtained

$$V_0 = \sqrt{2gh}$$

The ball and the block collide completely inelastically, and the momentum is conserved during the collision process. After the collision, the common velocity

$$mV_0 = (M+m)V$$

$$V = \frac{m}{M+m}\sqrt{2gh}$$

Select the starting position of the wooden block as the zero gravitational potential energy point and assume that the spring compression amount before the collision is X_0. Then

$$kX_0 = Mg$$

Assuming that from the position of the spring when the collision starts to its maximum compression position, the compression amount of the spring is X, then the maximum compression amount of the spring is

$$\Delta X = X_0 + X$$

From the position where the ball collides with the wooden block to the maximum compression position of the spring, only the conservative force does work, and the mechanical energy of the system is conserved, so

$$\frac{1}{2}(M+m)V^2 + \frac{1}{2}kX_0^2 = \frac{1}{2}k(X+X_0)^2 - (M+m)gX$$

Solve the equations to get

$$X = \frac{mg}{k} + \frac{mg}{k}\sqrt{1+\frac{2kh}{(m+M)g}}$$

So the maximum compression amount of the spring

$$\Delta X = \frac{Mg}{k} + \frac{mg}{k} + \frac{mg}{k}\sqrt{1+\frac{2kh}{(m+M)g}}$$

(2) According to the analysis, the mechanical energy loss of the whole process is only

in the completely inelastic collision between the ball and the block. The mechanical energy loss of the system caused by the collision is

$$\Delta E = \frac{1}{2}mV_0^2 - \frac{1}{2}(m+M)V^2 = \frac{Mmgh}{m+M}$$

3.6 Conservation of energy

According to the conservation of mechanical energy, for a system with only conservative forces doing work, other forms of energy must increase by the equal value when the mechanical energy of the system decreases. When the mechanical energy of the system increases, there must be other forms of energy decreasing by the same value. As a result, the mechanical energy of the system is conserved.

In summarizing various natural phenomena, it is found that if a system is isolated and has no energy exchange with outside, various forms of energy in the system can be transformed into each other, or transferred from one object to another in the system. But the sum of these energies remains unchanged. If the total energy of the system increases (or decreases) under the action of external force, the external objects acting on the system must have the same amount of energy decrease (or increase) at the same time. That is to say, **energy can neither disappear nor be created and it can only be converted from one form to another; for an isolated system, no matter what kind of change process occurs, various forms of energy can be converted into each other, but the total energy of the system remains unchanged.** This conclusion is called the **conservation of energy**, one of the most universal laws in natural science that all natural phenomena must obey. The conservation of mechanical energy is a case of the **conservation of energy**.

The **conservation of energy** is one of the most important and basic laws in nature, applicable not only to macroscopic phenomena but also to molecules, atoms and the internal processes of atoms; not only to physics but also to various disciplines of natural science such as chemistry and biology.

3.7 Angular momentum and conservation of angular momentum

In physics, it is often encountered that objects rotate around a certain point, for example, the orbital motion of planets around the sun and the motion of electrons in atoms around the nucleus. In these problems, the concepts of momentum and kinetic energy cannot reflect all the motion of the particle. For example, astronomical observations show that the rotation speed of the earth around the sun is relatively great near the perihelion and small near the aphelion; Under the action of the sun's gravity, planets can move around the sun again and again, but they will not fall on the sun, while the satellites flying around the earth will fall to the ground after running for a period of time. These problems will be easy to explain if the concepts of angular momentum and the law of conservation are used. In the following we will introduce the concept of angular momentum, the theorem of angular momentum and the law of conservation of angular momentum on the basis of Newton's law of motion.

3.7.1 Angular momentum of a particle relative to a certain point

As shown in Figure 3.7.1, there is a particle with mass m, located at point P, the position vector is r, and the velocity is v (the momentum is $p = mv$); then the **angular momentum** of the particle relative to the origin is defined as

$$L = r \times p = r \times mv \tag{3.7.1}$$

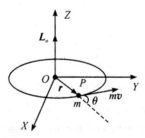

Figure 3.7.1 Angular Momentum of a particle relative to a certain point

The angular momentum of a particle L is a vector whose direction is perpendicular to the plane formed by r and p, and determined by the right-hand rule. When the four fingers turn from r to p through an angle θ smaller than 180°, the direction of the thumb is the direction of the angular momentum L.

According to the definition of the cross product, the magnitude of the angular momentum is

$$L = rp\sin\theta = rmv\sin\theta \tag{3.7.2}$$

In the International System of Units, the unit of angular momentum is $kg \cdot m^2 \cdot s^{-1}$.

It can be seen from the above definition that the angular momentum of a particle depends not only on its momentum, but also on its position vector relative to a fixed point. So the angular momentums of a particle relative to different points are different.

Notes about the concept of angular momentum:

(1) Angular momentum is an instantaneous quantity. The angular momentum of a particle at a certain moment is determined by its position vector and momentum at that moment.

(2) Angular momentum is a relative quantity. The angular momentum of a particle is related to the selection of the reference point in the inertial reference system. When the

same particle selects different reference points, the angular momentum is also different.

(3) Angular momentum is a physical quantity which describes the rotational state of a particle. As long as the particle is moving, there must be an angular momentum of the particle relative to a certain reference point. It doesn't matter whether the particle is moving in a straight line or a curve. However, when the direction of the particle's momentum or its reverse extension passes through the reference point, its angular momentum relative to the reference point equals zero.

3.7.2 Angular momentum theorem for a particle relative to a fixed point

1. Torque

When a particle moves relative to a certain reference point O and is subjected to a force, the object's motion state will change, and the angular momentum with respect to the reference point will also change. Experiments and theories show that the change in angular momentum is related not only to the magnitude and direction of the force but also to the point of application of the force. To quantitatively describe the reasons for the change in angular momentum, the concept of torque must be introduced.

Assuming that a mass point P moves relative to the reference point O under the action of the force $\boldsymbol{F}$, the vector product of the position vector of the action point of the force and the action force is called the moment of the force relative to the reference point, which is represented by a symbol $\boldsymbol{M}$, that is,

$$\boldsymbol{M} = \boldsymbol{r} \times \boldsymbol{F} \tag{3.7.3}$$

Torque $\boldsymbol{M}$ is a vector whose direction is always perpendicular to the plane composed by $\boldsymbol{r}$ and the force and can be determined by the right-hand rule, as shown in Figure 3.7.2. According to the cross product rule, the magnitude of the torque can be obtained as

$$M = rF\sin\theta = Fr\sin\theta = Fd \tag{3.7.4}$$

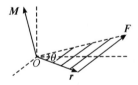

Figure 3.7.2 Torque for a fixed point

In the equation, θ is the angle between the position vector $\boldsymbol{r}$ and the force $\boldsymbol{F}$, and d is the perpendicular distance from the point to the line of the force, which is called the lever arm. Therefore, the magnitude of the torque is equal to the magnitude of the force multiplied by the lever arm.

The unit of torque is determined by the units of force and length. In the International System of Units, the unit of torque is N·m.

2. Angular momentum theorem for a particle

When a particle is subjected to a force, its momentum will change. According to Newton's second law, the rate of change of a particle's momentum with respect to time is numerically equal to the magnitude of the net external force acting on the particle. Similarly, when a particle is acted upon by a torque, the angular momentum of the particle will also change. The relationship between the torque of a particle and the rate of change of its angular momentum is studied below.

Taking the derivative at both sides of Equation (3.7.1) with respect to time t, we can get

$$\frac{d\boldsymbol{L}}{dt} = \boldsymbol{r} \times \frac{d(m\boldsymbol{v})}{dt} + \frac{d\boldsymbol{r}}{dt} \times m\boldsymbol{v} \tag{3.7.5}$$

Since $\boldsymbol{F} = m\dfrac{d\boldsymbol{v}}{dt}$ and $\boldsymbol{v} = \dfrac{d\boldsymbol{r}}{dt}$, Equation (3.7.5) can be written as

$$\frac{d\boldsymbol{L}}{dt} = \boldsymbol{r} \times \boldsymbol{F} + \boldsymbol{v} \times m\boldsymbol{v} \tag{3.7.6}$$

According to the property of cross product, $\boldsymbol{v} \times m\boldsymbol{v} = 0$, and $\boldsymbol{M} = \boldsymbol{r} \times \boldsymbol{F}$, then we have

$$\frac{d\boldsymbol{L}}{dt} = \boldsymbol{M} \tag{3.7.7}$$

The equation above indicates that the torque $\boldsymbol{M}$ of the net force acting on the particle with respect to a fixed reference point O is equal to the rate of change with time of the particle's momentum $\boldsymbol{L}$ with respect to the same point. This is known as the angular momentum theorem of a particle with respect to a point, also referred to as the differential form of the principle.

The angular momentum theorem is analogous to the principle of linear momentum, where torque $\boldsymbol{M}$ corresponds to force $\boldsymbol{F}$, and angular momentum $\boldsymbol{L}$ corresponds to linear momentum $\boldsymbol{p}$. The angular momentum theorem for a particle with respect to a fixed point can be rewritten as the following form:

$$d\boldsymbol{L} = \boldsymbol{M} dt \tag{3.7.8}$$

On the right side the product of the torque acting on the particle and time represents the cumulative effect of the torque acting on the particle in an infinitely small time interval. The above equation shows that the differential of the angular momentum of a particle relative to a fixed point O is equal to the cumulative effect of the torque over the time interval relative to point O.

If the torque $\boldsymbol{M}$ varies with time from t_1 to t_2, the integral form of the theorem of angular momentum of a particle relative to a fixed point O can be obtained as

$$\int_{L_1}^{L_2} d\boldsymbol{L} = \boldsymbol{L}_2 - \boldsymbol{L}_1 = \int_{t_1}^{t_2} \boldsymbol{M} dt \tag{3.7.9}$$

In the equation, $\boldsymbol{L}_1$ and $\boldsymbol{L}_2$ are the angular momentum of the particle at times t_1 and t_2 respectively, and $\int_{t_1}^{t_2} \boldsymbol{M} dt$ is the integral of the torque $\boldsymbol{M}$ over the time interval from t_1 to

t_2, which is called angular impulse and denoted as

$$H = \int_{t_1}^{t_2} M \, dt$$

Equation (3.7.9) shows that the increment of the particle's angular momentum with respect to point O over a certain time interval is equal to the cumulative effect of the torque over this time interval.

The unit of angular impulse is determined by the unit of moment and the unit of time. In the International System of Units, the unit of angular impulse is N·m·s, and the dimensional formula of angular impulse is the same as that of angular momentum.

3. Conservation of angular momentum

When the net torque on a particle $M_O = 0$, from Equation (3.7.8) of the angular momentum theorem, we can get

$$L_O = \text{constant vector} \quad (3.7.10)$$

The equation above shows that if the net torque acting on a particle is zero, then the particle's angular momentum remains constant which is known as the particle's angular momentum conservation law.

When applying the particle's angular momentum and angular momentum conservation laws, the following should be noted:

(1) The net torque and angular momentum of the particle depend on the choice of the reference point. When applying the laws, the same fixed reference point in the same inertial reference frame must be used.

(2) The angular momentum of the particle with respect to the reference point O is conserved only when the net torque M on the particle with respect to O is zero. Additionally, if the net torque M in a particular direction is zero, then the angular momentum of the particle with respect to O in that direction is conserved.

(3) Angular momentum is a vector, and the conservation of angular momentum implies that the magnitude and direction of the angular momentum both remain constant.

(4) The angular momentum law is derived from the second law of Newton, which only applies to inertial reference frames. Therefore, the laws only apply to inertial reference frames.

(5) The conservation laws of momentum, energy, and angular momentum are the three major conservation laws in physics, and they apply not only to macroscopic systems but also to microscopic ones.

4. Conservation of angular momentum in central motion

The motion of planets around a star belongs to a class of motion known as "central motion". The planets move around the sun in elliptical paths, with the sun located at one of the foci of the ellipse. For any planet (e.g., the earth), the force acting on it is almost solely the sun's gravitational force, which is a central force because its line of action passes through the sun's center S. Thus, the motion of a particle under the influence of a central force is called central motion.

Figure 3.7.3 shows a schematic diagram of a planet in the solar system orbiting the sun in an elliptical path. If the center of the sun is chosen as the reference point, since the gravitational force acting on the planet from the sun is a central force and the position vector r of the planet with respect to the center of the sun is collinear with F, the torque on the planet with respect to the center of the sun is $M = r \times F \equiv 0$. Therefore, the angular momentum of the planet with respect to the center of the sun is conserved, i.e.,

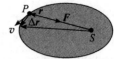

Figure 3.7.3 Schematic diagram of a planet orbiting around the sun

$$L = r \times (mv) = C \qquad (3.7.11)$$

According to the definition of angular momentum, the angular momentum of a planet relative to the center of the sun is always perpendicular to the planet's position vector r with respect to the center of the sun and its velocity vector v. The angular momentum L is a constant vector, so the position and velocity of the planet can only lie in the plane perpendicular to L. Therefore, the motion of a planet under the influence of a central force is confined to the plane, and the resulting trajectory is a planar curve. The plane of motion is determined by the planet's initial position vector and initial velocity.

Kepler's second law can be easily proved by the law of conservation of angular momentum. Assuming that the particle moves from point P to Q along the orbit in the time interval Δt and the corresponding position vector r changes to $r + \Delta r$, as shown in Figure 3.7.3, the area swept by the vector r in time Δt is approximately a sector area ΔA which is

$$\Delta A = \frac{1}{2} |r \times \Delta r| \qquad (3.7.12)$$

Divide both sides of the above equation by Δt, and take the limit to have

$$\frac{dA}{dt} = \lim_{\Delta t \to 0} \frac{\Delta A}{\Delta t} = \frac{1}{2} \lim_{\Delta t \to 0} \left| r \times \frac{\Delta r}{\Delta t} \right| = \frac{1}{2} |r \times v| \qquad (3.7.13)$$

The above equation can be further written as

$$\frac{dA}{dt} = \frac{1}{2m} |r \times mv| \qquad (3.7.14)$$

where m is the mass of the planet, and $\frac{dA}{dt}$ is the area swept by the vector r in the unit time, also known as the area velocity. According to the definition of the angular momentum, $|r \times mv|$ is the magnitude of the planet's angular momentum relative to the center of the sun, so Equation (3.7.14) can be rewritten as

$$\frac{dA}{dt} = \frac{1}{2m} L \qquad (3.7.15)$$

Because in any centric force field, the magnitude of a particle's angular momentum L is constant, in the universal gravitational field, the area swept by the planet's position vector per unit time is a constant, that is,

$$\frac{dA}{dt} = \text{Constant} \qquad (3.7.16)$$

which is known as **Kepler's second law**.

Example 3.7.1 As Figure 3..7.4 shows, a small ball with mass m is attached to one end of a rope that passes through a vertical tube, restricting the motion of the ball to a smooth horizontal plane. The ball initially moves in a circular path with radius r_0 around the center of the tube, with a velocity v_0. The rope is then pulled downwards to create a circular path with radius r for the ball.

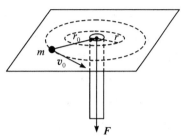

Figure 3.7.4 Figure of Example 3.7.1

Find the magnitude of the velocity v of the ball when it is at a distance r from the center of the tube, and the work done by the force $\boldsymbol{F}$ as the rope is shortened from r_0 to r.

Solution The force exerted on the ball by the rope always passes through the center point O, which is a central force, and the torque is always zero, so the angular momentum is conserved and we can get

$$mv_0 r_0 = mvr$$

Then

$$v = \frac{v_0 r_0}{r}$$

It can be seen that the speed increases and the kinetic energy increases, because the force $\boldsymbol{F}$ does work.

$$W = \frac{1}{2}mv^2 - \frac{1}{2}mv_0^2 = \frac{1}{2}mv_0^2 \left[\left(\frac{r_0}{r} \right)^2 - 1 \right]$$

 ## Scientist Profile

Qian Xuesen (1911 – 2009) who was born in Hangzhou, Zhejiang, was a Chinese aerodynamicist, member of the Chinese Academy of Sciences and the Chinese Academy of Engineering, and recipient of the Two Bombs, One Satellite Merit Medal. He served as a professor at MIT and Caltech in the United States and made significant contributions to the missile and space programs of both China and the US, earning him the nickname "Father of Chinese Missiles".

Qian was a pioneer and outstanding representative of Chinese aerospace technology industry, hailed as "Father of Chinese Space" and "King of Rockets". While studying and conducting research in the US, he collaborated with others to complete *A Review and Preliminary Analysis of Distant Rockets*, which laid the theoretical foundation for surface-to-surface missiles and sounding rockets. His work with Theodore von Kármán on supersonic flow theory also laid the groundwork for the development of aerodynamics. In October 1955, with the help of Premier Zhou Enlai's diplomatic efforts, Qian overcame various obstacles to return to China. In 1956, he was appointed the first director of Chinese first rocket and missile research institute. He led the completion of the "Jet and Rocket Technology Establishment" plan, participated in the development of short-and medium-range missiles and Chinese first artificial satellite, and directly led the use of intermediate-range missiles to carry out the "Two Bombs Combination" test. He also participated in the development of Chinese first interstellar aviation development plan and established engineering control theory and systems science. He made pioneering contributions in technical scientific fields such as aerodynamics, aeronautical engineering, jet propulsion, engineering control theory, and theoretical and applied research in physics and mechanics. He was the founder and advocate of modern mechanics and systems engineering theory and application research in China. In October 1991, the State Council and the Central Military Commission awarded Qian the honorary title of "National Outstanding Contribution Scientist".

While studying and working in the US, Qian never forgot his roots in China and he was a member of the Chinese nation. Upon learning of the establishment of the People's Republic of China, he unhesitatingly regarded his country's needs as his responsibility, giving up the generous treatment he received in the US and choosing to return to China. This choice was not only the highest responsibility of a scientist but also the highest mission of a Chinese descendant. His extraordinary life experience and outstanding achievements left a brilliant mark in the history of China, and even in world history, shining with dazzling light.

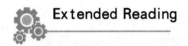 Extended Reading

Symmetry and conservation law

There are many theorems, laws, conservation laws and rules in various fields of physics, but their status is not the same. If you look at the laws and rules from the top of the whole building of physics, you'll find that they follow the framework: symmetry – conservation law – fundamental law – theorem – definition.

Symmetry is an understanding formed by observing the physical characteristics of objective things. Symmetry is regarded as an aesthetic principle in nature and is widely used in architecture, plastic arts and industrial arts. It was discovered in the late 19th century that certain symmetries of space and time are equivalent to the three conservation laws of mechanics. In 1918, a German woman scientist proposed the famous Noether theorem, which stated that if the law of motion has invariance under a certain transformation, there must be a corresponding conservation law. In short, a symmetry of the laws of physics corresponds to a conservation law. For example, the spatial translational symmetry of the law of motion leads to the law of conservation of momentum, the temporal translational symmetry leads to the law of conservation of energy, and the spatial rotational symmetry leads to the law of conservation of angular momentum. Noether's theorem, which related symmetries and conservation laws in classical physics, was later generalized to hold true in quantum mechanics as well.

In quantum mechanics and particle physics, some new internal degrees of freedom are introduced, and some new symmetries of abstract space and corresponding conservation laws are recognized. This law of conservation of parity, which had convinced physicists that it was undoubtedly universal, was carefully analyzed by Li Zhengdao and Yang Zhenning in 1956. They pointed out that there was no spatial inversion symmetry and conservation of parity in the motion of particles under weak action, and soon this conclusion was confirmed by Wu Jianxiong et al. From the historical development process, whether in classical physics or in modern physics, some important conservation laws were often recognized earlier than the general laws of motion. Laws of conservation of mass, conservation of energy, conservation of momentum, and conservation of charge were the first group of conservation laws. The establishment of these conservation laws provided clues and enlightenment for the later understanding of the general motion laws.

Discussion Problems

3.1 Internal force can change the kinetic energy of a particle system, but cannot change the momentum of the particle system. Why?

3.2 Someone says that in a certain motion process of a particle system, if the mechanical energy is conserved, then the momentum must also be conserved; or if the momentum is conserved, then the mechanical energy must also be conserved. Are these statements correct? Can you give an example of a motion process in which mechanical energy is conserved but momentum is not conserved, or momentum is conserved but mechanical energy is not conserved?

3.3 Can an object have energy without momentum? Can it have momentum without energy? Please give an example.

3.4 "Because the vector sum of all internal forces acting on all particles in a particle system is always equal to zero, internal force cannot change the total kinetic energy of the particle system." Is this statement correct? Can you give some examples of internal forces that can change the total kinetic energy of a particle system?

3.5 Try to determine whether the mechanical energy of the system is always conserved in the following processes:

(1) The satellite orbits the earth along an elliptical orbit (Ignore the effects of air resistance and the forces of other celestial bodies);

(2) The upper end of a spring is fixed, a heavy object is suspended from the lower end, and the object vibrates near its equilibrium position (Ignore the air resistance);

(3) An object falls freely from the air and falls into a sandpit.

3.6 If the mutual interaction force between two particles acts along the line connecting them and its magnitude depends on the distance between them, that is $f_1 = f_2 = f(r)$, this force is called a central force. Gravitational force is a kind of central force. Is it correct to say that any central force is a conservative force?

Problems

3.7 As shown in Figure T3-1, a small ball vibrates under the action of a spring force. The spring force $F = -kx$, and the displacement $x = A\cos\omega t$, where k, A and ω are constants. What is the impulse of the spring force on the ball during the time interval from $t=0$ to $t=\pi/2\omega$?

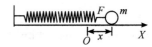

Figure T3-1 Figure of Problem 3.7

3.8 A ball weighing 0.3 kg and moving horizontally at a speed of 20 m·s^{-1} is hit by a stick, and flies vertically to a height of 10 m. How much impulse does the stick give to the ball? Assuming the contact time between the ball and the stick is 0.02 s, what is the average impulsive force on the ball?

3.9 As shown in Figure T3-2, three objects A, B and C, each with a mass of M, are placed on a smooth horizontal tabletop, with B and C leaning against each other and connected by a thin rope of length 0.4 m. In its initial state the rope is not taut. Another rope is connected to the other side of B and passed over

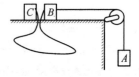

Figure T3-2 Figure of Problem 3.9

a fixed pulley across the edge of the table and connected to A. The friction on the pulley shaft can be ignored, and the length of the rope is constant. After A and B start moving, how long does it take for C to start moving? What is the velocity of C when it starts moving(Take $g = 10$ m·s^{-2})?

3.10 A particle of mass m is tethered to one end of a thin rope, and the other end of the rope is fixed. The particle moves in a circle with radius r on a rough horizontal surface. Suppose the particle's initial speed is v_0, and it changes to $v_0/2$ after one revolution.

(1) How much work is done by the frictional force?

(2) What is the sliding friction coefficient?

(3) How many revolutions the particle moves before it stops?

3.11 As shown in Figure T3-3, a steel ball m_1 of mass 1.0 kg is suspended from one end of a rope with length 0.8 m, and the other end of the rope is fixed at point O. After pulling the rope to the horizontal position, the ball is released from rest. At the lowest point, it collides with a steel block m_2 of mass 5.0 kg elastically. Find the maximum height that the steel ball can reach after the collision.

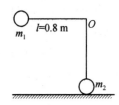

Figure T3-3 Figure of Problem 3.11

3.12 A particle of mass m slides down from the top of a circular arc groove with radius R and central angle $\pi/2$, as shown in Figure T3-4. Neglecting all friction, find

(1) the velocities of the particle and the groove when the particle just leaves the bottom of the groove;

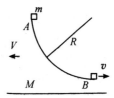

Figure T3-4 Figure of Problem 3.12

(2) the work done by the particle on the groove as it slides from point A to point B;

(3) the pressure on the groove when the particle reaches point B.

3.13 A particle of mass m is located at (x_1, y_1), and its velocity is $\boldsymbol{v} = v_x \boldsymbol{i} + v_y \boldsymbol{j}$. The particle is influenced by a force in the negative X direction. Find the angular momentum of the particle relative to the origin.

3.14 Halley's Comet orbits the sun in an elliptical path. When it is the closest to the sun at $r_1 = 8.75 \times 10^{10}$ m, its velocity v_1 is 5.46×10^4 m·s^{-1}. And its velocity v_2 is 9.08×10^2 m·s^{-1} when it is the farthest from the sun. Find its distance r_2 from the sun when it is the farthest (The sun is located at one focus of the elliptical orbit).

3.15 Assuming that the earth is a homogeneous sphere with mass M and radius R, try to estimate the minimum energy required to completely disassemble the earth.

Challenging Problems

3.16 A collision is called a central collision when the velocities of two objects before and after the collision lie along a straight line passing through their centers of mass. If the mechanical energy is not conserved during such a collision, it is called an inelastic collision. Newton summarized a collision law from the experimental results: The separation velocity $v_2 - v_1$ of two balls after collision is proportional to the approach velocity $v_{10} - v_{20}$ before collision, and the ratio depends on the material properties of the two balls, which can be formulated as

$$e = \frac{v_2 - v_1}{v_{10} - v_{20}}$$

e is called the coefficient of restitution.

(1) Derive the velocity formula for an inelastic elastic collision;

(2) Calculate the mechanical energy lost during an inelastic central collision between two particles.

3.17 The rocket is a kind of spacecraft launched into space by utilizing the recoil force generated by the gas ejected after the fuel is burned.

(1) What is the formula for its flight speed if the rocket is flying in free space without being affected by any external forces such as gravity or air resistance?

(2) What are the formulae for its flight speed, height, and acceleration within not too large a height if the rocket is launched vertically upwards from the surface of the earth with a fuel burning rate of α (Air resistance is neglected)?

(3) Assuming that the mass M_0 of the rocket before launch is 2.8×10^6 kg, the burning rate α of the fuel is 1.20×10^4 kg·s^{-1}, and the speed u of the gas ejected relative to the rocket is 2.90×10^3 m·s^{-1}. What are the features of the curves of height, speed, and acceleration changing with time within the 100 seconds after the rocket is launched? What values do they ultimately reach?

3.18 China's first artificial satellite orbits the earth along an elliptical orbit, with the center of the earth located at one of the foci of the ellipse. The radius R_e of the earth is 6.378×10^6 m, the minimum height h_1 of the artificial satellite above the ground (the perigee) is 4.39×10^5 m, and the maximum height h_2 (the apogee) is 2.384×10^6 m. The speed v_1 of the satellite at the perigee is 8.10×10^3 m·s^{-1}. Draw the orbit of the satellite, and calculate the speed v_2 of the satellite at the apogee and the period T of its motion.

Chapter 4

Fundamentals of Rigid Body Mechanics

In the previous chapters, we focused on the translational motion of a particle—an ideal model whose size and shape can be neglected. However, when we talk about some problems such as the rotation of motor rotors, the spin of shells, the rolling of wheels and the balance of bridges, the shape and size of the objects often play a very important role and should be taken into consideration. This makes these problems much more complicated than translational motions. Fortunately, in many cases, the deformation of these objects is very small, making their shapes and sizes unchanged during the motions. Therefore, to facilitate the study and grasp the essential characteristics of these problems, an ideal model named "rigid body" will be introduced.

In this chapter, we will discuss the fundamentals of rigid body mechanics, including the definition of a rigid body, the rotation law, the kinetic energy law, conservation of angular momentum, and applications of rigid body rotation about a fixed axis.

4.1 Kinematics of rotation about a fixed axis

4.1.1 Definition of a rigid body

In general, the shape and size of any object will change when it is subjected to an external force. For example, when cars go across the bridge, the pier will be compressed and deformed, and the bridge body will be bent and deformed. Nevertheless, such kind of deformation is very small and has little influence on the mechanical study. Hence, an ideal model "rigid body" is introduced to simplify the mechanical problems. **A body is rigid only if its shape and size are unchanged when it is under an external force.** An object is made up of a large number of particles, so a rigid body can also be defined as a body where, under the action of a force, the distance between all the particles that build up it remains unchanged. For example, when studying the mechanical problems of a rotating flywheel, the flywheel can be regarded as a rigid body. The rigid body is a very useful ideal model in mechanics although there is no real rigid body in daily life.

4.1.2 Basic motion of a rigid body

The motion of a rigid body is generally complicated, but there are two basic forms: translation and rotation. In consequence, a general motion of a rigid body can always be considered as a combination of rotation and translation.

As shown in Figure 4.1.1, **during a motion of a rigid body, if the lines joining any two points in the body always remain parallel to that joining their initial positions, this motion is called translation.** According to the characteristics of translation of a rigid body, it can be proved that all points in the body have the same instantaneous velocity and acceleration; therefore, the motion of the entire rigid body can be represented by the motion of any point of the body. Usually, the center of mass of a body is chosen as the representative point and all the external forces acting on the body can be regarded as acting only on the center of mass. As a result, the study of the translation of a rigid body can be simplified to the study of particle motion governed by Newton's second law.

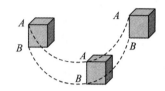

Figure 4.1.1 Translation of a rigid body

On the other hand, during the motion of a rigid body, if all the particles of the body make circular paths around a straight line, this motion is called rotation, and the straight line is called the axis of rotation. The axis may be fixed or may change its direction relative to the reference system. Here, we focus on the rotation about a fixed axis such as the rotation of a door about its hinge and the rotation of a motor rotor about its central vertical axis.

4.1.3 Description of rotation about a fixed axis

When studying the rotation of a rigid body about a fixed axis, it is common to define the plane perpendicular to the rotation axis as the rotation plane, and the intersection of the rotation plane and the rotation axis is the rotation center. In this way, any mass element of the rigid body will make a circular motion around the rotation center in the rotation plane passing through the mass element. Therefore, the fixed axis rotation of a rigid body is essentially the circular motion of each mass element on the rigid body around its rotation center in its rotation plane.

It is easy to find that, when a rigid body rotates about a fixed axis, although mass elements with different rotation radii on the body have different displacements, velocities, and accelerations, their rotation angles, angular velocities, and angular accelerations relative to their rotation center are the same during the same period. Therefore, it is more convenient to describe the rotation of a rigid body about a fixed axis using angular quantities including angular position, angular displacement, angular velocity and angular acceleration. The relationships between angular quantity and linear quantity we have learned are also applicable when describing the rotation of a rigid body about a fixed axis.

As shown in Figure 4.1.2, when a rigid body rotates about the Z axis, we can select one arbitrary particle P in the body as a reference point, and the circular motion of this particle P is then representative of the rotation of the entire rigid body. The rotation plane passing through point P crosses the axis at point O, which is set to be the rotation center. If we establish a polar coordinate system centered at point O and introduce a ray OX as the polar axis, then the angle θ starting from the polar axis OX to the polar diameter OP can denote the angular position of point P.

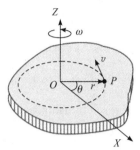

Figure 4.1.2 Rotation of a rigid body about a fixed axis

When a rigid body rotates about a fixed axis, its angular position changes with time and is a single-valued continuous function of time, so $\theta = \theta(t)$ **is defined as the motion function of rotation of a rigid body about a fixed axis.**

During a period of time Δt, the increment of the angular position $\Delta \theta = \theta(t + \Delta t) - \theta(t)$ is the **angular displacement**.

To describe the speed of rotation, **angular velocity** ω can be expressed by

$$\omega = \lim_{\Delta t \to 0} \frac{\Delta \theta}{\Delta t} = \frac{d\theta}{dt} \tag{4.1.1}$$

Meanwhile, angular acceleration β is used to reflect the change of the angular velocity, and its definition is as follows:

$$\beta = \lim_{\Delta t \to 0} \frac{\Delta \omega}{\Delta t} = \frac{d\omega}{dt} = \frac{d^2\theta}{dt^2} \tag{4.1.2}$$

According to the relationship between the angular quantity and linear quantity,

$$v = r\omega$$

$$a_\tau = \frac{dv}{dt} = r\beta$$

$$a_n = \frac{v^2}{r} = r\omega^2$$

in which r is the distance between the mass element and its rotation center.

It is worth noting that the angular displacement, angular velocity, and angular acceleration are vectors with both magnitude and direction. If we specify the counterclockwise rotation as the positive reference direction, then all the angular quantities θ, $\Delta\theta$, ω and β can be regarded as algebraic quantities with a positive sign indicating counterclockwise rotation and a negative sign indicating clockwise rotation, respectively.

Example 4.1.1 A flywheel with a radius of 0.1 m rotates about a fixed horizontal axis passing through its center, and the motion function is $\theta = t^3 + 3t + 5$ rad. Find

(1) the angular velocity at $t=1$ s;

(2) the linear velocity, tangential acceleration and normal acceleration of a point on the edge of the flywheel at $t=1$ s.

Solution (1) According to Equation (4.1.1),

$$\omega=\frac{d\theta}{dt}=3t^2+3$$

When $t=1$ s, $\omega=6$ rad·s^{-1}.

(2) According to Equation (4.1.2), $\beta=\dfrac{d\omega}{dt}=6t$.

From the relationship between the angular quantity and linear quantity, when $t=1$ s, the linear velocity is

$$v=r\omega=0.1\times6=0.6 \text{ m·s}^{-1}$$

the tangential acceleration is

$$a_\tau=r\beta=0.1\times6=0.6 \text{ m·s}^{-2}$$

and the normal acceleration is

$$a_n=r\omega^2=0.1\times6^2=3.6 \text{ m·s}^{-2}$$

4.2　Fundamentals of dynamics of rigid body rotation about a fixed axis

Force is the reason for changing the translational state, while torque is the reason for changing the rotational state. In this section, we will discuss the dynamic law of the rotation of a rigid body about a fixed axis.

4.2.1　Torque

A stationary object with a fixed rotating axis may rotate under the action of an external force. The rotation of the object depends not only on the magnitude of the force, but also on the acting point and direction of the force. Therefore, it is necessary to study the effect of torque on a rotation problem.

In Section 3.7 Chapter 3, we defined the torque of a force relative to a reference point O and here, we should discuss the torque relative to an axis.

As shown in Figure 4.2.1 (a), suppose that the Z axis is the rotation axis of the rigid body, the force $\boldsymbol{F}$ acts on point P and is in the rotation plane, point O is the rotation center, and the position vector from O pointing to P is $\boldsymbol{r}$, then the torque of $\boldsymbol{F}$ relative to the Z axis is defined as

$$M_z=rF\sin\theta \qquad (4.2.1)$$

in which r is the magnitude of the position vector $\boldsymbol{r}$, F is the magnitude of the force $\boldsymbol{F}$ and θ is the angle rotated counterclockwise from $\boldsymbol{r}$ to $\boldsymbol{F}$ when observed facing the Z axis. M_z is an algebraic value with its sign determined by the angle θ. The positive and negative signs of M_z indicate that the torque is in the positive and negative direction of the Z axis, respectively.

In Figure 4.2.1(a), OB is perpendicular to the action line of the force $\boldsymbol{F}$, referring to the vertical distance between the Z axis and the force $\boldsymbol{F}$, so OB is named as the moment arm and expressed by the letter d.

Since $d = |r\sin\theta|$, Equation (4.2.1) can also be written as
$$M_z = \pm Fd \tag{4.2.2}$$
In this way, the torque of a force $\boldsymbol{F}$ relative to the Z axis is the product of the magnitude of the force and the moment arm.

The force in the rotation plane has been discussed above. If the force $\boldsymbol{F}$ has an angle with the rotation plane, then the force $\boldsymbol{F}$ can be decomposed into two components, one along the rotation plane and the other perpendicular to the rotation plane, as shown in Figure 4.2.1 (b), i.e., $\boldsymbol{F} = \boldsymbol{F}_\perp + \boldsymbol{F}_{/\!/}$. $\boldsymbol{F}_{/\!/}$ is parallel to the axis and its torque is perpendicular to the axis, which can not change the rotation state of any rigid body. On the contrary, $\boldsymbol{F}_\perp$ is perpendicular to the axis and its torque $M_z = rF_\perp \sin\theta$ is parallel to the axis, which can make the rigid body rotate about the Z axis. As a result, this torque M_z is called rotation torque, specifically referring to the torque produced by the force in the rotation plane. For the convenience of discussion, unless otherwise stated, all the forces we will discuss in the following are in the rotation plane.

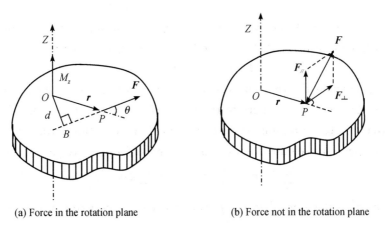

(a) Force in the rotation plane   (b) Force not in the rotation plane

Figure 4.2.1 Torque relative to an axis

If there are multiple forces acting on a rigid body at the same time, the torque produced by the net force is the net torque. The net torque can also be calculated by the algebraic sum of torques produced by each force since each torque is an algebraic value with only a positive or negative sign.

4.2.2 Law of rotation (Newton's second law for rotation)

We regard the rigid body as a special system of particles. By using the familiar laws of particle motion and the definition of torque, we can deduce the law of rotation for the rotation of a rigid body about a fixed axis.

As shown in Figure 4.2.2, suppose that a rigid body rotates about a fixed axis Z, at a certain time, the angular velocity is ω, and the angular acceleration is β. If we select an arbitrary mass element Δm_i on the rigid body, the distance between this mass element and the axis is r_i, and the net external force acting on the mass element is $\boldsymbol{F}_i$ while the net internal force on it is $\boldsymbol{f}_i$, then the mass element Δm_i is doing a circular motion with the center of O and radius of r_i. Applying Newton's second law to the mass element Δm_i, we can get

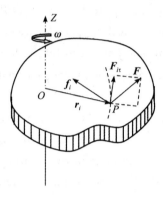

Figure 4.2.2 Law of rotation

$$\boldsymbol{F}_i + \boldsymbol{f}_i = \Delta m_i \boldsymbol{a}_i \qquad (4.2.3)$$

Project both sides of this equation onto the tangential and normal direction of the circular path of the mass element m_i, and we have

$$F_{i\tau} + f_{i\tau} = \Delta m_i a_{i\tau} \qquad (4.2.4)$$
$$F_{in} + f_{in} = \Delta m_i a_{in} \qquad (4.2.5)$$

The torque produced by F_{in} and f_{in} is zero since their extension lines pass through the axis, making no contribution to the rotation. Therefore, we only discuss the tangential equation.

Multiply both sides of Equation (4.2.4) by r_i and apply the relationship between angular quantity and linear quantity $a_{i\tau} = r_i \beta$, and the equation becomes

$$F_{i\tau} r_i + f_{i\tau} r_i = \Delta m_i r_i^2 \beta \qquad (4.2.6)$$

in which $F_{i\tau} r_i$ indicates the rotation torque relative to the Z axis produced by an external force acting on Δm_i and $f_{i\tau} r_i$ refers to the rotation torque relative to the Z axis produced by the internal force. Summing the above equation and considering that the angular acceleration of each mass element is the same, the following equation can be obtained

$$\sum_i F_{i\tau} r_i + \sum_i f_{i\tau} r_i = \left(\sum_i \Delta m_i r_i^2\right)\beta \qquad (4.2.7)$$

in which $\sum_i F_{i\tau} r_i$ is the sum of torques produced by all the external forces acting on the rigid body, called the net torque and expressed by M_z; $\sum_i f_{i\tau} r_i$ is the sum of torques produced by all the internal forces. All internal forces occur in pairs (Newton's third law), and each pair has the same magnitude, opposite direction and the same moment arm about the rotation axis. Therefore, the net internal torque is zero, that is $\sum_i f_{i\tau} r_i = 0$.

Let

$$J_z = \sum_i \Delta m_i r_i^2 \qquad (4.2.8)$$

be the rotational inertia of a rigid body, and Equation (4.2.7) can be written as

$$M_z = J_z \beta \qquad (4.2.9)$$

Equation (4.2.9) is called the law of rotation, indicating that the angular acceleration of a rigid body rotating about a fixed axis is proportional to the external torque acting on the rigid body and inversely proportional to the rotational inertia of the rigid body.

It should be noted that all physical quantities in the law of rotation are relative to the same rotation axis; the law of rotation describes the instantaneous effect of torque, and its role in rigid body mechanics is similar to that of Newton's second law in particle mechanics. It is obvious that the angular acceleration of a rigid body about a fixed axis is given by an equation having the same form as that for the linear acceleration of a particle $F = ma$. Comparing Equation (4.2.9) with Newton's second law in particle mechanics, $F = ma$, it can be seen that J_z has a similar position to m to describe the translational inertia of a particle, and then J_z should be a physical quantity describing the rotational inertia of a rigid body.

4.3 Calculation of rotational inertia

From Equation (4.2.9) which defines the rotational inertia J_z, we can know that, if a rigid body is made up of discrete particles, the rotational inertia can be calculated by

$$J_z = \sum_i m_i r_i^2 \qquad (4.3.1)$$

Otherwise, if the mass of the body is continuously distributed, we must replace the sum in Equation (4.3.1) by an integral and the definition of the rotational inertia becomes

$$J_z = \int_m r^2 \, dm \qquad (4.3.2)$$

where r is the distance from dm to the axis. The SI unit of rotational inertia is kg · m².

Further analysis shows that the rotational inertia of a rigid body is related to three factors: ① the total mass of the rigid body; ② the mass distribution and geometry of the rigid body relative to the rotation axis; ③ the position of the rotation axis. Different rotation axis positions of the same object correspond to different rotational inertia.

For a rigid body with regular geometry, its rotational inertia can be calculated easily by the definition, but in general, it is very troublesome to directly calculate the rotational inertia because of the complex shape of the rigid body or the uneven mass distribution. At this time, the rotational inertia is often measured by experiments.

Table 4.3.1 gives the rotational inertia of some rigid bodies with simple shapes and indicated axis of rotation.

Table 4.3.1 Rotational inertia of some common rigid bodies

Hoop about central axis $I=MR^2$ (a)	Annular cylinder (or ring) about central axis $I=\frac{1}{2}M(R_1^2+R_2^2)$ (b)	Solid cylinder (or disk) about central axis $I=\frac{1}{2}MR^2$ (c)
Solid cylinder (or disk) about central diameter $I=\frac{1}{4}MR^2+\frac{1}{12}ML^2$ (d)	Thin rod about axis through center perpendicular to length $I=\frac{1}{12}ML^2$ (e)	Solid sphere about any diameter $I=\frac{2}{5}MR^2$ (f)
Thin spherical shell about any diameter $I=\frac{2}{3}MR^2$ (g)	Hoop about any diameter $I=\frac{1}{2}MR^2$ (h)	Slab about perpendicular axis through center $I=\frac{1}{12}M(a^2+b^2)$ (i)

There are two simple theorems very helpful in obtaining the rotational inertia of a body, which are the parallel-axis and perpendicular-axis theorems.

1. Parallel-axis theorem

Figure 4.3.1 shows that Z_C is the axis passing through the center of mass of the rigid body and the rotational inertia of the body corresponding to this axis is J_C. If there is another axis Z parallel with Z_C, then it can be proved that the rotational inertia of the body relative to the Z axis is

$$J_Z = J_C + md^2 \qquad (4.3.3)$$

where m is the mass of the body, and d is the vertical distance between the two axes. The equation above describes the parallel-axis theorem of rotational inertia, from which, we can find that the

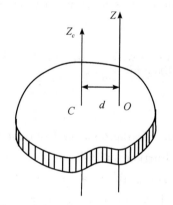

Figure 4.3.1 Parallel-axis theorem

rotational inertia of a rigid body reaches the minimum value (J_C) when the axis passes through the center of mass, while rotational inertias relative to any other axis Z (J_Z) should be larger than J_C, that is $J_Z > J_C$.

By the parallel-axis theorem, it is very easy to obtain the rotational inertia of a rod about a perpendicular axis passing through its endpoint

$$J = J_c + md^2 = \frac{1}{12}mL^2 + m\left(\frac{L}{2}\right)^2 = \frac{1}{3}mL^2$$

Similarly, the rotational inertia of a uniform disk about a perpendicular axis passing through one of its edge points is

$$J = J_c + mR^2 = \frac{1}{2}mR^2 + mR^2 = \frac{3}{2}mR^2$$

2. Perpendicular-axis theorem

As Figure 4.3.2 shows, a thin disk is located in the XY plane. Select point O at the intersection of the two vertical axes as the coordinate origin, and the OZ axis is perpendicular to the plate surface and the rotational inertia of the thin disk about the OZ axis is

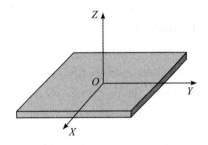

Figure 4.3.2   Perpendicular-axis theorem

$$J_z = \sum \Delta m_i r_i^2 = \sum \Delta m_i (x_i^2 + y_i^2)$$

where x_i and y_i are the vertical distances from the element Δm_i to the Y axis and X axis, respectively. Since $\sum \Delta m_i x_i^2 = J_y$ and $\sum \Delta m_i y_i^2 = J_x$, the equation above becomes

$$J_z = J_x + J_y \qquad (4.3.4)$$

This equation states that the sum of the rotational inertia of a sheet rigid body about any two perpendicular axes in the plane is equal to its rotational inertia about an axis passing through their point of intersection and perpendicular to the plane. The perpendicular-axis theorem can be applied only to planar rigid bodies, that is, two-dimensional rigid bodies, or rigid bodies of uniform thickness that can be neglected compared to the other dimensions.

Example 4.3.1   Figure 4.3.3 shows a uniform rod of mass m and length L. Calculate its rotational inertia about the Z axis perpendicular to the rod and passing through its one endpoint.

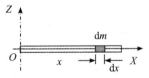

Figure 4.3.3   Figure of Example 4.3.1

Solution   Choose the intersection of the Z axis and the rod as the coordinate origin O, and build the OX axis along the rod. Select dx as a mass element located at position x, and then the mass of the element is

$$dm = \frac{m}{L}dx$$

The distance between the element dm and the Z axis is x, so the rotational inertia of mass element dm about the Z axis is

$$dJ_z = x^2 dm$$

As a result, the total rotational inertia of the entire rod about the Z axis is

$$J_z = \int_0^L x^2 \frac{m}{L} dx = \frac{1}{3} mL^2$$

Discussion: If the axis passes through the middle point of the rod, the corresponding rotational inertia is $J_z = \frac{1}{12} mL^2$, which indicates that the magnitude of the rotational inertia is related to the position of the rotation axis.

Example 4.3.2 Figure 4.3.4 shows a uniform disk with mass m and radius R. Find its rotational inertia about the symmetry axis Z.

Solution In this case, we choose a thin ring of radius r and thickness dr; then its area is $dS = 2\pi r \, dr$ and its mass is $dm = \sigma dS$, in which $\sigma = \frac{m}{\pi R^2}$

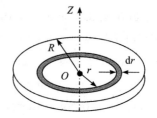

Figure 4.3.4 Figure of Example 4.3.2

is the surface mass density. Consequently, the rotational inertia of the ring is

$$dJ_z = r^2 dm = 2\pi \sigma r^3 dr$$

And the rotational inertia of the entire disk is

$$J_z = \int r^2 dm = 2\pi \sigma \int_0^R r^3 dr = \frac{\pi}{2} \sigma R^4$$

If the surface mass density $\sigma = \frac{m}{\pi R^2}$ is substituted, it becomes

$$J_z = \frac{1}{2} mR^2$$

Example 4.3.3 Figure 4.3.5 shows a uniform thin disk of mass m and radius R has a negligible thickness. Try to find the rotational inertia about any one of its diameters.

Solution Select the center of the disk as the coordinate origin O, and build the coordinate system with the X and Y axes located in the disk. From the analysis of the symmetry of the disk, we can get $J_x = J_y$; meanwhile, according to the perpendicular-axis theorem, there is

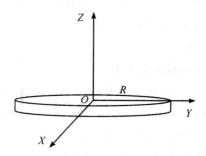

Figure 4.3.5 Figure of Example 4.3.3

$$J_z = J_x + J_y = 2J_x$$

We can know from Example 4.3.2 that $J_z = \frac{1}{2} mR^2$; therefore, the rotational inertia about the diameter OX is

$$J_x = \frac{1}{4} mR^2$$

Chapter 4 Fundamentals of Rigid Body Mechanics

4.4 Application of the law of rotation

The problem of rotation of a rigid body about a fixed axis is usually studied by the "isolation method", and the solution steps are similar to those taken in using Newton's second law.

Example 4.4.1 As shown in Figure 4.4.1, the pulley is a uniform disk of mass m and radius R, and the light rope passing over the pulley connects two objects of mass m_1 and m_2, respectively. Suppose that the light rope is nonretractable, there is no relative sliding between the rope and the pulley, and the friction at the axis of the pulley can be ignored. Find the tension in the rope and the angular acceleration of the pulley.

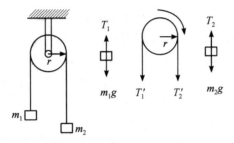

Figure 4.4.1 Figure of Example 4.4.1

Solution There is no relative sliding between the rope and the pulley, so there must be static friction between them, which drives the pulley to rotate. Hence, the tensions in two sections of the rope on both sides of the pulley could not be equal.

Firstly, apply the "isolation method", and choose mass m_1, mass m_2, the pulley and the contact part between the pulley and the rope as the research object, respectively. Let the acceleration of mass m_2 be vertically down. Because the rope is nonretractable, the accelerations of m_1 and m_2 should be the same in magnitude. Choose the clockwise direction as the positive direction of the angular acceleration β of the pulley.

From the law of rotation, we can write
$$T_2 R - T_1 R = J_z \beta \tag{1}$$
For mass m_1, according to Newton's second law, there is
$$T_1 - m_1 g = m_1 a \tag{2}$$
For mass m_2, according to Newton's second law, we have
$$m_2 g - T_2 = m_2 a \tag{3}$$
From the relationship between the angular and linear quantity, we can get
$$a = R\beta \tag{4}$$
The pulley can be regarded as a uniform disk, so its rotational inertia is

$$J_z = \frac{1}{2}mR^2 \qquad (5)$$

Solving Equations (1)~(5) simultaneously, we can get

$$\beta = \frac{2(m_2 - m_1)}{[2(m_1 + m_2) + m]R}g$$

$$T_1 = \frac{(4m_2 + m)m_1}{2(m_1 + m_2) + m}g$$

$$T_2 = \frac{(4m_1 + m)m_2}{2(m_1 + m_2) + m}g$$

As can be seen from this result, $T_2 \neq T_1$. The reason is that the mass of the pulley is not zero, so $T_2 \neq T_1$ is the necessary condition to make the pulley rotate. Another point that should be noted is that the static friction between the rope and the pulley is the reason why there is no relative sliding between them and is the source of power to rotate the pulley.

Example 4.4.2 A circular plate with radius R is placed flat on the horizontal desktop, and the friction coefficient between the plate and the horizontal desktop is μ, as shown in Figure 4.4.2. At a certain time, the plate starts to rotate with an angular velocity ω_0 about a fixed axis passing through its center and perpendicular to the plate surface. After how many revolutions will it stop (It is known that the rotational inertia of the plate about this axis is $J = \frac{1}{2}mR^2$, where m is the mass of the plate)?

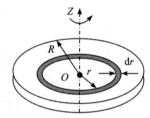

Figure 4.4.2 Figure of Example 4.4.2

Solution Choose a thin ring of radius r and thickness dr, and its area is $dS = 2\pi r\,dr$ and its mass is $dm = \sigma dS$, in which $\sigma = \frac{m}{\pi R^2}$ is the surface mass density. The friction force exerted on this ring is

$$df = \mu\, dmg = 2\pi\sigma\mu gr\, dr$$

So the torque of the friction force on the ring is

$$dM = -r\,df = -2\pi\sigma\mu gr^2\, dr$$

The total torque of the friction force on the entire circular plate is

$$M = \int dM = -\int_0^R 2\pi\sigma\mu gr^2\, dr = -\frac{2}{3}\mu mgR$$

The angular acceleration of the plate can be obtained by the law of rotation

$$\beta = \frac{M}{J_z} = -\frac{4\mu g}{3R}$$

When the plate stops, its angular velocity is $\omega = 0$. Suppose that the plate rotated $\Delta\theta$ before its stop, there should be

$$\omega^2 - \omega_0^2 = 2\beta\Delta\theta$$

Consequently,

$$\Delta\theta = \frac{\omega^2 - \omega_0^2}{2\beta} = \frac{3R\omega_0^2}{8\mu g}$$

The number of rotations is

$$N = \frac{\Delta\theta}{2\pi} = \frac{3R\omega_0^2}{16\pi\mu g}$$

4.5 Kinetic energy and work in rotational motion

4.5.1 Work done by torque

In Figure 4.5.1, a rigid body rotates about the Z axis. If an external force $\boldsymbol{F}_i$ in the rotation plane acts on point P, the point will do circular motion around point O when the rigid body rotates about the fixed axis passing through point O. The position vector of point P is $\boldsymbol{r}_i$. During the process that the body rotates $d\theta$, the displacement of point P is $d\boldsymbol{r}$. According to the definition of work, the work done by the external force $\boldsymbol{F}_i$ is dA_i, which can be expressed by

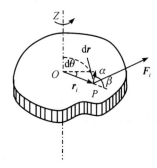

Figure 4.5.1 Work done by torquef

$$dA_i = \boldsymbol{F}_i \cdot d\boldsymbol{r} = F_i |d\boldsymbol{r}| \cos\alpha$$

where α is the angle between $\boldsymbol{F}_i$ and the displacement $d\boldsymbol{r}$. Let $F_i \cos\alpha = F_{i\tau}$ and $|d\boldsymbol{r}| = r_i d\theta$, and we can obtain

$$dA_i = F_{i\tau} r_i d\theta = M_{iz} d\theta \tag{4.5.1}$$

where M_{iz} is the torque of the external force $\boldsymbol{F}_i$ relative to the rotation axis Z, so dA_i is also the work done by the external torque. Equation (4.5.1) indicates that the elemental work done by the external torque equals the product of the external force torque and angular displacement.

If the rigid body rotates an angle displacement $d\theta$ under the action of multiple external forces, the total work dA of the net external torque is equal to the algebraic sum of the work of each external torque, that is

$$dA = \sum_i dA_i = \sum_i M_{iz} d\theta = M_z d\theta \tag{4.5.2}$$

where $\sum_i M_{iz} = M_z$ is the algebraic sum of the torques on the rigid body, named as the net external torque. The elemental work done by the net external torque is equal to the product of the net external torque and angular displacement.

As a result, the total work done by the net external torque during the process that the rigid body rotates from the angular position θ_1 to the angular position θ_2 is

$$A = \int_{\theta_1}^{\theta_2} M_z \, d\theta \tag{4.5.3}$$

It should be pointed out here that the rigid body is a special particle system, where the relative displacement of a pair of interacting internal forces is zero, so the work done by a pair of internal forces is zero. We only need to consider the work of the external torque.

4.5.2 Kinetic energy of rotation

A rigid body can be regarded as a system composed of many particles, so the kinetic energy of rotation of a rigid body is the sum of the kinetic energy of every particle on the rigid body. Let a rigid body rotate about a fixed axis Z with an angular velocity ω, and then each particle of the body should do circular motion in their rotation plane with the same angular velocity ω. Let Δm_i be the mass of any one mass element, r_i away from the axis, and then its linear velocity is $v_i = r_i \omega$ and its kinetic energy is

$$E_{ki} = \frac{1}{2} \Delta m_i v_i^2$$

Taking the relationship between the linear and angular quantities, $v_i = r_i \omega$, into consideration, the kinetic energy for rotation of the entire rigid body is

$$E_k = \sum_i E_{ki} = \sum_i \frac{1}{2} \Delta m_i v_i^2 = \frac{1}{2} \left(\sum_i \Delta m_i r_i^2 \right) \omega^2 = \frac{1}{2} J_z \omega^2$$

In summary, the kinetic energy for the rotation of a rigid body is

$$E_k = \frac{1}{2} J_z \omega^2 \tag{4.5.4}$$

The equation above shows that the kinetic energy of rotation is half of the product of the rotational inertia and the square of the angular speed ω, which has a very similar form to the kinetic energy of the particle.

4.5.3 Kinetic energy theorem of rigid body rotation about a fixed axis

Rewrite the law of rotation as

$$M_z = J_z \beta = J_z \frac{d\omega}{dt} = J_z \frac{d\omega}{d\theta} \frac{d\theta}{dt} = J_z \omega \frac{d\omega}{d\theta}$$

and move $d\theta$ to the left side of the equation

$$M_z \, d\theta = J_z \omega \, d\omega = d\left(\frac{1}{2} J_z \omega^2 \right)$$

where, $M_z \, d\theta = dA$ is the work done by the net external torque during the rotation of $d\theta$, and can be written as

$$dA = d\left(\frac{1}{2} J_z \omega^2 \right) \tag{4.5.5}$$

This equation states that the work done by the net external torque of the rigid body is equal to the differential of the kinetic energy of the rotation of the rigid body about a fixed axis.

If the rigid body rotates from the initial state $\theta=\theta_1$, $\omega=\omega_1$ to the final state $\theta=\theta_2$, $\omega=\omega_2$ under the effect of the external torque, the integral on both sides of Equation (4.5.5) gives

$$\int_{\theta_1}^{\theta_2} M_z d\theta = \int_{\omega_1}^{\omega_2} d\left(\frac{1}{2}J_z\omega^2\right)$$

that is

$$A = \frac{1}{2}J_z\omega_2^2 - \frac{1}{2}J_z\omega_1^2 \qquad (4.5.6)$$

where $A = \int_{\theta_1}^{\theta_2} M_z d\theta$ is the work done by the net external torque of the rigid body during the process that the body's angular velocity changes from ω_1 to ω_2. In other words, **the work done by the net external torque on the rigid body is equal to the increment of the kinetic energy of the rigid body, which is called the kinetic energy theorem of rotation of a rigid body about a fixed axis.**

If a rigid body is subjected to a conservative force, the concept of potential energy can also be introduced. For a system containing rigid bodies, if only conservative forces do work in the process of motion, the mechanical energy of the system is conserved.

Example 4.5.1 A uniform rigid thin rod of length l rotates in the vertical plane about the smooth horizontal fixed axis at its end A, as shown in Figure 4.5.2. Now release the rod from the horizontal position ($\theta_1=0$, $\omega_1=0$). Find out the work done by gravity until the rod rotates to the position of $\theta_2=\frac{\pi}{6}$ and the angular speed of the rod at this position.

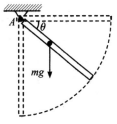

Figure 4.5.2　Figure of Example 4.5.1

Solution As shown in the figure, when the displacement of the rod is $d\theta$, the work done by the torque of gravity is

$$dA = M_z d\theta = mg\frac{l}{2}\cos\theta d\theta$$

During the process that the rod rotates from the position $\theta_1=0$ to $\theta_2=\frac{\pi}{6}$, the total work done by the torque of gravity is

$$A = \int_0^{\frac{\pi}{6}} mg\frac{l}{2}\cos\theta d\theta = \frac{mgl}{4}$$

The angular velocity of the rod at the horizontal position is $\omega_1=0$, and when it arrives at $\theta_2=\frac{\pi}{6}$, the angular velocity is ω. According to the kinetic energy theorem of rotation of rigid body about a fixed axis, there is

$$\frac{mgl}{4} = \frac{1}{2}J_z\omega^2$$

Since the rotational inertia of the rod about the horizontal axis is

$$J_z = \frac{1}{3}ml^2$$

the angular velocity should be

$$\omega = \sqrt{\frac{3g}{2l}}$$

In this case, only gravity works during the rotation of the rod, hence it is also suitable to solve the angular velocity by the conservation of mechanical energy. These two methods should give the same result.

Example 4.5.2 Figure 4.5.3 shows that a light rope is wound around the fixed pulley A, and an object C is hung on the rope after the rope passes over another pulley B. Pulleys A and B can be regarded as uniform disks of radius R_1 and R_2 and with mass m_1 and m_2, respectively. The friction of the axle can be ignored and there is no relative sliding between the light rope and the two pulleys. Find the speed of the object C at the time it falls to h from rest.

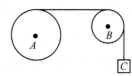

Figure 4.5.3 Figure of Example 4.5.2

Solution Take A, B and C as a system. The work done by the tension in the rope is zero during the motion, so there is only the work done by gravity, making the mechanical energy of the system conserved. Choose the highest position of the object C as the zero reference point of potential energy, at which, both the kinetic energy and the potential energy of the system are zero, so the total mechanical energy is zero. At the final state, the mechanical energy of the system includes the kinetic energy of A, B and C, as well as the potential energy of C. Use ω_1 and ω_2 to represent the angular speed of A and B respectively, and v to refer to the speed of C, and then we can get

$$0 = \frac{1}{2}J_1\omega_1^2 + \frac{1}{2}J_2\omega_2^2 + \frac{1}{2}m_3v^2 - m_3g \tag{1}$$

The rotation inertias of the pulleys A and B are respectively

$$J_1 = \frac{1}{2}m_1R_1^2, \quad J_2 = \frac{1}{2}m_2R_2^2 \tag{2}$$

Since there is no relative sliding between the light rope and the two pulleys, the speed of the object C is equal to the speed of the point at the edge of the pulley. The relationship between the linear quantity and the angular quantity gives

$$v = R_1\omega_1 = R_2\omega_2 \tag{3}$$

Combining Equations (1)~(3) yields

$$v = 2\sqrt{\frac{m_3g}{m_1+m_2+2m_3}}$$

This problem can also be solved by other two methods. The first one is to find the acceleration of the object C, and consequently get its speed at point h through integral. The second one is to apply the kinetic energy theorem.

Chapter 4 Fundamentals of Rigid Body Mechanics 97

4.6 Angular momentum of a rigid body and conservation of angular momentum

4.6.1 Angular momentum of a rigid body

As shown in Figure 4.6.1, when a rigid body rotates with an angular velocity ω about the Z axis, mass elements on the body do circular motion in the rotation plane with the same angular velocity about the Z axis. Suppose that a mass element has the mass of Δm_i and is a distance r_i away from the axis, then the angular momentum of this mass element about the Z axis is $L_{zi} = \Delta m_i r_i^2 \omega$. Consequently, the angular momentum of the

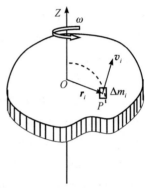

Figure 4.6.1 Angular momentum of a rigid body

entire rigid body about the Z axis is the sum of the angular momentums of all the particles on the body, i.e.

$$L_z = \sum_i \Delta m_i r_i^2 \omega = J_z \omega \qquad (4.6.1)$$

It is called the angular momentum of a rigid body about a fixed axis.

4.6.2 Angular momentum theorem

Since the rotation inertia of a rigid body is a constant value during the rotation of the body, the law of rotation Equation (4.2.9) can be rewritten as

$$M_z = J_z \frac{d\omega}{dt} = \frac{d(J_z \omega)}{dt} = \frac{dL_z}{dt} \qquad (4.6.2)$$

where $L_z = J_z \omega$. Separate the variables and take the integral from the initial state of θ_1 and ω_1 to the final state of θ_2 and ω_2

$$\int_{\theta_1}^{\theta_2} M_z \, dt = \int_{\omega_1}^{\omega_2} dL_z \qquad (4.6.3)$$

The result is

$$\int_{\theta_1}^{\theta_2} M_z \, dt = J_z \omega_2 - J_z \omega_1 \qquad (4.6.4)$$

where $\int_{\theta_1}^{\theta_2} M_z \, dt$ is the moment of impulse of the net external force about the Z axis on the rigid body. Equation (4.6.4) indicates that the increment of that angular momentum of a rigid body is equal to the moment of impulse of the net external force, which is the law of angular momentum of rotation of a rigid body about a fixed axis.

4.6.3 Conservation of angular momentum

If the net external torque M_z on the rigid body is always to be zero, that is,

$$\frac{dL_z}{dt} = \frac{d(J_z\omega)}{dt} = 0$$

This could result in

$$L_z = J_z\omega = \text{constant} \qquad (4.6.5)$$

The equation shows that the angular momentum of a rigid body could remain constant when the net external torque on the body is zero, and this is called the law of conservation of angular momentum.

It can be proved that the angular momentum of a deformable object can also be conserved when it moves with the net external torque being zero. That is, the law of conservation of angular momentum is suitable for not only a rigid body but also a deformable object. For example, ballet dancers and divers can make many beautiful movements by changing their angular speed through altering their rotational inertia. Just like the law of conservation of momentum and the law of conservation of energy, the law of conservation of angular momentum is one of the universal laws in nature. It is applicable not only to the macroscopic fields including celestial bodies, but also to the microscopic fields such as atoms and nuclei.

4.6.4 Application of conservation of angular momentum in engineering technology

The law of conservation of angular momentum has a wide application in engineering technology such as helicopters and gyroscopes.

When the main propeller of a helicopter rotates, the torque of the accelerated main propeller is the internal torque to the system, while the sum of it and the internal torque acting on the fuselage is zero, so the internal torque has no effect on the angular momentum of the system. When the torque of the air resistance on the main propeller is ignored, the external torque is zero at this time, so the angular momentum of the system is conserved. If the angular acceleration of the main propeller increases, the fuselage will rotate in the opposite direction to offset the increased angular momentum due to the acceleration of the main propeller, so that the total angular momentum of the system remains unchanged. In order to prevent the rotation of the fuselage, a small propeller needs to be installed at the tail to generate an additional torque which offsets the torque generated by the main propeller, as shown in Figure 4.6.2.

Figure 4.6.2 Helicopter

The structure of the gyroscope is shown in Figure 4.6.3. The rotor, inner ring and outer ring can rotate freely around their respective rotation axes, and the three rotating axes are orthogonal to the center of mass of the rotor. In the case of ignoring friction and air resistance, this design ensures that the external torque of the rotor is zero and the angular momentum of the rotor is conserved.

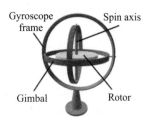

Figure 4.6.3   Gyroscope

The rotational inertia of the gyroscope does not change with time. If the rotation axis of the gyroscope is pointed to a certain direction and the rotor rotates about its own axis of symmetry at a high angular speed ω, no matter how to change the orientation of the frame, the spatial orientation of the spindle remains unchanged. According to this characteristic of gyroscopes, if they are installed on ships, aircraft, missiles, or spacecraft, they can point out the direction of these devices relative to a certain direction in space, correct their possible direction deviation in operation at any time, and play a role in navigation. When the gyroscope is used together with the camera on the mobile phone, it can play an anti-shake role, which will greatly improve the camera capability of the mobile phone; In all kinds of flying games, shooting games, and other sensors, the displacement of players' hands is completely monitored by gyroscopes, making it possible to achieve various game operation effects.

Example 4.6.1   A uniform rod of length l and mass m_1 is suspended on the horizontal smooth fixed axis O at one end, as shown in Figure 4.6.4. The rod swings freely from the horizontal position without initial velocity to the vertical position and makes a completely elastic collision with a block with mass of m_2. After the collision, the rod still swings

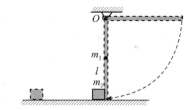

Figure 4.6.4   Figure of Example 4.6.1

in the original direction, and the block slides on the horizontal plane until it stops after a certain distance. The friction coefficient between the block and the horizontal plane is μ. Find the sliding distance of the block on the horizontal plane.

Solution   This problem can be divided into three simple processes:

(1) Select the system composed of the rod and the earth as the research object. In the process that the rod rotates from the horizontal position to the vertical position but before it collides with the block, only gravity torque works, so the mechanical energy of the system is conserved. Let the position where the center of mass of the rod is located when the rod is at the horizontal state be the zero point of potential energy. Let the angular speed of the rod be ω when it is at the vertical position. According to the conservation of mechanical energy, there is

$$0 = -\frac{1}{2} m_1 g l + \frac{1}{2} J_z \omega^2 \tag{1}$$

The rotational inertia of the rod is

$$J_z = \frac{m_1 l^2}{3} \tag{2}$$

(2) Select the system composed of the rod and the block as the research object. During the completely elastic collision, the net torque coming from the axis is zero, so the angular momentum and the energy of the system is conserved. Let the angular speed of the rod after collision be ω' and the speed of the block be v, and there are

$$J_z \omega = J_z \omega' + l m_2 v \tag{3}$$

$$\frac{1}{2} J_z \omega^2 = \frac{1}{2} J_z \omega'^2 + \frac{1}{2} m_2 v^2 \tag{4}$$

(3) Select the object block as the research object. Let the object block slide on the horizontal plane for a distance s after the collision, during which, the kinetic energy of the block changes due to the work done by the friction force. Applying the kinetic energy theorem of the particle yields

$$-\mu m_2 g s = 0 - \frac{1}{2} m_2 v^2 \tag{5}$$

Combining Equations (1)~(5) results in

$$s = \frac{6 m_1^2 l}{(m_1 + 3 m_2)^2 \mu}$$

Example 4.6.2 A rotary table with mass of M and radius of R can rotate about a vertical axis without friction. The initial angular speed of the rotary table is zero. Now, a person with a mass of m moves on the rotary table along the circumference with radius r from rest, as shown in Figure 4.6.5. Find the angle the rotary table has turned relative to the ground by the time that the person walks around the rotary table once and returns to the original position.

Figure 4.6.5 Figure of Example 4.6.2

Solution Suppose that the rotary table rotates θ relative to the ground when the person walks around the rotary table once and returns to the original position. Take the system composed of the person and the rotatory table as the research object. The external force acting on the system includes gravity and support force, making the net external torque zero, so the angular momentum of the system is conserved, that is

$$m r^2 \omega_{m\text{Ground}} - J_z \omega_{M\text{Ground}} = 0$$

From the relationship between the relative angular speeds

$$\omega_{m\text{Ground}} = \omega_{mM} - \omega_{M\text{Ground}}$$

we can obtain

$$\omega_{mM} = \frac{m r^2 + J_z}{m r^2} \omega_{M\text{Ground}}$$

Taking the integral at both sides gives

$$\int_0^t \omega_{mM}\,\mathrm{d}t = \int_0^t \frac{mr^2 + J_z}{mr^2}\omega_{MGround}\,\mathrm{d}t$$

that is

$$2\pi = \frac{mr^2 + J_z}{mr^2}\theta$$

Therefore, the angle that the rotary table rotates relative to the ground is

$$\theta = \frac{mr^2}{mr^2 + J_z}2\pi$$

Here, one thing worth noting is that the law of conservation of angular momentum is only applicable to the inertial reference system, so the rotary table cannot be selected as the reference system.

Scientist Profile

Qian Weichang (1912 – 2010), a famous Chinese dynamicist, applied mathematician, educator, and social activist, was one of the founders of modern mechanics in China. He successively served as the president and vice president of many famous universities in China, and was elected as vice chairman of the National Committee of the Chinese People's Political Consultative Conference for four consecutive terms.

In 1931, Qian Weichang was admitted to the Department of History of Tsinghua University. After the September 18th Incident, Qian Weichang resolutely quit literature and transferred to the Department of Physics, determined to study for the country and use the knowledge for the country. In 1940, Qian Weichang stayed in Canada and majored in elasticity. In 1942, he received a degree of Doctor from the University of Toronto. In May, 1946, Qian Weichang returned to serve the motherland and started to work as a professor in the Department of Machinery of Tsinghua University, and a professor of Peking University and Yanjing University.

Qian Weichang was engaged in mechanical research for a long time, and made outstanding contributions to plate and shell problems, generalized variational principles, analytical solutions of ring shells, and macro font coding of Chinese characters. In his early years, he cooperated with his mentor Singh to study the intrinsic theory of plates and shells, creating a new direction of plate and shell theory, which attracted the attention of the international academic community. The "parameter perturbation method" proposed by him not only solved the problem of large deflection deformation of circular thin plates proposed by von Karman in 1910, but also could be widely used to solve various non-linear partial differential equations, and was called "Qian's perturbation method" by Soviet scholars. He worked on the generalized variational principle, theoretically clarified the relationship between the variational principle and the variational constraints, proposed systematically eliminating the variational constraints by the Laplace multiplier method, and widely applied the generalized variational principle to various theoretical and practical problems of solid mechanics, fluid mechanics, heat transfer, electrics, vibration, fracture mechanics and general mechanics.

Qian Weichang created the first mechanics major in Chinese universities, recruited the first batch of mechanics graduate students, and published China's first monograph *Elasticity*. He participated in the preparation of the Institute of Mechanics and the Institute of Automation of the Chinese Academy of Sciences. In the 1970s, he founded the professional group of rational mechanics and mathematical methods in mechanics in the Chinese Society of Mechanics. In 1980, he founded the earliest academic journal *Applied Mathematics and Mechanics* in China, which promoted the international academic exchange of mechanical research results. Qian Weichang made important contributions to the cause of mechanics in China and the development of the Chinese Society of Mechanics.

Extended Reading

From cat falling and turning to sports biology

Hold up a cat with both hands, make it four feet up, and then suddenly let go. The cat can turn over in the air and fall safely with four feet down. It is reported that a cat fell from the 32nd floor, and only its chest and one tooth were slightly damaged. This familiar phenomenon, which has not been reasonably explained by the principle of mechanics for decades, has become a famous "cat case" in the development of mechanics.

As early as the end of the 19th century, the Frenchman Gulong tried to explain this fact with the principle of conservation of angular momentum. According to him, in the process of falling, the cat first contracts its forelimbs, stretches its hind legs, and then rotates its front half. Because the moment of inertia of the front half relative to the longitudinal axis is less than that of the rear half, the angle of rotation of the front half is larger than that of the rear half in the opposite direction at the same time. Then, the cat stretches out its forelimbs, contracts its hind legs, and turns its back half. According to the same principle, the back half also turns a larger angle. The result is that although the angular momentum is always zero, the cat as a whole can still rotate in one direction. This explanation can be called the theory of "open limbs". Although it conforms to the principle of mechanics, it does not conform to reality. This kind of opening and closing movement of the limbs was not observed at all during the falling process of the cat.

In the 1940s, the Soviet physicist kazynski and others put forward the "tail theory", that is, the cat can maintain momentum conservation by rapidly rotating its tail in one direction and turning its body in the opposite direction around its longitudinal axis.

However, due to the great difference between the mass of the cat's trunk and tail, it is obviously impossible to make the trunk rotate 180° in 1/8 seconds, which requires the speed of the tail almost catch up with that of the propeller of the aircraft. In 1960, the British physiologist MacDonald experimented with a cat with its tail cut off, and the cat could still turn over deftly in the air. Therefore, this "tail turning" theory is also untenable.

In 1969, Kane and Scheer from the USA observed the high-speed photographic photos of the cat falling and turning over. As shown in Figure ER 1, the spine of the cat bends in all directions (front, right, rear, left) in turn during the falling process, much like the conical movement of the waist when people do gymnastics. When the front half of the cat's body moves this way, the rear half of the body will rotate in the opposite direction, and when the front half of the body finishes one conical movement, the whole body will rotate 180° in the opposite direction. According to this process, the physical model of "double rigid body system" is established. Computer simulation results are completely consistent with the reality. The movement of the cat is the superposition of the turning movement of the whole body with the rear half of the body, and the reverse rotation of the front half of the body relative to the rear half of the body. The angular momentum of the superposition of these two opposite rotating is conserved. Kane's theory has gradually been accepted.

Figure ER 1 Photos of a cat's falling and turning

These discussions about cats have given us great enlightenment, enabling us to explain and design human motion, which gradually gave birth to the discipline of sports biomechanics. Sports biomechanics is a science that studies the laws of human mechanical movement in sports and other human sports. It deals with the kinematic, physiological and anatomical characteristics of various sports movements, as well as the internal and external causes of sports and their interactions, and points out the keys to sports movements, so as to guide training and improve sports level. Usually, high-speed photography is done for sports movements, followed by data processing and mechanical analysis, which is very important in modern sports technology and aerospace technology.

In sports, many actions of the human body are completed in the flying stage. Because it is only affected by gravity, the momentum moment of the human body relative to the center of mass is conserved in the flight stage. From the mechanical interpretation of cat somersault, it can be seen that the human body can also achieve air rotation through correct movements. The forward somersault with half twist movement in diving can be regarded as one of such examples. As shown in Figure ER 2, when leaving the springboard, the diver leans forward slightly to make the reaction force of the springboard pass behind the center of gravity, so as to obtain an initial moment of momentum for the following forward somersault. The moment of momentum is parallel to the horizontal axis of the human body, so the human body cannot rotate around the vertical axis. But then the athlete bends (and then stretches the arm) to cause the difference in the moment of inertia of the upper and lower body relative to the longitudinal axis. The upper and lower bodies are respectively driven around the longitudinal axis, which can realize the 360° rotation before jumping into the water.

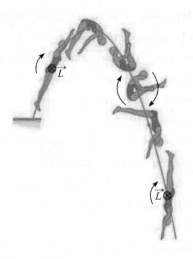

Figure ER 2 Movements of a diver

During space flight, astronauts are in a state of weightlessness. They must wear shoes with hooks to walk on the mesh floor, or use foot covers on the spacecraft to fix themselves in a certain position. However, sometimes, astronauts have to complete various twist tasks in the floating state. At this time, the actions of astronauts will be very different from the routine actions on the ground. For example, if astronauts want to achieve 180° rotation around the longitudinal axis in the static floating state, it will not be possible to use the method of backward rotation on the ground, because the momentum moment of the body is zero, and the rotation of any part of the body will be accompanied by the reverse rotation of another part. In order to enable astronauts to successfully complete various tasks and ensure the stability of their bodies in the weightless floating state, the aerospace department has designed a set of standard actions to train astronauts after complex mechanical research. As long as astronauts raise their arms and make their hands move in circles above their heads, their trunks will slowly rotate in the opposite direction; When the arm movement stops, the rotation of the trunk also stops, so they can turn any angle around the longitudinal axis. Considering the convenience of arm movement and in order to improve the efficiency of rotation, they can also make a conical movement of arms on both sides of the body in the same rotation direction. At this time, the trunk will rotate in the opposite direction. In order to further improve the rotation efficiency, they can use legs with larger mass to replace arms: kick legs apart, then make a half cone movement with the right leg to the right and the left leg to the left, and then retract legs into an upright state.

Discussion Problems

4.1 What factors are related to the rotational inertia of a rigid body? Is it right to say that "a certain rigid body has a certain rotational inertia"?

4.2 A rigid body with a fixed axis is subjected to two forces. When the resultant force of these two forces is zero, is their resultant torque about the axis zero? When the resultant torque of these two forces about the axis is zero, is the resultant force necessarily zero?

4.3 What is the difference between the conservation of momentum and the conservation of angular momentum in a system? Someone says that "the momentum must be conserved in the process of collision". Is this right?

4.4 Some vectors are determined relative to a certain point (or axis), and some vectors are independent of the selection of a fixed point (or axis). Which category do the following types of vectors fall into? (1) position vector; (2) displacement; (3) velocity; (4) momentum; (5) angular momentum; (6) force.

Problems

4.5 The angular position θ of a rotation runner about a fixed axis is a variable of time with the relationship of $\theta = \dfrac{t^3}{3} + \dfrac{t^2}{2} + t + 5$, where all quantities are in international units. Calculate the angular velocity and angular acceleration of the runner at time t.

4.6 There is an engine flywheel that is braked. At the beginning, the angular acceleration of the flywheel is $\beta = -6t$, where all quantities are in SI units, and the flywheel has zero initial condition $\omega_0 = 0$, $\theta_0 = 0$. Try to find the angular velocity and angular position of the flywheel after that.

4.7 The flywheel with mass $m = 60$ kg and radius $r = 0.25$ m, rotates about the horizontal central axis O, and the speed is 900 R · min^{-1}. Now, use a light brake shoe to exert the braking force F in the vertical direction at one end of the brake rod to slow down the flywheel. The dimension of the brake rod is shown in Figure T4-1, the friction coefficient between the brake shoe and flywheel is $\mu = 0.4$, and the rotational inertia of the flywheel can be calculated as a homogeneous disc.

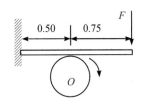

Figure T4-1 Figure of Problem 4.7

(1) For $F = 100$ N, how long does the flywheel need to stop? How many revolutions has the flywheel made during this period?

(2) How much braking force F is needed to halve the flywheel speed within 2 s?

4.8 As shown in Figure T4-2, A and B are two identical fixed pulleys with the light rope. From pulley A hangs an object with mass M, and pulley B is under tension F with the magnitude of $F = Mg$. Try to find the angular accelerations of pulley A and pulley B. The friction between the pulley and the axle is negligible.

4.9 For a uniform thin rod with a mass of M and a length of L, one end B is placed on the edge of the table and the other end is held in a horizontal position by hand, as shown in Figure T4 - 3. Now suddenly release the end A, and calculate the acceleration of the center of mass C of the thin rod at this moment and the angular acceleration (around the end B).

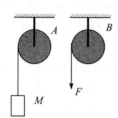
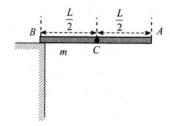

Figure T4 - 2　Figure of Problem 4.8　　Figure T4 - 3　Figure of Problem 4.9

4.10 There is an inhomogeneous rod with length l, which can swing freely in the vertical plane about the smooth axis O perpendicular to the paper surface. The linear mass density at point P on it is $\lambda = 2 + 3x$, as shown in Figure T4 - 4. Now release the rod from the horizontal position and find out the work done by gravity during the rod turning to the vertical position.

4.11 As shown in Figure T4 - 5, two uniform thin rods with equal length and mass rotate about the smooth horizontal axis O_1 and O_2 respectively. When they rotate 90° respectively, calculate the speeds of endpoints A and B.

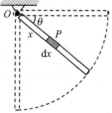
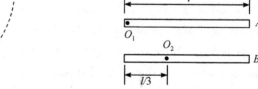

Figure T4 - 4　Figure of Problem 4.10　　Figure T4 - 5　Figure of Problem 4.11

4.12 A uniform thin rod with a mass of M and a radius of R is placed horizontally on a horizontal table with a sliding friction coefficient of μ, and it can rotate about the axis passing through the center and perpendicular to the rod. At the beginning, the rod rotates at an angular speed ω_0. Find

(1) the friction torque on the rod during the rotation;

(2) the work done by the friction torque from the beginning to the rest of the rod.

4.13 Two skaters are walking opposite to each other along two parallel lines 1.5 m apart. The mass of them is $m_A = 60$ kg, $m_B = 70$ kg, and their speeds are $v_A = 7$ m·s^{-1}, $v_B = 6$ m·s^{-1}, respectively. When the two skaters are the closest, they hold hands with each other and start to make a circular motion around the center of mass, keeping the distance between them 1.5 m. At that time, calculate

(1) the total angular momentum of the system formed by these skaters about the vertical axis passing through the center of mass;

(2) the angular velocity of the system;

(3) the total kinetic energy before and after the two skaters joining hands. Is the energy conserved during this process?

4.14 As shown in Figure T4-6, the radius of the pulley is r, the rotational inertia is J, the stiffness coefficient of the spring is k, and the mass of the object is m. It is assumed that the object is stationary at the beginning and the spring does not extend, the rope and pulley do not slide relatively during the falling process of the object, and the friction between the axles is ignored. Try to find

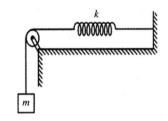

Figure T4-6 Figure of Problem 4.14

(1) the speed of the object when the falling distance of the object is l;

(2) the maximum distance the object can fall.

4.15 A uniform disc-shaped platform with radius R and mass m_1 rotates with a constant angular velocity ω_0 about the smooth vertical axis passing through the center O. A person with mass m_2 stands at the axle of the platform, as shown in Figure T4-7. If this person walks from the axis to the edge of the platform and stops there, what is the angular velocity of the platform?

4.16 A uniform thin bar with a length of l and a mass of m moves in a smooth horizontal plane at a speed v_0 perpendicular to the direction of the bar length and collides completely inelastically with a fixed smooth fulcrum O located $l/4$ away from the center of the rod, as shown in Figure T4-8. Calculate the angular velocity of the rod rotating about the axis at point O perpendicular to the plane of the rod after the collision.

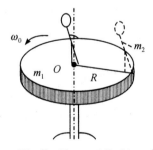

Figure T4-7 Figure of Problem 4.15

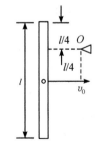

Figure T4-8 Figure of Problem 4.16

Challenging Problems

4.17 As shown in Figure T4-9, the uniform rectangular thin plate has a mass of m, a length of a, and a width of b, and can rotate about the vertical side. When it rotates, its initial angular velocity is ω_0 and it is subject to air resistance. The direction of the air resistance is perpendicular to the plate surface, the air resistance of each small area is proportional to the product of the nth ($n \geq 0$) power of its area and the speed and the scale coefficient is k_n ($k_n > 0$). Calculate the angular velocity and angle of the thin plate. When n takes different positive numbers, what are the characteristics of the relationship curves about the angular velocity and the angle versus time?

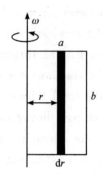

Figure T4-9 Figure of Problem 4.17

4.18 As shown in Figure T4-10, a uniform rod with a length of $2l$ and a mass of m naturally falls from its vertical position on a smooth horizontal plane. What is the relationship of the angular velocity and angular acceleration of the rod and the velocity and acceleration of the center of mass with the angle θ? What is the relationship between angle and time? What is the relationship of the angular velocity and angular acceleration of the rod and the velocity and acceleration of the center of mass with the time t?

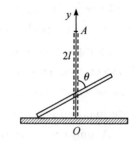

Figure T4-10 Figure of Problem 4.18

4.19 Two skaters are walking opposite to each other along two parallel lines 1.5 m apart. The mass of them is $m_A = 50$ kg, $m_B = 70$ kg, and their speeds are $v_1 = 10$ m/s, $v_2 = 8$ m/s respectively. When the two skaters are the closest, they hold hands with each other and start to make a circular motion around the center of mass, keeping the distance between them 1.5 m, as shown in Figure T4-11. Calculate

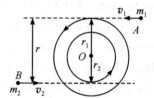

Figure T4-11 Figure of Problem 4.19

(1) the velocity of the center of mass;
(2) the angular velocity of the skaters about the center of mass;
(3) the force between the two skaters.
(4) Is the kinetic energy conserved throughout the motion (The friction from the ground is negligible)?

Chapter 5

Mechanical Oscillation

Oscillation is one of the most common forms of motion in nature, engineering technology and daily life, such as the swaying of branches in the wind, the undulation of waves, the swing of a pendulum, the beating of a heart, the vibration of a spring oscillator and the vibration of the bridge caused by a crossing train. Periodicity is the most obvious character of oscillation. Generally speaking, the repeated variation around a fixed value of any physical quantity (position vector and kinetic energy of an object, alternating current and voltage, strength of an electric or a magnetic field in an oscillating circuit) can be called oscillation. Among different vibration phenomena, the simplest and most basic vibration is simple harmonic vibration. Although the physical nature of vibrations is different, the mathematical description is the same. Therefore, studying the law of mechanical vibration is helpful to understand the law of other kinds of vibration. The Fourier transform shows that any complex vibration can be regarded as a combination of multiple simple harmonic vibrations. Therefore, the study of simple harmonic vibration is the basis of complex vibration. In this chapter, we will mainly discuss the basic properties and laws of simple harmonic vibration.

5.1 Simple harmonic motion

5.1.1 Definition of simple harmonic motion

One end of a spring with negligible mass is fixed and the other end is connected to a block as in Figure 5.1.1. Suppose the spring is always within the elastic range during the movement of the particle, then the system formed by the spring and the block is called a spring oscillator. When the spring is at its original length, the net force on the block is zero. If

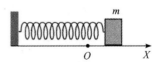

Figure 5.1.1 Spring oscillator

the spring is stretched or compressed, an elastic force pointing to the equilibrium position is generated, forcing the block to return to the equilibrium position. Under the action of this elastic force, the block will do reciprocating motion near its equilibrium position.

Now let us quantitatively analyze the small vibration of the spring oscillator. The mass of the block is m. The stiffness coefficient of the spring is k. Take the stable equilibrium position of the spring at its original length as the origin of coordinates, and the displacement of the block from its equilibrium position is x. Ignore the air resistance during the vibrator motion. According to Hooke's law, the force on the block can be expressed as

$$\boldsymbol{F} = -kx\boldsymbol{i} \tag{5.1.1}$$

The negative sign indicates that the direction of the elastic force is opposite to the displacement direction of the oscillator and always points to the equilibrium position in the motion. The force is proportional to the displacement of the oscillator, and is called the linear restoring force. If the resultant force $\boldsymbol{F}$ is proportional to and in the opposite direction of the displacement x from the equilibrium position, the motion of the object under the action of this restoring force is called simple harmonic vibration. This relation shows the mechanical characteristics of simple harmonic vibration. A body is in simple harmonic motion, if the net force on the body satisfies Equation (5.1.1).

According to Newton's second law, $\boldsymbol{F} = m\boldsymbol{a}$

$$m \frac{d^2 x}{dt^2} = -kx \tag{5.1.2}$$

If $\omega^2 = \dfrac{k}{m}$,

$$\frac{d^2 x}{dt^2} + \omega^2 x = 0 \tag{5.1.3}$$

Equation (5.1.3) is a second-order homogeneous differential equation with constant coefficients. According to advanced mathematics, the general solution to the differential equation is

$$x = A\cos(\omega t + \varphi) \tag{5.1.4}$$

where A and φ are constants, and ω is angular frequency which depends on the nature of the vibration system. Therefore, its kinematics equation can be written in terms of cosine (or sine) and this is why we call this kind of motion simple harmonic motion.

As can be seen from the above analyses, Equation (5.1.1) reflects the dynamic characteristics of the spring oscillator in the vibration process, which is the dynamic equation of simple harmonic oscillation. Equations (5.1.3) and (5.1.4) describe the motion of the vibration process, which are the kinematics equations of simple harmonic motion. The laws of a physical process are governed by the dynamic characteristics of the motion in a system. When the physical quantity describing the motion state of the system satisfies Equation (5.1.3) or Equation (5.1.4), this object is in simple harmonic motion.

Take the derivative of Equation (5.1.4) with respect to time, and the expressions of velocity and acceleration of simple harmonic motion can be obtained

$$v = \frac{dx}{dt} = -\omega A \sin(\omega t + \varphi) \tag{5.1.5}$$

$$a = \frac{dv}{dt} = -\omega^2 A \cos(\omega t + \varphi) \tag{5.1.6}$$

It can be seen that the velocity and acceleration of an object in simple harmonic motion also change periodically with time, as shown in Figure 5.1.2.

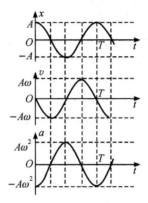

Figure 5.1.2 Displacement, velocity and acceleration curves of simple harmonic motion with time

Example 5.1.1 A pendulum is an ideal vibrating system, and it consists of a small ball of mass m and an inelastic, massless string with the length l that is fixed at one end. The ball is suspended from the other end of the string, as in Figure 5.1.3. The ball is moved a little bit away from the equilibrium position, and the angular displacement of the oscillation θ is small ($\theta < 5°$). If air resistance is ignored, please prove the simple pendulum is in simple harmonic motion.

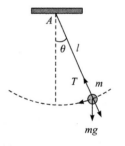

Figure 5.1.3 Figure of Example 5.1.1

Solution The forces exerted on the pendulum are the gravity mg, and the tension of the rope T as shown in Figure 5.1.3. Take the counterclockwise direction as the positive direction of the angular displacement. When the cycloid makes an angle of θ with the vertical direction, the component of the resultant force on the pendulum along the tangent of the arc is the component of gravity in this direction.

$$F_\tau = -mg \sin\theta$$

Because the angular displacement of the oscillation θ is small ($\theta < 5°$), $\sin\theta \approx \theta$ and

$$F_\tau = -mg\theta$$

This expression shows the similar form to Equation (5.1.1). Therefore, in the case of very small angular displacement, the vibration of simple pendulum is simple harmonic motion.

Example 5.1.2 A homogeneous thin rod with length l and mass m can rotate freely in the vertical plane around the axis O at one end, as shown in Figure 5.1.4. Try to prove that when the rod rotates at a small angle, the motion of the rod is simple harmonic motion.

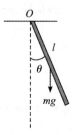

Figure 5.1.4　Figure of Example 5.1.2

Solution The resulting external torque on the rod is the gravitational torque. As shown in the figure, when the thin rod deviates from the equilibrium position by an angle θ, suppose the anticlockwise direction is positive, the moment of gravity on the rod is

$$M = -mg\frac{l}{2}\sin\theta$$

The negative sign indicates that the direction of the gravitational torque is opposite to the direction of the angular displacement. For a small angle rotation, $\sin\theta \approx \theta$. Then

$$M = -mg\frac{l}{2}\theta$$

According to the law of rotation $M = J\beta$

$$J\frac{d^2\theta}{dt^2} + mg\frac{l}{2}\theta = 0$$

If $\omega = \sqrt{\dfrac{mgl}{2J}}$,

$$\frac{d^2\theta}{dt^2} + \omega^2\theta = 0$$

The above equation has the similar form to Equation (5.1.3), so the rotation of the rod at a small angle is a simple harmonic motion.

In order to describe oscillation, some characteristic quantities are needed.

1. **Amplitude** A

In Equation (5.1.4), since the absolute value of cosine cannot be greater than 1, the absolute value of x cannot be greater than A. The object vibrates between $-A$ and $+A$. The absolute value of the maximum displacement (or angular displacement) of a body away from its equilibrium position is called the amplitude of the motion. The amplitude takes a positive value, it gives the range of motion of the particle and its magnitude is generally determined by the initial conditions.

2. **Period (or cycle)** T **and frequency** f

When an object makes simple harmonic motion, the time required to complete a full vibration is called the period of simple harmonic motion. After a period, the vibration state repeats exactly according to the periodicity of the cosine function

$$x = A\cos[\omega(t+T) + \varphi] = A\cos(\omega t + \varphi)$$

The minimum value of time that satisfies this equation should be $\omega T = 2\pi$; therefore

$$T = \frac{2\pi}{\omega} \tag{5.1.7}$$

The number of vibrations completed in unit time is called the frequency of simple harmonic vibration, and its magnitude is equal to the reciprocal of the period.

$$f = \frac{1}{T} = \frac{\omega}{2\pi} \tag{5.1.8}$$

In the SI system, the unit of frequency is the Hertz (Hz).

According to Equation (5.1.7), the angular frequency can be expressed as

$$\omega = 2\pi f \tag{5.1.9}$$

For the spring oscillator, $\omega = \sqrt{\frac{k}{m}}$. The period and frequency of the spring oscillator are respectively

$$T = 2\pi \sqrt{\frac{m}{k}} \tag{5.1.10}$$

and

$$f = \frac{1}{2\pi} \sqrt{\frac{k}{m}} \tag{5.1.11}$$

The mass m and stiffness coefficient k are the inherent properties of the spring oscillator system. It means that the period and frequency of the spring oscillator depend entirely on its own properties. Therefore, they are often referred to as the natural period and natural frequency.

3. Phase $\omega t + \varphi$ and phase constant φ

From Equations (5.1.4), (5.1.5) and (5.1.6), when the amplitude A and angular frequency ω are known, the position, velocity and acceleration of the vibrating body at any time t are completely determined by $\omega t + \varphi$. $\omega t + \varphi$ is called the phase of the motion related to the motion state and described by an angle. The constant φ is called the **phase constant** (or phase angle, initial phase), which depends on the displacement and velocity of the object at time $t = 0$. Because the period of the cosine function is 2π, the phase corresponds to the vibration state varying in the range from 0 to 2π. The vibration state repeats for every 2π change in phase.

Phase is not only a very important concept in simple harmonic motion, but also widely used in wave, optics, electrical engineering, wireless communication technology and so on.

Suppose there are two simple harmonic oscillations, and they can be expressed as $x_1 = A_1 \cos(\omega_1 t + \varphi_1)$ and $x_2 = A_2 \cos(\omega_2 t + \varphi_2)$ respectively. Their phase difference is $\Delta \varphi = (\omega_2 t + \varphi_2) - (\omega_1 t + \varphi_1)$. When $\omega_1 = \omega_2$, the phase difference is $\Delta \varphi = \varphi_2 - \varphi_1$ which is the difference of phase constant irrespective of time. When $\Delta \varphi$ is an even multiple of π, the vibration states of two vibrating objects are exactly the same, and such two vibrations are said to be in phase. And when $\Delta\varphi$ is an odd multiple of π, one of them reaches a positive maximum displacement while the other reaches a negative maximum displacement. Such two vibrations are said to be out of phase.

If $\Delta\varphi=(\varphi_2-\varphi_1)>0$, the phase of the second vibration is said to be $\Delta\varphi$ ahead of the phase of the first vibration (or the first vibration lags $\Delta\varphi$ behind the second vibration).

For a simple harmonic motion, if A, ω and φ are known, the vibration is completely determined. Therefore, these three quantities are called the three characteristic quantities describing simple harmonic motion.

For a simple harmonic motion, how to determine the amplitude and initial phase? ω is determined by the nature of the system itself. Constants A and φ can be determined by the initial conditions of the vibration system. When $t=0$, displacement and velocity are x_0 and v_0 respectively. According to Equation (5.1.5) and Equation (5.1.6)

$$x_0 = A\cos\varphi \qquad (5.1.12)$$
$$v_0 = -A\omega\sin\varphi \qquad (5.1.13)$$

Thus

$$A = \sqrt{x_0^2 + \left(\frac{v_0}{\omega}\right)^2} \qquad (5.1.14)$$

$$\tan\varphi = -\frac{v_0}{\omega x_0} \qquad (5.1.15)$$

It can be seen from the above two equations that the amplitude and initial phase of simple harmonic motion are determined by the initial conditions. The quadrant of the initial phase can be determined by the directions of x_0 and v_0. When $x_0>0$ and $v_0<0$, φ is in the first quadrant. When $x_0<0$ and $v_0<0$, φ is in the second quadrant. When $x_0<0$ and $v_0>0$, φ is in the third quadrant. When $x_0>0$ and $v_0>0$, φ is in the fourth quadrant.

Example 5.1.3 An object vibrates in simple harmonic motion along the OX axis. The amplitude is 0.12 m. The period is 2 s. When $t=0$ s, the coordinate of the object is 0.06 m, and the direction of the object is the positive X direction. Find

(1) the initial phase of the vibration;

(2) the coordinate, velocity, and acceleration of the object when $t=0.5$ s.

Solution The horizontal right direction is chosen as the positive direction of the OX axis, and the kinematics equation of the object is

$$x = A\cos(\omega t + \varphi)$$

(1) According to what is given, $A=0.12$ m and $\omega=\frac{2\pi}{T}=\pi$ rad·s^{-1}.

When $t=0$ s, $x=0.06$ m and $v_0>0$. Then

$$x_0 = 0.06 = 0.12\cos\varphi$$

$$\cos\varphi = \frac{1}{2}$$

Then, $\varphi=\frac{\pi}{3}$ or $\varphi=\frac{5\pi}{3}$. When $t=0$ s, $v_0>0$ and thus $\varphi=\frac{5\pi}{3}$, so the kinematics equation

of the object is

$$x = 0.12\cos\left(\pi t + \frac{5\pi}{3}\right)$$

(2) The coordinate, velocity and acceleration of the object at any time are respectively

$$x = 0.12\cos\left(\pi t + \frac{5\pi}{3}\right)$$

$$v = \frac{\mathrm{d}x}{\mathrm{d}t} = -0.12\pi\sin\left(\pi t + \frac{5\pi}{3}\right)$$

$$a = \frac{\mathrm{d}v}{\mathrm{d}t} = -0.12\pi^2\cos\left(\pi t + \frac{5\pi}{3}\right)$$

When $t = 0.5$ s, the coordinate, velocity and acceleration of the object are respectively

$$x_{0.5} = 0.12\cos\left(\pi \times 0.5 + \frac{5\pi}{3}\right) = 0.104 \text{ m}$$

$$v_{0.5} = -0.12\pi\sin\left(\pi \times 0.5 + \frac{5\pi}{3}\right) = -0.188 \text{ m} \cdot \text{s}^{-1}$$

$$a_{0.5} = -0.12\pi^2\cos\left(\pi \times 0.5 + \frac{5\pi}{3}\right) = -1.03 \text{ m} \cdot \text{s}^{-2}$$

 5.1.2 Simple harmonic motion and uniform circular motion

In order to understand the physical significance of quantities such as amplitude, phase and angular frequency of simple harmonic motion (SHM), an easy way to model SHM is by considering uniform circular motion. As show in Figure 5.1.5, a vector **A** with length A starts from the origin O. When time is 0 s, the angle between the vector **A** and the coordinate axis OX is φ. **A** rotates counterclockwise around the origin O with an angular velocity ω. The endpoint M of **A** will draw a circle in the plane. At time t, the angle between **A** and OX is $\omega t + \varphi$. The coordinate of point P which is the projection of the end point M on the OX axis is

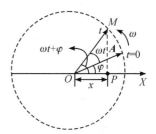

Figure 5.1.5 SHM and uniform circular motion

$$x = A\cos(\omega t + \varphi) \tag{5.1.16}$$

which is the equation of simple harmonic motion. It can be seen that the motion of the projection point P on the OX axis is the simple harmonic motion. Using uniform circular motion to describe simple harmonic motion is very intuitive.

The length of the vector **A** is equal to the amplitude of the simple harmonic motion. When time is 0 s, the angle between the vector **A** and OX is the phase constant. The angular velocity of the rotation vector is equal to the angular frequency of simple harmonic motion.

Example 5.1.4 A particle makes a simple harmonic motion along the OX axis with an amplitude A and a period T, as shown in Figure 5.1.6.

(1) When $t=0$, the particle is at $x_0=\dfrac{A}{2}$ and in the negative X direction. What is the phase constant?

(2) If the particle starts at $x=A$, what is the minimum time required for it to pass through the equilibrium position for the second time?

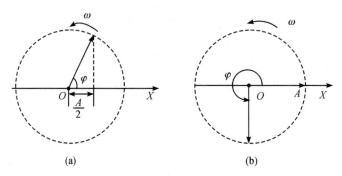

Figure 5.1.6 Figure of Example 5.1.4

Solution (1) When $t=0$, the displacement of the particle is $x_0=\dfrac{A}{2}$. So, the angle between vector **A** and OX axis is $\dfrac{\pi}{3}$ or $-\dfrac{\pi}{3}$. Because the particle is in the negative X direction, the phase constant should be $\dfrac{\pi}{3}$.

(2) The particle starts at $x=A$. When it passes through the equilibrium position the second time, the vector rotates from $\varphi=0$ to $\varphi=\dfrac{3\pi}{2}$. So $\Delta\varphi=\dfrac{3\pi}{2}$. The angle frequency is $\omega=\dfrac{2\pi}{T}$. Thus, the time to pass through the equilibrium position for the second time is $\Delta t=\dfrac{\Delta\varphi}{\omega}=\dfrac{3}{4}T$.

Example 5.1.5 The vibration curve of simple harmonic motion of a particle is shown in Figure 5.1.7, Find the vibration equation of the particle.

Solution Suppose the function of motion for this SHM is $x=A\cos(\omega t+\varphi)$. From Figure 5.1.7, the amplitude is 2 cm. When $t=0$ s, the displacement is $x_0=-\dfrac{A}{2}$. From the phasor notation $\varphi=\pm\dfrac{2\pi}{3}$; for the next time, the particle moves to the negative X direction. Then, $\varphi=\dfrac{2\pi}{3}$.

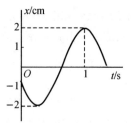

Figure 5.1.7 Figure of Example 5.1.5

The phasor rotates at ω from $t=0$ to $t=1$ s, $\omega\Delta t=\dfrac{4\pi}{3}$. Then $\omega=\dfrac{4\pi}{3}$.

Thus, the function of motion for this SHM is

$$x = 2\cos\left(\frac{4\pi}{3}t + \frac{2\pi}{3}\right) \text{ cm}$$

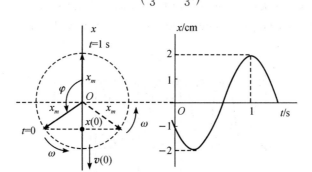

Figure 5.1.8 Figure of Example 5.1.5

5.2　Energy in simple harmonic motion

When a system makes simple harmonic motion, the energy of the vibration system includes kinetic energy and potential energy. Now we still use the spring oscillator as an example to illustrate the energy of a vibration system. The mass of the oscillator is m. The stiffness coefficient of the spring is k. The vibration equation is $x = A\cos(\omega t + \varphi)$. Then the velocity of the oscillator at any time is $v = -\omega A\sin(\omega t + \varphi)$. Therefore, the kinetic energy of the spring oscillator is

$$E_k = \frac{1}{2}mv^2 = \frac{1}{2}m\omega^2 A^2 \sin^2(\omega t + \varphi) \tag{5.2.1}$$

For $\omega^2 = \dfrac{k}{m}$ the kinetic energy can also be expressed as

$$E_k = \frac{1}{2}kA^2 \sin^2(\omega t + \varphi) \tag{5.2.2}$$

The original strength of the spring is the zero point of elastic potential energy. The elastic potential energy of the spring vibrator is

$$E_p = \frac{1}{2}kx^2 = \frac{1}{2}kA^2 \cos^2(\omega t + \varphi) \tag{5.2.3}$$

Then, the mechanical energy of the spring oscillator is

$$E = E_k + E_p = \frac{1}{2}kA^2 \tag{5.2.4}$$

It can be seen that the total energy of the simple harmonic motion system is proportional to the square of the amplitude, and the mechanical energy is conserved in the vibration process. This is because in the process of vibration, only the elastic force of the spring does work, and the elastic force of the spring is conservative. This means that amplitude not only describes the motion range of simple harmonic motion, but also reflects the magnitude of the energy of the vibrating system. Figure 5.2.1 shows the variation curves of kinetic energy, potential energy and total energy with time when the initial phase is zero.

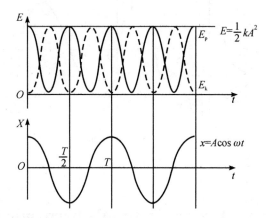

Figure 5.2.1 Curves of kinetic energy, potential energy and total energy of a harmonic oscillator over time

In the process of vibration, the kinetic energy and potential energy change periodically with time, respectively. When the kinetic energy is zero, the potential energy reaches the maximum. And when the potential energy is zero, the kinetic energy reaches the maximum. The kinetic energy and potential energy are converted to each other, but the total mechanical energy remains unchanged.

The average kinetic energy of the spring oscillator over a period is

$$\bar{E}_k = \frac{1}{T}\int_0^T E_k dt = \frac{1}{T}\int_0^T \frac{1}{2}kA^2 \sin^2(\omega t + \varphi) dt = \frac{1}{4}kA^2 \tag{5.2.5}$$

The average potential energy of the spring oscillator over a period is

$$\bar{E}_p = \frac{1}{T}\int_0^T E_p dt = \frac{1}{T}\int_0^T \frac{1}{2}kA^2 \cos^2(\omega t + \varphi) dt = \frac{1}{4}kA^2 \tag{5.2.6}$$

The average kinetic energy and potential energy of simple harmonic motion in a period are equal, which are half of the total energy.

These conclusions are of general significance to any simple harmonic vibration system.

Example 5.2.1 A horizontal spring oscillator with mass m has the equation of motion $x = 2\cos\left(\frac{\pi}{2}t\right)$ cm.

(1) When does the vibrator have the maximum kinetic energy?

(2) What is the coordinate of the oscillator when its kinetic energy is equal to the potential energy?

Solution (1) The velocity of the oscillator is

$$v = \frac{dx}{dt} = -\pi \sin\left(\frac{\pi}{2}t\right)$$

The kinetic energy of the oscillator is

$$E_k = \frac{1}{2}mv^2 = \frac{1}{2}m\pi^2 \sin^2\left(\frac{\pi}{2}t\right)$$

When the kinetic energy is maximum, $\sin^2\left(\frac{\pi}{2}t\right) = 1$

$$\frac{\pi}{2}t = (2n+1)\frac{\pi}{2} \quad (n=0, 1, 2, \cdots)$$

$$t = 2n+1 \quad (n=0, 1, 2, \cdots)$$

(2) When kinetic energy and potential energy are equal

$$\frac{1}{2}m\pi^2 \sin^2\left(\frac{\pi}{2}t\right) = \frac{1}{2}m\pi^2 \cos^2\left(\frac{\pi}{2}t\right)$$

$$\sin^2\left(\frac{\pi}{2}t\right) = \cos^2\left(\frac{\pi}{2}t\right)$$

$$\tan^2\left(\frac{\pi}{2}t\right) = 1$$

Then,

$$\frac{\pi}{2}t = (2n+1)\frac{\pi}{4} \quad (n=0, 1, 2, \cdots)$$

Substitute it into the equation of motion and we get

$$x = \pm\sqrt{2} \text{ cm}$$

5.3 Combination of simple harmonic motions

Simple harmonic motion(SHM) is the most simple and basic vibration. Any complex vibration can be regarded as a combination of simple harmonic motions. The basic knowledge of the composition of vibrations has a wide range of applications in acoustics, optics and radio technology. For example, when two sound waves reach a certain point at the same time, the mass element of air at this point will participate in both vibrations caused by these two waves. Then the motion of the mass element is actually a combination of two vibrations. In practice, the combination of vibrations is generally complicated, which depends on the frequencies, amplitudes and phases of these individual vibrations. Here we discuss the combination of two simple harmonic vibrations in the same direction.

  5.3.1 Combination of two SHMs on the same line and with the same frequency

Suppose that a particle simultaneously participates in two simple harmonic motions with the same direction (along the X axis) and frequency. The vibration equations of the two SHMs are $x_1 = A_1 \cos(\omega t + \varphi_1)$ and $x_2 = A_2 \cos(\omega t + \varphi_2)$, respectively. The displacement of the combined vibration is

$$x = x_1 + x_2 = A_1\cos(\omega t + \varphi_1) + A_2\cos(\omega t + \varphi_2)$$
$$= A\cos(\omega t + \varphi) \quad (5.3.1)$$

where A is the resultant amplitude, and φ is the resultant initial phase.

$$A = \sqrt{A_1^2 + A_2^2 + 2A_1A_2\cos(\varphi_2 - \varphi_1)} \quad (5.3.2)$$

$$\tan\varphi = \frac{A_1\sin\varphi_1 + A_2\sin\varphi_2}{A_1\cos\varphi_1 + A_2\cos\varphi_2} \quad (5.3.3)$$

So, the combination of two simple harmonic motions with the same direction and frequency are still a simple harmonic motion.

Figure 5.3.1 shows the combination by the phasor diagram. The vector sum of individual phasors represents the phasor of the resultant oscillation. The projections of vectors $\boldsymbol{A}_1$ and $\boldsymbol{A}_2$ onto the X axis correspond to x_1 and x_2, respectively. The angles between OX and the vectors are φ_1 and φ_2. The two vectors rotate at the same angular velocity ω. The parallelogram will remain the same shape in the rotating process. By vector superposition, the combination of $\boldsymbol{A}_1$ and $\boldsymbol{A}_2$ is

$$\boldsymbol{A} = \boldsymbol{A}_1 + \boldsymbol{A}_2 \quad (5.3.4)$$

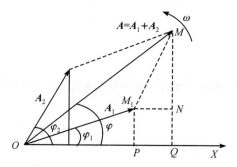

Figure 5.3.1 Phasor diagram for the combination of two SHMs

Meanwhile, the resultant vector rotates at the same angular speed with the constant length A. The projection of $\boldsymbol{A}$ onto the X axis, i.e., the resultant vibration has the equation

$$x = A\cos(\omega t + \varphi) \quad (5.3.5)$$

In the triangle $\triangle OQM$, we can get the length of $\boldsymbol{A}$ using the Pythagorean theorem

$$A = \sqrt{A_1^2 + A_2^2 + 2A_1A_2\cos(\varphi_2 - \varphi_1)} \quad (5.3.6)$$

And the angle between $\boldsymbol{A}$ and OX

$$\tan\varphi = \frac{A_1\sin\varphi_1 + A_2\sin\varphi_2}{A_1\cos\varphi_1 + A_2\cos\varphi_2} \quad (5.3.7)$$

When $\Delta\varphi = \varphi_2 - \varphi_1 = 2k\pi$ $(k = 0, \pm 1, \pm 2, \cdots)$,

$$A = \sqrt{A_1^2 + A_2^2 + 2A_1A_2} = A_1 + A_2 \quad (5.3.8)$$

The resultant amplitude equals the sum of the amplitudes of the two partial vibrations, i.e., the maximum.

When $\Delta\varphi=\varphi_2-\varphi_1=(2k+1)\pi$ $(k=0, \pm1, \pm2, \cdots)$,
$$A=\sqrt{A_1^2+A_2^2-2A_1A_2}=|A_1-A_2| \qquad (5.3.9)$$

The resultant amplitude reaches the minimum, which equals the absolute value of the difference between the amplitudes of the two partial vibrations. If $A_1=A_2$, $A=0$ and the particle remains at rest.

When $\Delta\varphi$ gets any other values, A will range between A_1+A_2 and $|A_1-A_2|$.

Example 5.3.1 A particle participates in two simple harmonic vibrations simultaneously. Their oscillation equations are $x_1=4\cos\left(\frac{\pi}{3}t+\frac{\pi}{6}\right)$ and $x_2=3\cos\left(\frac{\pi}{3}t+\frac{\pi}{2}\right)$ respectively. Find the oscillation equation of the particle.

Solution The oscillation equation of the particle is the resultant oscillation of the two harmonic oscillations. The resultant amplitude and phase are

$$A=\sqrt{A_1^2+A_2^2+2A_1A_2\cos(\varphi_2-\varphi_1)}=\sqrt{4^2+3^2+2\times 4\times 3\times \cos\left(\frac{\pi}{2}-\frac{\pi}{6}\right)}=\sqrt{37}\,\text{cm}\approx 6.08\,\text{cm}$$

$$\tan\varphi=\frac{A_1\sin\varphi_1+A_2\sin\varphi_2}{A_1\cos\varphi_1+A_2\cos\varphi_2}=\frac{4\sin\frac{\pi}{6}+3\sin\frac{\pi}{6}}{4\cos\frac{\pi}{6}+3\cos\frac{\pi}{6}}=\frac{5}{6}\sqrt{3}$$

$$\varphi=\arctan\frac{5}{6}\sqrt{3}\approx 0.96$$

The oscillation equation of the particle is

$$x=6.08\cos\left(\frac{\pi}{3}t+0.96\right)\,\text{cm}$$

Example 5.3.2 A particle participates in two simple harmonic vibrations simultaneously. One of the partial oscillations is $x_1=3\cos(\omega t)$ cm. The resultant oscillation equation is $x=3\sqrt{3}\sin(\omega t)$ cm. Find the second partial oscillation equation.

Solution The resultant oscillation equation can be written as

$$x=3\sqrt{3}\cos\left(\omega t-\frac{\pi}{2}\right)\,\text{cm}$$

Figure 5.3.2 shows the vector diagram of vibration composition when $t=0$ s. The amplitude of the second partial oscillation is $A_2=6$ cm by the Pythagorean theorem and the constant phase is $\varphi_2=-\frac{2\pi}{3}$.

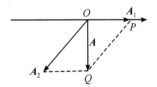

Figure 5.3.2 Figure of Example 5.3.2

Thus, the second partial oscillation equation is

$$x_2=6\cos\left(\omega t-\frac{2\pi}{3}\right)\,\text{cm}$$

5.3.2 Combination of two SHMs on the same line and with different frequencies

The vibration equations of two SHMs on the same line and with different frequencies are respectively

$$x_1 = A_1 \cos(\omega_1 t + \varphi_1)$$
$$x_2 = A_2 \cos(\omega_2 t + \varphi_2)$$

The displacement of the combined vibration is

$$x = x_1 + x_2$$

Because the angular frequencies of these two simple harmonic motions are different (suppose $\omega_1 > \omega_2$), the phase difference is $\Delta\varphi = (\omega_2 - \omega_1)t + \varphi_2 - \varphi_1$ and depends on time. So, the magnitudes of the resultant A and resultant phase $\omega t + \varphi$ also vary with time. This means that the projection of the resultant A is not SHM.

Suppose $\varphi_1 = \varphi_2 = 0$ and $A_1 = A_2 = A_0$, then

$$x = x_1 + x_2 = A_0 \cos(\omega_1 t) + A_0 \cos(\omega_2 t) \tag{5.3.10}$$

Using $\cos\alpha + \cos\beta = 2\cos\dfrac{\alpha-\beta}{2}\cos\dfrac{\alpha+\beta}{2}$,

$$x = 2A_0 \cos\dfrac{\omega_1 - \omega_2}{2}t \cos\dfrac{\omega_1 + \omega_2}{2}t \tag{5.3.11}$$

If $A = 2A_0 \cos\dfrac{\omega_1 - \omega_2}{2}$,

$$x = A\cos\dfrac{\omega_1 + \omega_2}{2}t \tag{5.3.12}$$

When $\omega_2 \approx \omega_1$, we get $\omega_2 + \omega_1 \gg \omega_2 - \omega_1$. Equation (5.3.12) can be regarded as a periodic vibration with varying amplitude $\left|2A_0 \cos\dfrac{\omega_1 - \omega_2}{2}\right|$ and angular velocity $\dfrac{\omega_1 + \omega_2}{2}$. $\left|2A_0 \cos\dfrac{\omega_1 - \omega_2}{2}\right|$ changes within $0 \sim 2A_0$. We can find that the amplitude of the resultant vibration changes periodically with time, which means the resultant amplitude will appear strong or weak. This phenomenon is called beat. Figure 5.3.3 shows the combination of the beat phenomena. The period of the absolute value of the cosine function is π. Thus, the period of varing amplitude τ is $\tau = \left|\dfrac{2\pi}{\omega_1 - \omega_2}\right|$ and frequency of varying amplitude is

$$f_{\text{beat}} = \left|\dfrac{\omega_1 - \omega_2}{2\pi}\right| = |f_2 - f_1| \tag{5.3.13}$$

which is the number of amplitude changes per unit time, called beat frequency. Its value equals the frequency difference of two partial vibrations.

The beat phenomenon has many important applications in technology, such as the superheterodyne radio and beat-frequency oscillator that produces very low frequency electromagnetic oscillation. In addition, musicians use beat phenomenon to tune instruments. Simply make an instrument sound together with the standard frequency and tune it until the beat disappears, and the instrument's frequency will be the same as the standard frequency.

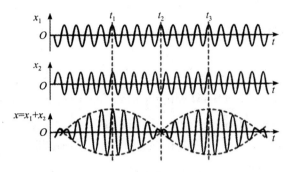

Figure 5.3.3 Beat

 ### 5.3.3 Combination of two SHMs with the same frequency and perpendicular to each other

Suppose a particle simultaneously participates in two simple harmonic motions with the same frequency and in the directions of X and Y axes, respectively. Their corresponding oscillation equations are $x = A_1 \cos(\omega t + \varphi_1)$ and $y = A_2 \cos(\omega t + \varphi_2)$ respectively. Eliminating time t from the above two equations, we can get the trajectory equation of the combined oscillation.

$$\frac{x^2}{A_1^2} + \frac{y^2}{A_2^2} - \frac{2xy}{A_1 A_2}\cos(\varphi_2 - \varphi_1) = \sin^2(\varphi_2 - \varphi_1) \tag{5.3.14}$$

We can see that the trajectory of the combined vibration of two simple harmonic motions with the same frequency and perpendicular to each other is generally elliptical. The shape and orientation of the ellipse depend on the phase difference $\varphi_2 - \varphi_1$ and amplitude of two partial SHMs.

(1) When $\varphi_2 - \varphi_1 = 0$, Equation (5.3.14) becomes

$$y = \frac{A_2}{A_1} x \tag{5.3.15}$$

It shows that the path of the particle is a straight line with a slope $\frac{A_2}{A_1}$ in the first and third quadrants, which indicates that the combination of two SHMs perpendicular to each other and having the same frequency and phase is a SHM along a straight line with a slope of $\frac{A_2}{A_1}$ as Figure 5.3.4(a) shows.

(2) When $\varphi_2 - \varphi_1 = \pi$, Equation (5.3.14) becomes

$$y = -\frac{A_2}{A_1} x \tag{5.3.16}$$

It shows a straight line with a slope $-\dfrac{A_2}{A_1}$ in the second and fourth quadrants. The resultant oscillation is a SHM along a straight line with a slope of $-\dfrac{A_2}{A_1}$, as Figure 5.3.4 (b) shows.

In the two cases above, at any given time, the displacement of the particle from its equilibrium position is

$$r = \sqrt{x^2 + y^2} = \sqrt{A_1^2 + A_2^2}\cos(\omega t + \varphi) \qquad (5.3.17)$$

Thus, the resultant oscillation is a SHM with the same frequency as the partial oscillation and the amplitude is $\sqrt{A_1^2 + A_2^2}$. The difference is that the line in Case (1) goes through the first and third quadrants, while the line in Case (2) goes through the second and fourth quadrants.

(3) When $\varphi_2 - \varphi_1 = \dfrac{\pi}{2}$, the partial oscillation in the Y direction is $\dfrac{\pi}{2}$ ahead of the partial oscillation in the X direction. Equation (5.3.14) becomes

$$\frac{x^2}{A_1^2} + \frac{y^2}{A_2^2} = 1 \qquad (5.3.18)$$

The trajectory of the combined vibration is an ellipse with X and Y axes as its two axes, as shown in Figure 5.3.4 (c). When $\varphi_2 - \varphi_1 = \dfrac{\pi}{2}$, or $\varphi_2 = \varphi_1 + \dfrac{\pi}{2}$, the oscillation equation of the partial vibration in the Y direction is $y = A_2 \cos\left(\omega t + \varphi_1 + \dfrac{\pi}{2}\right)$. When $\omega t + \varphi_1 = 0$, $x = A_1$ and $y = 0$. When $\omega t + \varphi_1$ is slightly greater than zero, x is slightly smaller than A_1, $\omega t + \varphi_1 + \dfrac{\pi}{2}$ is slightly greater than $\dfrac{\pi}{2}$, and $y = A_2 \cos\left(\omega t + \varphi_1 + \dfrac{\pi}{2}\right) < 0$. The particle will be in the fourth quadrant. It means that the particle moves clockwise along an ellipse over time. When $A_1 = A_2$, the particle moves clockwise along a circle.

(4) When $\varphi_2 - \varphi_1 = -\dfrac{\pi}{2}$, the partial oscillation in the Y direction is $\dfrac{\pi}{2}$ behind the partial oscillation in the X direction. According to the same analytical method with Case (3), the particle moves anticlockwise along an ellipse over time, as Figure 5.3.4(d) shows.

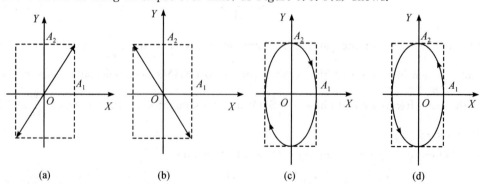

Figure 5.3.4 Combination of two SHMs with the same frequency and perpendicular to each other

(5) When $\varphi_2 - \varphi_1$ equals any other value, the trajectory of the combined oscillation is generally an inclined ellipse. The direction and magnitude of the long and short axes and the direction of motion are determined by the amplitude and phase difference of the oscillation.

5.3.4 Combination of two SHMs with different frequencies and perpendicular to each other

In general, because the phase difference is not a constant value, the combination of two SHMs with different frequencies and perpendicular to each other is more complex. If the frequency ratio between two partial oscillations is an integer, the resultant trajectory curve is closed which is called the Lissajous figure as shown in Figure 5.3.5. Using the Lissajous figure, we can obtain the unknown frequency of an oscillation from an oscillation with a known frequency. This is the method commonly used in radio technology to determine the frequency of oscillation. If the frequency ratio between two partial oscillations is not an integer, the resultant trajectory curve is not closed. The resultant motion will no longer be periodic.

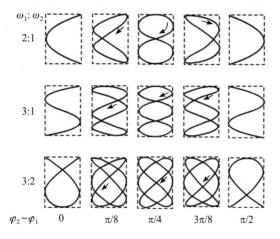

Figure 5.3.5　The Lissajous figure

5.4　Damped and forced oscillation

5.4.1 Damped oscillation

The simple harmonic oscillation discussed earlier is just an ideal case, that is, the vibration of an object under ideal conditions without any resistance during the movement:

The vibration is of equal amplitude, and the energy of the vibration system is conserved during the vibration process. In actual vibration, vibration systems are always subject to resistance. Therefore, the energy obtained at the beginning of the vibration system gradually decreases during the vibration process due to the continuous work done to overcome the resistance, until the vibration stops. This kind of vibration whose amplitude decreases continuously due to the resistance on the vibration system is called damped oscillation.

There are generally two ways in which the energy of a vibrating system is reduced due to damping. One is that due to the frictional resistance on the vibration system, part of the energy is gradually converted into the energy of molecular thermal motion through friction, which is called frictional damping, such as vibration of a spring vibrator in air. The other is that the vibration system causes the vibration of the adjacent particles, so that the energy of the vibration system is radiated to the surrounding, which is called radiation damping. For example, the vibration of a tuning fork not only consumes energy due to friction, but also reduces energy due to radiating sound waves.

The law of actual damped vibration is more complicated. In the following we take the spring vibrator as an example to discuss the effect of damping on vibration. Due to the existence of friction, the spring vibrator is also affected by resistance in addition to elastic force. Experiments show that when the oscillator velocity is not too large, it can be considered that the resistance is proportional to the velocity, but the direction is opposite to the velocity of the vibrator.

$$f = -\gamma v = -\gamma \frac{dx}{dt} \tag{5.4.1}$$

where γ is damping constant. Under the combined action of elastic force and resistance, the dynamic equation of the spring oscillator becomes

$$m \frac{d^2 x}{dt^2} = -kx - \gamma \frac{dx}{dt} \tag{5.4.2}$$

If $\omega_0 = \sqrt{\frac{k}{m}}$ which is the natural angular frequency of the oscillator without damping and $\beta = \frac{\gamma}{2m}$ which is the damping factor, Equation (5.4.2) becomes

$$\frac{d^2 x}{dt^2} + 2\beta \frac{dx}{dt} + \omega_0^2 = 0 \tag{5.4.3}$$

This is a second-order linear homogeneous differential equation with constant coefficients. Generally, different damping factors result in different solutions to Equation (5.4.3). The solution to this equation is discussed below in three cases.

1. Underdamped

If $\beta < \omega_0$, the solution to Equation (5.4.3) is

$$x = A_0 e^{-\beta t} \cos(\omega t + \varphi_0) \qquad (5.4.4)$$

where $\omega = \sqrt{\omega_0^2 - \beta^2}$ and A_0 and φ_0 are determined by the initial conditions. The curve of displacement versus time during damped vibration is shown in Figure 5.4.1. The dotted line in the figure indicates that the amplitude of damped vibration decays exponentially with time. It is no longer a periodic motion in the strict sense. The quasi-period of damped vibration is

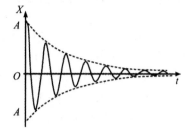

Figure 5.4.1 Underdamped vibration curve

$$T = \frac{2\pi}{\omega} = \frac{2\pi}{\sqrt{\omega_0^2 - \beta^2}} \qquad (5.4.5)$$

This period is greater than the natural period of the simple harmonic vibration of the same system without resistance, which is caused by the slow movement due to resistance.

2. Overdamping

If $\beta > \omega_0$, the solution to Equation (5.4.3) is

$$x = c_1 e^{-(\beta - \sqrt{\beta^2 - \omega_0^2})t} + c_2 e^{-(\beta + \sqrt{\beta^2 - \omega_0^2})t} \qquad (5.4.6)$$

where c_1 and c_2 are constants. At this time, the vibration system no longer reciprocates, but slowly returns to the equilibrium position. Curve B in Figure 5.4.2 is the vibration curve at overdamping, where the spring oscillator is doing damped motion instead of vibrating.

3. Critically damped

If $\beta = \omega_0$, the solution to Equation (5.4.3) is

$$x = (c_1 + c_2 t) e^{-\beta t} \qquad (5.4.7)$$

where c_1 and c_2 are constants. At this time, the vibrator cannot generate vibration, and quickly returns to the equilibrium position from the maximum displacement. Curve A in Figure 5.4.2 is the vibration curve when critical damping occurs. In the case of critical damping, the spring vibrator performs a damped motion instead of vibrating.

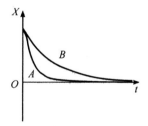

Figure 5.4.2 Critically damped (A) and overdamping vibration (B) curves

In engineering technology, in order to control the motion of a system, the value of damping is often controlled for various needs. For example, a high-sensitivity instrument requires the pointer to return to the equilibrium position quickly and without vibration in order to read as quickly as possible, in which case a critical damping state is needed.

5.4.2 Forced oscillation

If there is no constant replenishment of energy to a vibrating object, vibration will eventually stop due to the damping effect. In order to obtain stable vibration, a periodic external force is usually applied to the vibration system. The vibration of a vibrating system under the action of periodic external force is called forced oscillation. This periodic external force is called driving force.

Still take the spring oscillator as an example to discuss the motion under the action of the driving force at underdamping. The driving force can be simply expressed as

$$F = F_0 \cos(\omega t) \qquad (5.4.8)$$

where F_0 is the amplitude of the driving force, and ω is the angular frequency of the driving force. The dynamic equation of forced vibration can be written as

$$m \frac{d^2 x}{dt^2} = -kx - \gamma \frac{dx}{dt} + F_0 \cos(\omega t) \qquad (5.4.9)$$

If $\omega_0 = \sqrt{\frac{k}{m}}$, $\beta = \frac{\gamma}{2m}$, and $f_0 = \frac{F_0}{m}$,

$$\frac{d^2 x}{dt^2} + 2\beta \frac{dx}{dt} + \omega_0^2 x = f_0 \cos(\omega t) \qquad (5.4.10)$$

This is a second-order linear non-homogeneous differential equation with constant coefficients whose solution is

$$x = A_0 e^{-\beta t} \cos(\sqrt{\omega_0^2 - \beta^2}) + A \cos(\omega t + \varphi) \qquad (5.4.11)$$

The first term is damped vibration, which rapidly decays over time and reflects the transient behavior of forced vibration independent of the driving force. The second term represents a vibration with stable amplitude. After some time, the first term decays to be negligible, so the vibration equation when the forced vibration is stable is

$$x = A \cos(\omega t + \varphi) \qquad (5.4.12)$$

Substitute Equation (5.4.12) into Equation (5.4.10) to obtain

$$A = \frac{f_0}{\sqrt{(\omega_0^2 - \omega^2)^2 + 4\beta^2 \omega^2}} \qquad (5.4.13)$$

$$\tan \varphi = -\frac{2\beta \omega}{\omega_0^2 - \omega^2} \qquad (5.4.14)$$

These results are independent of the initial vibration conditions but depend on the properties of the vibrating system, damping force and the magnitude and properties of the driving force.

5.4.3 Resonance

The amplitude of forced vibration is related to the amplitude and frequency of the driving force, as well as the natural frequency and resistance of the vibration system.

Figure 5.4.3 shows the relationship between the amplitude A of the forced vibration with different damping constants and the angular frequency of the driving force. It can be seen from the figure that when the driving force frequency is a definite value, the amplitude reaches the maximum value, and this phenomenon is called resonance. From Equation (5.4.13), the angular frequency corresponding to the maximum value of the amplitude can be obtained by using the method of finding the extreme value

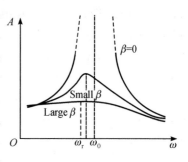

Figure 5.4.3 Resonance

$$\omega_r = \sqrt{\omega_0^2 - 2\beta^2} \tag{5.4.15}$$

The maximum amplitude is

$$A_r = \frac{f_0}{2\beta\sqrt{\omega_0^2 - \beta^2}} \tag{5.4.16}$$

From Equation (5.4.15), when the damping is small $\beta \ll \omega_0$ and the frequency of the driving force is approximately equal to the natural frequency of the vibration system, the displacement amplitude reaches the maximum value, that is, the resonance phenomenon occurs.

When the resonance phenomenon occurs, the direction of the driving force is always consistent with the moving direction of the object, so in the whole cycle, the driving force always does positive work to the system, the energy supplied to the system is the largest, and the vibration amplitude of the object is also the largest. At this time, the system can fully absorb the energy of the external excitation source, and this mechanism is called resonance absorption of energy.

5.4.4 Application of the resonance phenomenon in engineering technology

The resonance phenomenon has important significance in engineering technology. For example, radios and televisions change their natural frequency by changing the parameters of the tuning circuit so that the circuit will resonate with the signal emitted by the radio station or the TV station. In this way, radios and televisions can receive programs that people are interested in. The human ear canal is actually a resonance system with a resonance frequency of 3200 Hz to 3500 Hz. Therefore, the human ear is most sensitive to the sound in this frequency range. Medically, diseases can be diagnosed by using the magnetic resonance phenomenon of atomic nuclei.

Resonance theory is also used in noise control. Converting the energy of noise into other forms through thin plate resonance can eliminate noise. For example, the plywood parapet in the auditorium, rehearsal hall and piano room is the thin plate resonance sound absorption structure. In the structure periphery of thin plywood, hard fiberboard, wood or metal plate is fixed on the frame close to the wall, and a certain thickness of air layer is left between the plate and the wall. When the sound wave hits the thin plate, it causes the plate to vibrate, so that the thin plate is bent and deformed. Due to the friction between the thin plate and the fixed fulcrum and the internal friction of the thin plate itself, the energy of the vibration is converted into heat. If the frequency of the incident wave is equal to the natural frequency of the system, the system will resonate, the amplitude will reach the maximum value at this time, and the corresponding noise energy consumption will also be the largest, achieving the purpose of noise control. Through theoretical calculation, the natural frequency of the thin plate resonance structural system is $f_0 = \frac{1}{2\pi}\sqrt{\frac{\gamma p}{\sigma D}}$ where γ is the gas specific heat ratio, p is the atmospheric pressure, σ is the sheet quality areal density, and D is the distance between the sheet and the wall. It can be seen that the natural frequency of the system is related to the depth D of the cavity and the quality areal density of the sheet. When the thickness of the practical wood sheet sound-absorbing structure is generally 3 - 6 mm and the thickness of the cavity is 30 - 100 mm, the resonance frequency of this system falls within 100 - 300 Hz, so does its frequency of noise control. Since the low frequency wave is closer to the natural frequency of the thin plate, the sound absorption frequency is mainly in the low frequency range. The effective sound absorption range can be improved by combining multiple layers of the thin plate or adding an elastic material layer to the cavity.

Resonance also has a downside. Due to resonance, the vibration of a system is very strong, which will cause damage to the system. For example, resonance generated by an operating machine can affect the accuracy of the machine; resonance caused by high winds or earthquakes can break bridges, snap airplane wings, or damage buildings. In 1940, the Tacoma Bridge in the United States collapsed due to resonance caused by high winds in Figure 5.4.4. Although the wind speed at that time was less than 1/3 of the design wind speed limit, the accident occurred because the actual anti-resonance strength of the bridge did not reach the design value. In 2020, the Humen Bridge in Guangdong shook abnormally in the breeze, and slightly undulating waves appeared on the bridge deck. According to the expert group, the reason for the vibration of the Humen Bridge this time was that the water horses continuously set along the bridge rails changed the aerodynamic shape of the steel box girder. Under the specific wind environment of that day, a Karman vortex street was formed on the downstream side of the main girder of the box body of the bridge. The vibration frequency of this vortex street was just close to the natural frequency of the bridge itself, resulting in the phenomenon of vortex vibration. Therefore, when constructing high-rise buildings, large bridges and extra-large bridges in different regions of the world, in order to ensure the safety of the buildings, designers should analyze wind vibration as one of the main factors and design different anti-wind-vibration structures and anti-resonance dampers.

Figure 5.4.4   Resonance phenomenon

Scientist Profile

Zhou Peiyuan (1902 – 1993), born in Yixing County, Jiangsu Province, was a famous hydrodynamist, theoretical physicist, educator and social activist. He was an academician of Chinese Academy of Sciences and one of the founders of modern mechanics and theoretical physics.

Zhou Peiyuan graduated from Tsinghua University in 1924, and in 1927, he studied at the California Institute of Technology in the United States and obtained a doctorate. He was the first Chinese doctoral student to graduate from the California Institute of Technology. In 1928, he went to the University of Leipzig, Germany, to conduct research on quantum mechanics under the guidance of Heisenberg, and in 1929, he studied at the Technical School of Zurich, Switzerland, under the guidance of Professor Pauli. After returning to China in 1929, he served as a professor at Tsinghua University, Southwest Associated University and Peking University. He was provost of Tsinghua University, president of Peking University, vice president of the Chinese Academy of Sciences, chairman of the Chinese Association for Science and Technology, vice chairman of the World Association for Science and Technology, president of the China Association for the Promotion of International Science and Technology, and chairman of the Chinese Physical Society.

Zhou Peiyuan's academic achievements focused mainly on the study of the theory of gravity in Einstein's general relativity and the turbulence theory in fluid mechanics. In general relativity, Zhou Peiyuan was committed to finding the definite solution to the gravitational field equation and applied it to the study of cosmology. He added the strict harmonic condition as a physical condition to the gravitational field equation and obtained a series of static, steady state and cosmic solutions. He guided the graduate students to carry out the comparison experiment of the speed of light parallel and vertical to the ground, and obtained the relative difference between the two in the order of 10^{-11} for the first time in the world, which had a great impact on the understanding of Einstein's theory of gravity. In the field of turbulence theory, two methods for solving turbulence motion were proposed, which attracted wide attention in the world and led to the formation of a school of "turbulence model theory". This had a profound influence on the study of fluid mechanics, especially turbulence theory. For the first time in the world, he confirmed the law of turbulence energy decay from the initial stage to the later stage and the theoretical results of the Taylor turbulence microscale diffusion law by experiment.

Zhou Peiyuan was engaged in higher education for more than 60 years, and his students were all over the world. Among his early students, Wang Zhuxi, Peng Huanwu, Lin Jiaqiao and Hu Ning all became famous scientists. In the process of teaching, he accumulated rich experience in teaching and running schools, and formed his own teaching style and school running ideology.

Extended Reading

Introduction to nonlinear vibration

Vibration is a physical phenomenon that exists widely in physics and technical science, such as vibration of buildings and machines, electromagnetic vibration in radio technology and optics, self-excited vibration in control systems and tracking systems and acoustic vibration. These seemingly very different phenomena can all be unified into vibration theory through vibration equations. The oscillation law is determined by the properties of various forces acting on a system, or determined by the following equation

$$m\frac{d^2x}{dt^2} + \beta\frac{dx}{dt} + kx = f(t) \tag{1}$$

where m is the vibration mass, x is the vibration displacement, $\beta\frac{dx}{dt}$ is the damping force, kx is the elastic restoring force, and $f(t)$ is the periodic disturbance force. Because the elastic force and damping force are both linear functions, Equation (1) is a second-order linear inhomogeneous differential equation. Such a system is called a linear vibration system. If either or both of the elastic force and damping force is or are nonlinear functions, the vibration equation becomes a nonlinear differential equation

$$m\frac{d^2x}{dt^2} + f_1(x) + f_2\left(\frac{dx}{dt}\right) + kx = f(t) \tag{2}$$

In this case the system is called a nonlinear vibration system.

The study of nonlinear vibration has experienced a long history of development. Galileo, the founder of modern physical science, carried out pioneering research on vibration problems. He discovered the isochronism of simple pendulums and used his free fall formula to calculate the period of simple pendulums. In the seventeenth century,

Huygens noticed the deviation of the pendulum's large swing from isochronism and the synchronization of two frequencies close to the clock, which was the earliest record of the nonlinear vibration phenomenon. The theoretical study of rigorous nonlinear vibrations began in the late nineteenth century, with the theoretical foundation laid by Poincaré. He opened up a whole new direction in the study of vibration problems, namely qualitative theory. In a series of papers from 1881 to 1886, Poincaré discussed the classification of singularities in second-order systems, introduced the concept of limit cycles and established the criterion for the existence of limit cycles, and defined singularities and the exponents of limit cycles. A special and important aspect of qualitative theory is stability theory, the earliest result of which was the stability criterion for the equilibrium position of conservative systems established by Lagrange in 1788. In 1892, Lyapunov gave a strict definition of stability and proposed a direct method to study stability problems. In terms of approximate analytical methods for nonlinear vibrations, Poisson proposed the basic idea of the perturbation method when he studied simple pendulum vibration in 1830, Lindstedt solved the problem of the duration term of the perturbation method in 1883, and Poincaré established the mathematical basis of the perturbation method in 1892. In 1920, Van der Pol proposed the basic idea of slow variable coefficient method when studying the nonlinear oscillation of electron tubes. In 1934, Krylov and Bogolyubov developed it into the average method suitable for general weak nonlinear systems. In 1955, Mitropolsky extended this method to unsteady systems and finally formed the KBM method.

The study of nonlinear vibration gave people a new understanding of the mechanism of vibration. It was recognized that in addition to free vibration and forced vibration, there was another kind of vibration, namely self-excited vibration. In 1926 Van der Poel studied the self-excited vibration of triode tube circuit, and in 1933 Baker's work showed that dry friction with energy input would cause self-excited vibration. The study of nonlinear vibration also helped people to recognize a new form of motion, chaotic vibration, whose discovery and research opened up an active new field, and the discipline of nonlinear vibration entered a new stage of development.

The main task of nonlinear vibration theory is to study the periodic vibration laws (variation laws of amplitude, frequency, and phase) of various vibration systems or to find periodic solutions, as well as to study the stability conditions for periodic solutions. From the perspective of engineering technology, its task is to study how to reduce the vibration of a system or make effective use of the vibration, so that the system has reasonable structural forms and parameters. When studying a vibration system, its damping force and elastic force can sometimes be linearized, and in other cases its nonlinear properties must be considered. Whether to consider the nonlinear properties of the force depends on the nature of the research problem and the required precision. There are various nonlinear factors in the actual mechanical system, such as nonlinearity of electric field force, magnetic field force and gravitational force, kinematic nonlinearity of normal acceleration and the Coriolis acceleration, material nonlinearity of nonlinear constitutive relation and elastic large deformation. Therefore, the vast majority of the vibration systems in engineering practice are nonlinear systems.

Since there is no universally effective accurate solution method for nonlinear differential equations, and the mathematical theory of linear ordinary differential equations has been perfected, replacing nonlinear systems with linear systems is an effective method commonly used in engineering, but only within a certain range. When the nonlinear factor is strong, the results obtained by the linear theory not only have too large errors, but also cannot be used for self-excited vibration, parametric vibration, multi-frequency response, superharmonic and subharmonic vibration, internal resonance, jump phenomena and synchronization phenomena. The above-mentioned practical phenomena appear more and more frequently in modern engineering technology. As early as 1940, the collapse of the Tacoma suspension bridge in the United States due to vibration caused by wind load is a typical example of damage caused by nonlinear vibration. Therefore, it is necessary to develop nonlinear vibration theory, study the methods for nonlinear system analysis and calculation, and explain the physical nature of various nonlinear phenomena to analyze and solve practical nonlinear vibration problems in engineering technology.

The purpose of nonlinear vibration theory research is to determine the qualitative characteristics and quantitative laws of system motion under different parameters and initial conditions based on the mathematical model of nonlinear vibration systems. The mathematical models of nonlinear vibrations are usually nonlinear differential equations. Unlike linear differential equations, there is no universally valid solution for nonlinear differential equations. Therefore, compared with linear vibration systems, it is difficult to obtain accurate analytical solutions for nonlinear vibration systems. For practical nonlinear vibration problems in engineering, in addition to experimental methods, the commonly used theoretical research methods can be divided into geometric methods, numerical methods and analytical methods.

The application of nonlinear vibration systems can be divided into the application of the system with smooth nonlinear restoring force, the application of the system with piecewise linear nonlinearity, the application of the system with hysteretic nonlinear force, the application of self-excited vibration, the application of nonlinear vibration system with shock, the application of nonlinear wave system, the application of slowly varying parameter system, the application of frequency capture principle, the application of bifurcation solution, the application of chaos, etc.

Discussion Problems

5.1 When a spring vibrator performs simple harmonic vibration in the horizontal direction and vertical suspension, is the frequency the same? If it is placed along a smooth slope, will it still vibrate in simple harmonic motion? Has the vibration frequency changed?

5.2 Pull a simple pendulum to make it deviate from the equilibrium position by a small angle θ, and then release it with no initial velocity to make it swing. Is the angle θ the initial phase angle?

5.3 Are the following statements correct?

(1) All periodic motions are simple harmonic motions;

(2) In simple harmonic motion, the period is proportional to the amplitude;

(3) In simple harmonic motion, the total energy is proportional to the square of the amplitude.

5.4 Is the composite vibration of two simple harmonic vibrations in the same direction and with the same frequency a simple harmonic motion? If yes, what is its frequency? What factors does the amplitude depend on?

5.5 What is resonance? What are the conditions for resonance?

Problems

5.6 The vibration equation of a simple harmonic oscillator composed of a small ball with a mass of 10×10^{-3} kg and a light spring is $x = 0.1\cos(8\pi t + \dfrac{2\pi}{3})$, where the unit of t is the second (s) and the unit of x is the meter (m). Find

(1) the angular frequency, period, amplitude and initial phase of the simple harmonic motion;

(2) the maximum value of velocity and acceleration of the ball;

(3) the maximum restoring force.

5.7 A block-spring simple harmonic oscillator vibrates along the X axis with amplitude A and period T. Find the initial phase and function of motion when the block at time $t = 0$ is

(1) at $x_0 = -A$;

(2) at the equilibrium position and starts to move to the $+X$ direction;

(3) at $x = \dfrac{A}{2}$ and starts to move to the $-X$ direction;

(4) at $x = -\dfrac{A}{\sqrt{2}}$ and starts to move to the $+X$ direction.

5.8 An object with mass $m = 10 \times 10^{-3}$ kg vibrates in simple harmonic motion with amplitude of $A = 24$ cm and period of $T = 4.0$ s. When $t = 0$, the object is at $x = 24$ cm. Find

(1) the position of the object and the magnitude and direction of the force acting on the object when $t = 0.5$ s;

(2) the shortest time required for the object to move from the initial position to $x = 12$

cm;

(3) the total energy of the object at $x=12$ cm.

5.9 An object of mass M is suspended from the lower end of a light spring, and the elongation of the spring is 9.8×10^{-2} m. The object is made to vibrate up and down, and the downward direction is specified as the positive direction.

(1) If the object is at a position of 8.0×10^{-2} m above the equilibrium position when $t=0$ and starts to move downward from rest, find the function of motion;

(2) If the object is at the equilibrium position when $t=0$ and starts to move at a speed of 0.60 m·s^{-1} upward, find the function of motion.

5.10 An object with a mass of 0.25 kg vibrates in simple harmonic motion under the action of elastic force, and the coefficient of stiffness $k = 25$ N·m^{-1}. If it starts to vibrate with potential energy of 0.6 J and kinetic energy of 0.2 J, find

(1) the amplitude;

(2) the displacement when the kinetic energy is exactly equal to the potential energy;

(3) the velocity of the object when passing the equilibrium position.

5.11 Figure T5-1 shows the curve of $x(t)$ of a simple harmonic motion. Find the motion function.

5.12 A spring-block oscillator is placed on a smooth horizontal surface and rests at the original length of the spring. A bullet with mass m and velocity v_0 is injected horizontally into it, and the system starts simple harmonic vibration (as shown in Figure T5-2). If the moment when the vibration starts is selected as the initial time $t=0$, find

(1) the initial phase φ;

(2) the amplitude A;

(3) the function of motion.

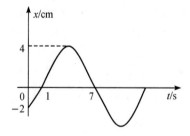
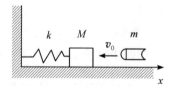

Figure T5-1 Figure of Problem 5.11   Figure T5-2 Figure of Problem 5.12

5.13 The following two simple harmonic motions are on the same line: $x_1 = 0.05\cos(10t + \frac{3}{5}\pi)$ and $x_2 = 0.06\cos(10t + \frac{1}{5}\pi)$. Find the amplitude, initial phase and motion function of the combination of these two simple harmonic motions by the phasor method and algebraic method respectively.

5.14 Figure T5-3 shows the x-t curves of two simple harmonic motions with the same frequency and amplitude. Find

(1) the phase difference of the two simple harmonic motions;
(2) the motion function of the combination of these two simple harmonic motions.

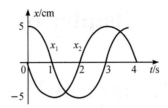

Figure T5-3 Figure of Problem 5.14

5.15 The motion functions of three simple harmonic motions with the same direction and frequency are respectively $x_1 = 0.08\cos(314t + \frac{\pi}{6})$, $x_2 = 0.08\cos(314t + \frac{\pi}{2})$ and $x_3 = 0.08\cos(314t + \frac{5\pi}{6})$. Find

(1) the resultant angular frequency, amplitude, initial phase and motion function of their combination;

(2) the shortest time required for the combined vibration to move from the initial position to $x = \frac{\sqrt{2}}{2}A$ (A is the combined vibration amplitude).

Challenging Problems

5.16 (1) A light rod of length l is connected to a small ball of mass m to form a simple pendulum. Neglecting friction and air resistance, find the relationship between the period and angular amplitude of the simple pendulum.

(2) Demonstrate the animation of the simple pendulum vibration and compare it with the simple harmonic vibration.

(3) While the degree of the pendulum angular amplitude is from 1° to 7° (the interval is 1°), compare the angular position and angular velocity of the simple pendulum motion with the simple harmonic motion. Carry out the same comparisons while the degree of the pendulum angle amplitude is from 30° to 180° (in intervals of 30°).

5.17 A cylindrical rigid body with mass m, radius r, and moment of inertia J_C can roll without sliding on an arc with radius R, and form a circular arc roll. Find the period of the roll motion. Demonstrate the motion of rolling pendulums with the same mass and radius but different moment of inertia (for example, solid and hollow cylinders).

Chapter 6

Mechanical Wave

The study of mechanical waves developed along with the study of sound waves. Sound is one of the earliest physical phenomena studied by human beings, and the early acoustic research work in the world is mainly in music. The three-point profit and loss method（三分损益法）which is the earliest method to specify mathematical natures of music in ancient China is recorded in *Guanzi Diyuan*（《管子·地员》）. In ancient Greece, Pythagoras also proposed a similar natural law, using strings as the basis. The problem of sound propagation had been noticed for a long time. In 1635, someone used the sound of gunfire from a distance to measure the speed of sound, and the method had been continuously improved since then. In 1738, scientists from the Paris Academy of Sciences measured the sound of cannons and found that the speed of sound in the air at 0℃ was 332 m·s^{-1}. In 1827, the Swiss physicist Daniel and French mathematician Sturm conducted experiments on Lake Geneva and found that the speed of sound in water was 1435 m·s^{-1}. This was a remarkable achievement when the only "acoustic instrument" was human ears at that time. In fact, sound waves are one kind of mechanical waves, and the laws that sound waves satisfy are the same as those that mechanical waves follow. This chapter mainly studies the theory of the formation of mechanical waves and the basic principles and laws that they satisfy.

6.1 Generation and basic characteristics of a mechanical wave

6.1.1 Formation of a mechanical wave

To generate mechanical waves, first of all, a mechanical vibration object is needed as the wave source. Secondly, an elastic medium is also necessary for the propagation of mechanical waves.

Without a wave source or elastic medium, mechanical waves cannot be generated. For example, the generation of seismic waves has a vibrational source, and the earth itself is the medium. The generation of sound waves requires a sound generator, and air is the medium. The generation of water waves has a vibration source, and water is the medium. The loud sound produced by a nuclear explosion inside the sun is inaudible on earth for lack of an elastic medium. Of course, not all waves need a medium. For example, electromagnetic waves can propagate freely in a vacuum without a medium.

The elastic medium is composed of an infinite number of continuous mass elements, and there are elastic forces between these mass elements, which can also produce relative motion. When any mass element in the elastic medium leaves the equilibrium position due to vibration, the adjacent mass element will generate an elastic restoring force to make it vibrate near the equilibrium position. At the same time, due to Newton's third law, this mass element will also exert an elastic restoring force on the adjacent mass element, so that the latter also vibrates around its equilibrium position. In this way, the vibration is transmitted from the wave source to the surroundings from near to far at a certain speed, thereby forming a mechanical wave. It should be noted that each mass element only vibrates near its equilibrium position and does not move forward in the propagation direction.

6.1.2 Transverse and longitudinal waves

If the vibration direction of the mass element and the propagation direction of a wave are perpendicular to each other, we call the wave a transverse wave (also called shear wave). For example, hold one end of a rope in your hand and shake it up and down, as shown in Figure 6.1.1. It can be observed that the state of up-and-down vibration at this end propagates along the rope, and

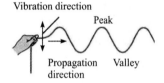

Figure 6.1.1 Transverse wave

the vibration direction of the mass element in the rope is perpendicular to the direction of wave propagation. During the propagation of transverse waves, the mass element arrives at the peak (maximum positive displacement) and the valley (maximum negative displacement) in turn. So the waveform of a transverse wave is characterized by the directional movement of the peaks and valleys.

Take a transverse wave as an example to illustrate its formation process. As shown in Figure 6.1.2, when $t=0$, the mass elements in the medium are in equilibrium. When the mass element at the wave source, i.e., point 1 starts to vibrate with a period T, it drives the adjacent mass element at point 2 to do harmonic vibration with the same period. When $t=T/4$, the vibration propagates to the mass element at point 3 and it begins to do harmonic vibration. When $t=T/2$, the vibration is transmitted to the mass element at point 5. When $t=T$, the mass element at point 1 returns to the equilibrium position after one cycle, and the vibration is transmitted to the mass element at point 9. In this way, the vibration of a mass element in the elastic medium propagates from near to far, forming a mechanical wave.

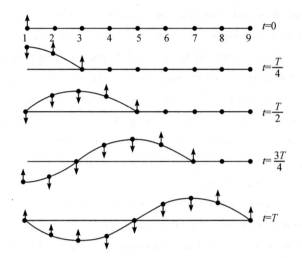

Figure. 6.1.2 Time periodicity of a wave

If the vibration direction of the mass element and the propagation direction of the wave are parallel to each other, we call the wave a longitudinal wave. Take a spring as an example. Fix one end of it to the wall, and hold the other end in your hand to quickly push and pull. As shown in Figure 6.1.3, the left and right vibration state at the left end of the spring propagates along the spring to the fixed end, and the vibration direction of each part of the spring is parallel to the direction of wave propagation. In the process of longitudinal wave propagation, the mass elements present a state of sparse or dense state, so the characteristic of a longitudinal wave's waveform is the directional movement of the sparse and dense state.

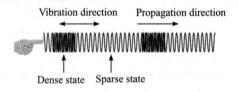

Figure 6.1.3 Longitudinal wave

When vibration propagates in an elastic solid medium, one layer of medium will translate laterally relative to another layer of medium, thereby generating transverse force. Such tangential elastic force cannot be generated in gas and liquid, so transverse waves can only be transmitted in the solid medium. When a longitudinal wave is formed in an elastic medium, the medium needs to be compressed or stretched. That is, a volume change occurs. Solid, liquid, and gas can all undergo volume change, so the longitudinal wave can propagate in all of them. Transverse waves and longitudinal waves are just a simple classification of waves. Some waves are neither pure transverse waves nor pure longitudinal waves, but they can be regarded as the superposition of transverse waves and longitudinal waves.

6.1.3 Wave line and wave surface

In order to describe the propagation of waves in space, the concepts of wave line and wave surface are introduced from a geometrical view. Make some lines with arrows in the direction of wave propagation, and they are called wave lines. The wave surface refers to the surface formed by the points of the same vibration phase in the medium during the wave propagation process, also known as the in-phase surface. The headmost wave surface is called the wavefront. In an isotropic medium, the wave line is always perpendicular to the wave surface and points to the direction that the vibration phase lags behind. According to the geometry of the wave surface, waves can be divided into plane waves, spherical waves, and cylindrical waves. From Figure 6.1.4, the wave lines of a plane wave are straight lines perpendicular to the wave surface and parallel to each other. From Figure 6.1.5, the wave lines of a spherical wave are a family of rays radiating radially and centered on the wave source. Both plane waves and spherical waves are ideal situations. Spherical waves in localized regions far from the source can be considered as plane waves.

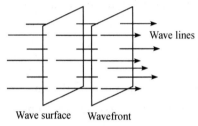

Figure 6.1.4　Wavefront and wavelines of a plane wave

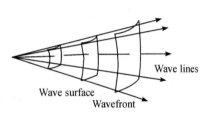

Figure 6.1.5　Wavefront and wavelines of a spherical wave

6.1.4　Characteristic physical quantities describing a wave

1. Wave speed

Wave is the propagation of vibrational states. The distance that the vibration state travels per unit time is called the wave speed, denoted by u. The wave speed is also called the phase speed because the propagation of the wave is actually the propagation of the vibration phase. For mechanical waves, the wave speed depends on the properties of the medium.

It can be shown that the wave speed of the transverse wave in a tensioned rope or string is

$$u = \sqrt{\frac{T}{\mu}} \qquad (6.1.1)$$

where T is the tension in the rope or string, and μ is the mass per unit length of the rope or string.

In a solid, the wave speed of transverse and longitudinal waves can be expressed as follows.

For transverse waves

$$u_\perp = \sqrt{\frac{G}{\rho}} \qquad (6.1.2)$$

For longitudinal waves

$$u_{//}=\sqrt{\frac{E}{\rho}} \tag{6.1.3}$$

where G and E are the transverse elastic modulus and Young's modulus in the medium, respectively, and ρ is the density of the medium. For the same solid medium, there is generally $G < E$, so

$$u_{\perp} < u_{//}$$

Only longitudinal waves can propagate in liquids and gases with a velocity of

$$u=\sqrt{\frac{B}{\rho}} \tag{6.1.4}$$

where B is the bulk modulus of the liquid or gas, and ρ is the mass density of the liquid or gas.

For an ideal gas, the propagation of sound waves is approximated as an adiabatic process. According to the molecular kinetic theory and thermodynamics, the sound speed formula in an ideal gas can be deduced as

$$u=\sqrt{\frac{\gamma p}{\rho}}=\sqrt{\frac{\gamma RT}{M_{mol}}} \tag{6.1.5}$$

where M_{mol} is the molar mass of the gas, γ is the specific heat ratio of the gas, p is the pressure of the gas, T is the thermodynamic temperature of the gas, ρ is the density of the gas, and R is the universal gas constant.

If the vibration of the wave source is periodic, the propagation of the wave in space has both time and spatial periodicity.

2. Wavelength

The length of a complete waveform (the shape of the wave) is called the wavelength (denoted as λ) which is the distance between two adjacent vibrational mass elements with a phase difference of 2π in the wave propagation direction, as shown in Figure 6.1.6. The

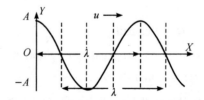

Figure 6.1.6 Spatial periodicity of a wave

distance between two adjacent peaks or between two adjacent valleys on a transverse wave is one wave length. The distance between the corresponding points of two adjacent sparse parts or two dense parts on the longitudinal wave is also a wavelength. On a wave line, two points with a distance of one wavelength vibrate exactly the same, so the wavelength characterizes the spatial periodicity of the wave.

3. Period and frequency

The time it takes for a wave to travel the distance of one wavelength is called the period, denoted by T. The reciprocal of the period is called the frequency, which is the number of complete waves that pass through a point per unit time, denoted by $\nu = 1/T$. From Figure 6.1.6, when the wave source makes a complete vibration, the wave advances one wavelength. So the period of the wave source vibration is equal to the wave period, and the frequency of the wave is also the frequency of the wave source. Since each mass element in the medium repeats the vibration of the wave source in turn, the time required for any mass element in the medium to complete a full vibration is also the period of the wave. At the same time, the time for a complete waveform to pass through a fixed point in the medium is also equal to the period of the wave. From the above analysis, it can be seen that the period of the wave reflects the time period of the wave. In the wave propagation, the wave speed links the time periodicity of the wave with the spatial periodicity. If the mass element vibrates ν times per unit time, the wave advances ν wavelengths forward. In other words, the distance traveled is $\nu\lambda$ which equals the wave speed.

$$u = \frac{\lambda}{T} = \nu\lambda \tag{6.1.6}$$

6.2 Wave function of a plane simple harmonic wave

In order to quantitatively describe a wave, a wave function is often needed to describe how the displacement of each mass element in the medium changes with time. Generally speaking, since the vibration of each mass element in the medium is very complicated, the wave is also very complicated. But it can be proved that any complex wave can be regarded as the superposition of several simple harmonic waves. In a simple harmonic wave, the wave source and all the elements in the medium are in simple harmonic vibration. If the wavefront of a simple harmonic wave is a plane, it is called a plane simple harmonic wave.

6.2.1 Wave function of a plane simple harmonic wave

A plane simple harmonic wave propagates in the positive direction of the X axis, as shown in Figure 6.2.1. Since the phases of the mass elements in the plane perpendicular to the X axis are all the same, the vibration states of all the mass elements on any phase plane can be described by the vibration state of the mass element at the intersection of the plane and the X axis. Therefore, the vibration of the mass elements in the whole

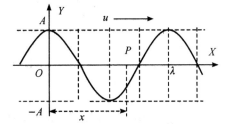

Figure 6.2.1 Plane simple harmonic wave

medium can be simplified to the vibration of the mass element on the X axis. Suppose that the amplitude of the mass element at the origin is A, the angular frequency is ω and the initial phase is φ, the vibration equation of the mass element at the origin is

$$y_O = A\cos(\omega t + \varphi)$$

In order to find the displacement of any mass element at any time during the wave propagation, take an arbitrary point P on the OX axis, with its coordinate x. Obviously, when the vibration propagates from O to P, the mass element at P will repeat the vibration at O. Because the time taken for the vibration to travel from O to P is $\frac{x}{u}$, the vibration at point P lags behind the vibration at point O by time $\frac{x}{u}$. Therefore, the displacement of the mass element at point P at time t is equal to the displacement of the mass element at point O at time $t - \frac{x}{u}$. From this, the displacement of the mass element at P at time t can be written as

$$y_P = A \cos\left[\omega\left(t - \frac{x}{u}\right) + \varphi\right] \quad (6.2.1)$$

Similarly, when the wave propagates in the negative direction of X, the displacement of the mass element at P at time t is

$$y_P = A \cos\left[\omega\left(t + \frac{x}{u}\right) + \varphi\right] \quad (6.2.2)$$

Combining Equations (6.2.1) and (6.2.2), we can get

$$y_P = A \cos\left[\omega\left(t \mp \frac{x}{u}\right) + \varphi\right] \quad (6.2.3)$$

The above expressions represent the displacement of the mass element at any point (x away from the origin) on the wave line at any time t. "$-$" indicates that the wave propagates in the positive direction of the X axis. "$+$" indicates that the wave propagates in the negative direction of the X axis. Using $\omega = 2\pi\nu$ and $u = \nu\lambda$, Equation (6.2.3) can also be written as

$$y(x, t) = A \cos\left[2\pi\left(\nu t \mp \frac{x}{\lambda}\right) + \varphi\right] \quad (6.2.4)$$

$$y(x, t) = A \cos\left[2\pi\left(\frac{t}{T} \mp \frac{x}{\lambda}\right) + \varphi\right] \quad (6.2.5)$$

$$y(x, t) = A \cos\left[\omega t + \varphi \mp \frac{2\pi}{\lambda}x\right] \quad (6.2.6)$$

Equations (6.2.3), (6.2.4), (6.2.5) and (6.2.6) are collectively referred to as the wave function of the plane harmonic wave.

 ### 6.2.2 Physical meaning of the wave function

From the wave function of the plane simple harmonic wave, it can be seen that when the wave characteristics, such as the angular frequency, wave speed, and amplitude of the wave, are determined, the wave function contains only two variables: coordinate x and time t.

When $x = x_0$, the wave function $y = y(x, t)$ is just a function of t, that is, $y = y(t)$. The wave function now represents the displacement of the mass element at x_0 deviating from the equilibrium at different times. So the wave function becomes the vibration equation of the mass element at x_0. Besides, from Equation (6.2.3), the phase of the mass element at x_0 is behind that of the origin and the phase difference is $\Delta\varphi = \frac{2\pi}{\lambda}x_0$. The larger the value of x_0 is, the more the phase lags. Therefore, in the propagation direction, the vibration phase of each mass element lags behind in turn. The phase of the mass element at $x=\lambda$ lags behind the phase of the mass element at $x=0$ by 2π. For the cosine function, the vibration curves of these two points are exactly the same, indicating that the wavelength reflects the periodicity of the wave in space.

When $t=t_0$, the wave function $y=y(x,t)$ is just a function of x, that is, $y=y(x)$. The wave function now represents the displacement of each mass element on the wave line relative to its equilibrium position and is the waveform equation at time t_0. At the same instant, the phases of two mass elements respectively at distances x_1 and x_2 from the origin O are different. According to Equation (6.2.3), it can be concluded that the phases of the two mass elements are respectively

$$\varphi_1 = \omega t_0 + \varphi - \frac{2\pi}{\lambda}x_1$$

$$\varphi_2 = \omega t_0 + \varphi - \frac{2\pi}{\lambda}x_2$$

The phase difference between the vibrations of the two mass elements is

$$\Delta\varphi = \frac{2\pi}{\lambda}(x_2 - x_1) = \frac{2\pi}{\lambda}\delta \qquad (6.2.7)$$

where $\delta = x_2 - x_1$ is wave path difference. The waveform curve at $t=0$ is the same as that at $t=T$, indicating that the period reflects the periodicity of the wave in time.

When both x and t change, the wave function $y = y(x,t)$ represents the displacement of all the mass elements on the wave line at any time, which is the vibration of each mass element. Then the wave function also reflects the propagation of the waveform as shown in Figure 6.2.2.

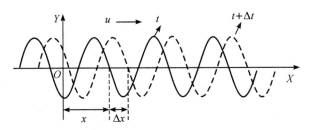

Figure 6.2.2　Wave propagation

The solid line represents the waveform at time t, and the dashed line represents the waveform at time $t + \Delta t$. The vibration state (the phase) travels along the wave line a distance $\Delta x = u\Delta t$, and the entire waveform also travels a distance Δx, so the speed of the wave is the speed at which the waveform propagates forward, and the wave function also describes the propagation of the waveform.

Example 6.2.1 A continuous transverse wave propagates in the positive direction of the X axis at a frequency of 25 Hz. The distance between the centers of adjacent dense parts on the wave line is 24 cm, and the maximum displacement of a mass element is 3 cm. When $t=0$ s, the displacement of the mass element at the origin is zero and it moves in the positive direction of the Y axis. Find

(1) the vibration equation of the mass element at the origin;
(2) the expression of the wave function;
(3) the waveform equation when $t=1$ s;
(4) the vibration equation of the mass element at $x=0.24$ m;
(5) the phase difference of mass element vibrations at $x=0.12$ m and $x=0.36$ m.

Solution (1) From what is given, $A=0.03$ m, $\omega=2\pi\nu=50\pi$ rad·s^{-1}, and the phase constant of the mass element at the origin is $-\dfrac{\pi}{2}$ according to the relationship between SHM and uniform circular motion. The vibration equation of the mass element at the origin is

$$y = 0.03\cos\left[50\pi t - \frac{\pi}{2}\right] \text{ m}$$

(2) The wave moves to the positive direction of the Y axis, so the wave function is

$$y = 0.03\cos\left[50\pi t - \frac{2\pi}{\lambda}x - \frac{\pi}{2}\right] \text{ m}$$

The wavelength is 0.24 m, so the wave function is

$$y = 0.03\cos\left[50\pi t - \frac{25\pi}{3}x - \frac{\pi}{2}\right] \text{ m}$$

(3) When $t=1$ s, the waveform equation is

$$y = 0.03\cos\left[50\pi - \frac{25\pi}{3}x - \frac{\pi}{2}\right] = 0.03\cos\left[\frac{99}{2}\pi - \frac{25\pi}{3}x\right] \text{ m}$$

(4) The vibration equation of the mass element at $x=0.24$ m is

$$y = 0.03\cos\left[50\pi t - \frac{25\pi}{3}\times 0.24 - \frac{\pi}{2}\right] = 0.03\cos\left[50\pi t - \frac{5\pi}{2}\right] \text{ m}$$

(5) The phase difference of mass element vibrations at $x=0.12$ m and $x=0.36$ m

$$\Delta\varphi = \varphi_2 - \varphi_1 = \left(50\pi t - \frac{25\pi}{3}x_2 - \frac{\pi}{2}\right) - \left(50\pi t - \frac{25\pi}{3}x_1 - \frac{\pi}{2}\right) = \frac{25\pi}{3}(x_1 - x_2)$$

If $x_1 = 0.12$ m and $x_2 = 0.36$ m, the phase difference is

$$\Delta\varphi = \frac{25\pi}{3}(x_1 - x_2) = -2\pi \text{ rad}$$

Example 6.2.2 A mechanical wave propagates in the positive direction along the OX axis. The waveform at $t=0$ s is shown in Figure 6.2.3(a). The wave speed is 10 m·s^{-1}. The wavelength is 2 m. Find

(1) the wave function;

(2) the vibration equation of the mass element at point P;

(3) the minimum time required for point P to return to the equilibrium position.

Solution (1) $\lambda = 2$ m, $u = 10$ m/s, $T = \lambda/u = 0.2$ s and $\omega = 2\pi/T = 10\pi$ rad·s^{-1}. From Figure 6.2.3 (a), the amplitude is $A = 0.1$ m. When $t = 0$ s, $y_0 = A/2$ and $v_0 < 0$. Thus the phase constant is $\varphi = \pi/3$. The vibration equation of the mass element at the origin is

$$y_O = 0.1\cos\left(10\pi t + \frac{\pi}{3}\right) \text{ m}$$

The wave function is

$$y = 0.1\cos\left[10\pi\left(t - \frac{x}{10}\right) + \frac{\pi}{3}\right] \text{ m}$$

(2) When $t = 0$ s, $y_P = -A/2$ and $v_P < 0$; the phase constant is $\varphi = 2\pi/3$. The vibration equation of the mass element at point P is

$$y_P = 0.1\cos\left(10\pi t + \frac{2\pi}{3}\right) \text{ m}$$

(3) According to (2) we can draw Figure 6.2.3 (b), and then the angle that the rotation vector should turn while returning to the equilibrium position from point P is

$$\Delta\varphi = \frac{\pi}{3} + \frac{\pi}{2} = \frac{5\pi}{6} \text{ rad}$$

So the minimum time required is

$$\Delta t = \frac{\Delta\varphi}{\omega} = \frac{1}{12} \approx 0.83 \text{ s}$$

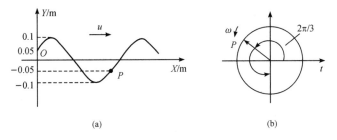

Figure 6.2.3　Figure of Example 6.2.2

6.3　Wave energy and energy flux density

6.3.1　Wave energy

When a wave propagates in an elastic medium, each mass element in the medium vibrates near its respective equilibrium position. So its kinetic energy has a certain value.

At the same time, the relative positions of the mass elements in the medium also change, and there is internal stress (additional internal force per unit area). So there is elastic potential energy. The following is an example of a transverse wave propagating on a rope from which to derive the energy expression of the wave.

Assume that a plane simple harmonic wave propagates along a rope with density ρ and cross-sectional area S. Taking the propagation direction of the wave as the positive direction of the OX axis, and the vibration direction of the rope as the Y axis, the wave can be expressed as

$$y = A\cos\left[\omega\left(t - \frac{x}{u}\right) + \varphi_0\right] \tag{6.3.1}$$

Take a mass element of length Δx at x on the rope, and the volume of the element is $\Delta V = S\Delta x$ and the mass of the element is $\Delta m = \rho \Delta V$.

The vibration velocity is

$$v = \frac{dy}{dt} = -A\omega\sin\left[\omega\left(t - \frac{x}{u}\right) + \varphi_0\right] \tag{6.3.2}$$

The kinetic energy of the mass element is

$$E_k = \frac{1}{2}\Delta m v^2 = \frac{1}{2}\rho \Delta V \omega^2 A^2 \sin^2\left[\omega\left(t - \frac{x}{u}\right) + \varphi_0\right] \tag{6.3.3}$$

As shown in Figure 6.3.1, in the process of wave propagation, the mass element has displacement not only in the Y direction, but also in the X direction from the original length Δx to Δl with the elongation $\Delta l - \Delta x$. When the amplitude of the wave is very small, the tensions T_1 and T_2 on both ends of the mass element can be approximately regarded as equal, $T = T_1 = T_2$. During the elongation of the mass element, the work done by the tension is equal to the potential energy of the mass element

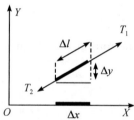

Figure 6.3.1 Mass element deformation during wave propagation

$$E_p = T(\Delta l - \Delta x)$$

When Δx is small, we have

$$\Delta l = \sqrt{(\Delta x)^2 + (\Delta y)^2} = \Delta x\left[1 + \left(\frac{\Delta y}{\Delta x}\right)^2\right]^{1/2} \approx \Delta x\left[1 + \left(\frac{\partial y}{\partial x}\right)^2\right]^{1/2}$$

Using the Taylor expansion and omitting higher-order terms,

$$\Delta l \approx \Delta x\left[1 + \left(\frac{\partial y}{\partial x}\right)^2\right]$$

Therefore, the potential energy of the mass element is

$$E_p = T(\Delta l - \Delta x) = \frac{1}{2}T\left(\frac{\partial y}{\partial x}\right)^2 \Delta x \tag{6.3.4}$$

Take the first derivative of the wave function with respect to x,

$$\frac{dy}{dx} = A\frac{\omega}{u}\sin\left[\omega\left(t - \frac{x}{u}\right) + \varphi_0\right] \tag{6.3.5}$$

The mass per unit length is $\mu = \rho S$ and the wave speed is $u = \sqrt{\dfrac{T}{\mu}}$, so $T = \rho S u^2$. Substitute Equation (6.3.5) and $T = \rho S u^2$ into Equation (6.3.4), and the potential energy expression of the mass element is

$$E_p = \frac{1}{2}\rho \Delta V \omega^2 A^2 \sin^2\left[\omega\left(t - \frac{x}{u}\right) + \varphi_0\right] \tag{6.3.6}$$

The total mechanical energy of the mass element is

$$W = E_k + E_p = \rho \Delta V \omega^2 A^2 \sin^2\left[\omega\left(t - \frac{x}{u}\right) + \varphi_0\right] \tag{6.3.7}$$

It can be seen from Equations (6.3.3) and (6.3.6) that, during wave propagation the kinetic energy and potential energy of a certain volume element at any time have the same value, and their variation laws with time are also the same. When the kinetic energy is the maximum, the potential energy reaches the maximum. When the kinetic energy is zero, the potential energy is also zero. The total energy of the mass element varies periodically with time t. The above results are quite different from those of a single simple harmonic oscillator. When the potential energy of a single simple harmonic oscillator is maximum, the kinetic energy is zero. When the kinetic energy is maximum, the potential energy is zero. But the sum of the two is a constant, that is, the mechanical energy is conserved. In the process of wave propagation, the phase of the kinetic energy and potential energy of each mass element changes synchronously, and the mechanical energy of each mass element in the medium is not conserved. For example, in Figure 6.3.2, as a simple harmonic wave passes through a mass

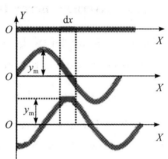

Figure 6.3.2 Schematic diagram of mass element deformation

element with mass m and length L it does simple harmonic oscillation. When the mass element moves to the equilibrium position $y = 0$, its speed and kinetic energy reach the maximum. It shows maximum deformation, so the elastic potential energy reaches the maximum. When the mass element moves to the maximum displacement $y = y_m$, its speed and kinetic energy is zero. There is no deformation, so the elastic potential energy is zero.

It can be seen that the mechanical energy of the line element is the largest at the equilibrium position, and the mechanical energy is zero at the maximum displacement. As a result, the mechanical energy is not conserved. The reason is that in the process of wave propagation, each mass element of the medium continuously exchanges energy with its adjacent mass elements, and the process of wave propagation is also the process of energy propagation.

It should also be noted that the results discussed above for transverse waves are also valid for longitudinal waves. Although the specific expressions of kinetic energy and potential energy are different for longitudinal waves, the conclusion that kinetic energy and potential energy have the same value is generally true. Equation (6.3.7) shows that the energy of the wave is related to the volume of the mass element under consideration. Thus, we define the energy of a unit volume medium as energy density, which is used to represent the distribution of wave energy in the medium, usually represented by w.

$$w = \frac{\Delta W}{\Delta V} = \rho \omega^2 A^2 \sin^2\left[\omega\left(t - \frac{x}{u}\right) + \varphi_0\right] \tag{6.3.8}$$

The average value of the energy density in a period is the average energy density, usually denoted by $\bar{w}$.

$$\bar{w} = \frac{1}{T}\int_0^T w \, dt = \frac{1}{2}\rho\omega^2 A^2 \tag{6.3.9}$$

The above formula indicates that, for a plane simple harmonic wave propagating in an isotropic homogeneous medium, the average energy density of the wave is independent of time and location. It is proportional to the density of the medium, the square of the amplitude, and the square of the angular frequency.

6.3.2 Average energy flux density vector

In Figure 6.3.3, let S be a cross section perpendicular to the direction of wave propagation. Energy flow is the energy passing through a section of a medium in unit time. Energy flow through this section is $uS\rho\omega^2 A^2 \sin^2\left[\omega\left(t - \frac{x}{u}\right) + \varphi_0\right]$. Since the energy flow changes periodically, the average over one cycle is usually taken, which is called the average energy flow. Thus the average energy flow through this section is $\bar{w}uS$.

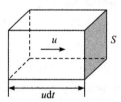

Figure 6.3.3 Relationship between energy density and energy flux

The average energy flux per unit time through a unit area perpendicular to the direction of wave propagation is the average energy flux density, or wave intensity, denoted by I.

$$I = \frac{\bar{w}uS}{S} = \bar{w}u$$

Usually, I is regarded as a vector, and its direction is the propagation direction of the wave, that is, the direction of the wave speed.

$$\boldsymbol{I} = \frac{\bar{w}uS}{S} = \bar{w}\boldsymbol{u} = \frac{1}{2}\rho\omega^2 A^2 \boldsymbol{u} \tag{6.3.10}$$

The vector $\boldsymbol{I}$ is called the average energy flux density vector, also known as the Poynting vector, and its unit is watt · meter^{-2} (W · m^{-2}). When a wave of a certain frequency propagates in an isotropic medium, the intensity of the wave is proportional to the square of amplitude.

6.3.3 Amplitudes of plane and spherical waves

From Equation (6.3.10), the energy flux density of the wave(or wave intensity) is related to the amplitude, so the change of the amplitude can be studied with the help of Equation (6.3.10) and the concept of energy conservation.

1. Plane wave

Suppose a plane wave propagates in a homogeneous medium with a speed u, as shown in Figure 6.3.4. S_1 and S_2 are two plane surfaces with the same area and both perpendicular to the wave lines. Assume that the medium does not absorb the energy of the wave. According to the conservation of energy, the energy passing through the S_1 and S_2 planes in one cycle is equal.

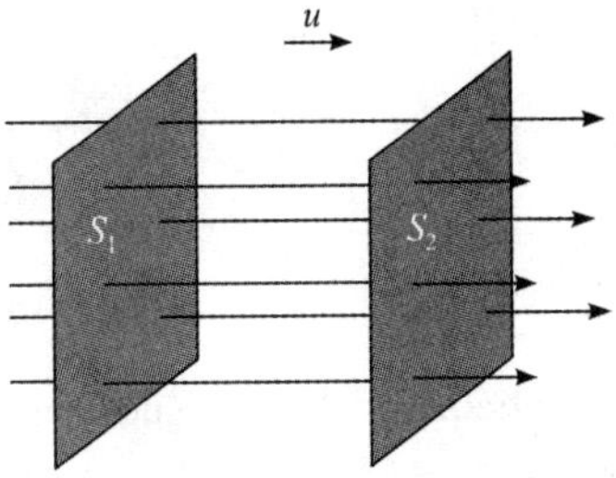

$$I_1 S_1 T = I_2 S_2 T$$

Figure 6.3.4 Plane wave

According to Equation (6.3.10), we can get

$$\frac{1}{2}\rho\omega^2 A_1^2 u S_1 T = \frac{1}{2}\rho\omega^2 A_2^2 u S_2 T$$

For plane waves, $S_1 = S_2$, and thus $A_1 = A_2$. It means that the amplitude of the plane simple harmonic wave remains unchanged when propagating in a homogeneous medium that does not absorb energy.

2. Spherical wave

Choose two spherical surfaces S_1 and S_2 with the wave source as the center and of radius r_1 and r_2 respectively, as shown in Figure 6.3.5. Under the condition that the medium does not absorb the energy of the wave, according to the conservation of energy, the energy passing through the two spherical surfaces in one cycle is

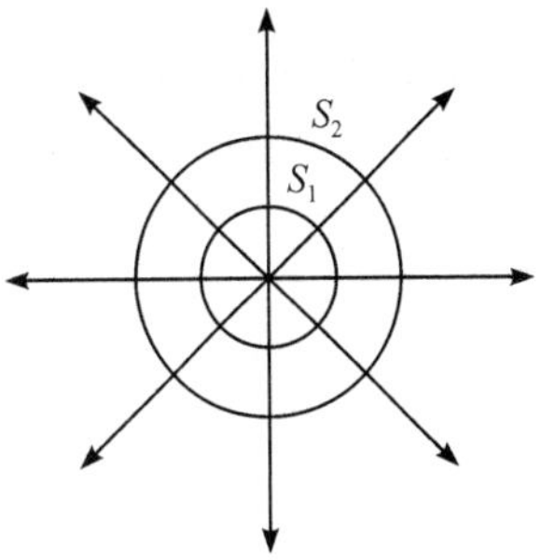

$$I_1 S_1 T = I_2 S_2 T$$

For spherical waves, $S = 4\pi r^2$, and then

Figure 6.3.5 Spherical wave

$$\frac{1}{2}\rho\omega^2 A_1^2 u 4\pi r_1^2 T = \frac{1}{2}\rho\omega^2 A_2^2 u 4\pi r_2^2 T$$

Therefore

$$\frac{A_1}{A_2} = \frac{r_1}{r_2}$$

That is

$$A \propto \frac{1}{r}$$

It can be seen that the amplitude of the spherical wave is inversely proportional to the distance from the point to wave source. Since the vibration phase of the spherical wave with r in the medium is similar to that of the plane wave, the wave function of the spherical simple harmonic wave can be written as

$$y = \frac{A_0}{r} \cos\left[\omega\left(t - \frac{x}{u}\right) + \varphi_0\right]$$

The constant A_0 in the formula can be determined according to the amplitude of a certain wave surface and the corresponding spherical radius.

6.4 Huygens' principle

Theoretical and experimental results prove that when a wave propagates in an isotropic homogeneous medium, the shapes of the wavefront and the wave surface remain unchanged, and the wave line also remains a straight line. But when a wave encounters an obstacle during its propagation, or when it propagates from one medium to another medium, the shapes of the wavefront and the direction of wave propagation change. As shown in Figure 6.4.1, when a planar water wave passes through an obstacle with a small hole, a spherical wave appears behind the small hole and the original wavefront and wave surface change, as if the small hole is a new wave source. Waves emitted from the hole is called wavelets. By summarizing such phenomenon, the Dutch physicist Huygens proposed a principle in 1690, which is called Huygens' principle. In the propagation of waves, every point on a wavefront is the source of spherical wavelets, and the wavelets emanating from different points mutually interfere. The envelopes of these wavelets form the wavefront.

Figure 6.4.1 Diffraction of a water wave passing through a slit

Huygens' principle applies to propagation of any wave, whether it is a mechanical wave or an electromagnetic wave, and whether the medium in which the wave propagates is uniform or not. As long as the wavefront at a certain moment is known, the wavefront at the next moment can be determined by geometrical mapping according to this principle. Thus, the problem of wave propagation direction is solved to a large extent. In the following we take spherical waves and plane waves as examples to illustrate the application of Huygens' principle. As shown in Figure 6.4.2 (a), a plane wave propagates to the right with a speed u, and the wavefront at time t is the plane S_1. After time Δt, every point on plane S_1 can be regarded as the source of spherical wavelets according to Huygens' principle. Draw a number of hemispherical wavelets with points on S_1 as the center and r as the radius. The superposition of these spherical wavelets forms the new wavefront plane S_2. Figure 6.4.2 (b) is a spherical wave centered at O. Assuming that the spherical wave velocity is u, and the wavefront at time t is a sphere S_1 of radius R_1. Every point on plane S_1 can be regarded as the source of new wavelets according to Huygens' principle. With each point on S_1 as the center and $r=u\Delta t$ as the radius, draw a hemispherical wavelet, and the envelope of these spherical surfaces is wavefront S_2 which is a sphere centered at O and with radius $R_2=R_1+u\Delta t$.

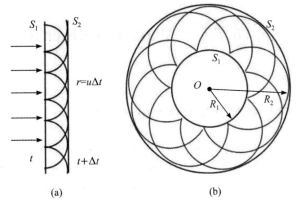

Figure 6.4.2 Using the Huygens' principle to make a new wavefront

Huygens' principle can also qualitatively explain the phenomenon of wave diffraction. When a wave encounters an obstacle in the process of propagation, its propagation direction is deflected around the obstacle or aperture, which is called wave diffraction. The diffraction phenomenon shown in Figure 6.4.1 can be explained by Huygens' principle. Similar to the method shown in Figure 6.4.2, when the wavefront reaches the small hole, each point at the small hole becomes the source of wavelets, and the envelopes of these wavelets form a new wavefront. It can be found that the diffracted wavefront is different from the original planar wavefront, and the wavefront bends near the edge, that is, the wave bypasses the obstacle and continues to propagate.

Huygens' principle can explain not only the propagation of waves in a medium and the diffraction of waves, but also the reflection and refraction of waves at the interface of two media as shown in Figure 6.4.3.

It should be pointed out that since the Huygens principle does not explain the intensity distribution of wavelets, it can only solve the problem of the propagation direction of the wave. In fact, diffracted waves have different intensities in all directions. Later, Fresnel made an important supplement to the Huygens principle, forming the Huygens-Fresnel principle, which has important applications in wave optics.

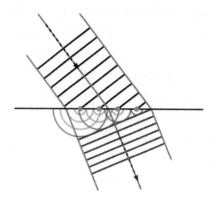

Figure 6.4.3 Refraction of a wave at the interface of two media

6.5 Wave interference

6.5.1 Principle of superposition of waves

It is common for two or more waves to pass through the same area at the same time. After observation and research, the following two laws have been concluded:

(1) Several waves meet in a certain area and then separate during the propagation process. The propagation of each wave is the same as that before they meet, and still maintains its original characteristics (frequency, wavelength, vibration direction, etc.) and continues in the original propagation direction. That is, waves do not interfere with each other, which is called the independence of wave propagation. For example, when several people are speaking or several instruments are playing at the same time, we can distinguish the voice of each person or the sound of different instruments. Electromagnetic waves also have this independence. For example, there are many wireless electromagnetic waves in the air at the same time, and we can receive the broadcast of a certain radio station at will.

(2) In the meeting area, the vibration of the mass element at any place is the result of vibrations caused by each wave alone. That is, at any moment, the displacement of the mass element at this place is the vector sum of the displacements caused by each wave alone. This law is called the principle of superposition of waves.

6.5.2 Interference of waves

In general, the problem that several waves meet and superimpose in space is very complicated. Only one of the simplest and most important wave superposition cases is discussed here, which is, the superposition of two waves with the same frequency, vibration direction, and a constant phase difference. If waves satisfy these three conditions, we can regard them as coherent, and the wave sources are called coherent wave sources. When two coherent waves are superimposed, the resultant vibration at each point where the two waves meet maintains a constant amplitude, and the resultant vibration is always strengthened at some locations and weakened at some other locations. This phenomenon is called wave interference, another important feature of waves which can help to judge whether a certain motion has wave property.

Suppose there are two coherent wave sources S_1 and S_2. Their vibration equations are respectively

$$y_1 = A_1 \cos(\omega t + \varphi_1)$$
$$y_2 = A_2 \cos(\omega t + \varphi_2)$$

The two waves meet at point P at distances r_1 and r_2 respectively from their wave sources, as shown in Figure 6.5.1. Then the vibrations caused by the two waves at point P are

$$y_1 = A_1 \cos\left(\omega t + \varphi_1 - \frac{2\pi}{\lambda} r_1\right)$$

$$y_2 = A_2 \cos\left(\omega t + \varphi_2 - \frac{2\pi}{\lambda} r_2\right)$$

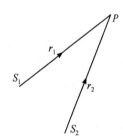

Figure 6.5.1 Wave interference

Since the vibration directions of these two partial vibrations are the same, according to the synthesis rule of vibration with the same direction and frequency, the motion of point P is still a simple harmonic vibration, and the vibration equation is

$$y = y_1 + y_2 = A\cos(\omega t + \varphi)$$

where A is the amplitude of the resultant vibration.

$$A^2 = A_1^2 + A_2^2 + 2A_1 A_2 \cos\Delta\varphi \qquad (6.5.1)$$

where $\Delta\varphi$ is the phase difference of the bipartite vibration at point P.

$$\Delta\varphi = \left(\omega t + \varphi_2 - \frac{2\pi}{\lambda} r_2\right) - \left(\omega t + \varphi_2 - \frac{2\pi}{\lambda} r_2\right) = (\varphi_2 - \varphi_1) - \frac{2\pi}{\lambda}(r_2 - r_1) \qquad (6.5.2)$$

where $\varphi_2 - \varphi_1$ is the initial phase difference between the two coherent wave sources and $\frac{2\pi}{\lambda}(r_2 - r_1)$ is the phase difference caused by the difference in the path difference of the two waves. For a given point P in space, the initial phase difference $\varphi_2 - \varphi_1$ of the two coherent wave sources is certain, and the path difference $\delta = r_2 - r_1$ is also certain. Therefore the phase difference $\Delta\varphi$ is constant. From Equation (6.5.1), it can be seen that there will be different constant amplitude values for different points in space.

According to Equation (6.5.2), in the position where the phase difference satisfies

$$\Delta\varphi = (\varphi_2 - \varphi_1) - \frac{2\pi}{\lambda}(r_2 - r_1) = \pm 2k\pi \quad (k = 0, 1, 2, \cdots) \qquad (6.5.3)$$

the amplitude reaches the maximum $A_{max}=A_1+A_2$. That is, at the positions where the phase difference is zero or an even multiple of π, the vibration is always strengthened, which is called constructive interference.

Similarly, when the phase difference satisfies

$$\Delta\varphi=(\varphi_2-\varphi_1)-\frac{2\pi}{\lambda}(r_2-r_1)=\pm(2k+1)\pi \ (k=0, 1, 2, \cdots) \qquad (6.5.4)$$

the resultant amplitude is minimum $A_{min}=|A_1-A_2|$. That is, at the positions where the phase difference is an odd multiple of π, the vibration is always weakened, which is called destructive interference.

If the initial phases of the two wave sources are the same $\varphi_2=\varphi_1$, then $\Delta\varphi$ is only determined by the path difference $\delta=r_2-r_1$. The above condition is simplified to

$$\delta=r_2-r_1=\pm k\lambda \ (k=0, 1, 2, \cdots) \qquad (6.5.5)$$

$$\delta=r_2-r_1=\pm\frac{(2k+1)}{2}\lambda \ (k=0, 1, 2, \cdots) \qquad (6.5.6)$$

The above two equations show that when the waves emitted by two coherent wave sources with the same initial phase are superimposed in space, for points where the wave path difference is equal to zero or an integer multiple of the wavelength, the interference is strengthened while for points where the path difference is equal to an odd multiple of a half wavelength, the interference is weakened. The phenomenon of interference has a wide range of applications in optics, acoustics, modern physics, and many engineering disciplines.

Example 6.5.1 A and B are two coherent wave sources. The mutual distance is 30 m and the amplitude is the same. A lags behind B in phase by π, the speed of the two waves is 400 m·s^{-1}, and the frequency is 100 Hz. Find the position of each point on the line connecting A and B that is at rest due to interference.

Solution Choose point P for analysis. The distance between point P and wave source A is r_1, and the distance from wave source B is r_2. According to what is given, the wavelengths of the two waves are $\lambda=\dfrac{u}{f}=4$ m.

If point P is located on the left of point A, as shown in Figure 6.5.2 (a), then the path difference of the two waves at point P is

$$\delta=r_2-r_1=30 \text{ m}$$

From Equation (6.5.2), it can be concluded that the phase difference between the two waves at point P is

$$\Delta\varphi=\pi-\frac{2\pi}{\lambda}\delta=-14\pi$$

Similarly, if point P is located on the right side of point B, as shown in Figure 6.5.2 (b), the wave path difference is

$$\delta=r_2-r_1=-30 \text{ m}$$

and the phase difference is

$$\Delta\varphi = \pi - \frac{2\pi}{\lambda}\delta = 16\pi$$

It can be seen that if P is at any point on the left side of point A or on the right side of point B, the condition of constructive interference will be satisfied. Therefore, there will be no points at rest on both sides of A and B due to interference.

If point P is located between the two wave sources A and B, as shown in Figure 6.5.2(c), the path difference of the two waves at point P is

$$\delta = r_2 - r_1 = 30 - 2r_1 \text{ m}$$

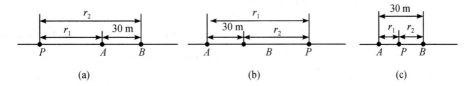

Figure 6.5.2 Figure of Example 6.5.1

To satisfy the condition of destructive interference, the phase difference is

$$\Delta\varphi = \pi - \frac{2\pi}{\lambda}\delta = -14\pi + \pi r_1 = \pm(2k+1)\pi$$

that is

$$r_1 = 14 \pm (2k+1) \quad (k=1, 2, 3, \cdots)$$

Since point P is between the two wave sources of A and B, the points at rest due to interference are $r_1 = 1, 3, 5, \cdots, 25, 27, 29$ m from point A.

Example 6.5.2 As shown in Figure 6.5.3, S_1 and S_2 are two coherent wave sources in the same medium. The amplitude is 5 cm, and the frequency is 100 Hz. The oscillation of S_1 is ahead of that of S_2. When S_1 is the peak, S_2 is exactly the valley. The wave speed is 10 m · s^{-1}. Assuming that the oscillations of S_1 and S_2 are both perpendicular to the paper surface, try to find the result of the interference when the two waves emitted by them reach point P.

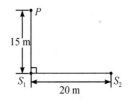

Figure 6.5.3 Figure of Example 6.5.2

Solution From the figure, $S_1P = 15$ m, $S_1S_2 = 20$ m and then $S_2P = \sqrt{15^2 + 20^2} = 25$ m. $\varphi_1 - \varphi_2 = \pi$. The wavelength is $\lambda = \frac{u}{f} = 0.1$ m. The phase difference is

$$\Delta\varphi = (\varphi_2 - \varphi_1) - \frac{2\pi}{\lambda}(S_2P - S_1P) = -201\pi$$

which satisfies the condition of destructive interference. So the resultant amplitude is $A = |A_1 - A_2|$ at point P.

6.6 Standing wave

A standing wave is a special interference phenomenon, which is formed by the superposition of two coherent waves with the same amplitude, the same frequency, and the same vibration direction, but propagating in opposite directions of the same straight line.

6.6.1 Standing wave experiment

Figure 6.6.1 is a schematic diagram of an experimental setup for observing standing waves. Figure 6.6.2 shows the standing wave pattern. The A end of the string is attached to a tuning fork, and the B end is attached to a weight through a pulley in order to tight the string. We can treat a standing wave as a special interference of two coherent waves with opposite propagation directions. When the tuning fork vibrates, the resulting wave propagates to the right along the string. After the wave arrives at the wedge tip at the B end, a reflected wave is generated. In this way, the incident wave propagating to the right and the reflected wave propagating to the left satisfy the coherence condition on the string line. If the length of AB has some special values, then we can find that the string AB vibrates in segments. Some points on the string AB keep still, while the points between two still points vibrate with different amplitudes at various positions. This is the standing wave.

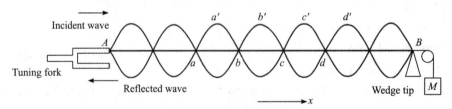

Figure 6.6.1 Schematic diagram of a standing wave experimental setup

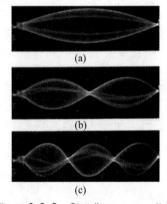

Figure 6.6.2 Standing wave pattern

6.6.2 Standing wave equation

The formation of standing waves can be quantitatively studied by the principle of wave superposition. There are two plane simple harmonic waves with the same amplitude, the same frequency, the same vibration direction and opposite propagation directions. If their amplitude is denoted by A, their frequency is f, and the initial phase is 0, their wave functions can be written, respectively, as

$$y_1 = A\cos 2\pi \left(ft - \frac{x}{\lambda}\right), \quad y_2 = A\cos 2\pi \left(ft + \frac{x}{\lambda}\right)$$

According to the principle of wave superposition, the wave function of the resultant standing wave should be

$$y = y_1 + y_2 = A\cos 2\pi \left(ft - \frac{x}{\lambda}\right) + A\cos 2\pi \left(ft + \frac{x}{\lambda}\right)$$

Using trigonometric relations, it can be simplified as

$$y = 2A\cos \frac{2\pi x}{\lambda} \cos 2\pi ft \tag{6.6.1}$$

where $\cos 2\pi ft$ is the cosine function of time t which shows that after the standing wave is formed, all the mass elements are in simple harmonic vibration with the same frequency. $2A\cos \frac{2\pi x}{\lambda}$ is the cosine function of the coordinate x which means the amplitude is dependent on the coordinate x. We can think of it as a simple harmonic vibration with the amplitude $\left|2A\cos \frac{2\pi x}{\lambda}\right|$ of the mass element at x.

1. Amplitude characteristics of a standing wave

From Equation (6.6.1), the amplitude of a standing wave can be expressed as $A_x = \left|2A\cos \frac{2\pi x}{\lambda}\right|$. The amplitude is always zero when x satisfies

$$\frac{2\pi x}{\lambda} = (2k+1)\frac{\pi}{2} \quad (k=0, \pm 1, \pm 2, \cdots)$$

which is

$$x = (2k+1)\frac{\lambda}{4} \quad (k=0, \pm 1, \pm 2, \cdots)$$

These points are the nodes of the standing wave. The distance between two adjacent nodes is

$$x_{k+1} - x_k = [2(k+1)+1]\frac{\lambda}{4} - (2k+1)\frac{\lambda}{4} = \frac{\lambda}{2}$$

The distance between two adjacent wave nodes is half a wavelength.

The amplitude is the largest, when x satisfies the following expression

$$\frac{2\pi x}{\lambda} = k\pi \quad (k=0, \pm 1, \pm 2, \cdots)$$

which is

$$x = k\frac{\lambda}{2} \quad (k=0, \pm 1, \pm 2, \cdots)$$

These points are the antinodes of the standing wave. The distance between two adjacent antinodes is

$$x_{k+1} - x_k = (k+1)\frac{\lambda}{2} - k\frac{\lambda}{2} = \frac{\lambda}{2}$$

The distance between two adjacent antinodes is also half a wavelength.

It can be seen from the above discussion that the amplitude of the mass element vibration at the node is zero, and it is always in a static state. The amplitude of mass element vibration at the antinode is the largest, which is equal to $2A$. The amplitudes of other mass element vibrations are between zero and the maximum value, the distance between two adjacent nodes or two adjacent antinodes is half a wavelength, and the distance between the adjacent antinode and node is $\lambda/4$. That is, the antinodes and nodes are alternately arranged at equal distances, as shown in Figure 6.6.1 or Figure 6.6.2.

2. Phase characteristics of a standing wave

$2A\cos\frac{2\pi x}{\lambda}$ is positive or negative for different values of x. If we take a segment between two adjacent nodes, $2A\cos\frac{2\pi x}{\lambda}$ of each point in each segment has the same sign while the signs of two adjacent segments are opposite. This shows that the vibration phase of each mass element in the same segment of a standing wave is the same, while the vibration phase of each mass element in two adjacent segments is opposite. Therefore, each mass element on the same segment reaches the maximum vibration displacement in the same direction at the same time, and passes through the equilibrium position in the same direction at the same time. The mass elements on both sides of the node reach the positive and negative maximum vibration displacement in the opposite directions at the same time, and pass through the equilibrium position in the opposite directions at the same time. That is, the standing wave is a segmental vibration, each segment vibrates synchronously as a whole, and the adjacent segments vibrate in opposite directions.

3. Half-wave loss

In the experiment shown in Figure 6.6.1, the wave is reflected at the fixed point B, forming a node at the reflection end. If the end is free, point B is an antinode. In general, whether a node or an antinode is formed at the interface between two media is related to the type of wave, the properties of the two media and the incident angle. Theories and experiments show that when a wave is perpendicularly incident from one elastic medium to the second elastic medium and the product of the mass density and the wave speed of the second medium is larger than that of the first one, that is $\rho_2 v_2 > \rho_1 v_1$, then a node appears on the interface. The first medium is called the wave-sparse medium, and the second medium is called the wave-dense medium. Therefore, when the wave is perpendicularly incident from the wave-sparse medium to the wave-dense medium, the reflected wave forms a node at the interface of the media. On the contrary, when the wave is reflected from the wave-dense medium back to the wave-sparse medium, the reflected wave forms an antinode at the interface.

When a wave is reflected from the wave-dense medium back to the wave-sparse medium, a node is formed at the reflective surface. It means that the phase of the incident wave and the reflected wave are opposite, that is, a sudden phase change of π occurs. This phase change π corresponds to the phase difference between two points separated by half a wavelength, which is equivalent to adding (or losing) a half-wavelength wave path. This phenomenon of sudden phase change π is usually called half-wave loss.

4. Energy of a standing wave

A standing wave is formed by the superposition of two waves with the same wave intensity while propagating in opposite directions, so the wave intensity vector sum in the medium is zero, that is, the energy flux density is zero. There is no unidirectional propagation of energy or wave form. When the vibration displacement of each point between two nodes reaches their respective positive or negative maximum value, the kinetic energy at each point is zero. The relative deformation near the node is the largest and the potential energy has the maximum value, while the relative deformation near the antinode is the smallest and the potential energy has a minimum value. When the points between two wave nodes pass through the equilibrium position in the same direction, the relative deformation everywhere in the medium is zero, the potential energy is zero, the kinetic energy is the maximum at the antinode due to the maximum vibration speed, and the kinetic energy becomes smaller as the points become closer to the node. At other moments, kinetic energy and potential energy exist at the same time. Obviously, for each mass element, only kinetic energy and potential energy convert to each other, and there is no directional energy propagation.

5. Conditions for the formation of a standing wave on a string

It can be seen from the above analysis that for a string with fixed ends, it is possible to form a standing wave on the string only when the string length l is equal to an integer multiple of half a wavelength.

$$l = n\frac{\lambda}{2} \quad (n=1, 2, 3, \cdots)$$

or

$$f = \frac{u}{\lambda} = \frac{nu}{2l} \quad (n=1, 2, 3, \cdots) \tag{6.6.2}$$

where u is the wave speed. The relationship expressed by Equation (6.6.2) is the standing wave condition, which is widely used in quantum mechanics, acoustics, laser principle, atomic physics and other disciplines. The standing wave vibration mode when $n=1$ is called the fundamental mode or the first harmonic wave, the standing wave vibration mode when $n=2$ is called the second harmonic wave, and the standing wave vibration mode when $n=3$ is called the third harmonic wave and so on. The set of all possible vibration modes is called the harmonic wave series, and n is called the order of the nth harmonic wave. Figures 6.6.2 (a), (b) and (c) are the vibration patterns of the first, second and third harmonic wave of a string standing wave, respectively.

Example 6.6.1 As shown in Figure 6.6.3, there is a right-propagating plane harmonic wave $y = A\cos 2\pi\left(\dfrac{t}{T}-\dfrac{x}{\lambda}\right)$ which is reflected by the vertical interface at point P, $L = 5\lambda$ away from the coordinate origin O. Suppose there is a half-wave loss at the reflection point. The amplitude of the reflected wave is approximately equal to the amplitude of the incident wave. Try to find

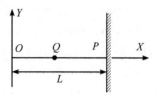

Figure 6.6.3   Figure of Example 6.6.1

(1) the expression of the reflected wave;

(2) the expression of the standing wave;

(3) the coordinates of each node and antinode between the origin O and the reflection point P.

Solution (1) The vibration equation of the incident wave at point P is

$$y = A\cos 2\pi\left(\dfrac{t}{T}-\dfrac{L}{\lambda}\right) = A\cos 2\pi\left(\dfrac{t}{T}-\dfrac{5\lambda}{\lambda}\right) = A\cos 2\pi\dfrac{t}{T}$$

Since the wave has a half-wave loss when reflected at point P, the vibration equation of the reflected wave at point P is

$$y = A\cos\left(2\pi\dfrac{t}{T}+\pi\right)$$

A mass element Q is arbitrarily chosen in the traveling direction of the reflected wave, with its coordinate x, and the phase of the vibration of the mass element at point Q lags behind that of the vibration of the mass element at point P, $\varphi_P - \varphi_Q = \dfrac{2\pi}{\lambda}(L-x) = 10\pi - \dfrac{2\pi}{\lambda}x$. The phase at point Q is

$$\varphi_Q = 2\pi\dfrac{t}{T}+\pi-\left(10\pi-\dfrac{2\pi}{\lambda}x\right) = 2\pi\dfrac{t}{T}+\dfrac{2\pi}{\lambda}x-9\pi$$

The vibration equation of the mass element at point Q is

$$y = A\cos\left(2\pi\dfrac{t}{T}+\dfrac{2\pi}{\lambda}x-9\pi\right) = A\cos\left(2\pi\dfrac{t}{T}+\dfrac{2\pi}{\lambda}x-\pi\right)$$

So the expression for the reflected wave is

$$y_{ref} = A\cos\left(2\pi\dfrac{t}{T}+\dfrac{2\pi}{\lambda}x-\pi\right)$$

(2) The expression of the standing wave is

$$y_s = y + y_{ref} = A\cos 2\pi\left(\dfrac{t}{T}-\dfrac{x}{\lambda}\right) + A\cos\left(2\pi\dfrac{t}{T}+\dfrac{2\pi}{\lambda}x-\pi\right)$$

$$y_s = 2A\sin\frac{2\pi}{\lambda}x \sin 2\pi\frac{t}{T}$$

(3) At the node, the amplitude is 0, or $\left|\sin\frac{2\pi}{\lambda}x\right|=0$

$$\frac{2\pi}{\lambda}x = k\pi \quad (k=0, \pm 1, \pm 2, \cdots)$$

$$x = k\frac{\lambda}{2} \quad (k=0, \pm 1, \pm 2, \cdots)$$

As a result, the coordinates of the nodes between the origin O and the reflection point P are

$$x = 0, \frac{\lambda}{2}, \lambda, \frac{3\lambda}{2}, 2\lambda, \frac{5\lambda}{2}, 3\lambda, \frac{7\lambda}{2}, 4\lambda, \frac{9\lambda}{2}, 5\lambda$$

At antinodes, the amplitude is the largest $\left|\sin\frac{2\pi}{\lambda}x\right|=1$

$$\frac{2\pi}{\lambda}x = (2k+1)\frac{\pi}{2} \quad (k=0, \pm 1, \pm 2, \cdots)$$

$$x = (2k+1)\frac{\lambda}{4} \quad (k=0, \pm 1, \pm 2, \cdots)$$

The coordinates of the antinodes between the origin O and the reflection point P are

$$x = \frac{\lambda}{4}, \frac{3\lambda}{4}, \frac{5\lambda}{4}, \frac{7\lambda}{4}, \frac{9\lambda}{4}, \frac{11\lambda}{4}, \frac{13\lambda}{4}, \frac{15\lambda}{4}, \frac{17\lambda}{4}, \frac{19\lambda}{4}$$

6.7 The Doppler effect

6.7.1 The Doppler effect

In daily life, when an approaching train whistles, the pitch of the whistle heard will be higher; as the train moves away, the whistle heard is at a lower pitch. This is because there is a relative motion between the wave source and the observer, so that the frequency of the wave received by the observer is not the same as the frequency emitted by the wave source, which is known as the Doppler effect. This phenomenon was first discovered by the Austrian physicist Doppler in 1842, and in 1845 the Dutch meteorologist Barrott arranged a team of trumpeters standing on a speeding open train to play musical instruments. A change in pitch was detected by him on the platform, which is one of the most interesting experiments in the science history. The Doppler effect applies not only to sound waves, but also to electromagnetic waves, including microwaves, radio waves, and visible light. Here we take the sound wave as an example for discussion, and the whole medium (air) in which the sound wave propagates as the reference frame.

For simplicity, only the case that the wave source and the receiver (or observer) move collinearly is discussed here. A sonic source S and a receiver R move along the line connecting them. u_S is the velocity of the wave source relative to the air, u_R is the velocity of the receiver relative to the air, and u is the velocity of the wave in the medium (speed of sound). The vibration frequency of the wave source is ν_0. The frequency received by the receiver is ν_R. Three cases are discussed below.

(1) The Doppler effect when the wave source does not move and the observer moves at a uniform speed relative to the medium.

As shown in Figure 6.7.1, the wave source is stationary in the medium and the receiver R is moving towards the wave source at a speed relative to the medium. According to the velocity composition theorem, the velocity of the wave relative to the receiver is $u' = u + u_R$. Since the wavelength remains the same, the number of complete wavelengths received by the receiver per unit time is

$$\nu_R = \frac{u'}{\lambda} = \frac{u + u_R}{\lambda} = \frac{u + u_R}{u} \nu_0 \tag{6.7.1}$$

Therefore, the frequency received by the observer when moving towards the wave source is $\dfrac{u + u_R}{u}$ times the frequency of the wave source.

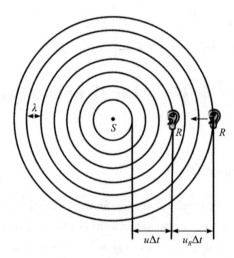

Figure 6.7.1 The Doppler effect when the receiver is moving

When the receiver moves away from the wave source, the velocity of the wave relative to the receiver is $u' = u - u_R$. So the frequency measured by the observer is

$$\nu_R = \frac{u - u_R}{u} \nu_0 \tag{6.7.2}$$

At this time, the received frequency is lower than the frequency of the wave source.

(2) The Doppler effect when the receiver does not move and the wave source moves at a uniform speed relative to the medium.

Figure 6.7.2 is the ripple diagram of the water wave when the swan swims. The wavelength in front of the swan becomes shorter, and the wavelength in the back becomes longer when the receiver R is still in the medium. The wave source S moves towards the receiver with a velocity relative to the medium. During time T, the wavefront A_1 has moved forward by a distance uT to A_1', and the distance moved by the wave source is $u_S T$. Therefore, in the direction of motion of the wave source S, according to the definition of wavelength, the distance between two adjacent wavefronts A_1 and A_2 in time T is $\lambda' = uT - u_S T$ which is the wavelength of the wave. So the frequency received by the receiver should be

$$\nu_R = \frac{u}{\lambda'} = \frac{u}{uT - u_S T} = \frac{u}{u - u_S} \nu_0 \tag{6.7.3}$$

It can be seen that the frequency received by the receiver is larger than the frequency of the wave source.

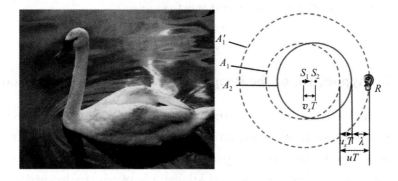

Figure 6.7.2 The Doppler effect when the wave source is moving

When the wave source moves away from the receiver, similarly the frequency received by the receiver can be obtained as

$$\nu_R = \frac{u}{u + u_S} \nu_0 \tag{6.7.4}$$

At this time, the frequency received by the receiver is smaller than the frequency of the source.

(3) The Doppler effect when both the observer and the wave source move at different velocities relative to the medium.

Based on the above analysis, it can be concluded that when the wave source and the receiver move towards each other, the frequency received by the receiver is

$$\nu_R = \frac{u + u_R}{u - u_S} \nu_0 \tag{6.7.5}$$

When the source and receiver are away from each other, the frequency received by the receiver is

$$\nu_R = \frac{u - u_R}{u + u_S} \nu_0 \tag{6.7.6}$$

The above discussions are about the motion of the wave source and the receiver on the connecting line of the two. If the motion of the receiver and the wave source is not on the connecting line, the result can be understood as the velocity component along the connecting line.

Example 6.7.1 Two trains are moving towards each other at the speed of 72 km · h^{-1} and 54 km · h^{-1} respectively, and the first train emits a 600 Hz whistle sound. If the speed of sound is 340 m · s^{-1}, what are the frequencies of the sound the observer on the second train hears before and after the trains meet, respectively?

Solution Let the speed of the whistling train be $u_S = 72$ km · h^{-1} = 20 m · s^{-1} and the speed of the train receiving the whistle be $u_R = 54$ km · h^{-1} = 15 m · s^{-1}. Then, the frequency received before the encounter is

$$\nu_1 = \frac{u + u_R}{u - u_S} = \frac{340 + 15}{340 - 20} \times 600 = 665 \text{ Hz}$$

The frequency received after the two trains meet is

$$\nu_2 = \frac{u - u_R}{u + u_S} = \frac{340 - 15}{340 + 20} \times 600 = 541 \text{ Hz}$$

6.7.2 Application of the Doppler effect in engineering technology

In nature, bats rely on emitting ultrasonic waves and then detecting reflected waves for navigation and foraging. After the ultrasound is emitted from the bat's nostril, it will be reflected back into the bat's ear if it encounters a moth. The relative motion of the bat and the moth causes a difference of several kilohertz between the frequency the bat hears and the frequency it emits, and the bat relies on this difference to determine its relative velocity to the moth. Meanwhile, when moths hear the bat's ultrasound, they avoid the direction of the ultrasound so that the bat can't hear the echo.

The Doppler effect of sound waves can also be used for medical diagnosis, which is what we usually call color Doppler ultrasound. Checking the blood movement states of the heart and blood vessels and understanding the speed of blood flow can be achieved by transmitting ultrasound. The ultrasonic oscillator generates a high-frequency constant-amplitude ultrasonic signal, which stimulates the transmitter probe to generate continuous ultrasonic waves, which are emitted to the cardiovascular organs of the human body, resulting in the Doppler effect. When the ultrasonic beam meets the moving organs and blood vessels, the reflected signal is accepted by the transducer. Then the blood flow velocity can be calculated according to the difference between the reflected wave and the transmitted frequency, and the direction of blood flow can be determined according to the increase or decrease of the reflected wave frequency.

In mobile communication, the frequency becomes higher when the mobile station moves to the base station, and the frequency becomes lower when it moves away from the base station, so we must fully consider the Doppler effect in mobile communication. In daily life, our moving speed is small and will not bring a very large frequency offset, but because the speed of the aircraft is very high, we must fully consider the Doppler effect in satellite mobile communication. In order to avoid this influence in communication, we have to take various technical considerations into account, thus increasing the complexity of mobile communication.

In traffic monitoring, the system transmits ultrasonic or electromagnetic waves with a known frequency to the moving vehicle and at the same time measures the frequency of the reflected wave, and the speed of the vehicle can be known according to the change in the frequency of the reflected wave. Monitors equipped with the Doppler speedometers are sometimes installed just above the road, taking pictures of the vehicle's registration number while the speed is measured.

Weather radars based on the Doppler effect also play an important role in weather warnings. The weather radar emits electromagnetic waves into the air, and the electromagnetic waves are scattered after encountering water vapor condensation in the atmosphere. By measuring the high frequency difference between the received signal and the transmitted signal, the speed of the scatterer relative to the radar can be determined. It is of great significance to studying the formation of precipitation, analyzing small and medium-scale weather systems, and warning against severe convective weather, etc.

The application of the Doppler effect is very extensive, and it is also widely used in the microelectronics industry, aviation test technology, research on the motion of celestial bodies, etc.

6.7.3 Shockwave

When the velocity of a wave source towards a receiver v_S is greater than the velocity of the wave u, according to Equation (6.7.5), the frequency received by the receiver is less than zero, and the formula is not applicable. The reason is that at any moment the wave source itself will exceed the wavefront it emits, and there cannot be any wave in front of the wave source, as shown in Figure 6.7.3.

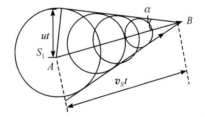

Figure 6.7.3 The Mach cone and shock wave

Let the point wave source move from A to B within time t, and the distance is $AB = v_S t$. At the same time, the wave from the wave source at A propagates ut. As a result, the tangent planes of the wavefronts form a cone, and the half-width of the vertex angle is α. Then

$$\sin\alpha = \frac{u}{v_S}$$

With the passage of time, each wavefront continues to expand, and the cone surface also continues to expand. This V-shaped wave formed by a point wave source is called a shockwave. The envelope surface of the shockwave is conical, which is called the Mach cone. $M=\dfrac{u}{v_S}$ is called the Mach number.

Ultrasonic waves are generated when aircraft and shells fly at supersonic speeds. Figure 6.7.4 shows the shockwave generated by the "Super Hornet" fighter jet at the New York Air Show. The reason why this wave can be "seen" is that the air pressure inside the shockwave suddenly decreases, causing water molecules in the air to condense and form a fog. The accumulation of shockwave fronts causes sudden changes in pressure, density and temperature on both sides of the cone, so the shockwave often produces huge destructive force.

Figure 6.7.4　Shockwave generated by aircraft

Scientist Profile

Deng Jiaxian (1924 - 1986) was born in Huaining County, Anhui Province. He was a famous nuclear physicist and academician of the Chinese Academy of Sciences.

Deng Jiaxian was the main organizer and leader of China's nuclear weapons research and development, and was awarded the medal of "Two Bombs". In the research of the atomic bomb and hydrogen bomb, Deng Jiaxian led the basic theoretical research on detonation physics, fluid mechanics, equation of state, neutron transport, etc., completed the theoretical scheme of the atomic bomb, and participated in the guidance of the nuclear detonation simulation test. After the success of the atomic bomb test, Deng Jiaxian organized forces to explore the design principle of the hydrogen bomb and select technical ways. He led and personally participated in the development and experiment of China's first hydrogen bomb in 1967.

Summary of Theoretical Research on China's First Atomic Bomb, co-written by Deng Jiaxian and Zhou Guangzhao, is a seminal fundamental work on the theoretical design of nuclear weapons. It summarizes the research achievements of hundreds of scientists. This work not only plays a guiding role in the theoretical design in the future, but also serves as a textbook for the training of scientific researchers. Deng Jiaxian also made important contributions to the study of the equation of state at high temperature and high pressure. In order to train young researchers, he wrote many lectures on electrodynamics, plasma physics, and spherical detonation wave theory. Even after serving as the dean, he began to write *Quantum field theory* and *Group theory* in his spare time.

Deng Jiaxian was an excellent representative of Chinese intellectuals. For the prosperity of the motherland and the development of national defense scientific research, he was willing to be an unknown hero and struggled for decades. He, often at critical moments and regardless of personal safety, appeared in the most dangerous post, which fully embodied his noble selfless dedication. He made a remarkable contribution to the development of China's nuclear weapons. It was not until after his death that people learned of his deeds.

Extended Reading

Ultrasonic suspension

Ultrasonic levitation technology is the use of sound radiation pressure generated by high-intensity ultrasonic fields to achieve the suspension of objects. Compared with pneumatic suspension, electromagnetic suspension, and superconducting suspension, acoustic suspension can suspend any material object with good stability, and does not produce significant thermal effect. In 1886, the German scientist Kundt first proposed the phenomenon of ultrasonic suspension, and floated charged dust with acoustic suspension in a resonant tube. Since then, physicists around the world had tried to suspend other samples. Most notably, the Canadian physicist King in 1934 suspended rigid spheres and revealed that ultrasonic suspension is a nonlinear phenomenon under high intensity conditions.

The experimental principle of ultrasonic levitation is to let the vertical levitation force overcome the gravity of the suspended object, and at the same time produce a horizontal positioning force to fix the object in a certain position (or vibrate left and right within a certain range). The essence of the sound wave used in ultrasonic levitation technology is the standing wave. The sound pressures on both sides of the node of the standing wave are opposite, the combined sound pressure is zero, and the object is not affected by the levitation force. Therefore, if the object is not acted upon by other forces, it will stop at the node, but if the gravity is not negligible, the object will stop at the point below the node where the sound pressure is upward, as shown in Figure ER 1. The directional force in the horizontal direction is provided by the sound field stimulated by the cylindrical resonator. Because standing waves have multiple nodes of sound pressure, several objects can be suspended at the same time. The standing wave suspension instrument is generally composed of an ultrasonic transmitter, a transducer, an amplitude transformer, an emission end, a reflection end, a quartz tube and a tuning mechanism, as shown in Figure ER 2. In order to obtain a relatively large suspension force, ultrasonic waves with a frequency of 20 to 50 kHz are usually used in experiment, and in this frequency range, the performance of piezoelectric ultrasonic transducers is just superior.

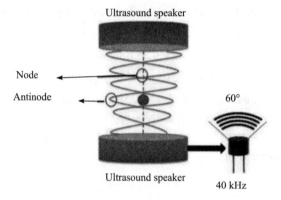

Figure ER 1　Acoustic levitation

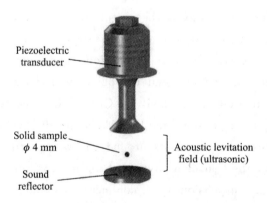

Figure ER 2　Standing wave suspension instrument

The ultrasonic suspension experiment has requirements on the size of the object. Since the distance between adjacent nodes is half a wavelength, the size of the suspended object in the direction of sound wave propagation cannot exceed half of its wavelength. If the suspended object is to be stabilized, only the range of 1/6 above and below the node can be used as the range of the gathering object, that is, the size of the object cannot exceed 1/3 of the wavelength. In general, the larger the frequency of the sound wave, the smaller the wavelength and the smaller the size of the object that can be suspended. Because the ultrasonic levitation force $F = -\dfrac{5}{6}\pi \rho_0^2 a^3 \sin \theta$ is proportional to the cubic diameter of the object and the gravity is also proportional to the cubic diameter of the object, the size of the levitation capacity is only related to the energy of the sound field and the density of the object and has no relationship with the size of the object. Therefore, the shape of the suspended item does not affect the stability of ultrasonic suspension, and the greater the density of the object, the greater the sound intensity required for suspension. For example, the suspension of a drop of water needs to reach 160 dB, while the suspension of the higher density metal tungsten needs to reach 172 dB.

In recent years, ultrasonic suspension has been widely used in protein crystallization, liquid alloy condensation, droplet kinetics, biochemical analysis, and even the drying of colloidal droplets. It can also provide a container-free environment on the ground to simulate space conditions. The research group of Academician Wei Bingbo of Northwestern Polytechnical University studied the dynamic mechanism of the rapid growth of dendrites and eutectic in the melt of deep supercooled alloy by using ultrasonic suspension technology, revealed the coupling effect of microgravity and deep supercooling conditions on the rapid solidification process, explored the pattern of change of the thermal physical properties of deep supercooled alloy melt, and measured the physical properties of liquid density and surface tension.

Discussion Problems

6.1 What is a wave surface? What are the differences between a wave surface and a wavefront? What is the connection between a wavefront and a wave line?

6.2 When someone writes the function of a wave traveling in the positive direction of the X axis, he thinks that the wave propagates from point O to point P and the vibration at point P is x/u later than that at point O. Therefore, the phase of point O at time t should pass to point P at time $t+x/u$. The function of the plane simple harmonic wave should be $y = A\cos\left[\omega\left(t+\dfrac{x}{u}\right)+\varphi_0\right]$. Do you think that's right and why?

6.3 The wave function of a plane simple harmonic wave is $y=A\cos\left[\omega\left(t-\dfrac{x}{u}\right)+\varphi_0\right]$. What do $\dfrac{x}{u}$ and φ_0 stand for? If the wave function is written as $y=A\cos\left(\omega t-\dfrac{\omega}{u}x+\varphi_0\right)$, what does $\dfrac{\omega}{u}x$ stand for?

6.4 As shown in Figure T6-1, it could be a standing wave or a traveling wave at some moment. The wavelength is λ. If it is a standing wave, what is the phase difference between points A and B? If it is a traveling wave, what is the phase difference between points A and B?

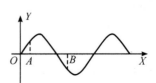

Figure T6-1 Figure of Problem 6.4

6.5 In a standing wave, there is no "running" of the waveform and no propagation of energy. Does this phenomenon contradict the definition of a wave? How to explain it?

6.6 What is the Doppler effect?

Problems

6.7 It's known that the wave equation of a wave is $y = 5 \times 10^{-2} \sin(10\pi t - 0.6x)$ m.

(1) Find the wavelength, frequency, wave speed and direction of propagation;

(2) Explain the meaning of the wave equation when $x = 0$ m, and draw the graph.

6.8 A planar simple harmonic wave propagates forward along the X axis with a velocity of $u = 0.2$ m·s^{-1} in a medium. The vibration equation at point A ($x = 0.05$ m) is $y = 0.03\cos\left(4\pi t - \dfrac{\pi}{2}\right)$. Find

(1) the wave equation;

(2) the vibration equation at point P ($x = -0.05$ m).

6.9 A planar simple harmonic wave propagates along the negative X axis with a velocity of $u = 10$ m·s^{-1}. The vibration curve of particle at point $x = 0$ m is shown in Figure T6-2. Find the wave equation of this wave.

6.10 The waveform of a planar simple harmonic wave at $t = 0.25$ s is shown in Figure T6-3. Find the wave equation of this wave.

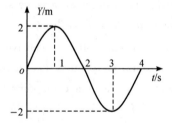
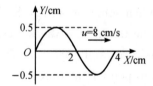

Figure T6-2 Figure of Problem 6.9   Figure T6-3 Figure of Problem 6.10

6.11 The vibration period, wavelength and amplitude of a simple harmonic wave are $T = 0.5$ s, $\lambda = 10$ m and $A = 0.1$ m, respectively. The displacement of the wave source vibration is exactly the maximum value in the positive direction at $t = 0$ s. The wave source is on the coordinate origin, and the wave propagates in the positive direction of the OX axis. Find

(1) the wave expression;

(2) the displacement of the particle at $x_1 = \lambda/4$ when $t_1 = T/4$;

(3) the vibration velocity of the particle at $x_1 = \lambda/4$ when $t_2 = T/2$.

6.12 The vibration expression of the plane wave source is $y_0 = 6.0 \times 10^{-2} \sin \dfrac{\pi}{2} t$ m. The wave velocity is 2 m·s^{-1}. Find the vibration equation of the particle 5 m away from the wave source and the phase difference between the particle and the wave source.

6.13 A mechanical wave propagates forward along the OX axis. The waveform at $t=0$ s is shown in Figure T6-4. The wave velocity is 10 m · s^{-1} and the wavelength is 2 m. Find

(1) the wave equation;

(2) the vibration equation at point P;

(3) the minimum time required for point P to return to the equilibrium position.

6.14 As shown in Figure T6-5, S_1 and S_2 are two coherent wave sources. Both the amplitudes are A. The distance between S_1 and S_2 is 1/4 the wavelength, and the phase of S_1 is $\pi/2$ ahead of that of S_2. Find the amplitude of the resultant wave at each point outside S_1 and S_2 and on the line connecting S_1 and S_2.

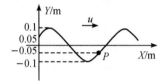
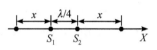

Figure T6-4 Figure of Problem 6.13   Figure T6-5 Figure of Problem 6.14

6.15 Two coherent point wave sources S_1 and S_2 have the amplitude A, the initial phase difference $\varphi_1 - \varphi_2 = \dfrac{3\pi}{2}$, and the wavelength λ. If the vibration amplitude of the medium mass element at each point outside S_1 and on the connection line between S_1 and S_2 is $2A$, regardless of wave attenuation, try to find

(1) the ratio of the wave intensity at each point outside S_1 and on the line connecting S_1 and S_2 to the wave intensity when a single wave source exists;

(2) the distance between S_1 and S_2 expressed by the wavelength λ.

6.16 The expression of an incident wave is $y_1 = A\cos 2\pi \left(\dfrac{t}{T} + \dfrac{x}{\lambda}\right)$. Reflection occurs at $x=0$ m, and the reflection point is a free end. Find

(1) the expression of the reflected wave;

(2) the expression of the resultant standing wave.

6.17 Two waves propagate on a very long string, and their expressions are $y_1 = 6\cos 2\pi(0.02x - 8t)$ and $y_2 = 6\cos 2\pi(0.02x + 8t)$ respectively. Find

(1) the frequency, wavelength and wave velocity of each wave;

(2) the position of the node;

(3) the position where the amplitude is the maximum.

6.18 The speed of sound in the air is 330 m · s^{-1}. A train is traveling at a speed of 30 m · s^{-1}, and the frequency of the whistle on the train is 600 Hz. What are the frequencies heard by a stationary observer when he is in front of the train and when the train passes him respectively? If the observer is moving towards the train with a velocity of

$10 \text{ m} \cdot \text{s}^{-1}$, what are the frequencies of the sounds he hears at the two positions respectively?

6.19 The frequency of the sound source is 1080 Hz, and it moves to the right at a speed of $30 \text{ m} \cdot \text{s}^{-1}$ relative to the ground. On its right there is a reflective surface moving to the left at a speed of $65 \text{ m} \cdot \text{s}^{-1}$ relative to the ground. The speed of sound in the air is $330 \text{ m} \cdot \text{s}^{-1}$. Find

(1) the wavelength of the sound emitted by the sound source in the air;

(2) the frequency and wavelength of the reflected sound.

Challenging Problems

6.20 The frequency of a train whistle is f_0. When the train goes through a crossing with its whistle, a passerby is at a distance d from the track. What is the frequency of the train heard by the passerby at the crossing? The other train is moving towards the former train. Both trains have the same speed. The distance between the two trains is s. What is the frequency of the former train heard by the driver of the latter train?

Part 2   Thermology

Thermology is a discipline that studies the law of thermal motion of matter and its application. It involves the influence of thermal motion on the macroscopic properties of matter and the transformation of the matter state, as well as the mutual transformation between thermal motion and other forms of motion of matter. The thermal theory is not only applied in the field of physics, but also widely adapted in chemistry, biology, meteorology, astrophysics and other natural science fields, and is a basic theory with universal significance.

According to the research methods, thermology can be studied by macroscopic and microscopic theory. The macroscopic theory of thermal motion of matter is referred to as thermodynamics. It summarizes the basic laws of thermal phenomena based on experimental facts and studies the process of thermal motion from a macroscopic point of view without considering the microscopic motion of molecules. The microscopic theory of thermal motion of matter is referred to as statistical physics. Based on the microscopic structure of matter and the mechanical laws followed by each molecule, it studies the relationship between the macroscopic quantities and the corresponding average of microscopic quantities by statistical methods, so as to reveal the microscopic nature of thermal phenomena.

It can be seen that although thermodynamics and statistical physics study the same object, they use different methods. The theory of statistical physics can be verified by thermodynamics, and the macroscopic properties of matter studied by thermodynamics can also be understood via the analysis of statistical physics. Therefore, they complement each other, so that we can have a deeper understanding of the law of thermal motion of matter.

The basic concepts and methods of gas dynamic theory and thermodynamics are described by taking gas as a research object in this part.

Chapter 7

Kinetic Theory of Gas

The kinetic theory of gas is a classical microstatistical theory established in the mid-19th century, which mainly studies the thermal phenomena of gas. Gas is composed of a large number of molecules. The molecules show irregular thermal motion, and the movement of molecules follows the classical Newtonian mechanics. The collective behavior of a large number of molecules can be investigated by statistical method, and the macroscopic thermal properties and the gas laws, such as pressure, temperature, equations of state, internal energy, specific heat and transport process, can be explained from a quantitative microscopic point of view. The kinetic theory of gas reveals its macroscopic thermal properties and the microscopic nature of thermodynamic process and gives the relationship between them.

From the 17th century to the early 19th century, the development of kinetic theory of gas was in the embryonic period. In 1678, Hooke proposed that gas pressure was the result of a large number of gas molecules colliding with the wall, in 1738 Bernoulli derived the pressure formula and explained the Boyer law, and in 1744 Lomonosov proposed that heat was the expression of molecular motion.

In the mid-19th century, the kinetic theory of gas had a major development. Clausius, Maxwell and Boltzmann were the founders. In 1858, Clausius proposed the concept of mean free path of gas molecules and derived the relevant formula. In 1860, Maxwell pointed out that the frequent collisions of gas molecules do not make their velocities converge but make them reach a stable distribution, and derived the speed distribution and velocity distribution of gas molecules in equilibrium. In 1868, Boltzmann introduced the gravitational field in the Maxwell speed distribution, and then proposed the statistical concept of non-equilibrium state. After that, the statistic theories of equilibrium state and non-equilibrium state were combined with each other and formed the statistical mechanics.

7.1 Microscopic characteristics of thermal motion of a gas system

The kinetic theory of gas studies the law of thermal phenomena from the microstructure of gas. So the microscopic characteristics of thermal motion of molecules are introduced at first.

7.1.1 Microscopic characteristics of a gas system

(1) A gas system is composed of a large number of microscopic particles (molecules or atoms).

Experiments show that the number of molecules or atoms contained in 1 mole of any substance is $N_A = 6.0221367(36) \times 10^{23}$ mol^{-1}. N_A is called Avogadro's constant.

(2) The molecules (or atoms) inside a gas system are in perpetual motion. This motion is irregular, and their intensities are related to the temperature of the gas.

The direction and speed of each molecule is constantly changing as the molecules collide with each other. Since the molecules are moving in different directions, each molecule may be hit in different directions. Thus the motion trajectory of each molecule is chaotic. The higher the temperature, the more violent the random movement of molecules. Because the random motion of molecules is related to temperature, it is usually called the **thermal motion of molecules**.

(3) There is an interaction between molecules (or atoms).

There are complex interactions between molecules (or atoms) in gas. The relationship between the force f and the distance r of two molecules can be approximately described by the curve shown in Figure 7.1.1. When the distance between two molecules is small ($r < r_0$), the interaction force appears repulsive, and the repulsive force increases rapidly with the decrease of distance. When the distance is large ($r > r_0$), the interaction force behaves as an attractive

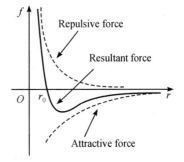

Figure 7.1.1 Interaction between two molecules

force, and the attractive force gradually approaches zero with the increase of distance.

The interaction between molecules tends to bring molecules together to form a certain regular distribution in space, while the thermal motion of molecules tends to break up this regular arrangement by dispersing them. Because of these two opposite effects, molecules behave in different ways at different temperatures, making the substance exist with different states, such as ice, water, and vapour.

A gas system is composed of a large number of molecules which have irregular thermal motion and interactions between each other. The motion of molecules follows the classical Newtonian mechanics. According to the above microscopic characteristics of gas molecules, the collective behavior of a large number of molecules can be investigated by the statistical average method for studying the macroscopic thermal properties and laws of gas.

7.1.2 Statistical law of thermal motion of gas molecules

The greatest characteristic of molecular thermal motion is disorder. If we trace the motion of a molecule, we will find that the speed and direction of its motion are constantly changing, and it is difficult to find any regularity. The number of microscopic particles that make up macroscopic objects is large, and collisions between these particles are frequent. Every collision could change the magnitude and direction of the velocity of the molecules. If we use classical mechanics to deal with the collision of molecules, we must study the motion of each molecule, that is, we must solve a large number of dynamic equations, which is quite difficult. However, although the movement of a single molecule is chaotic, there is a certain law governing the movement of a large number of molecules from the view of statistics. This law comes from the collection of a large number of accidental events, which is called the **statistical law**. Gas molecules in equilibrium are subject to the following statistical assumptions.

(1) If gravity and other external forces are ignored, each molecule has an equal probability of appearing at any position in the container. For a large number of molecules, the number of molecules in any unit volume is equal at any time, and the molecular number density can be obtained

$$n = \frac{dN}{dV} = \frac{N}{V}$$

That is, if the gas is in equilibrium, the density of a large number of gas molecules in random thermal motion is equal everywhere.

(2) The probability of the molecule moving in any direction is the same.

According to this assumption, the statistical averages of the velocity components of a large number of molecules are equal in all directions.

In the rectangular coordinate system $OXYZ$, the velocity component in the positive direction of the axis is positive, the velocity component in the negative direction of the axis is negative, and the arithmetic mean of each component of the molecular velocity is

$$\bar{v}_i = \frac{\sum v_i}{N} \quad (i = x, y, z)$$

N is the total number of molecules, and the sum symbol represents the sum of the components of all molecular velocities in the i direction. Then there is

$$\bar{v}_x = \bar{v}_y = \bar{v}_z = 0$$

The mean of the squared molecular velocity components is

$$\overline{v_i^2} = \frac{\sum v_i^2}{N} \quad (i = x, y, z)$$

and

$$\overline{v_x^2} = \overline{v_y^2} = \overline{v_z^2}$$

For every molecular, $v^2 = v_x^2 + v_y^2 + v_z^2$. Thus, $\overline{v^2} = \overline{v_x^2} + \overline{v_y^2} + \overline{v_z^2}$, and then

$$\overline{v_x^2} = \overline{v_y^2} = \overline{v_z^2} = \frac{1}{3}\overline{v^2}$$

7.2 Pressure of ideal gas

Ideal gas is the simplest thermodynamic model, which exhibits some common properties of real gases in a certain range, so this model has been widely studied in the kinetic theory of gas. According to the microscopic model of the ideal gas, the pressure formula of the ideal gas is derived by a statistical method, and the statistical significance of the pressure of the ideal gas is discussed.

7.2.1 Microscopic model of ideal gas

The ideal gas microscopic model is actually the result of idealization and abstraction of real gas under the condition that the pressure is not large and the temperature is not low. The microscopic model is as follows:

(1) The linearity of the gas molecules is much smaller than the average distance between the molecules, and the molecules are regarded as particles;

(2) The interaction forces between molecules, as well as the forces between molecules and the container wall, are ignored, except for the forces produced during collision;

(3) Collisions between molecules and between molecules and the wall of the container are completely elastic, meaning that the kinetic energy of the gas molecules is not lost due to collisions.

7.2.2 Pressure formula of ideal gas

Hooke proposed that the gas pressure is the result of a large number of gas molecules colliding with the wall. According to this physical concept, every molecule in the container is in random motion, and there are constant collisions between molecules and between molecules and the wall of the container. For a molecule, each time it collides with the wall, the location of the collision and the magnitude of the impulse it gives to the wall are accidental. However, when taking a large number of molecules as a whole subject, there are many molecules colliding with the wall at any given time, and these collisions, from a macroscopic perspective, exert constant and continuous forces on the wall, creating a constant pressure. This is very similar to the rain falling on the umbrella, in which case a large number of dense raindrops falling on the umbrella will make us feel a continuous downward pressure.

To sum up, the pressure of the gas on the container wall is the result of a large number of constant collisions between the gas molecules and the container wall. The pressure formula of ideal gas in equilibrium is derived by using the above ideal gas model and statistical hypothesis in what follows.

A container of any shape has a volume of V, in which a certain amount of ideal gas with molecular mass m and in equilibrium is stored. The total number of the gas molecules is N, so the number of molecules in the unit volume is $n = \dfrac{N}{V}$.

We divide all the molecules into several groups according to their speed, that is, molecules in one group have the same speed. Let the number of molecules in a group be ΔN_i and the speed of these molecules be v_i. The number of molecules with speed v_i per unit volume in the container is $\Delta n_i = \dfrac{\Delta N_i}{V}$. Since the gas is in equilibrium, the pressure on the wall is equal everywhere, so it is sufficient to study only the pressure on any small area of the wall.

Choose the rectangular coordinate system $OXYZ$ and take a small element area dA on the wall so that it is perpendicular to the X axis, as shown in Figure 7.2.1. First, consider the collisions within the area dA of all molecules with speed v_i in time dt. Make a slanted column with dA as the base, v_i as the axis and $v_{ix}dt$ as the height, as shown in Figure 7.2.1. The volume of this slanted column is $v_{ix}dt\,dA$. Among all the molecules with speed of v_i, only the molecules in the above oblique cylinder can collide with dA in dt, and the number of molecules that can collide is $\Delta n_i v_{ix} dt\,dA$.

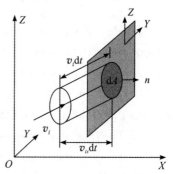

Figure 7.2.1 Diagram of ideal gas pressure formula derivation

Since the collision is completely elastic, the speed component of the molecule in the X direction changes from v_{ix} to $-v_{ix}$. The change of momentum in the X direction is $-mv_{ix} - mv_{ix} = -2mv_{ix}$. The impulse exerted on dA in dt is $\Delta n_i v_{ix} dt\,dA(-2mv_{ix})$.

Considering that only gas molecules with $v_{ix} > 0$ can collide with dA and the probability of molecular thermal motion occurring in any direction is equal in equilibrium, the number of gas molecules with $v_{ix} > 0$ accounts for half of the total number of molecules. The magnitude of the total impulse on dA in dt is

$$dI = \sum_{v_{ix}>0} 2\Delta n_i m v_{ix}^2 \, dt \, dA = \frac{1}{2} \sum_i 2\Delta n_i m v_{ix}^2 \, dt \, dA \qquad (7.2.1)$$

So the force exerted by these molecules on dA is

$$dF = \frac{dI}{dt} = \sum_i \Delta n_i m v_{ix}^2 \, dA \qquad (7.2.2)$$

The pressure of the gas on the container wall is

$$p = \frac{dF}{dA} = \sum_i \Delta n_i m v_{ix}^2 = m \sum_i \Delta n_i v_{ix}^2 \qquad (7.2.3)$$

According to the definition of average value, $\dfrac{\sum_i \Delta N_i v_{ix}^2}{N} = \dfrac{\sum_i V \Delta n_i v_{ix}^2}{N} = \overline{v_x^2}$, and then

$$p = mn\overline{v_x^2} \qquad (7.2.4)$$

According to the assumption that molecules have equal chances of moving in any direction and $\overline{v_x^2} = \overline{v_y^2} = \overline{v_z^2} = \dfrac{1}{3}\overline{v^2}$, we have

$$p = \frac{1}{3} nm\overline{v^2} \qquad (7.2.5)$$

Introduce the mean translational kinetic energy of the molecule $\overline{\varepsilon}_t = \dfrac{1}{2} m\overline{v^2}$, and

$$p = \frac{2}{3} n\overline{\varepsilon}_t \qquad (7.2.6)$$

which is the pressure formula of ideal gas, and is one of the basic formulas of the kinetic theory of gas.

7.2.3 Statistical significance and microscopic nature of the pressure formula

According to the ideal gas pressure formula, the pressure of an ideal gas is related to the number of molecules n per unit volume and the average translational kinetic energy of the molecules $\overline{\varepsilon}_t$. The larger the product of n and $\overline{\varepsilon}_t$, the larger the pressure p. Pressure is a state parameter that describes the state of a gas and is a macroscopic quantity that can be measured directly. The molecular number density of a gas n, is a microscopic quantity and a statistical average. The average translational kinetic energy of gas molecules $\overline{\varepsilon}_t$ is also a statistical average of a microscopic quantity. However, the statistical average of microscopic quantities cannot be directly measured by experimental methods. Thus, the pressure formula for an ideal gas relates the macroscopic quantity p which describes the properties of the gas to the statistical average n and $\overline{\varepsilon}_t$ which are the microscopic quantities.

The ideal gas pressure formula shows that, when $\overline{\varepsilon}_t$ is constant, the larger the number of molecules n in the unit volume, the larger the number of molecules colliding with the unit area wall in the unit time and the larger the pressure on the wall. On the other hand, when the molecular number density n of an ideal gas is constant, a larger $\overline{\varepsilon}_t$ results in a larger pressure p. The reason is that a larger $\bar{\varepsilon}_t$ indicates a larger root-mean-square rate of the molecule and a larger average velocity of the molecule, which means more collisions in the unit time and greater average impulse force on the wall during collisions.

Pressure has statistical significance and is the statistical average of the microscopic quantity of a large number of gas molecules, and its value is equal to the average force of gas molecules on the unit wall area, which is the microscopic nature of pressure.

7.3 Microscopic interpretation of temperature

Temperature is one of the most basic concepts in heat, and in this section we will derive the temperature formula from the gas pressure formula, clarify the microscopic nature of temperature, and explain the statistical significance of temperature.

For an ideal gas, let the mass of a molecule be m, the number of molecules of the gas be N, the total mass of the gas be M, and the molar mass of the gas be M_{mol}. Thus, $M = Nm$ and $M_{mol} = N_A m$. According to the equation of state for an ideal gas $pV = \dfrac{M}{M_{mol}} RT$,

$$pV = \frac{M}{M_{mol}} RT = \frac{Nm}{N_A m} RT = \frac{N}{N_A} RT \tag{7.3.1}$$

So

$$p = \frac{N}{V} \frac{R}{N_A} T \tag{7.3.2}$$

Define $n = N/V$ as the number of molecules per unit volume, that is, the molecular number density and introduce $k = R/N_A = 1.38 \times 10^{-23}$ J·K^{-1}, called the Boltzmann constant, and we have

$$p = nkT \tag{7.3.3}$$

The above equation is another expression of the ideal gas equation of state, showing the relationship between the macroscopic quantity p and the number of molecules per unit volume n and the macroscopic quantity temperature T.

Comparing the pressure formula of the kinetic theory of gases Equation (7.2.6) with the ideal gas equation of state Equation (7.3.3), we can get

$$T = \frac{2}{3k} \bar{\varepsilon}_t \tag{7.3.4}$$

Equation (7.3.4) can be rewritten as

$$\bar{\varepsilon}_t = \frac{3}{2} kT \tag{7.3.5}$$

This shows that the average translational kinetic energy of an ideal gas molecule depends only on the temperature and is proportional to the thermodynamic temperature. At the same temperature, the average translational kinetic energy of all gas molecules is the same. Equation (7.3.4) is a basic equation of the kinetic theory of gas molecules, which is called the temperature formula of the kinetic theory of gas. The important physical significance of the temperature formula is that it reveals the microscopic nature of temperature: The temperature of a gas indicates the intensity of the random thermal motion made by a large number of molecules in the gas, that is, temperature is a measure of the average translational kinetic energy of molecules. The higher the temperature of the gas, on average, the more intense the thermal motion of the molecules inside the gas and the greater the average translational kinetic energy of the molecules.

The temperature formula reflects the relationship between the macroscopic quantity T of a system composed of a large number of molecules and the statistical average of the microscopic quantity $\bar{\varepsilon}_t$. Therefore, temperature, like pressure, is also the collective performance of a large number of molecules doing random thermal motion. So temperature is of statistical average significance. For a system composed of a single molecule or a small number of molecules, it is not suitable to say how high its or their temperature is, that is, for a single molecule or a system composed of a small number of molecules, the concept of temperature is meaningless.

When two systems with different temperatures reach thermal equilibrium through thermal contact, the macroscopic performance during this process is that net energy is transferred from the system with high temperature to the system with low temperature until the temperatures of the two systems are equal; the microscopic nature of it is that the molecules from each of the systems exchange their energy through collisions, resulting in energy redistribution. If the two systems have the same mean translational kinetic energy of molecules, that is, their temperatures are equal, then the two systems reach thermal equilibrium no matter how many molecules are contained in each of the two systems.

As can be seen from Equation (7.3.4), with the decrease of temperature, the average translational kinetic energy of gas molecules will decrease. When the thermodynamic temperature $T=0$ K, $\bar{\varepsilon}_t=0$ J, indicating that the random thermal motion of the ideal gas molecules will stop, which contradicts experimental facts. In fact, the thermal motion of molecules is never stopped, that is, it is impossible to reach absolute zero temperature through any finite process. Before the temperature reaches the absolute zero, a gas has become a liquid or a solid, and the gas kinetic theory model based on the laws of classical mechanics is no longer applicable. At this time, the laws of quantum mechanics play a major role, and the statistical formula of temperature is not valid.

According to the definition of the average translational kinetic energy of an ideal gas molecule $\bar{\varepsilon}_t = m\overline{v^2}/2$ and Formula (7.3.5), we can get

$$\sqrt{\overline{v^2}} = \sqrt{\frac{3kT}{m}} = \sqrt{\frac{3RT}{M_{\text{mol}}}} \tag{7.3.6}$$

$\sqrt{\overline{v^2}}$ represents the square root of molecular speed squared average of a large number of gas molecules, called the **root-mean-square speed of gas molecules**, which represents the statistical average of the microscopic quantity of gas molecules. The above equation shows that the root-mean-square speed of gas molecules is proportional to the square root of the thermodynamic temperature of the gas and inversely proportional to the square root of the molecular or molar mass of the gas. The higher the temperature and the smaller the mass of the gas molecules (or molar mass), the greater the root-mean-square speed of the molecules.

Example 7.3.1 There are N different ideal gases with the same temperature mixed and stored in the same container. Prove that the pressure of the mixed gas is equal to the sum of the partial pressures of the component gases.

Proof According to Equation (7.3.5), the average translational kinetic energy of various gas molecules with the same temperature is equal.

$$\bar{\varepsilon}_{t1} = \bar{\varepsilon}_{t2} = \cdots = \bar{\varepsilon}_{tN} = \bar{\varepsilon}_t = \frac{3}{2}kT$$

Let the number of molecules containing various gases per unit volume be $n_1, n_2, \cdots, n_N$. The number of molecules of the mixed gas per unit volume is $n = n_1 + n_2 + \cdots + n_N$. So the pressure of the mixed gas is

$$p = \frac{2}{3}n\bar{\varepsilon}_t = \frac{2}{3}(n_1 + n_2 + \cdots + n_N)\bar{\varepsilon}_t$$

$$= \frac{2}{3}n_1\bar{\varepsilon}_{t1} + \frac{2}{3}n_2\bar{\varepsilon}_{t2} + \cdots + \frac{2}{3}n_N\bar{\varepsilon}_{tN}$$

$$= p_1 + p_2 + \cdots + p_N$$

This is Dalton's law of partial pressure where $p_1 = \frac{2}{3}n_1\bar{\varepsilon}_{t1}$, $p_2 = \frac{2}{3}n_2\bar{\varepsilon}_{t2}$, $\cdots$, $p_N = \frac{2}{3}n_N\bar{\varepsilon}_{tN}$, representing the partial pressures of various gases respectively.

Example 7.3.2 Hydrogen gas is stored in a container at pressure of one atmosphere and temperature of 27℃. Find

(1) the number of molecules per unit volume in the container;

(2) the mass density of hydrogen;

(3) the root-mean-square speed of hydrogen molecules;

(4) the average translational kinetic energy of hydrogen molecules.

Solution (1) According to Equation (7.3.3), the number of molecules per unit volume is

$$n = \frac{p}{kT} = \frac{1.013 \times 10^5}{1.38 \times 10^{-23} \times 300} = 2.45 \times 10^{25} \text{ m}^{-3}$$

(2) Let the mass of hydrogen be M and the volume be V, and the mass density of hydrogen is

$$\rho = \frac{M}{V} \tag{1}$$

The number of molecules per unit volume is

$$n = \frac{\frac{M}{M_{mol}} N_A}{V} = \frac{M N_A}{M_{mol} V} \quad (2)$$

The mass density of hydrogen can be obtained by combining Equations (1) and (2)

$$\rho = \frac{M_{mol}}{N_A} n = \frac{2 \times 10^{-3}}{6.02 \times 10^{23}} \times 2.45 \times 10^{25} = 8.14 \times 10^{-2} \text{ kg} \cdot \text{m}^{-3}$$

(3) The root-mean-square speed of hydrogen molecules is

$$\sqrt{\overline{v^2}} = \sqrt{\frac{3RT}{M_{mol}}} = \sqrt{\frac{3 \times 8.31 \times 300}{2 \times 10^{-3}}} = 1.93 \times 10^3 \text{ m} \cdot \text{s}^{-1}$$

(4) The average translational kinetic energy of hydrogen molecules is

$$\overline{\varepsilon_t} = \frac{3kT}{2} = \frac{3}{2} \times 1.38 \times 10^{-23} \times 300 = 6.21 \times 10^{-21} \text{ J}$$

7.4 Energy equipartition theorem

In the previous discussion, we only studied the average translational kinetic energy of the ideal gas molecules, during which, the ideal gas molecule was treated as a particle. When an ideal gas molecule is composed of two or more atoms, it has a certain size and internal structure. In addition to the overall translational motion of the molecule, the atoms of the molecule may also vibrate and rotate. These forms of motion correspond to a certain amount of energy, so the total energy of molecular thermal motion includes the energy of these three forms.

When studying the energy of gases, we divide ideal gas molecules into monatomic, diatomic and polyatomic molecules. In order to calculate the average energy of molecular thermal motion by a statistical method, the concept of the degree of freedom is first introduced.

 ### 7.4.1 Energy equipartition theorem

The degree of freedom is a physical quantity that describes the degree of freedom of movement of an object. It refers to the number of independent variables needed to determine the spatial position of an object.

For a gas, its molecule can be monatomic, diatomic, or polyatomic. For an ideal gas composed of monatomic molecules, the monatomic molecule can be regarded as a particle, and its position in space can be determined by only three coordinates (x, y, z) in the Cartesian coordinate system, as shown in Figure 7.4.1. So the degree of freedom of a single atomic gas molecule is 3.

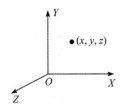

Figure 7.4.1 Monatomic molecule

The average translational kinetic energy of a monatomic molecule is

$$\overline{\varepsilon}_t = \frac{1}{2}m\overline{v^2} \qquad (7.4.1)$$

For a monatomic molecule with only translational kinetic energy, the expression for its average energy is

$$\overline{\varepsilon} = \overline{\varepsilon}_t = \frac{1}{2}m\overline{v_x^2} + \frac{1}{2}m\overline{v_y^2} + \frac{1}{2}m\overline{v_z^2} \qquad (7.4.2)$$

When a gas is in equilibrium, the probability of molecules moving in any direction is equal, that is

$$\overline{v_x^2} = \overline{v_y^2} = \overline{v_z^2} = \frac{1}{3}\overline{v^2} \qquad (7.4.3)$$

According to Equation (7.3.5), we can get

$$\frac{1}{2}m\overline{v_x^2} = \frac{1}{2}m\overline{v_y^2} = \frac{1}{2}m\overline{v_z^2} = \frac{1}{6}m\overline{v^2} = \frac{1}{3}\overline{\varepsilon}_t = \frac{1}{2}kT \qquad (7.4.4)$$

Equation (7.4.4) shows that each translational degree of freedom has the same average translational kinetic energy, whose value is $\frac{1}{2}kT$. Although this conclusion is deduced from the translational motion, Boltzmann generalized it by hypothesizing that, for an ideal gas system in thermal equilibrium at temperature T, every degree of freedom of the molecule has the same average kinetic energy $\frac{1}{2}kT$, which is called the **energy equipartition theorem according to degree of freedom.**

The degree of freedom of translation, rotation and vibration are represented by t, r and v, respectively. The average kinetic energy of the molecule can be expressed as

$$\overline{\varepsilon} = (t+r+v)\frac{1}{2}kT \qquad (7.4.5)$$

If i is the total number of the degree of freedom of the molecule, $i = t+r+v$ and the average kinetic energy of the molecule is

$$\overline{\varepsilon} = \frac{i}{2}kT$$

For monatomic molecules, there is $t = 3$, $r = 0$, and $v = 0$, so $\overline{\varepsilon} = \frac{3}{2}kT$.

For ideal gases composed of diatomic molecules, we only discuss rigid diatomic molecular gases, that is, the distance between two atoms in a gas molecule is fixed. As a consequence, the gas molecule can be viewed as a system in which two atoms are connected by a massless thin rod. The molecular motion can be viewed as a superposition of the translation of the center of mass C and the rotation through the center of mass C in the X, Y, and Z directions. Take the X axis as the direction of the connection of two atoms in the molecule, as shown in Figure 7.4.2. To determine the position of the molecule in space, firstly, a translational degree of freedom of 3 ($t = 3$) is needed to point out the position of the center of mass, and then three azimuths α, β and γ are required to determine the orientation of the thin rod. Because $\cos^2\alpha + \cos^2\beta + \cos^2\gamma = 1$ and only two of the three azimuths are independent, a rigid diatomic molecule has a rotational degree of freedom of 2 ($r=2$). According to the equipartition principle, its average kinetic energy is

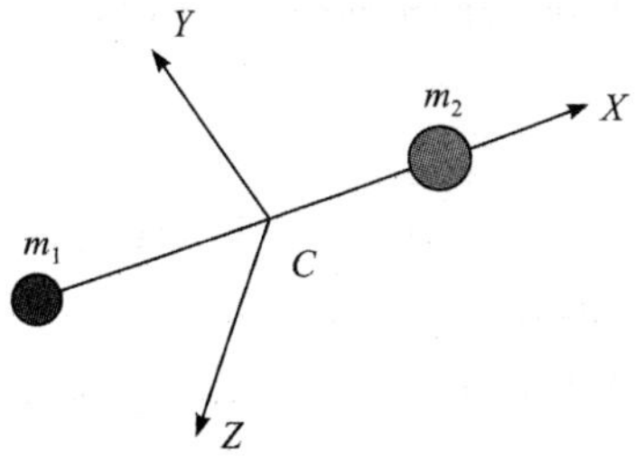

Figure 7.4.2 Rigid diatomic molecule

$$\bar{\varepsilon}_k = \frac{1}{2}(t+r)kT \qquad (7.4.6)$$

where $t=3$ and $r=2$, and then

$$\bar{\varepsilon} = \frac{5}{2}kT$$

For rigid molecules with many atoms, if the molecular structure is linear (e.g. CO_2), as with a rigid diatomic molecule, then it has $t=3$, $r=2$, $v=0$, and $\bar{\varepsilon}=\frac{5}{2}kT$. If it's a nonlinear rigid molecule, we need to add an independent coordinate to determine the rotation about the axis, that is $t=3$, $r=3$, $v=0$, and $\bar{\varepsilon}=\frac{6}{2}kT=3kT$.

The energy equipartition theorem according to degree of freedom can be derived from general statistical theory and is an important conclusion in classical statistical physics. For an individual molecule in a gas, its energy of each degree of freedom is not necessarily equal. But for a large group of molecules, due to the frequent collisions between molecules, energy is exchanged and transferred between molecules and degrees of freedom. Therefore, when the gas reaches equilibrium, energy is evenly distributed to each degree of freedom. The energy equipartition theorem according to degree of freedom applies not only to gases, but also to liquids and solids.

7.4.2 Internal energy of ideal gas

In a thermodynamic system, the sum of the energy of molecular thermal motion is called the internal energy of the system. For an ideal gas, the force between molecules is negligible, so the potential energy is zero, and the internal energy of the system is the sum of the kinetic energy of all the molecules in the system. If the total number of molecules in an equilibrium ideal gas system is N at temperature T, then the internal energy can be expressed as

$$E = N\bar{\varepsilon} = N\frac{i}{2}kT \qquad (7.4.7)$$

Internal energy of 1 mole of an ideal gas is

$$E_{mol} = N_A \bar{\varepsilon} = N_A \frac{i}{2}kT = \frac{i}{2}RT \qquad (7.4.8)$$

where N_A is the Avogadro constant and R is the universal gas constant.

Internal energy of ν moles of an ideal gas is

$$E = \nu \frac{i}{2} RT \tag{7.4.9}$$

which shows that the internal energy of an ideal gas with a certain mass in equilibrium depends only on the temperature of the system.

When the temperature of a gas changes ΔT, the internal energy changes

$$\Delta E = \frac{M}{M_{mol}} \frac{i}{2} R \Delta T = \nu \frac{i}{2} R \Delta T \tag{7.4.10}$$

Example 7.4.1 There is a rigid diatomic ideal gas at 0℃. Try to find

(1) the molecular average translational kinetic energy;

(2) the molecular average rotational kinetic energy;

(3) the molecular average kinetic energy;

(4) the molecular average energy;

(5) the internal energy of 0.5 moles of the gas.

Solution It is given that the ideal gas is a rigid diatomic molecule, so its degree of freedom is $i=5$, of which the translational degree of freedom is 3 and the rotational degree of freedom is 2. According to the energy equipartition theorem,

(1) the average translational kinetic energy is

$$\overline{\varepsilon}_t = \frac{3}{2} kT = \frac{3}{2} \times 1.38 \times 10^{-23} \times 273 = 5.65 \times 10^{-21} \text{ J}$$

(2) the average rotational kinetic energy is

$$\overline{\varepsilon}_r = \frac{2}{2} kT = \frac{2}{2} \times 1.38 \times 10^{-23} \times 273 = 3.77 \times 10^{-21} \text{ J}$$

(3) the average kinetic energy

$$\overline{\varepsilon}_k = \frac{5}{2} kT = \frac{5}{2} \times 1.38 \times 10^{-23} \times 273 = 9.42 \times 10^{-21} \text{ J}$$

(4) the average energy is

$$\overline{\varepsilon} = \overline{\varepsilon}_k = 9.42 \times 10^{-21} \text{ J}$$

(5) the internal energy of 0.5 moles of the gas

$$E = \nu \frac{i}{2} RT = \frac{1}{2} \times \frac{5}{2} \times 8.31 \times 273 = 2.84 \times 10^3 \text{ J}$$

7.5 Law of the Maxwell speed distribution

Gas molecules are in random thermal motion with constantly changing speed due to collision, hence, the magnitude and direction of the velocity of a given molecule at a moment are entirely accidental. However, for a large number of molecules as a whole, the speed distribution of molecules under certain conditions follows a certain statistical law—the speed distribution law of gas.

The statistical law of the distribution of gas molecules according to speed was first derived by Maxwell in 1859 on the basis of the probability theory, and in 1920 Stern confirmed the statistical law from experiments.

7.5.1 Function of speed distribution

At a certain temperature, the molecules of a gas in equilibrium are distributed according to speed and this statistical law is called the **speed distribution**. Let N be the total number of molecules in a system and dN be the number of molecules with speed values in the interval $v \sim v+dv$. So dN/N is the percentage of the number of molecules in the speed distribution to the total number of molecules. For a single molecule, we can introduce a function of speed distribution $f(v)$ which represents the probability that the molecular speed is in the interval $v \to v+dv$. Thus,

$$\frac{dN}{N} = f(v)dv \tag{7.5.1}$$

Therefore

$$f(v) = \frac{dN}{Ndv} \tag{7.5.2}$$

which represents the ratio of the number of molecules in the unit speed interval near the speed v to the total number of molecules in the system and is called the function of speed distribution. The physical meaning of this function of speed distribution is that, $f(v)$ is the ratio of the number of molecules distributed in the unit speed interval $\dfrac{dN}{dv}$ near the speed v to the total number of molecules, and it also represents the probability that the speed of any molecule occurs in a unit speed interval near the speed v. Figure 7.5.1 shows the relationship between $f(v)$ and v, which is called the **speed distribution curve of gas molecules**. It graphically displays the speed distribution of gas molecules.

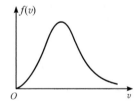

Figure 7.5.1 Speed distribution curve

By integrating Equation (7.5.2) over velocity, we get the percentage of the number of molecules in all speed intervals (equal to the total number of molecules N) to the total number of molecules N, which is equal to 1.

$$\int_0^N \frac{dN}{N} = \int_0^\infty f(v)dv = 1 \tag{7.5.3}$$

which is called the **normalization condition of the function of speed distribution**. It also means that the probability of a molecule's speed being located in all speed intervals is 1.

7.5.2 Law of the Maxwell speed distribution

In 1859, the British physicist Maxwell used the method of statistical physics to theoretically derive the statistical law of the speed distribution of ideal gas molecules in equilibrium.

$$\frac{dN}{N} = 4\pi \left(\frac{m}{2\pi kT}\right)^{\frac{3}{2}} e^{-\frac{mv^2}{2kT}} v^2 dv \tag{7.5.4}$$

where T is the temperature of the system, m is the mass of a molecule, and k is called the Boltzmann constant. The relationship between k and the ideal gas universal constant R as well as Avogadro's constant N_A is $k = \frac{R}{N_A} = 1.38 \times 10^{-23}$ J·K^{-1}. Compare Equations (7.5.1) and (7.5.4), and we get

$$f(v) = 4\pi \left(\frac{m}{2\pi kT}\right)^{\frac{3}{2}} e^{-\frac{mv^2}{2kT}} v^2 \tag{7.5.5}$$

which is called the **function of the Maxwell speed distribution**. It can be seen that this function is only related to the type of gas and temperature.

Figure 7.5.2 shows the speed distribution curves of nitrogen at different temperatures. The curves start at the origin, rise as the speed increases, pass a maximum point, then fall as the speed increases, and finally tend to the horizontal axis. These curves show that the speed of the gas molecules can be any possible value greater than zero. Meanwhile, the ratio of the number of

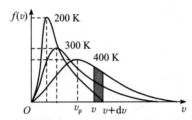

Figure 7.5.2 Speed distribution curves of Nitrogen at different temperatures

molecules with a relative high or low speed to the total number of molecules is small, and the ratio of the number of molecules with a moderate speed is much larger.

In a speed distribution curve, the area of a narrow strip under the curve in any speed interval $v \sim v+dv$ represents the ratio of the number of molecules whose speeds are located in this speed interval to the total number of molecules $\frac{dN}{N}$, and also represents the probability that the speed of a molecule falls in the speed interval $v \sim v+dv$. The entire area under the speed distribution curve represents the ratio of the sum of all molecules in the speed interval to the total number of molecules which should be equal to 1. It is the normalization condition of the function of speed distribution.

7.5.3 Three statistical speeds of speed distribution

The speed corresponding to the maximum value of the speed distribution curve in Figure 7.5.2 is called the most probable speed v_p. If the entire speed range is divided into many equivalent speed intervals and each speed interval is dv, then the ratio of the number of molecules in the speed interval $v_p \sim v_p + dv$ to the total number of molecules is the largest, which means that the probability that a molecule's speed is in the rate interval $v_p \sim v_p + dv$ is the highest. v_p is the most probable speed and can be found by the function of speed distribution.

$$\frac{df(v)}{dv} = 0 \tag{7.5.6}$$

By substituting Equation (7.5.5) into the above expression we can get

$$v_p = \sqrt{\frac{2kT}{m}} = \sqrt{\frac{2RT}{M_{mol}}} \approx 1.41\sqrt{\frac{RT}{M_{mol}}} \tag{7.5.7}$$

which shows that v_p decreases with the increase of molecular weight and increases with the increase of temperature.

Using the function of speed distribution, two other commonly used statistical average speeds can be obtained: the average speed and the root-mean-square speed.

The average speed is

$$\bar{v} = \frac{\int v \, dN}{N} = \int_0^\infty v f(v) \, dv \tag{7.5.8}$$

By substituting Equation (7.5.5) into the above equation, the average speed of an ideal gas molecule in equilibrium can be obtained

$$\bar{v} = \sqrt{\frac{8kT}{\pi m}} = \sqrt{\frac{8RT}{\pi M_{mol}}} \approx 1.59\sqrt{\frac{RT}{M_{mol}}} \tag{7.5.9}$$

In a similar way,

$$\overline{v^2} = \frac{\int v^2 \, dN}{N} = \int_0^\infty v^2 f(v) \, dv \tag{7.5.10}$$

By substituting Equation (7.5.5) into the above equation, the root-mean-square speed of an ideal gas molecule in equilibrium can be obtained

$$\sqrt{\overline{v^2}} = \sqrt{\frac{3kT}{m}} = \sqrt{\frac{3RT}{M_{mol}}} \approx 1.73\sqrt{\frac{RT}{M_{mol}}} \tag{7.5.11}$$

It can be seen that the values of these three speeds v_p, $\bar{v}$ and $\sqrt{\overline{v^2}}$ are proportional to $\sqrt{T}$ and inversely proportional to $\sqrt{m}$. Meanwhile, $v_p < \bar{v} < \sqrt{\overline{v^2}}$, as Figure 7.5.3 shows.

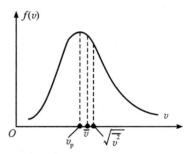

Figure 7.5.3 Three statistical speeds

When the gas type is given, the speed distribution curve is only related to temperature. When the temperature of the gas increases, the number of molecules with a smaller speed decreases, while the number of molecules with a larger speed increases and the most probable speed increases, so the peak of the speed distribution curve moves to the right side. The function of speed distribution should satisfy the normalization condition and the total area under the curve should be identical to 1, so the curve becomes flatter as the temperature increases. As can be seen from Figure 7.5.4, the higher the temperature, the smaller the value $f(v_p)$. Meanwhile, since the total area under the curve is equal to 1, the curve will become lower and flatter when the temperature rises, and expand to the region with higher speed.

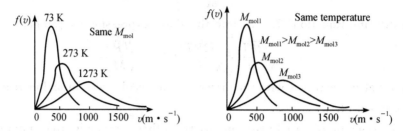

Figure 7.5.4 Speed distribution curves of gas molecules at different temperatures and molecular weights

The speed distribution curve at a constant temperature is only related to the type of gas. For different kinds of gases, since the most probable speed v_p is inversely proportional to the square root of the molecular mass of the gas $\sqrt{m}$, gas with a smaller molecular mass has a larger most probable speed. When the molecular mass of the gas decreases, the most probable speed of the gas increases, and the peak of the speed distribution curve moves towards the side with the higher speed. Due to the normalization condition of the function of speed distribution, the total area under the curve should be equal to 1, so the curve becomes flat as the gas molecular mass decreases.

It should be emphasized that the formulas of three statistical speeds given above are only the results calculated by the Maxwell speed distribution. For different speed distribution functions, the calculation formulas of the three statistical speeds may be different from the above and need to be derived again.

Example 7.5.1 There are N particles and the functions of speed distribution are

$$\begin{cases} f(v) = \dfrac{av}{v_0} & (0 \leqslant v \leqslant v_0) \\ f(v) = a & (v_0 \leqslant v \leqslant 2v_0) \\ f(v) = 0 & (v > 2v_0) \end{cases}$$

(1) Draw the speed distribution curve and find the constant a;

(2) Find the number of particles in the speed interval $(1.5v_0, 2v_0)$;

(3) Find the average particle speed.

Solution (1) The speed distribution curve is shown in Figure 7.5.5.

By normalization conditions

$$\int_0^\infty f(v) \, dv = 1$$

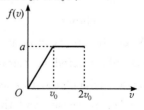

Figure 7.5.5 Figure of Example 7.5.1

$$\int_0^{v_0} a\,\frac{v}{v_0}\mathrm{d}v + \int_{v_0}^{2v_0} a\,\mathrm{d}v = 1$$

Then

$$a = \frac{2}{3v_0}$$

(2) The number of particles within the speed interval $(1.5v_0, 2v_0)$ is

$$\Delta N = N\int_{1.5v_0}^{2v_0} f(v)\,\mathrm{d}v = N\int_{1.5v_0}^{2v_0} a\,\mathrm{d}v = N\int_{1.5v_0}^{2v_0} \frac{2}{3v_0}\mathrm{d}v = \frac{N}{3}$$

(3) The average particle speed is

$$\bar{v} = \int_0^\infty v f(v)\,\mathrm{d}v = \int_0^{v_0} v\,\frac{av}{v_0}\mathrm{d}v + \int_{v_0}^{2v_0} av\,\mathrm{d}v = \frac{11}{9}v_0$$

7.6　Law of the Boltzmann distribution

7.6.1　Law of the Maxwell velocity distribution

The distribution of molecular speed discussed above does not take into account the direction of molecular velocity. To find the velocity distribution of molecules, it is necessary to find out the ratio of the number of molecules $\mathrm{d}N_v$ distributed in a volume element $\mathrm{d}v_x\mathrm{d}v_y\mathrm{d}v_z$ near the velocity $\boldsymbol{v}$ to the total number of molecules in the velocity space, and then the function of velocity distribution is defined as

$$f(\boldsymbol{v}) = \frac{\mathrm{d}N_v}{N\mathrm{d}v_x\mathrm{d}v_y\mathrm{d}v_z} \tag{7.6.1}$$

which represents the proportion of the number of molecules in the unit velocity space volume near the velocity v to the total number of molecules, that is, the velocity probability density, also known as the function of velocity distribution of gas molecules.

In 1859, Maxwell first derived the law of velocity distribution of an ideal gas

$$\frac{\mathrm{d}N_v}{N} = \left(\frac{m}{2\pi kT}\right)^{\frac{3}{2}} \mathrm{e}^{-\frac{m}{2kT}(v_x^2+v_y^2+v_z^2)}\mathrm{d}v_x\mathrm{d}v_y\mathrm{d}v_z \tag{7.6.2}$$

Then the Maxwell velocity distribution function is

$$f(\boldsymbol{v}) = \left(\frac{m}{2\pi kT}\right)^{\frac{3}{2}} \mathrm{e}^{-\frac{m}{2kT}(v_x^2+v_y^2+v_z^2)} \tag{7.6.3}$$

7.6.2　Law of the Boltzmann distribution

The law of Maxwell distribution is the law of velocity distribution of ideal gas molecules in thermal equilibrium without external force, or when the external force field can be ignored. Since there is no external force field, the distribution of molecules according to spatial position is uniform, that is, the molecular number density n is the same everywhere in a container. When there is a conservative external force (gravity field, electric field, etc.), the distribution of gas molecules in various spatial positions is no longer uniform, and the molecular number density is different at different positions.

Boltzmann extended the Maxwell velocity distribution to the case of an ideal gas in a conservative force field. ① In the external field, the kinetic energy $\dfrac{mv^2}{2}$ of molecules should be replaced by the total energy $E = E_k + E_p$ in Equation (7.6.2). ② The particles should be distributed not only according to the velocity interval $v_x \sim v_x + dv_x$, $v_y \sim v_y + dv_y$, and $v_z \sim v_z + dv_z$, but also according to the position interval $x \sim x + dx$, $y \sim y + dy$, and $z \sim z + dz$. Using the above-mentioned two generalizations as well as the probability theory, the following formula is derived

$$dN' = n_0 e^{-\frac{E_p}{kT}} dx\,dy\,dz \tag{7.6.4}$$

where dN' is the number of molecules in the volume element $dx\,dy\,dz$ and n_0 is the number density of molecules when $E_p = 0$. Equation (7.6.4) can also be rewritten in the following form

$$n = \frac{dN'}{dx\,dy\,dz} = n_0 e^{-\frac{E_p}{kT}} \tag{7.6.5}$$

Equation (7.6.5) is the Boltzmann distribution of molecular number density, which is the distribution of molecular number density according to potential energy. In the Boltzmann derivation, the following assumption is used that the number of molecules $dN' = n\,dx\,dy\,dz$ in any macroscopic volume element $dx\,dy\,dz$ is still a large number, this volume element still contains various velocities, and the molecules in this volume element obey the Maxwell velocity distribution. Let the number of molecules whose speeds are located in $v_x \sim v_x + dv_x$, $v_y \sim v_y + dv_y$, and $v_z \sim v_z + dv_z$ be dN in the macroscopic volume element $dx\,dy\,dz$. According to the Maxwell distribution, there are

$$\frac{dN}{dN'} = f(v)\,dv_x\,dv_y\,dv_z$$

$$dN = dN' f(v)\,dv_x\,dv_y\,dv_z$$

$$dN = n_0 e^{-\frac{E_p}{kT}} \left(\frac{m}{2\pi kT}\right)^{\frac{3}{2}} e^{-\frac{m}{2kT}(v_x^2 + v_y^2 + v_z^2)} dv_x\,dv_y\,dv_z\,dx\,dy\,dz$$

$$dN = n_0 \left(\frac{m}{2\pi kT}\right)^{\frac{3}{2}} e^{-\frac{E_k + E_p}{kT}} dv_x\,dv_y\,dv_z\,dx\,dy\,dz \tag{7.6.6}$$

Equation (7.6.6) is the Boltzmann distribution in which molecules are distributed by both speed interval and position interval. The Boltzmann distribution diagram is shown in Figure 7.6.1. If $E = E_k + E_p$, then

$$dN = n_0 \left(\frac{m}{2\pi kT}\right)^{\frac{3}{2}} e^{-\frac{E}{kT}} dv_x\,dv_y\,dv_z\,dx\,dy\,dz \tag{7.6.7}$$

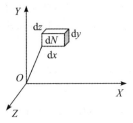

Figure 7.6.1 Diagram of the Boltzmann distribution

The Boltzmann distribution describes the energy distribution of ideal gas molecules in thermal equilibrium when they are subjected to non-negligible conservative external force or conservative external force field. In intervals of equal width, if $E_1 > E_2$, then the number of particles with high energy dN_1 is smaller than the number of particles with low energy dN_2, which is $dN_1 < dN_2$, or particles preferentially occupy states with low energy. This is an important result of the law of Boltzmann distribution. It should be noted that the law of the Boltzmann distribution applies to molecules, atoms, and Brownian particles, but not to systems composed of electrons and photons.

7.6.3 Application of the law of the Boltzmann distribution in engineering technology

The law of the Boltzmann distribution is a basic principle of statistical mechanics. It is applicable to the thermal equilibrium state of particles of any substance under the action of any conservative force field. It is not only of great scientific significance but also of extensive engineering application value.

The technique of isotope separation, which is widely used in nuclear reactions, biochemistry and geochemistry, is essentially the use of the law of the Boltzmann distribution. The isotope gas molecules are placed into a centrifuge cylinder with radius R, which rotates at a constant angular speed ω, so the molecules inside the cylinder are in a constant inertial centrifugal force field. In thermal equilibrium, the density of the gas is $n = n_0 e^{\frac{m\omega^2 r^2}{2kT}}$ distributed according to the Boltzmann distribution. It can be seen that molecules of different mass have different densities at the same radius, which allows them to be separated. At present, the nuclear industry uses this method to separate and produce the isotope ^{235}U.

Using the law of the Boltzmann distribution, we can find the active medium for laser manufacturing. The active medium is a medium in which particle population inversion occurs between two energy levels and the light base of a specific frequency is amplified under certain external conditions. It is not possible for every substance to realize population inversion between any two energy levels, so we must find the condition for it. According to the fact that the distribution of particle population at different energy levels depends on the law of the Boltzmann distribution in thermal equilibrium, a function of speed based on the Boltzmann distribution can be used to judge the condition for particle population inversion.

Although the Boltzmann distribution is derived from thermodynamics and statistical physics, it is popularized for applications in fluid mechanics, machine learning, and artificial intelligence. The Boltzmann distribution reflects the corresponding relationship between energy and probability, and this relationship is applied to machines' learning the sampling distribution in stochastic neural network algorithms, such as the Boltzmann machines, constrained Boltzmann machines, and deep Boltzmann machines.

7.7 Mean free path of a gas molecule

The average speed of thermal motion of gas molecules at room temperature is about $10^2 \sim 10^3$ m·s^{-1}. Judging by this speed, the processes of diffusion and heat conduction in the gas seem to proceed very quickly. However, this is not the case, and the mixing of gases (diffusion process) proceeds rather slowly. The reason for this is that when a molecule moves from one place to another, it constantly collides with other molecules, which causes the molecule to move along circuitous lines, as shown in Figure 7.7.1. Therefore, the speed of gas diffusion, heat conduction and other processes are related to the frequency of molecular collision.

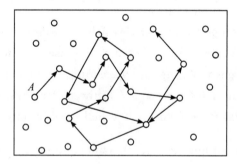

Figure 7.7.1 Schematic diagram of molecular motion trajectory

In frequent collisions with other molecules, the distance and time interval traveled by a molecule between successive collisions are various. It is impossible to find every distance and time interval but we can deal with this problem statistically. The distance that a molecule passes freely between two adjacent collisions is called the **free path**. The average free path between successive collisions is called the **mean free path of a molecule**, denoted as $\bar{\lambda}$. The average number of collisions between a molecule and other molecules in unit time is called the **average collision frequency of the molecule**, denoted as $\bar{Z}$. If the average speed of the molecules is expressed as $\bar{v}$, then

$$\bar{\lambda} = \frac{\bar{v}}{\bar{Z}} \tag{7.7.1}$$

In order to derive the expression of the mean collision frequency $\bar{Z}$, suppose that a gas molecule can be regarded as a rigid sphere with a certain diameter d and the other molecules are stationary. If molecule A moves at an average relative speed $\bar{u}$ and collides with another molecule B, then the distance between the centers of molecule A and B is equal to the effective diameter d of molecule A. Therefore, if a tortuous cylinder is made with the motion trajectory of the center of A as the axis and the effective diameter d as the radius, all molecules with their centers in the cylinder will collide with A, as shown in Figure 7.7.2. The cross-sectional area of this cylinder $\sigma = \pi d^2$ is called the collision cross section of the molecule.

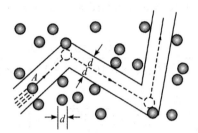

Figure 7.7.2 Schematic diagram of the molecular collision process

In time dt, the distance traveled by molecule A is $\bar{u} dt$ and the swept volume of the cylinder is $\sigma \bar{u} dt$. If n is the number of molecules per unit volume of the gas, then the number of molecules in a cylinder of volume $\sigma \bar{u} dt$ is $n\sigma \bar{u} dt$ which is the number of collisions between molecule A and other molecules in dt. So, the average number of collisions per unit time between molecule A and other molecules is

$$\bar{Z} = \frac{n\sigma \bar{u} dt}{dt} = n\sigma \bar{u} \tag{7.7.2}$$

Using the law of the Maxwell velocity distribution, it can be proved that the relationship between the average relative speed of gas molecules and the average speed is

$$\bar{u} = \sqrt{2}\, \bar{v} \tag{7.7.3}$$

Then the mean collision frequency is

$$\bar{Z} = \sqrt{2}\, n\sigma \bar{v} = \sqrt{2}\, \pi d^2 n \bar{v} \tag{7.7.4}$$

The above equation shows that the average collision frequency is proportional to the molecular number density, the average molecular speed, and the square of the molecular diameter.

Using the above formula and Equation (7.7.1), the mean free path can be obtained

$$\bar{\lambda} = \frac{1}{\sqrt{2}\, n\sigma} = \frac{1}{\sqrt{2}\, \pi d^2 n} \tag{7.7.5}$$

which shows that the mean free path is inversely proportional to the collision cross section and molecular number density, and has nothing to do with the mean molecular speed. Using the ideal gas equation of state $p=nkT$, the mean free path can be expressed as

$$\bar{\lambda}=\frac{kT}{\sqrt{2}\sigma p} \tag{7.7.6}$$

It should be noted that actual molecules are generally not spheres, and the interactions between molecules are complex. During the above derivation, we regard the gas molecule as a rigid ball with a diameter of d and consider the collision as an elastic collision. Therefore, the molecular diameter obtained by Equations (7.7.4) and (7.7.5) can only be considered as the effective diameter of the molecule under the above conditions.

Example 7.7.1 The effective diameter of a nitrogen molecule is 3.8×10^{-10} m. Try to find the mean free path, the mean collision frequency, and the mean time interval between two successive collisions of nitrogen molecules in the standard state.

Solution The standard state is a standard atmosphere at zero degrees Celsius.

The mean free path is

$$\bar{\lambda}=\frac{1}{\sqrt{2}\pi d^2 n}=\frac{kT}{\sqrt{2}\pi d^2 p}=\frac{1.38\times 10^{-23}\times 273}{\sqrt{2}\pi(3.8\times 10^{-10})^2\times 1.013\times 10^5}=5.80\times 10^{-8}\text{ m}$$

The average speed is

$$\bar{v}=\sqrt{\frac{8RT}{\pi M_{\text{mol}}}}=\sqrt{\frac{8\times 8.31\times 273}{\pi\times 28\times 10^{-3}}}=4.54\times 10^2 \text{ m}\cdot\text{s}^{-1}$$

According to $\bar{\lambda}=\dfrac{\bar{v}}{\bar{Z}}$

$$\bar{Z}=\frac{\bar{v}}{\bar{\lambda}}=7.83\times 10^9\text{ s}^{-1}$$

The average time interval between two successive collisions is

$$\bar{\tau}=\frac{\bar{\lambda}}{\bar{v}}=\frac{1}{\bar{Z}}=1.28\times 10^{-10}\text{ s}$$

Example 7.7.2 The linear dimension of a vacuum tube is 10^{-2} m and the vacuum degree is 1.33×10^{-3} Pa. Let the effective diameter of the air molecule be 3.8×10^{-10} m. When temperature is 27℃, find

(1) the molecular density of the air in the vacuum tube;

(2) the mean free path;

(3) the average collision frequency.

Solution According to $p=nkT$, the molecular number density in the vacuum tube can be obtained

$$n=\frac{p}{kT}=\frac{1.33\times 10^{-3}}{1.38\times 10^{-23}\times 300}=3.21\times 10^{17}\text{ m}^{-3}$$

The mean free path of the molecule is

$$\bar{\lambda}=\frac{1}{\sqrt{2}\pi d^2 n}=\frac{1}{\sqrt{2}\pi\times(3\times 10^{-10})^2\times 3.2\times 10^{17}}=7.82\text{ m}$$

This $\bar{\lambda}$ is much larger than the dimension 10^{-2} of the vacuum tube. This means that, actually, the air molecules are not likely to collide with each other, but only collide with the wall of the tube. Therefore, the mean free path should be the dimension of the vacuum tube. That is $\bar{\lambda}=10^{-2}$ m, not $\bar{\lambda}=7.82$ m.

So, the average collision frequency is

$$\bar{Z}=\frac{\bar{v}}{\bar{\lambda}}=\frac{1}{\bar{\lambda}}\sqrt{\frac{8RT}{\pi M_{mol}}}=\frac{1}{10^{-2}}\sqrt{\frac{8\times 8.31\times 300}{3.14\times 29\times 10^{-3}}}=4.7\times 10^{4}\ s^{-1}$$

Example 7.7.3 Find the average number of collisions of hydrogen molecules in a standard state in one second. The effective diameter of a hydrogen molecule is 2×10^{-10} m.

Solution The average speed of hydrogen molecules is

$$\bar{v}=\sqrt{\frac{8RT}{\pi M_{mol}}}$$

According to the pressure of an ideal gas $p=nkT$, the mean free path is $\bar{\lambda}=\frac{1}{\sqrt{2}\pi d^{2}n}$. The average number of collisions of molecules is $\bar{Z}=\frac{\bar{v}}{\bar{\lambda}}$.

Then, the average number of collisions of a hydrogen molecule in one second is

$$\bar{Z}=4\sqrt{\frac{R\pi}{M_{mol}T}}\frac{d^{2}p}{k}=8.12\times 10^{9}\ s^{-1}$$

Scientist Profile

Ludwig Edward Boltzmann (1844 – 1906), an Austrian physicist, was one of the founders of thermal and statistical physics.

He developed Maxwell's theory of molecular motion and extended the law of Maxwell speed distribution to polyatomic molecules and to situations where external forces act. In studying the law of non-equilibrium transport process, he introduced a function H defined by the non-equilibrium molecular distribution function, and obtained the famous H theorem. The H theorem could give the rate of entropy increase in the process from non-equilibrium to equilibrium, and it was the first time that the irreversibility or directivity of the macroscopic process had been proved by the microscopic theory of statistical physics.

He proposed the relationship between entropy and the number of microscopic states corresponding to the macroscopic state, i.e., the thermodynamic probability W, which is expressed as

$$S=k\ln W$$

which is known as Boltzmann's entropy formula. In the study of cavity heat radiation, he used the second law of thermodynamics to directly theoretically prove Stefan's experimental formula $M(T) = \sigma T^4$, which later became known as the Stefan-Boltzmann law, and later had great implications for Planck's blackbody radiation theory.

He was also an excellent teacher who was extremely strict with his students and never assumed authority. He often discussed problems with his students as equals. He believed that the greatest evil of scientific progress was self-isolation, and that science would only progress on the basis of discussion. On September 5, 1906, Boltzmann committed suicide due to the suffering of illness while on vacation in Italy. After his death, the formula $S = k \ln W$ was emblazoned on his tombstone.

Extended Reading

Paradoxes in the history of physics: four "divine beasts"

A paradox means that two contradictory conclusions can be derived from the same proposition. There are two kinds of paradoxes. One is called true paradoxes, which can't be solved, such as whether the chicken or the egg came first. The other is called the cognitive paradox, which may sound outrageous or highly counterintuitive at first but in fact can be solved by taking into account some factors missed previously. In the development of physics, scientists used thought experiments to create four animals representing the four most famous paradoxes: Zeno's turtle, Laplace's beast, Maxwell's demon, and Schrödinger's cat, which have been called the four great beasts of physics. They correspond to calculus, classical mechanics, thermodynamics and quantum mechanics respectively, which witness the development of physics from ancient times to the present.

The first paradox, Zeno's turtle is also called the Achilles Paradox which was proposed 2,500 years ago by the Greek philosopher Zeno. It was said that, Achilles, a hero who was good at running in ancient Greek mythology, raced with a turtle, the turtle was allowed to run 100 meters first, and Achilles chased after it. But when Achilles chased 100 meters, the turtle had already climbed 10 meters further; then Achilles chased 10 meters, and the turtle climbed another 1 meter; Achilles chased another meter, and the turtle climbed another 0.1 meter. In this way, the turtle was always ahead, and it could always create a distance between itself and Achilles. In ancient Greece, mathematics had not yet formed the concept of infinity, and the crux of this paradox was that the sum of infinite numbers is not necessarily infinite. Two thousand years later, the mathematical master Leibniz and the scientific master Newton established the "calculus" theory, which solved the space-time continuity with the "limit" in calculus, allowing Achilles to catch up with Zeno's turtle.

The second paradox is Laplace's beast created by Laplace in 1814 based on Newton's theory of classical mechanics. Laplace, like Newton, was a proponent of determinism, and he proposed that if a beast was willing to move its fingers and eyes to record the exact position and momentum of every atom in the universe, then it could use Newton's simple formula to calculate the past and future of the universe in an instant. Laplace's basic theory is that by understanding the state of motion of matter at the previous moment, we can deduce the state of motion at the next moment, and after determining the state of motion of every particle in the entire universe, we can deduce the state of motion at the next moment. Today's quantum mechanics completely negates the possibility of humans predicting the future, and the all-powerful Laplace beast finally exits the stage of physics. Because quantum has an uncertainty effect, it can only be described by probability rather than accurately measured, which is the law of nature in the microcosmic world, and therefore we can never predict which future will be realized.

In 1871, thermodynamics gave birth to a third paradox, Maxwell's demon. Maxwell is considered the greatest theoretical physicist after Newton and before Einstein. Maxwell came up with this thought experiment: There is a box filled with gas and a frictionless smooth valve inside. When the valve is closed, the container is split into two. A demon is inside the box and can manipulate the valve and keep the hot and cold atoms apart, so the box stays half hot and half cold and the entropy in the box works backwards without energy due to the smoothness of the valve. In the 1950s, the concept of information entropy was proposed. In order to achieve thermodynamic entropy reduction, it is necessary to obtain information about molecular motion, and it is impossible to obtain information without wasting energy. Therefore, Maxwell's demon cannot exist in an isolated system.

The fourth paradox was conceived in 1935 by one of the founders of quantum mechanics Schrödinger, that is Schrödinger's cat. There's a weird principle in quantum mechanics called superposition of states, which says that a microscopic particle can have an infinite number of states at the same time. Now put a cat and a radioactive atom in an airtight box, and once the atom decays, the particles released will trigger a switch and release a poison that will kill the cat. According to the superposition effect of states, the atom may be in both decayed and undecayed states at the same time, and the cat will be both dead and alive. Strangely, this superposition can persist as long as no one opens the box to take a look, and once someone opens the box, the cat will immediately choose one of the states (dead or alive) in which to appear in front of us. It seems that our look when opening the box decides whether the cat lives or dies. The key to solving this paradox is to explain why state superposition fails in the cat. When an atom is observed, there must be a photon or other particle colliding with this atom, which disturbs and destroys the superposition state, resulting in a collapse of the superposition state. Long before people open the box, the cat has collided countless times with the particles in the environment and the superposition state has collapsed, so we can never catch a dead and living cat.

Today, the four great beasts of physics have completed their respective missions. But physics will not stop. Today's frontier physics is still moving forward without stopping, and physicists are still working tirelessly to explore the unknown of the universe. New theories will also be slowly formed in the process of exploration, and there may be new concepts of physics similar to the four great beasts in the future.

Discussion Problems

7.1 In the process of deriving the ideal gas pressure formula, where is the microscopic model of an ideal gas used? Where is the equilibrium condition used? Where is the concept of statistical averages used?

7.2 What is the physical meaning of the function of speed distribution $f(v)$? Try to explain the physical meaning of each of the following quantities: (1) $f(v)dv$; (2) $nf(v)dv$; (3) $Nf(v)dv$; (4) $\int_0^v f(v)dv$; (5) $\int_0^\infty f(v)dv$; (6) $\int_{v_1}^{v_2} Nf(v)dv$ (n is the number of molecules density, and N is the total number of molecules in a system).

7.3 If a container of gas is moving with respect to a coordinate system, the velocities of the molecules in the container increase with respect to that coordinate system. Does the temperature increase?

7.4 Try to explain the physical meaning of each of the following quantities:
(1) $\frac{1}{2}kT$; (2) $\frac{3}{2}kT$; (3) $\frac{i}{2}kT$; (4) $\frac{M}{M_{mol}}\frac{i}{2}RT$; (5) $\frac{i}{2}RT$; (6) $\frac{3}{2}RT$.

7.5 For an ideal gas of a given mass, how do $\bar{Z}$ and $\bar{\lambda}$ of the molecule depend on temperature during isobaric expansion? When the volume stays the same and the temperature rises, how will $\bar{Z}$ and $\bar{\lambda}$ change?

Problems

7.6 It is known that the pressure on the wall of the gas with a temperature of 27℃ is 10^5 Pa. Find the number of molecules per unit volume in the gas.

7.7 The speed distribution of a system with N particles is shown in Figure T7-1. Find
(1) the expression of distribution function;
(2) the relationship between a and v_0;
(3) the number of particles with a speed between $1.5v_0$ and $2v_0$;
(4) the average speed of particles;
(5) the average particle speed between $0.5v_0$ and $1v_0$.

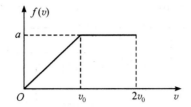

Figure T7 – 1 Figure of Problem 7.7

7.8 There is one bottle of oxygen and one bottle of hydrogen under the isobaric and isothermal condition and the oxygen volume is twice that of hydrogen. Find

(1) the ratio of molecular number density of oxygen to that of hydrogen;

(2) the ratio of the average speed of oxygen molecules to that of hydrogen molecules.

7.9 A cosmic ray particle with an energy of 1.6×10^{-7} J is shot into the neon tube. The neon tube contains 0.01 mole of neon. If the energy of ray particles is totally converted into the internal energy of neon, how much will the temperature of neon increase?

7.10 Oxygen is stored in a vessel with a volume of $V = 1.20 \times 10^{-2}$ m³. Its pressure is $p = 8.31 \times 10^5$ Pa and temperature is 300 K. Find

(1) the number of molecules n per unit volume;

(2) the average translational kinetic energy of the molecule;

(3) the internal energy of the gas.

7.11 What are the translational kinetic energy, rotational kinetic energy and internal energy of 1 mole of hydrogen at 27 ℃?

7.12 The effective diameter of nitrogen molecules is 3.8×10^{-10} m. Find the mean free path and the mean time interval of successive collisions in the standard state.

7.13 The vacuum degree of a vacuum tube is about 1.38×10^{-3} Pa (that is, 1.0×10^{-5} mmHg). Find the number of molecules per unit volume and the mean free path of molecules at 27 ℃ (the effective diameter of molecules $d = 3 \times 10^{-10}$ m).

Challenging Problems

7.14 Miller and Kusch did an experiment that more precisely proved the law of Maxwell's speed distribution. As shown in Figure T7 – 2, O is the vapor source, and thallium or potassium atoms escape through small holes to form rays. R is a cylinder made of aluminum alloy, and D is a detector. The whole instrument is placed in a high-vacuum container, and the cylinder can be rotated about the central axis. The length of the cylinder is $L = 20.40$ cm, the radius $r = 10.0$ cm, and many spiral grooves with a width $l = 0.0424$ cm are carved. The angle between the radii of the inlet and outlet of the groove is $\phi = 4.8°$. The intensity of atomic rays can be measured by the detector after passing through the groove. Try to explain the detection principle. Collect the experimental data near the theoretical curve and simulate the experimental results.

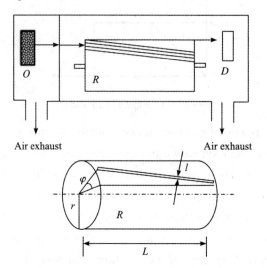

Figure T7 - 2   Figure of Problem 7.14

7.15 (1) Prove the law of molecular number density distribution according to height in the gravity field is

$$n = n_0 \exp\left(-\frac{mgz}{kT}\right)$$

where z is the height. The molecular mass of hydrogen, neon, nitrogen, oxygen and fluorine is 2, 20, 28, 32 and 38, respectively. What are the characteristics of the molecular number density distribution curves of hydrogen, neon, nitrogen, oxygen and fluorine according to height at 300 K? What are the characteristics of the molecular number density distribution curve of oxygen according to height at the temperatures ranging from 100 K to 400 K (with an interval of 50 K)?

(2) Use points to represent molecules and express the law of distribution of molecules according to height by the density of the points.

Chapter 8
Fundamentals of Thermodynamics

Human beings knew and applied thermal phenomena very early, but thermodynamics as a science has been studied quantitatively since the end of the 17th century. The development history of thermodynamics can be roughly divided into four stages. The first one is from the end of the 17th century to the middle of the 19th century. A large number of experiments and observations were accumulated during this period, steam engines were manufactured, and the nature of heat was studied and debated. However, the study of thermodynamics was still limited to the description of thermodynamic phenomena, and no mathematical description was introduced. The second stage is from the middle of the 19th century to the end of the 1870s. The first law of thermodynamics was established during this period, and the combination of the first law and Carnot's theory produced the second law of thermodynamics. The laws were fully theoreticalized, and the basic theoretical framework of thermodynamics was initially formed. At the same time, the kinetic theory of gas based on Newtonian mechanics also began to develop, but people did not understand the relationship between thermodynamics and kinetic theory of gas during this period. The third stage is from the end of the 1870s to the beginning of the 20th century. During this period Boltzmann first combined the theories of thermodynamics and molecular dynamics, proposed the ensemble theory, and gave birth to statistical thermodynamics. At the same time, he also proposed the idea of non-equilibrium statistical theory. The fourth stage is from the 1930s to the present. It is mainly the further development of nonequilibrium theory and the combination with quantum theory. A rich and huge theoretical system of modern thermophysics has been established and its ideas and research methods have penetrated into different fields of science.

The basis of thermodynamics is mainly some basic concepts and laws involved in the macroscopic theory of thermal phenomena. This chapter studies the properties and laws that thermodynamic systems satisfy during the process of state change from the perspective of energy conservation and conversion. This chapter deals with the first law of thermodynamics, the second law of thermodynamics, and the characteristics of thermodynamic processes investigated by these laws. Meanwhile, according to the second law of thermodynamics, this chapter discusses the conditions for the conversion between heat and work, the directionality of the thermodynamic process, Carnot's theorem, and the entropy change of the thermodynamic system.

8.1 Basic concepts of thermodynamics

8.1.1 Equilibrium state parameters

When studying thermodynamic problems, the macroscopic object studied is usually called the thermodynamic system, and other objects that interact with the thermodynamic system are called **the external environment (or environment)**. According to energy and mass transfer, the system is divided into the open system, isolated system, and closed system. A thermodynamic system with both energy and mass exchange with the environment is called an **open system**; a thermodynamic system without any interaction with the environment is called an **isolated system**; a thermodynamic system that exchanges energy with the environment without any mass exchange is called a **closed system.**

The equilibrium state means that various macroscopic properties inside a thermodynamic system do not change with time. For an isolated system, after a long enough time, the system will reach a state in which the macroscopic properties do not change with time. This can be regarded as an equilibrium state, which is a special state of a thermodynamic system. In the equilibrium state, the microcosmic particles that make up the system are still doing random motion, but their statistical average effect remains unchanged. Therefore, we usually call this dynamic equilibrium thermodynamic equilibrium.

In order to describe the equilibrium state of a thermodynamic system, we need to introduce several state parameters. Any object is composed of a large number of microscopic particles (molecules, atoms), and the physical quantities (mass, speed, energy, etc.) that describe the characteristics of these microscopic particles are usually called microscopic quantities. On the contrary, the physical quantities that describe the characteristics of macroscopic objects (pressure, temperature, volume, internal energy, etc.) are called macroscopic quantities. Macroscopic quantities are physical quantities that can be observed experimentally.

Temperature is one of the important state parameters to describe the thermal phenomenon of an object. It indicates the degree of coldness or heat of an object. The numerical representation of temperature is called a **temperature scale**. In thermodynamics, a thermodynamic temperature scale that does not depend on any material properties is used and is called **the absolute temperature scale,** and the temperature determined by this temperature scale is called the thermodynamic temperature or absolute temperature, expressed as T. Since 1960, it has been stipulated internationally that the thermodynamic temperature is a basic physical quantity, and its unit in the International System of Units is the Kelvin (K), usually called Kelvin temperature or thermodynamic temperature.

People often use the Celsius temperature scale in life and technology field, expressed by t (the unit is the degree, denoted as ℃), and the relationship between Celsius temperature and thermodynamic temperature is

$$t = T - 273.15 \tag{8.1.1}$$

In thermodynamics, our research object is the gas system, and only the macroscopic properties of the gas system in the equilibrium state can be described by a set of certain state parameters. For a certain amount of gas system, the volume (V), pressure (p), and temperature (T) can generally be used to characterize the equilibrium state of the gas. These three parameters are called the state parameters of the gas. If they satisfy a certain functional relationship, this relationship is called the state equation of the gas system.

8.1.2 State equation of ideal gas

For any gas system, its state equation is very complicated, so here we only discuss the state equation of ideal gas. A general gas system can be approximately regarded as an ideal gas when the temperature is not too low and the pressure is not too high. Based on many experiments, three laws have been summarized for a certain amount of ideal gas, namely Boyle's law, Gay-Lussac's law, and Charlie's law. These three laws can be summarized by a state equation, expressed as follows:

$$\frac{pV}{T} = C \tag{8.1.2}$$

where C is a constant.

Clapeyron studied the ideal gas system with mass M, pressure p, volume V and temperature T in the equilibrium state, whose state equation can be expressed as

$$pV = \frac{M}{M_{mol}} RT \tag{8.1.3}$$

where $R = 8.31 \, \text{J} \cdot \text{mol}^{-1} \cdot \text{K}^{-1}$ is the universal constant of gas, and this equation is also called the Clapeyron equation for ideal gas.

8.2 The first law of thermodynamics

8.2.1 Quasi-static process

A thermodynamic system changes from one state to another, and the system is said to undergo a **thermodynamic process**. During the process of system change, if each intermediate state experienced by the system is a non-equilibrium state, this process is called a non-quasi-static process; if each state experienced between the initial and final equilibrium states is infinitely close to an equilibrium state, then this process is called **quasi-static process**. Quasi-static process is an ideal process, and any state during the process can be regarded as an equilibrium state, so it can be described by a set of certain state parameters.

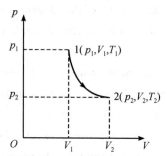

Figure 8.2.1 Process curve

Since the equilibrium state can be represented by a point in the parameter space, a quasi-static process can be represented by a continuous curve in the parameter space, and such a curve is called a **process curve**. The curve in Figure 8.2.1 shows that the system goes from equilibrium state 1 to equilibrium state 2 after a quasi-static process.

8.2.2 Work and heat in a quasi-static process

Changes in the state of a thermodynamic system can occur by exchanging energy with the environment, which can be achieved by doing work and exchanging heat. We first discuss the work done by a thermodynamic system in a quasi-static process.

Assume that the gas in a closed cylinder undergoes quasi-static expansion, the cross-sectional area of the piston is S, and the pressure of the gas in the cylinder is p, as shown in Figure 8.2.2. During the microelement displacement dl of the piston, the element work done by the gas on the piston is

$$dW = F\,dl = pS\,dl = p\,dV \qquad (8.2.1)$$

where $dV = S\,dl$ is the increment of gas volume. Although Equation (8.2.1) is derived from the special case in Figure 8.2.2, it can be generalized to any quasi-static process. Obviously, when the volume of the gas expands $dV > 0$, $dW > 0$ and the system does positive work on the environment. On the contrary, when the volume of the gas shrinks $dV < 0$, $dW < 0$ and the system does negative work on the environment.

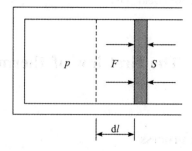

Figure 8.2.2 Schematic diagram of gas work

When the system undergoes a quasi-static process and the volume changes from V_1 to

V_2, the work done by the gas is

$$W = \int_{V_1}^{V_2} p \, dV \tag{8.2.2}$$

As shown in Figure 8.2.3, on the p-V diagram, the element work dW corresponds to the gray area $V \to V + dV$ under the process curve. Then, during the process of $V_1 \to V_2$, the total work done by the system on the environment is equal to the total area of $V_1 \to V_2$ under the process curve.

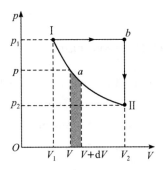

Figure 8.2.3 Process curve and work

If the system starts from the same initial state I, goes through two different quasi-static processes, I$\to a \to$II and I$\to b \to$II, and reaches the same final state II, the areas under the curves of the two processes are different, which means that in these two different thermodynamic processes, the system does different work on the environment. Therefore, work is a process quantity related to thermodynamic processes.

In the thermodynamic process, when a system exchanges heat with the environment, its state may change. The energy of this heat exchange is called **heat**, which is expressed by Q. For a given initial state and final state, heat transfer in different thermodynamic processes is generally different. So heat, like work, is a process quantity rather than a state quantity, and we cannot say how much heat a system has when it is in a certain state.

Both work and heat transfer can change the thermodynamic state, but they are different in nature. Work is accomplished through the macroscopic displacement of objects, which is the conversion between the energy of the orderly movement of the environment and the energy of the random thermal motion of the system molecules. Heat transfer is realized through the collision between molecules, which is the exchange of energy between the random movement of molecules inside and outside the system.

8.2.3 Internal energy in a quasi-static process

In Joule's mechanical equivalent of heat experiment, work is done on a system in various ways under adiabatic conditions, such as stirring and electric current heating. But as long as the initial state and final state of the system are fixed, no matter which method is used, the work required is the same. This shows that when a system changes from one state to another after an adiabatic process, the required work is only related to the initial and final states of the system, but has nothing to do with the specific process and method of doing work. Therefore, a thermodynamic system has a certain energy in a certain state, and this energy is only a single-valued function of the state, called **internal energy**, expressed by E. For a general gas system, the internal energy can be expressed as $E = E(p, V, T)$. For an ideal gas system, the Joule-Thomson experiment shows that the internal energy is only a function of temperature

$$E = E(T) \tag{8.2.3}$$

8.2.4 The first law of thermodynamics

In general, during system state changes, work and heat transfer often exist at the same time. Assuming that during a process a system absorbs heat Q from the environment and does work W on the environment and the internal energy of the system changes from the initial state E_1 to the final state E_2, then according to the law of energy conversion and conservation, we have

$$Q = E_2 - E_1 + W = \Delta E + W \tag{8.2.4}$$

The above formula is **the first law of thermodynamics**, which shows that in any thermodynamic process, the heat absorbed by the system from the environment is equal to the sum of the increase of the internal energy of the system and the work done by the system on the environment. In the formula, Q and W represent the heat absorbed by the system from the environment and the work done by the system, respectively, and ΔE represents the change of the internal energy between the initial state and final state. It is stipulated that $Q > 0$ means that the system absorbs heat and $Q < 0$ means that the system releases heat, $W > 0$ means that the system does positive work on the environment and $W < 0$ means that the environment does negative work on the system, and $\Delta E > 0$ means that the internal energy of the system increases and $\Delta E < 0$ means that the internal energy of the system decreases. If the system undergoes a small change, the first law of thermodynamics can be expressed as

$$dQ = dE + dW \tag{8.2.5}$$

Among them, dQ is the heat transfer from the environment to the system during the process, dW is the work done by the system on the environment, and dE is the change of the internal energy of the system.

The first law of thermodynamics can be rewritten as $\Delta E = Q - W$, which links the process quantity and the state quantity together, and shows that heat transfer and work have an equal status in the thermodynamic process. Both work and heat are process-related quantities, which can be used as a measure of the internal energy change within a system. In history, many people tried to develop a device that does not need any power and fuel but can continuously perform work on the environment. This device is called the **first type of perpetual motion machine**. The idea of building such a machine is impossible according to the first law of thermodynamics. Therefore, the first law of thermodynamics can be expressed as the impossibility of making the first type of perpetual motion machine.

8.3 Heat capacity

8.3.1 Definition of heat capacity

When the temperature rise is ΔT and the heat absorbed from the environment is ΔQ, the **heat capacity** of the system in this given process is defined as

$$C = \lim_{\Delta T \to 0} \frac{\Delta Q}{\Delta T} = \frac{dQ}{dT} \tag{8.3.1}$$

The heat capacity is represented by the symbol C, and its unit is $J \cdot K^{-1}$ in the International System of Units.

The heat capacity of a substance per unit mass is called the **specific heat capacity** c, and the unit is $J \cdot kg^{-1} \cdot K^{-1}$. The relationship between heat capacity and specific heat capacity is

$$C = Mc \tag{8.3.2}$$

The heat capacity of 1 mole of substance is called **molar heat capacity**, which is expressed in the symbol C_m, and its unit is $J \cdot mol^{-1} \cdot K^{-1}$ in the International System of Units.

The relationship between the heat capacity and the molar heat capacity is

$$C_m = \frac{M_{mol}}{M} C \tag{8.3.3}$$

Experiments have shown that the specific heat capacities of different substances have different values, and the specific heat capacity of a substance generally changes with temperature. But when the temperature change is not too large, the specific heat capacity can be approximately regarded as a constant. For a gas system with the given initial state and final state many processes can occur between them, during which the heat absorbed by the system from the environment may be unequal, so the heat capacity for different processes is not the same. This shows that the heat capacity depends not only on the structure of the system, but also on the specific process and is a process quantity. The heat capacities of an ideal gas system at constant volume and constant pressure are introduced respectively below.

8.3.2 Molar heat capacity at constant volume

For an ideal gas with mass m, under the condition that the volume V remains constant, according to the definition of heat capacity, the heat capacity at constant volume can be written as

$$C_V = \lim_{\Delta T \to 0} \frac{(\Delta Q)_V}{\Delta T} = \left(\frac{dQ}{dT}\right)_V \tag{8.3.4}$$

From the first law of thermodynamics for an isovolumetric process, $dQ = dE$ and the above equation can be further written as

$$C_V = \left(\frac{dQ}{dT}\right)_V = \frac{dE}{dT} \tag{8.3.5}$$

The heat capacity of 1 mole of a substance at constant volume is called **molar heat capacity at constant volume**, which is recorded as $C_{V,m}$

$$C_{V,m} = \frac{1}{v}C_V = \frac{1}{v}\left(\frac{dQ}{dT}\right)_V \tag{8.3.6}$$

8.3.3 Molar heat capacity at constant pressure

For an ideal gas with mass m, under the condition that the pressure p is kept constant, according to the definition of heat capacity, the heat capacity at constant pressure can be written as

$$C_p = \lim_{\Delta T \to 0} \frac{(\Delta Q)_p}{\Delta T} = \frac{dE}{dT} + \frac{p\,dV}{dT} \tag{8.3.7}$$

The internal energy of an ideal gas is only a single-valued function of temperature T, so

$$C_p = \frac{dE}{dT} + p\frac{dV}{dT} \tag{8.3.8}$$

The heat capacity of 1 mole of a substance at constant pressure is called the **molar heat capacity at constant pressure**, which is recorded as $C_{p,m}$. Obviously

$$C_{p,m} = \frac{1}{v}C_p = \frac{1}{v}\left(\frac{dQ}{dT}\right)_p \tag{8.3.9}$$

8.3.4 Relationship between molar heat capacity at constant pressure and molar heat capacity at constant volume

For 1 mole of ideal gas, it can be obtained from Equation (8.3.7) and the ideal gas state equation $pV = RT$ that

$$C_{p,m} = C_{V,m} + R \tag{8.3.10}$$

This equation is called **the Meyer formula**. It shows that the molar heat capacity at constant pressure of an ideal gas is R greater than the molar heat capacity at constant volume. The reason is that when the temperature increase is the same, the heat absorbed in the isovolumic process is all converted into the increase of internal energy, while the heat absorbed in the isobaric process not only converts into internal energy of the same magnitude as that in the isovolumic process but also does work externally.

For an ideal gas, the molar heat capacity at constant volume is

$$C_{V,m} = \frac{i}{2}R \qquad (8.3.11)$$

where i is the degree of freedom of the ideal gas. Thus, the molar heat capacity at constant pressure of an ideal gas is

$$C_{p,m} = C_{V,m} + R = \frac{i+2}{2}R \qquad (8.3.12)$$

Define the ratio of molar heat capacity at constant pressure to molar heat capacity at constant volume as **specific heat capacity ratio** γ, and then it can be expressed by

$$\gamma = \frac{C_{p,m}}{C_{V,m}} = \frac{i+2}{i} \qquad (8.3.13)$$

Table 8.3.1 lists the experimental and theoretical values of some gas molar heat capacities and specific heat capacity ratios.

Table 8.3.1 Experimental and theoretical data on the molar heat capacities and specific heat capacity ratios of some gases ($T=300$ K)

Molecular type	Gas	Theoretical value			Experimental value		
		$C_{V,m}/R$	$C_{p,m}/R$	γ	$C_{V,m}/R$	$C_{p,m}/R$	γ
Monatomic	He	1.5	2.5	1.67	1.5	2.5	1.67
	Ar						
Diatomic	H_2	2.5	3.5	1.4	2.45	3.46	1.41
	N_2				2.49	3.50	1.41
	CO				2.53	3.53	1.40
Polyatomic	CO_2	3	4	1.33	3.42	4.44	1.30
	H_2O				3.25	4.26	1.31
	CH_4				3.26	4.27	1.31

It can be seen from the above table that for monatomic and diatomic molecular gases, the theoretical values are in good agreement with the experimental values while for polyatomic molecular gases, the theoretical values are quite different from the experimental values. This difference shows that the ideal gas model can only approximately deal with gases composed of simple molecules (atoms).

The heat capacity at constant pressure and heat capacity at constant volume given by the classical gas kinetic theory are independent of temperature. However, the heat capacity measured in experiments varies with temperature. Figure 8.3.1 shows the relationship between the molar heat capacity at constant pressure of hydrogen and the temperature. It can be seen that the molar heat capacity at constant pressure changes with temperature in three obvious steps, which reflects the flaws of the classical theory. Only the quantum theory can explain these results correctly.

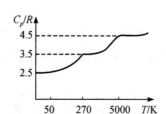

Figure 8.3.1 Molar heat capacity at constant pressure of hydrogen as a function of temperature

8.4 Application of the first law of thermodynamics

The first law of thermodynamics clarifies the interrelationship among internal energy, work, and heat in the process of state change of a thermodynamic system. Next, we apply the first law of thermodynamics together with the ideal gas state equation to the quasi-static isovolumetric, isobaric, isothermal and adiabatic processes of ideal gas.

8.4.1 Isovolumetric process

Suppose there is ideal gas with a certain mass in a closed container, and the number of moles is ν. The system changes from state (p_1, V_1, T_1) to state (p_2, V_1, T_2) after a quasi-static isovolumetric process, as shown in Figure 8.4.1. Since the volume remains constant during this process, the p-V curve corresponding to this process is a straight line parallel to the p axis, called the **isovolumetric line**. Any state parameter (p, V, T) in the process satisfies the process equation

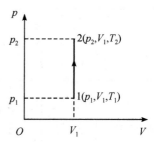

Figure 8.4.1 Isovolumetric process

$$V = C_1 \quad \text{or} \quad \frac{p}{T} = C_2 \tag{8.4.1}$$

where C_1 and C_2 are two constants and can be determined by a known state parameter in the process.

During the isovolumetric process, $dW = p\,dV = 0$.

So the system does work

$$W_V = 0$$

According to the first law of thermodynamics, the heat absorbed by the system is equal to the increment of internal energy. Assuming that the molar heat capacity at constant volume of the system $C_{V,m}$ is constant, we obtain

$$Q_V = \Delta E = \nu C_{V,m}(T_2 - T_1) = \frac{i}{2}\nu R(T_2 - T_1) \tag{8.4.2}$$

The heat absorbed by the system from the environment is all converted into the internal energy of the system.

8.4.2 Isobaric process

Suppose there is an ideal gas system with a certain mass, and the number of moles is ν. It changes from state 1 (p_1, V_1, T_1) to state 2 (p_1, V_2, T_2) after a quasi-static isobaric process, as shown in Figure 8.4.2. Since the pressure is constant, the p-V curve of the isobaric process is a straight line parallel to the V axis, called **the isobaric line**, and the corresponding process equation is

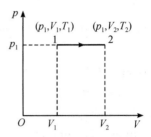

Figure 8.4.2 Isobaric process

$$p=C_1 \quad \text{or} \quad \frac{V}{T}=C_2 \qquad (8.4.3)$$

In the isobaric process, p is a constant, so the work done by the system on the environment is

$$W_p = \int_{V_1}^{V_2} p\,dV = p_1(V_2 - V_1) = \nu R(T_2 - T_1) \qquad (8.4.4)$$

Since internal energy is a state quantity and the internal energy of an ideal gas is only a function of temperature, the increment of internal energy in an isobaric process can be expressed as

$$\Delta E = \nu C_{V,\,m}(T_2 - T_1) = \frac{i}{2}\nu R(T_2 - T_1) \qquad (8.4.5)$$

According to the first law of thermodynamics, the heat absorbed by the system in the process can be obtained

$$Q_p = W_p + \Delta E = \nu C_{p,\,m}(T_2 - T_1) = \frac{i+2}{2}\nu R(T_2 - T_1) \qquad (8.4.6)$$

Part of the heat absorbed by the system from the environment is converted into internal energy of the system, and the other part is used to do work on the environment.

Example 8.4.1 Nitrogen is stored in a cylinder with a mass of 1.25 kg. Heat it slowly at the standard atmospheric pressure to raise the temperature by 1 K. Find the work W_p done when the gas expands, the increment of internal energy ΔE of the gas, and the heat Q_p absorbed by the gas (Nitrogen gas molecules are treated as rigid diatomic molecules, and the mass of the piston and its friction with the cylinder wall can be ignored).

Solution This process is an isobaric process, so the work done when the gas expands is

$$W_p = \int_{V_1}^{V_2} p\,dV = p_1(V_2 - V_1) = \nu R(T_2 - T_1) = \frac{m}{M_{\text{mol}}}R\Delta T = \frac{1.25}{0.028} \times 8.31 \times 1 = 371 \text{ J}$$

Because nitrogen is a diatomic molecule and its degree of freedom $i=5$,

$$C_{V,m} = \frac{i}{2}R = 20.8 \text{ J} \cdot \text{mol}^{-1} \cdot \text{K}^{-1}$$

The increment of gas internal energy is

$$\Delta E = \frac{m}{M_{mol}} C_{V,m} \Delta T = \frac{1.25}{0.028} \times 20.8 \times 1 = 929 \text{ J}$$

Therefore, the heat absorbed by the gas in this process is

$$Q_p = W_p + \Delta E = 1300 \text{ J}$$

Example 8.4.2 1 mole of monatomic molecular ideal gas changes from 0℃ to 100℃ after an isovolumic process and an isobaric process respectively. Try to find the heat absorbed in each process.

Solution The degree of freedom of monatomic molecular ideal gas $i=3$.

(1) Isovolumetric process

$$Q_V = \frac{m}{M_{mol}} C_{V,m}(T_2 - T_1) = 1 \cdot \frac{i}{2} R(T_2 - T_1) = \frac{3}{2} \times 8.31 \times 100 = 1.25 \times 10^3 \text{ J}$$

(2) Isobaric process

$$Q_p = \frac{m}{M_{mol}} C_{p,m}(T_2 - T_1) = 1 \cdot \frac{i+2}{2} R(T_2 - T_1) = \frac{5}{2} \times 8.31 \times 100 = 2.08 \times 10^3 \text{ J}$$

8.4.3 Isothermal process

Suppose there is ideal gas with a certain mass and the number of moles is ν. It changes from state $1(p_1, V_1, T_1)$ to state $2(p_2, V_2, T_1)$ after a quasi-static isothermal process, as shown in Figure 8.4.3. Since the temperature of the system remains constant, the p-V curve of the isothermal process is a hyperbola, which is called **the isothermal line**. The isotherm divides the p-V diagram into two areas, the temperature of the gas in the area above the isothermal line greater than T and the temperature of the gas in the area below the isothermal line less than T. Its process equation is

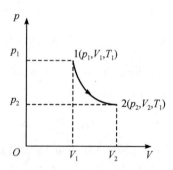

Figure 8.4.3 Isothermal process

$$T = C_1 \quad \text{or} \quad pV = C_2 \tag{8.4.7}$$

The internal energy of an ideal gas is only related to temperature, so the increase in internal energy is $\Delta E = 0$. According to the process equation $p = C_2/V = \nu RT_1/V$, the work done by the system in the isothermal process is

$$W_T = \int_{V_1}^{V_2} p \, dV = \int_{V_1}^{V_2} \frac{\nu RT_1}{V} dV = \nu RT_1 \ln \frac{V_2}{V_1} = \nu RT_1 \ln \frac{p_1}{p_2} \tag{8.4.8}$$

According to the first law of thermodynamics, the heat absorbed by the system is

$$Q_T = W_T = \nu RT_1 \ln \frac{V_2}{V_1} = \nu RT_1 \ln \frac{p_1}{p_2} \tag{8.4.9}$$

In the isothermal process, all the heat absorbed by the system from the environment is used to do work on the environment.

Example 8.4.3 Nitrogen gas with a mass of 2.8×10^{-3} kg, a pressure of 3 atm, and a temperature of 27℃ undergoes isothermal expansion to reduce the pressure to 1 atm. Find the change of internal energy of the system, the work done by the system on the environment, and the heat absorbed by the system during this process.

Solution Since this process is an isothermal process, the change of the internal energy of the system is $\Delta E = 0$ and the work done by the system on the environment is equal to the heat absorbed by the system.

Using the process equation of the isothermal process, we can get

$$p_1 V_1 = p_2 V_2$$

that is

$$\frac{V_2}{V_1} = \frac{p_1}{p_2} = 3$$

The work done by the system on the environment is equal to the heat absorbed by the system

$$Q_T = W_T = \nu R T_1 \ln \frac{V_2}{V_1} = \nu R T_1 \ln \frac{p_1}{p_2} = 0.1 \times 8.31 \times 300 \times \ln 3 = 274 \text{ J}$$

8.4.4 Adiabatic process

The adiabatic process is one during which the gas system does not exchange heat with the environment. There is no strict adiabatic process in nature, but an extremely fast one can be approximately regarded as an adiabatic process since the system has no time to communicate with the environment. For example, the combustion and explosion of mixed gas in the cylinder of an internal combustion engine, and the compression and expansion of the air during the propagation of the sound wave.

Suppose there is an ideal gas system with a certain mass and the number of moles is ν. It changes from state 1(p_1, V_1, T_1) to state 2 (p_2, V_2, T_2) after a quasi-static adiabatic process. Since in the adiabatic process the system never exchanges heat with the environment, that is $dQ = 0$, according to the first law of thermodynamics, there is

$$dE + p dV = 0 \qquad (8.4.10)$$

It shows that in the quasi-static adiabatic process, the work done by the environment on the system is completely converted into the internal energy of the system.

For an ideal gas,

$$dE = \nu C_{V,m} dT \qquad (8.4.11)$$

Substituting Equation (8.4.11) into Equation (8.4.10), we can get

$$\nu C_{V,m} dT + p dV = 0 \qquad (8.4.12)$$

On the other hand, the ideal gas state equation is

$$pV = \nu RT$$

Differentiate both sides of this equation at the same time, and we can get
$$p\,dV + V\,dp = vR\,dT \tag{8.4.13}$$
Combine Equation (8.4.12) and Equation (8.4.13), and eliminate dT to get
$$(C_{V,m} + R)p\,dV + C_{V,m}V\,dp = 0 \tag{8.4.14}$$
Divide both sides of the above expression by $C_{V,m}pV$ and use the definition of γ and Meyer's law, and we can get
$$\frac{dp}{p} + \gamma\frac{dV}{V} = 0 \tag{8.4.15}$$
This is the differential equation that the state parameters satisfy in the quasi-static adiabatic process of an ideal gas. Integrate both sides of the above formula to get
$$\ln p + \gamma \ln V = C \tag{8.4.16}$$
where C is the integral constant, and the above formula is often written as
$$pV^\gamma = C_1 \tag{8.4.17}$$
where C_1 is a constant. Using the ideal gas state equation, the relationship between T and V, and the relationship between p and T in the quasi-static adiabatic process can also be derived
$$TV^{\gamma-1} = C_2 \tag{8.4.18}$$
$$p^{\gamma-1}T^{-\gamma} = C_3 \tag{8.4.19}$$
C_2 and C_3 are two other constants. Equations (8.4.17), (8.4.18) and (8.4.19) are all quasi-static adiabatic process equations for ideal gas.

According to the adiabatic process equation, the adiabatic process curve can be drawn on the p-V diagram, as shown in Figure 8.4.4 and referred to as **the adiabatic line**.

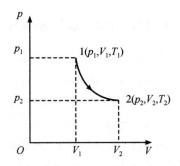

Figure 8.4.4 Adiabatic process

The gas system changes from the initial state (p_1, V_1, T_1) to the final state (p_2, V_2, T_2) after the adiabatic process, and the internal energy changes by
$$\Delta E = E_2 - E_1 = v\frac{i}{2}R(T_2 - T_1) \tag{8.4.20}$$

The work done by the system is
$$W = \int_{V_1}^{V_2} p\,dV = -\Delta E = -\frac{i}{2}vR(T_2 - T_1) \tag{8.4.21}$$

Using the adiabatic process Equation (8.4.17), the work done by the system on the environment during the adiabatic process can also be written as
$$W = \int_{V_1}^{V_2} p\,dV = \frac{p_1V_1 - p_2V_2}{\gamma - 1} \tag{8.4.22}$$

Figure 8.4.5 shows the quasi-static isothermal process curve and the adiabatic process curve of an ideal gas with a certain mass at the same time. Assuming that the isothermal line and the adiabatic line intersect at point A, then according to the process equations of the isothermal process and the adiabatic process, we can obtain their slopes at the point of intersection, which are

$$\left(\frac{dp}{dV}\right)_T = -\frac{p_A}{V_A} \tag{8.4.23}$$

$$\left(\frac{dp}{dV}\right)_Q = -\gamma\frac{p_A}{V_A} \tag{8.4.24}$$

Since $\gamma > 1$, the absolute value of the slope of the adiabatic line at the intersection is greater than the absolute value of the slope of the isothermal line, that is, the adiabatic line is steeper than the isothermal line.

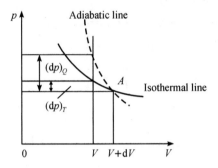

Figure 8.4.5 Isothermal line and adiabatic line

Example 8.4.4 Try to discuss whether the ideal gas absorbs heat or releases heat in the two processes Ⅰ and Ⅲ shown in Figure 8.4.6. It is known that process Ⅱ is an adiabatic process.

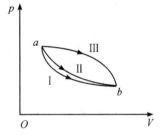

Solution As can be seen from the figure, the work of the three processes satisfies the relationship

$$0 < W_Ⅰ < W_Ⅱ < W_Ⅲ \tag{1}$$

Figure 8.4.6 Figure of Example 8.4.4

For the second adiabatic process, applying the first law of thermodynamics, we get

$$(E_b - E_a) + W_Ⅱ = 0$$

that is

$$-(E_b - E_a) = W_Ⅱ > 0 \tag{2}$$

It is known from the figure that the initial state and final state of the three processes are the same and the internal energy is a state quantity, so the increment of internal energy is the same.

For the first process, applying the first law of thermodynamics, we get

$$Q_Ⅰ = (E_b - E_a) + W_Ⅰ$$

Combining Equations (1) and (2), $Q_I < 0$ can be obtained, indicating that the system releases heat to the environment during the process.

For the third process, applying the first law of thermodynamics, we get

$$Q_{\text{III}} = (E_b - E_a) + W_{\text{III}}$$

Combining Equations (1) and (2), $Q_{\text{III}} > 0$ can be obtained, indicating that the system absorbs heat from the environment during the process.

Example 8.4.5 N_2 gas in the standard state with a mass of 0.014 kg doubles its volume after the following quasi-static processes: ① isobaric process; ② isothermal process; ③ adiabatic process. Try to find the increment of internal energy, work done on the environment, and heat absorbed by the system in each process (Nitrogen molecules are treated as rigid diatomic molecules).

Solution The degree of freedom of the nitrogen molecule is $i = 5$.

(1) Using the isobaric process equation, we can get

$$\frac{V_1}{T_1} = \frac{V_2}{T_2}$$

so the final temperature is

$$T_2 = \frac{V_2}{V_1} T_1 = 2 \times 273 = 546 \text{ K}$$

Then the increment of internal energy, work done on the environment, and heat absorbed by the system are respectively

$$\Delta E = \frac{i}{2} \nu R (T_2 - T_1) = \frac{5}{2} \times \frac{0.014}{0.028} \times 8.31 \times (546 - 273) = 2.84 \times 10^3 \text{ J}$$

$$W = p_1 (V_2 - V_1) = p_1 V_1 = \nu R T_1 = \frac{0.014}{0.028} \times 8.31 \times 273 = 1.13 \times 10^3 \text{ J}$$

$$Q = \Delta E + W = 3.97 \times 10^3 \text{ J}$$

(2) The increment of internal energy in the isothermal process is

$$\Delta E = 0$$

The work done on the environment and heat absorbed by the system are respectively

$$W = \nu R T_1 \ln \frac{V_2}{V_1} = \frac{0.014}{0.028} \times 8.31 \times 273 \times \ln 2 = 7.86 \times 10^2 \text{ J}$$

$$Q = W = 7.86 \times 10^2 \text{ J}$$

(3) $\gamma = \frac{C_{p,m}}{C_{V,m}} = \frac{i+2}{i} = \frac{7}{5}$. According to the adiabatic process equation, we can get

$$T_1 V_1^{\gamma-1} = T_2 V_2^{\gamma-1}$$

so the final temperature is

$$T_2 = \left(\frac{V_1}{V_2}\right)^{\gamma-1} T_1 = 206.9 \text{ K}$$

The heat absorbed in the adiabatic process is

$$Q = 0$$

The increase in internal energy of the system is

$$\Delta E = \frac{i}{2}\nu R(T_2 - T_1) = \frac{5}{2} \times \frac{0.014}{0.028} \times 8.31 \times (206.9 - 273) = -6.87 \times 10^2 \text{ J}$$

The work done on the environment is

$$W = -\Delta E = 6.87 \times 10^2 \text{ J}$$

8.5 Cycle process

In the development of thermodynamics, the research on heat engines is an indispensable and important part. When a heat engine is working, it needs to continuously convert heat into work, but this goal cannot be achieved relying on a single thermodynamic process; therefore, a cyclic process is necessary.

8.5.1 Cycle process

The working substance of a heat engine starts from a certain state, goes through a series of different state changes, and then returns to the original starting state. Such a complete thermodynamic process is called a **cyclic process**, or **cycle** for short. The substances that participate in the cycle process are called **working substances** or **working medium**. Since the internal energy does not change when the working medium returns to the initial state after a cycle process, the important feature of the cycle process is $\Delta E = 0$. According to the first law of thermodynamics, the net work done by the system on the environment during the cycle is equal to the net heat absorbed by the system. If each sub-process is a quasi-static process in the cycle experienced by the working medium, then the whole process is a quasi-static cyclic process. As shown in Figure 8.5.1, it is a closed curve in the p-V diagram, and the net work done by the system is equal to the area surrounded by the curve.

If the cycle proceeds in a clockwise direction, as shown in Figure 8.5.1, the net work done by the system is positive, and the cycle is a **positive cycle**. If the net work done by the system is negative, as shown in Figure 8.5.2, then the cycle is **a reverse cycle**. A machine with a working substance in a positive cycle is called a heat engine, such as a steam engine, an internal combustion engine, and a steam turbine; a machine with a working substance in a reverse cycle is called a refrigerating machine, such as a refrigerator and a refrigeration air conditioner.

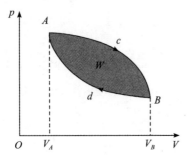
Figure 8.5.1 Positive cycle process curve

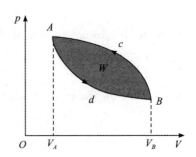
Figure 8.5.2 Reverse cycle process curve

8.5.2 Efficiency of a heat engine

The working medium of a heat engine does a positive cycle, and its internal energy remains unchanged after completing a cycle. According to the first law of thermodynamics, the net work done by the system in a cycle is

$$W = Q_1 - Q_2 \tag{8.5.1}$$

where Q_1 is the heat absorbed by the working medium from the high-temperature heat source in one cycle, and Q_2 is the heat released to the low-temperature heat source. The energy flow diagram is shown in Figure 8.5.3.

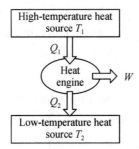

Figure 8.5.3 Schematic diagram of energy flow of a heat engine

An important physical quantity to measure the performance of a heat engine is the efficiency of the heat engine which refers to the ratio of the useful work done by the engine to the heat absorbed by the engine, which can be expressed as

$$\eta = \frac{W}{Q_1} = \frac{Q_1 - Q_2}{Q_1} = 1 - \frac{Q_2}{Q_1} \tag{8.5.2}$$

8.5.3 Coefficient of performance

In one cycle of a refrigerator, the environment does net work W on the working medium. The working medium absorbs heat Q_2 from the low-temperature heat source and releases heat Q_1 to the high-temperature heat source. The energy flow diagram is shown in Figure 8.5.4.

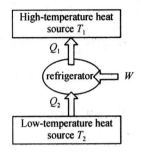

Figure 8.5.4 Schematic diagram of energy flow of a refrigerator

According to the first law of thermodynamics, there is
$$Q_1 = Q_2 + W \tag{8.5.3}$$
For a refrigerator, taking heat Q_2 from the low-temperature heat source is the working purpose of the working medium, and the external work on the working medium W is the price that must be paid. The working efficiency of a refrigerator is usually expressed by the coefficient of performance, which is defined as the ratio of the heat absorbed by the working medium from the low-temperature heat source in one cycle to the work done by the environment, expressed as
$$w = \frac{Q_2}{|W|} = \frac{Q_2}{Q_1 - Q_2} \tag{8.5.4}$$

Example 8.5.1 There are 2 moles of water vapor stored in the cylinder (as an ideal gas of rigid molecules), and the *abcda* cycle process is shown in Figure 8.5.5, where *a-b* and *c-d* are isovolumic processes, *b-c* is an isothermal process, and *d-a* is an isobaric process. Try to find

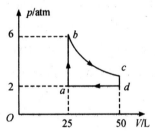

Figure 8.5.5 Figure of Example 8.5.1

(1) the work W_{da} done by water vapor in the process of *d-a*;

(2) the increment of internal energy ΔE_{ab} of water vapor in the process of *a-b*;

(3) the cycle efficiency.

Solution (1) The work done by water vapor in the process of *d-a* at constant pressure is
$$W_{da} = p(V_a - V_d) = -5.065 \times 10^3 \text{ J} < 0$$
So, the environment does work on the water vapor.

(2) The degree of freedom of the rigid polyatomic molecular ideal gas is $i = 6$, so the increase in the internal energy ΔE_{ab} of water vapor in the isovolumic process of *a-b* is
$$\Delta E_{ab} = \nu \left(\frac{i}{2}\right) R(T_b - T_a) = \left(\frac{6}{2}\right) V_a (p_b - p_a) = 3.039 \times 10^4 \text{ J}$$
According to the ideal gas state equation $p_b V_b = \nu R T_b$ and $V_b = V_a$
$$T_b = \frac{p_b V_a}{\nu R} = 914 \text{ K}$$

b-c is an isothermal process, and the work done by the gas is

$$W_{bc} = \nu R T_b \ln\left(\frac{V_c}{V_b}\right) = 1.05 \times 10^4 \text{ J}$$

In the whole cycle process, the net work done by the system and the heat absorbed by the system are respectively

$$W = W_{bc} + W_{da} = 5.44 \times 10^3 \text{ J}$$

$$Q_1 = Q_{ab} + Q_{bc} = \Delta E_{ab} + W_{bc} = 4.09 \times 10^4 \text{ J}$$

The cycle efficiency is

$$\eta = \frac{W}{Q_1} = 13\%$$

8.5.4 The Carnot cycle

1. The Carnot cycle

In order to improve the efficiency of the heat engine, in 1824 the French engineer Carnot proposed an ideal but theoretically important heat engine cycle—the Carnot cycle.

The Carnot cycle consists of two quasi-static isothermal processes and two quasi-static adiabatic processes, as shown in Figure 8.5.6. In the p-V diagram, there are two isothermal lines with temperatures T_1 and T_2 and two adiabatic lines, forming a closed curve. It can be seen that in the process of the Carnot cycle, the working medium only exchanges heat with two temperature constant heat sources, and the heat engine that completes the Carnot positive cycle is called **the Carnot heat engine**.

2. The Carnot heat engine

The Carnot positive cycle with ideal gas as working medium is shown in Figure 8.5.6, and each of its processes is as follows:

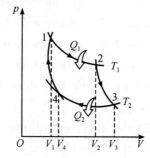

Figure 8.5.6 The Carnot positive cycle

1→2: The gas expands isothermally in contact with a high-temperature heat source of temperature T_1, and the volume increases from V_1 to V_2. The heat absorbed from a high-temperature heat source is

$$Q_1 = \frac{M}{M_{\text{mol}}} R T_1 \ln \frac{V_2}{V_1} \tag{8.5.5}$$

2→3: The gas is separated from the high-temperature heat source and expands adiabatically. The temperature drops to T_2 and the volume increases to V_3. There is no heat exchange in the process, but work is done on the environment.

3→4: The gas is in contact with the low-temperature heat source T_2 for isothermal compression, and the volume is reduced to V_4, so that state 4 and state 1 are located on the same adiabatic line. During the process, the environment does work on the gas, and the gas releases heat Q_2 to the low-temperature heat source with temperature T_2, which is

$$Q_2 = \frac{M}{M_{mol}} RT_2 \ln \frac{V_3}{V_4} \quad (8.5.6)$$

4→1: The gas is separated from the low-temperature heat source, and after adiabatic compression, it returns to the initial state 1 and completes a cycle. There is no heat exchange during the process, and the environment does work on the gas.

According to the definition of heat engine efficiency, the cycle efficiency of the Carnot heat engine with ideal gas as working medium can be obtained

$$\eta_C = 1 - \frac{Q_2}{Q_1} = 1 - \frac{T_2 \ln \frac{V_3}{V_4}}{T_1 \ln \frac{V_2}{V_1}} \quad (8.5.7)$$

Applying Equation (8.4.18) to the adiabatic processes 2→3 and 4→1, respectively, we have

$$T_1 V_2^{\gamma-1} = T_2 V_3^{\gamma-1}$$
$$T_1 V_1^{\gamma-1} = T_2 V_4^{\gamma-1} \quad (8.5.8)$$

Dividing the two equations, there is

$$\frac{V_2}{V_1} = \frac{V_3}{V_4} \quad (8.5.9)$$

Substituting it into Equation (8.5.7), we can get

$$\eta_C = 1 - \frac{T_2}{T_1} = \frac{T_1 - T_2}{T_1} \quad (8.5.10)$$

The above formula shows that to complete a Carnot cycle, there must be two heat sources with a constant high temperature and low temperature, respectively. The efficiency of the Carnot cycle is only related to the temperature of the two heat sources. The larger the difference between the temperature of the high temperature heat source and the low temperature heat source is, the higher efficiency of the Carnot cycle could be obtained. The efficiency of the Carnot cycle is always less than 1.

3. The Carnot refrigerator

The Carnot reverse cycle with an ideal gas as the working medium is shown in Figure 8.5.7. The gas is in contact with a low-temperature heat source, and the heat absorbed from the low-temperature heat source is

$$Q_2 = \frac{m}{M_{mol}} RT_2 \ln \frac{V_3}{V_4} \quad (8.5.11)$$

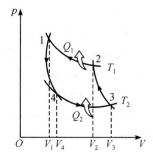

Figure 8.5.7 The Carnot reverse cycle

The amount of heat released by the gas to the high-temperature heat source is

$$Q_1 = \frac{m}{M_{mol}} RT_1 \ln \frac{V_2}{V_1} \quad (8.5.12)$$

The net work done during one cycle is
$$W = Q_1 - Q_2 \tag{8.5.13}$$
So, the coefficient of performance of the Carnot refrigerator is
$$w_C = \frac{Q_2}{|W|} = \frac{\frac{m}{M_{mol}} R T_2 \ln \frac{V_3}{V_4}}{\frac{m}{M_{mol}} R T_1 \ln \frac{V_2}{V_1} - \frac{m}{M_{mol}} R T_2 \ln \frac{V_3}{V_4}} \tag{8.5.14}$$
From the ideal gas adiabatic Equation (8.4.18), we can get
$$T_1 V_2^{\gamma-1} = T_2 V_3^{\gamma-1}$$
$$T_1 V_1^{\gamma-1} = T_2 V_4^{\gamma-1} \tag{8.5.15}$$
Dividing the above two equations, we get
$$\frac{V_2}{V_1} = \frac{V_3}{V_4} \tag{8.5.16}$$
Substituting the above expression into Equation (8.5.14), the coefficient of performance of the Carnot refrigerator is
$$w_C = \frac{Q_2}{|W|} = \frac{T_2}{T_1 - T_2} \tag{8.5.17}$$

It can be seen that the coefficient of performance of the Carnot refrigerator is only related to the temperature of the two heat sources. The larger the difference between the temperature of the high-temperature heat source and the temperature of the low-temperature heat source is, the smaller the coefficient of performance is. The coefficient of performance can be greater than 1.

Example 8.5.2 Imagine using the temperature difference between surface seawater and deep seawater to create a heat engine. It is known that the surface water temperature in tropical waters is about 25℃, and the water temperature at a depth of 300 meters is about 5℃.

(1) Find the efficiency of the Carnot heat engine operating between these two temperatures;

(2) When a power station works at the maximum theoretical efficiency, it obtains a mechanical power of 10^6 W. How fast will it discharge the waste heat?

(3) If both the mechanical work obtained by the power station and the waste heat discharged come from the heat released during the cooling of water from 25℃ to 5℃, how fast will the power station take the surface water at 25℃?

Solution (1) According to Equation (8.5.10), we can get
$$\eta_C = 1 - \frac{T_2}{T_1} = 6.7\%$$

(2) As $\eta_C = 1 - \frac{Q_2}{Q_1} = 1 - \frac{Q_2}{Q_2 + W}$, $Q_2 = \frac{W(1-\eta)}{\eta}$. According to what is given, the mechanical work obtained by the power station per second is $W = 10^6$ J, and then the waste heat discharged by the power station per second is

$$Q_2 = \frac{W(1-\eta)}{\eta} = \frac{10^6 \times (1-0.067)}{0.067} = 1.4 \times 10^7 \text{ J}$$

(3) $T_1 = 298$ K and $T_2 = 278$ K. According to what is given, $Q_1 = W + Q_2 = Cm(T_1 - T_2)$, so it can be obtained that the mass of water taken by the power station at 25℃ per second is

$$m = \frac{W + Q_2}{C(T_1 - T_2)} = \frac{10^6 + 1.4 \times 10^7}{4.18 \times 10^3 \times 20} = 1.8 \times 10^2 \text{ kg}$$

8.5.5 Application of the thermodynamic cycle process in engineering technology

A heat engine is the main equipment to convert heat energy into mechanical energy. The gasoline engine and diesel engine are two kinds of internal combustion engines commonly used in engineering. A cycle of an internal combustion engine is called the Otto cycle, and its working medium is a mixture of fuel and air, which uses the combustion heat of the fuel to generate huge pressure to perform work. Figure 8.5.8 is a schematic diagram of the structure of an internal combustion engine and its p-V diagram of the four-stroke cycle. Among them, ab is the process of adiabatic compression, bc is the isovolumetric process at the moment of fuel explosion caused by electric spark, cd is the process of adiabatic expansion doing work on the environment, and da is the isovolumetric process at the moment of opening the exhaust valve. The efficiency of the Otto cycle is

$$\eta_C = 1 - \left(\frac{V_2}{V_1}\right)^{\gamma - 1} \tag{8.5.18}$$

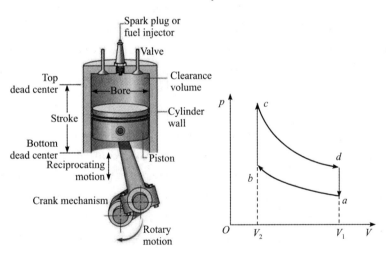

Figure 8.5.8 The Otto cycle of an internal combustion engine

It can be seen from Formula (8.5.18) that the efficiency of the Otto cycle is determined by the volume compression ratio of the cylinder. At present, most cars use reciprocating engines. In the engine cylinder, when the piston reaches the lowest point, the position at this time is called the bottom point, and the volume formed by the entire cylinder including the combustion chamber is the maximum volume. When the piston moves in the reverse direction and reaches the highest point, this position is called the top point, and the volume formed is the minimum volume of the entire piston. The compression ratio is the ratio of the maximum volume to the minimum volume. Therefore, the ratio of the total volume of the internal combustion engine cylinder to the volume of the combustion chamber is an important structural parameter of the internal combustion engine. The higher the compression ratio, the higher the thermal efficiency. However, with the increase of the compression ratio, the growth rate of the thermal efficiency becomes smaller and smaller. Meanwhile, the increase of the compression ratio results in the increase of the compression pressure and the maximum combustion pressure, so the mechanical efficiency of the internal combustion engine decreases. Too high a compression ratio of the gasoline engine may cause knocking. The compression ratio of modern diesel engines is generally 12 – 22, and the compression ratio of gasoline engines is usually in the region of 9 – 12.

A refrigerator is a machine that transfers heat by doing work on the environment. As shown in Figure 8.5.9, the compressor firstly compresses the low-temperature and low-pressure gaseous refrigerant (such as ammonia or Freon) to a pressure of 1 MPa, and the temperature rises above room temperature (ab: adiabatic compression process). Then the refrigerant enters the radiator and releases heat of Q_1, and gradually liquefies into the liquid reservoir (bc: isobaric compression process). In the following, the refrigerant expands and cools through the throttle valve (cd: adiabatic expansion process), and finally enters the freezer to absorb the heat Q_2 in the refrigerator and vaporizes from liquid to gas (da: isobaric expansion process). Then, it is sucked into the compressor again for the next cycle. It can be seen that the whole refrigeration process is that the compressor does work so that the refrigerant changes from a gaseous state to a liquid state, releasing Q_1, and then becomes gas again to absorb heat Q_2. By repeating this cycle, the purpose of cooling is achieved.

The commonly used refrigerant in the refrigerator is freon, but the discharge of freon into the atmosphere will lead to the decrease of the ozone content, so that creatures on the earth will be severely damaged by ultraviolet rays. In recent years, a semiconductor refrigerator has been invented which is completely different from ordinary refrigerators in terms of refrigeration principle. This type of refrigerator utilizes the Peltier effect to make a galvanic couple by n-type and p-type semiconductor materials. When a current flows through the semiconductor materials, in addition to generating irreversible Joule heat, heat absorption or release occurs at different joints. If the direction of current is reversed, the endothermic and exthermic joints will also be exchanged. At present, the temperature difference of commonly used semiconductor refrigerators can reach as much as 150℃. When the refrigeration capacity of this kind of refrigerator exceeds tens of liters, its efficiency is lower than that of a compression refrigeration refrigerator; but for small capacity, it is quite superior. Because this kind of refrigerator has the characteristics of no mechanical transmission, no refrigerant, and no noise, it is widely used in car refrigerators, USB refrigerators, and insulin refrigeration equipment.

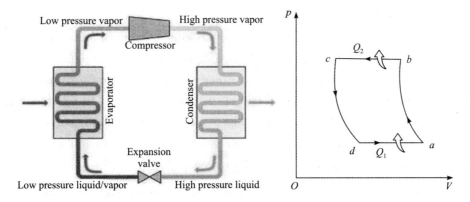

Figure 8.5.9   Refrigeration reverse cycle of a refrigerator

8.6   The second law of thermodynamics

The first law of thermodynamics states that energy is conserved in thermodynamic processes. However, when people study the working principle of heat engines, they find that not all thermodynamic processes satisfying energy conservation can be carried out in practice, and the actual thermodynamic processes can only proceed in a certain direction. The first law of thermodynamics does not describe the direction of system changes, and the second law of thermodynamics is about the directionality of natural processes.

8.6.1   Formulation of the second law of thermodynamics

1. The Kelvin formulation

At the beginning of the 19th century, steam engines were widely used in many fields, and how to improve the efficiency of heat engines became a research hotspot at that time. In 1851, Kelvin discovered that the work-heat conversion is irreversible while studying the working principle of the heat engine and the conversion of work to heat. His discovery can be expressed as: In a circulatory system, it is impossible to absorb heat from a single heat source and completely convert it into useful work without causing other changes. This formulation is called **the Kelvin formulation of the second law of thermodynamics**. It shows that the heat-work conversion process has a certain directionality.

In the Kelvin formulation, "circulatory system", "single heat source", and "without causing other changes" are the three key conditions. Although the isothermal expansion process of an ideal gas can completely convert the absorbed heat into work done on the environment, it is not a heat engine with cyclic action and instead causes other changes such as volume increase and pressure decrease.

2. The Clausius formulation

In 1850, when Clausius was studying the working principle of the refrigerator, he proposed that heat cannot be automatically transferred from a low-temperature object to high-temperature object without causing any other changes. This formulation is known as **the Clausius formulation of the second law of thermodynamics**. It shows that the heat conduction process also has directionality. The second law of thermodynamics can also be expressed as: The second type of perpetual motion machine is impossible to make.

3. Equivalence of the Kelvin and Clausius formulation

The Clausius formulation and Kelvin formulation of the second law of thermodynamics seem to be different, but in fact, both express the irreversibility of the natural macroscopic process and they are equivalent. Regarding the equivalence of these two expressions, we can use the method of proof by contradiction to prove it.

Firstly, assume that the Clausius formulation does not hold, that is, heat can be automatically transferred from a low-temperature object to a high-temperature object. Make heat Q_2 automatically transfer from the low-temperature source to the high-temperature source, as shown in Figure 8.6.1(a), and then use the Carnot heat engine to make it absorb heat Q_1 from the high-temperature heat source T_1, do work W on the environment, and release heat Q_2 to the low-temperature heat source T_2, $Q_2 = Q_1 - W$, as shown in Figure 8.6.2(b). The total effect is equivalent to the fact that the heat engine absorbs heat $Q_1 - Q_2$ from a high-temperature heat source and transforms all the heat into work, as shown in Figure 8.6.2(c), so the Kelvin formulation is also invalid.

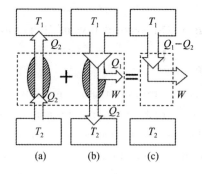

Figure 8.6.1 Assuming that the Clausius formulation does not hold

Secondly, assume that the Kelvin formulation does not hold. We can design a heat engine that converts all the heat Q absorbed by the working medium from the high-temperature heat source T_1 into work in one cycle, as shown in Figure 8.6.2(a); and use this work to drive a Carnot refrigerator working between the low-temperature heat source T_2 and the high-temperature heat source T_1 for it to absorb heat Q_2 from the source T_2 and release $Q_1 = W + Q_2 = Q + Q_2$ to the source T_1 in one cycle, as shown in Figure 8.6.2 (b). These two cycles can be regarded as a combined cycle, and the overall effect is that heat Q_2 is transferred from the source T_2 to the source T_1 without doing work on the environment in one cycle, as shown in Figure 8.6.2(c). Therefore, the Clausius formulation does not hold.

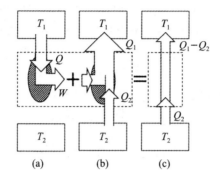

Figure 8.6.2　Assuming that the Kelvin formulation does not hold

The second law of thermodynamics states that spontaneous macroscopic processes in nature are directional or irreversible. The equivalence between the Kelvin formulation and the Clausius formulation shows that the irreversibility of the heat-work conversion process and the irreversibility of the heat conduction process in nature are interdependent.

In the state change process of a thermodynamic system, if the reverse process can repeat every state in the forward process without causing other changes, such a process is called a **reversible process**. Only some ideal processes, such as quasi-static isovolumetric, isothermal, isobaric and adiabatic processes, are reversible processes. Whether a thermodynamic process is reversible or not is closely related to whether the intermediate state experienced by the system is in equilibrium. Only when the process proceeds infinitely slowly, at the same time no friction, viscous force, or other dissipative force does work during the process and the whole process is composed of quasi-static processes infinitely close to the equilibrium state, is it a reversible process.

If the reverse process cannot repeat every state of the forward process without causing other changes, or the reverse process can repeat every state of the forward process but causes other changes at the same time, such a process is called an **irreversible process**. The actual macroscopic spontaneous processes in nature are all irreversible processes, such as the adiabatic free expansion process of gas and the mutual diffusion process of various gases.

 8.6.2　Carnot's theorem

Each process in the Carnot cycle is a quasi-static process, so the Carnot cycle is an ideal reversible cycle. Carnot's theorem, which is very important in the theory of heat

engines, was obtained when Carnot studied the cycle efficiency of heat engines. Its content is:

(1) All reversible heat engines working between the same high-temperature heat source and the same low-temperature heat source have equal efficiency regardless of the working substance;

(2) The efficiency of all irreversible heat engines working between the same high-temperature heat source and the same low-temperature heat source cannot be greater than that of the reversible heat engine.

If we choose a Carnot heat engine with an ideal gas as the working substance from reversible heat engines, then by Carnot's theorem we can get

$$\eta = 1 - \frac{Q_2}{Q_1} = 1 - \frac{T_2}{T_1} \tag{8.6.1}$$

Similarly, if η' is the efficiency of an irreversible heat engine, then by Carnot's theorem (2) we can get

$$\eta' \leqslant 1 - \frac{T_2}{T_1} \tag{8.6.2}$$

The equal sign in the formula is suitable for reversible heat engines, and the less than sign is suitable for irreversible heat engines.

Carnot's theorem points out the ways to improve the efficiency of a heat engine. Firstly, the actual irreversible heat engine should be as close as possible to the reversible heat engine, that is, the dissipation factors such as friction, air leakage and heat dissipation should be reduced. Secondly, increase the temperature difference between two heat sources as much as possible. Since the general heat engine always uses the surrounding environment as the low-temperature heat source, only increasing the temperature of the high-temperature heat source is feasible.

Another important theoretical significance of Carnot's theorem is that it can be used to define a temperature scale. From Equation (8.6.1), we can get

$$\frac{Q_2}{Q_1} = \frac{T_2}{T_1} \tag{8.6.3}$$

That is to say, in the Carnot cycle, the ratio of the heat absorbed by the working substance from the high-temperature heat source to the heat released to the low-temperature heat source is equal to the ratio of the temperatures of the two heat sources. Since this conclusion has nothing to do with the type of working substance, the ratio of the heat exchanged between any working substance undergoing the Carnot cycle and the high and low-temperature heat sources can be used to measure the temperatures of the two heat sources, or to define the temperatures of the two heat sources. If the triple point of water is taken as the fixed point of measuring temperature and its value is specified as 273.16, then the value of any temperature can be determined by the temperature ratio given by Equation (8.6.3). This method of measuring temperature was introduced by Kelvin and is called **the thermodynamic temperature scale.**

8.7 Statistical significance of the second law of thermodynamics and the principle of entropy increase

8.7.1 Statistical significance of the second law of thermodynamics

The second law of thermodynamics points out that all spontaneous processes related to thermal phenomena are irreversible processes, and the processes of heat-work conversion, heat conduction, and free expansion of gas are typical irreversible processes. From the point of view of molecular kinetic theory, the irreversibility of thermodynamic processes is determined by the random thermal motion of a large number of molecules, and the random motion of a large number of molecules follows statistical laws. Then, let us discuss the second law of thermodynamics from the view of statistics.

Take the free expansion process of an ideal gas as an example. As shown in Figure 8.7.1, suppose a container is divided into two parts A and B with equal volume. Room A is filled with some kind of gas, and room B is vacuumized. Now discuss the location distribution of the gas molecules in the container after the partition board is removed. Any molecule in the gas has an equal chance of appearing in the A room or in the B room, and the probabilities are both $\frac{1}{2}$. Suppose there are four molecules a, b, c, and d in room A. Then there are 16 ways of distributing the four molecules in the container after thermal motion, as shown in Table 8.7.1. The various possible distribution states of molecules in the two rooms A and B are called microscopic states.

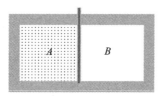

Figure 8.7.1 Free expansion of ideal gas

As can be seen from Table 8.7.1, there are $16=2^4$ microscopic states for the molecule distribution. Macroscopically, molecules are indistinguishable and we can only point out how many molecules there are in room A and room B but we cannot identify which molecules are in room A (or room B). Various possible distribution states with certain number of molecules in the two rooms A and B are called macroscopic states. In this way, the distribution of the four molecules in room A and room B has 5 macroscopic states. A macroscopic state corresponds to several microscopic states. Different macroscopic states correspond to different numbers of microscopic states. But the probability of each microstate appearing is the same. The probability that all four molecules automatically return to room A after thermal motion is $\dfrac{1}{16}=\dfrac{1}{2^4}$. Therefore, in the case where the total number of gas molecules is N, the probability that all the N molecules in the A room will automatically return to it after free expansion is $\dfrac{1}{2^N}$. Usually, the thermodynamic system contains a huge number of molecules, so the probability that all the molecules will automatically return to room A is too small to be observed. For example, for 1 mole of gas molecules, this probability is $\dfrac{1}{2^{6\times 10^{23}}} \approx 10^{-2\times 10^{23}}$. Hence, the free expansion of gas is an irreversible process.

Table 8.7.1 Distribution of four molecules in the container

Microstate		Macrostate		Microstate number	Probability
A	B	A	B		
$abcd$	0	4	0	1	$\dfrac{1}{16}$
abc abd acd bcd	d c b a	3	1	4	$\dfrac{1}{4}$
ab ac ad bc bd cd	cd bd bc ad ac ab	2	2	6	$\dfrac{3}{8}$
d c b a	abc abd acd bcd	1	3	4	$\dfrac{1}{4}$
0	$abcd$	0	4	1	$\dfrac{1}{16}$

There is only one microscopic state corresponding to the macroscopic state in which all four molecules are distributed in room A, and the probability of such occurrence is the smallest. And there are 6 microscopic states corresponding to the macroscopic state in which four molecules are evenly distributed in room A and room B, and the probability of such occurrence is the largest. From the above analysis, we can draw such a conclusion: **All the actual processes that take place in an isolated system that is not affected by the environment always change from a macroscopic state that contains a small number of microscopic states to a macroscopic state that contains a large number of microscopic states. This is the statistical significance of the second law of thermodynamics.**

8.7.2 Principle of entropy increase

The irreversibility of the spontaneous process indicates that an isolated system can never return to the initial state after changing from its initial state to its final state in any actual process. This reflects that there is a qualitative difference between the initial state and the final state, and it is this difference that determines the direction of the process. In order to describe this property of the state of a thermodynamic system and quantitatively illustrate the direction of the spontaneous process, we introduce the concept of entropy, denoted by S.

The number of microstates contained in a macrostate is called the thermodynamic probability of the macrostate, denoted by W. In 1877, Boltzmann gave a statistical explanation for the concept of entropy, pointing out that the entropy of the system in a certain macrostate is proportional to the logarithm of the thermodynamic probability W of the macrostate, namely

$$S = k \ln W \qquad (8.7.1)$$

where k is the Boltzmann constant, and the above formula is called the Boltzmann formula. The Boltzmann formula explains the statistical significance of entropy: The greater the thermodynamic probability (the greater the number of microstates corresponding to a certain macrostate), the greater the disorder of thermal motion in molecules (the greater the entropy). Therefore, **entropy is a measure of disorder of the microscopic particles in a system.**

Suppose that an isolated system changes from the macroscopic state with a thermodynamic probability of W_1 to the macroscopic state with a thermodynamic probability of W_2 after a spontaneous process. Because $W_2 > W_1$, the increase of entropy of the system is

$$\Delta S = S_2 - S_1 = k \ln W_2 - k \ln W_1 = k \ln \frac{W_2}{W_1} > 0$$

The above expression shows that all spontaneous processes of an isolated system proceed in the direction of increasing entropy, and when the equilibrium state is reached, the entropy of the system is the largest.

If a reversible process is carried out in an isolated system, it means that the thermodynamic probabilities of the two macroscopic states at the beginning and end of the process are equal, that is, $W_2 = W_1$, and then the entropy increase is

$$\Delta S = S_2 - S_1 = k \ln \frac{W_2}{W_1} = 0$$

The conclusion is that **the entropy of an isolated system never decreases**, i.e., $\Delta S \geqslant 0$, where the equal sign corresponds to a reversible process. This conclusion is called **the principle of entropy increase**. It gives the mathematical expression of the second law of thermodynamics and provides a reliable basis for judging the direction of the process.

It must be pointed out that the principle of entropy increase is only applicable to isolated systems. If the system is not isolated and exchanges matter or energy with the environment, then the process of system entropy decrease can occur.

Scientist Profile

Wang Zhuxi (1911 – 1983), born in Gong'an county, Hubei province, was a physicist, educator, and pioneer of Chinese thermodynamic statistical physics research.

Wang Zhuxi graduated from the Department of Physics of Tsinghua University in 1933, and received a Doctor of Philosophy degree from Cambridge University in 1938. In the same year, he returned to China and served as a professor of Tsinghua University at Southwest Associated University in Kunming. In 1955, he was elected as an academician of the Chinese Academy of Sciences.

Wang Zhuxi was engaged in the research of thermodynamics, statistical physics and mathematical physics, etc. He made a number of important achievements in the fields of turbulent wake theory, adsorption statistical theory, superlattice statistical theory, thermodynamic equilibrium and stability. He also composed the first batch of excellent theoretical physics textbooks such as *Thermodynamics* and *Introduction to Statistical Physics*, which laid the foundation for the establishment of China's theoretical physics teaching system. He was the editor-in-chief of *Acta Physica Sinica* for a long time, presided over the approval of Chinese physics terms, and made great contributions in promoting the research, dissemination, and exchange of physics in our country.

Wang Zhuxi taught in the physics departments of Tsinghua University and Peking University for more than 40 years. Several generations of Chinese physicists listened to his lectures, like Yang Zhenning and Li Zhengdao. He was not only a great scientist, but also a respected teacher and friend.

Extended Reading

Seeing the spirit of scientific exploration from the failed perpetual motion machine

A perpetual motion machine is an ideal device that can move continuously and perform work externally without consuming energy or with only one heat source. In history, many academic masters such as Da Vinci and Joule, and some inventors explored perpetual motion machines. The research ranged from the first type of perpetual motion machines, the second type of perpetual motion machines to the third type of perpetual motion machines. The Chinese Academy of Sciences proposed in the first issue of *Science Bulletin* that perpetual motion machines were not feasible, so that researchers should not continue its study. In the United States, so many people applied for patents on perpetual motion machines that the U.S. court ruled in 1990 that the patent department would no longer accept any patent applications for perpetual motion machines. In 1775, the French Academy of Sciences also issued that a patent involving perpetual motion machines could not be accepted. Although the idea of perpetual motion machines has not been realized, scientists' in-depth exploration has deepened people's understanding of the nature of heat and promoted the development of thermodynamics.

The first type of perpetual motion machine is the earliest one in history. It can be traced back to about 1200 AD. It was proposed by the Indian Bascara and spread throughout Europe. The most famous one was created by a man named Hennecau, as shown in Figure ER 1. Its principle is as follows. A round wheel is connected with several movable rods, and each rod is connected with a ball. When the round wheel rotates clockwise, the down rod and ball will move away from the center, and the down torque will increase; at the same time, the up rod and ball will approach the center, and the torque will gradually

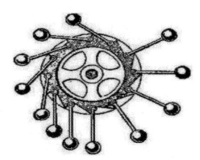

Figure ER 1 The first kind of perpetual motion machine

decrease. Hennecau hoped to rely on this principle to make the wheel go on forever. After that, Da Vinci also designed a similar perpetual motion machine. After numerous failures, Da Vinci concluded that the perpetual motion machine is impossible to realize. The first type of perpetual motion machine violates the law of energy conservation. Despite the failure, the research of scientists and some inventors promoted the continuous development of thermodynamics.

After countless failures in the research of the first type of perpetual motion machine, especially after the first law of thermodynamics was proposed, people no longer conducted research on the first type of perpetual motion machine. Consequently, some inventors asked whether it was possible to invent a machine that could absorb energy from the outside world, and then use the heat to do work to drive the machine to rotate. Hence the famous second type of perpetual motion machine in history. In 1881, the American Gamgee first designed the second type of perpetual motion machine, attempting to use seawater to vaporize liquid ammonia to provide thrust to the engine. Although the liquid ammonia was temporarily vaporized, there was no low-temperature heat source and the liquid ammonia could not be re-liquefied to carry out the circulation process, so it could not promote the continuous operation of the equipment. Later, after Clausius and Kelvin studied the Carnot cycle, the first law of thermodynamics, and the second law of thermodynamics, the idea of the second type of perpetual motion machine basically went bankrupt.

The first and the second laws of thermodynamics denied the production of the first and the second types of perpetual motion machines, that is, it is not feasible to do work on the environment by using no energy or absorbing energy from the environment. People then turned their attention to the cycle of matter and scientists conducted the Biosphere 2 experiment. In 1990, the Americans built a greenhouse of 1 hectare in the desert, and then according to the earth's biosphere, added various elements such as air, rivers, forests, animals, and people to the greenhouse. The greenhouse was called Biosphere 2, the third type of perpetual motion machine referred to most in current literature. Scientists hoped that the various elements in Biosphere 2 could realize self-circulation under the sun. But this type of perpetual motion machine violates a basic fact of physics, that is, materials cannot be completely recycled. Garbage will be produced in each cycle, that is, the substances participating in the cycle will be fewer and fewer, and the system will not be able to cycle and will finally collapse. In-depth analysis finds that Biosphere 2 also violates the second law of thermodynamics, but in a subtle way. The second law of thermodynamics is the principle of entropy increase, indicating not only the direction of energy conversion but also the direction of material conversion. Matter always changes from usable to unusable. Hence, Biosphere 2 can only end in failure.

To sum up, the idea of perpetual motion machines has lasted for nearly a thousand years in human history. Although some people are still studying perpetual motion machines, there has been no successful case. The refutation of the perpetual motion machine idea is conducive to people's correct understanding of both science and the world. Energy can neither be created nor destroyed; it can only be transformed from one form to another, or transferred from one object to another. The conversion and transfer of energy is also directional, for example heat can be spontaneously transferred from a hot object to a cold object, but not from a cold object to a hot object. On the other hand, it is precisely because of the imagination and enthusiasm of these enthusiasts that the development of thermodynamics has been promoted. Feng Duan, academician of the Chinese Academy of Sciences and a famous physicist, once said that in addition to commemorating the physicists who discovered the first and second laws of thermodynamics, we should also commemorate those who have studied perpetual motion machines.

Discussion Problems

8.1 Is it correct to say "How much heat does the system have" and "How much work does the system have"?

8.2 Why generally speaking, only in the quasi-static process, can work be calculated by $\int_{V_1}^{V_2} p\,dV$?

8.3 What are the characteristics of a gas in equilibrium? How is the equilibrium state of a gas different from the equilibrium state in mechanics?

8.4 Since the area enclosed by the closed curve in the p-V diagram during the cycle corresponds to the net work done by the working medium on the environment, the larger the area enclosed by the closed curve, the higher the efficiency of the cycle. Is it right?

8.5 Are ΔE, ΔT, A and Q positive or negative during the adb process in Figure T8 - 1. acb is an adiabatic process, and adb is a straight line in the p-V diagram.

8.6 A certain amount of ideal gas changes along a straight line from equilibrium state a to equilibrium state b in the V-T diagram, as shown in Figure T8 - 2. Does the pressure increase or decrease in the process? Is the process endothermic or exothermic?

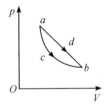
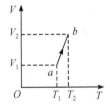

Figure T8 - 1 Figure of Problem 8.5   Figure T8 - 2 Figure of Problem 8.6

Problems

8.7 As shown in Figure T8 - 3, a system goes from state a to state c along abc with heat absorption of 350 J and work done on the environment of 126 J.

(1) In the process adc, the work done on the environment by the system is 42 J. How much heat is absorbed by the system?

(2) When the system returns to state a from state c along curve ac, the work done by the environment on the system is 84 J. Does the system absorb heat or release heat? During this process, how much heat is transferred between the system and the environment?

8.8 1 mol of oxygen changes from state 1 to state 2, and the processes experienced are shown in Figure T8-4. One process is along the 1→ m →2 path, and the other along the 1→2 straight line path. Try to find out the heat Q absorbed by the system, the work A done to the environment, and the change of internal energy $E_2 - E_1$ during each of the two processes.

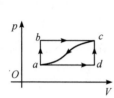

Figure T8-3 Figure of Problem 8.7

Figure T8-4 Figure of Problem 8.8

8.9 The volume of 2 moles of hydrogen at the pressure of 1.013×10^5 Pa and the temperature of 20℃ is V_0, so that it can reach the same state after the following two processes:

(1) keep the volume constant, heat it up to the temperature of 80℃, and then let it expand isothermally with the volume becoming twice the original volume;

(2) first make it expand isothermally to double the original volume, then keep the volume unchanged and raise the temperature to 80℃.

Try to calculate the heat absorbed, the work done by the gas, and the increase in internal energy in each process.

8.10 When the temperature of 0.01 m³ nitrogen is 300 K, it is compressed from 0.1 MPa (i.e., 1 atm) to 10 MPa. Try to find the volume, temperature, and work of nitrogen after isothermal and adiabatic compressions, respectively.

8.11 1 mole of ideal gas goes through the process shown in Figure T8-5, ab is a straight line, and the extension line passes through the origin O. Find the work done by the gas in process ab.

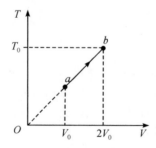

Figure T8-5 Figure of Problem 8.11

8.12 As shown in Figure T8-6, 1 mole of diatomic molecular ideal gas changes from state $A(p_1, V_1)$ to state $B(p_2, V_2)$ along the straight line shown in the p-V diagram. Try to find

(1) the increment of internal energy of the gas;
(2) the work done by the gas on the environment;
(3) the heat absorbed by the gas;
(4) the molar heat capacity of this process.

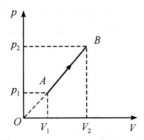

Figure T8 - 6   Figure of Problem 8.12

8.13   There is a heat engine cycle with an ideal gas as the working medium, as shown in Figure T8 - 7. Try to prove that its cycle efficiency is

$$\eta = 1 - \gamma \, \frac{\dfrac{V_1}{V_2} - 1}{\dfrac{p_1}{p_2} - 1}$$

8.14   As shown in Figure T8 - 8, it is a cycle process experienced by an ideal gas, where AB and CD are isobaric processes, BC and DA are adiabatic processes, and the temperatures of points B and C are T_2 and T_3, respectively. Find the cycle efficiency. Is this a Carnot cycle?

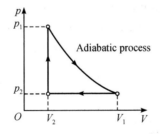

Figure T8 - 7   Figure of Problem 8.13   Figure T8 - 8   Figure of Problem 8.14

8.15   A Carnot heat engine works between two heat sources of 1000 K and 300 K. Try to calculate
(1) the efficiency of the heat engine;
(2) how much the temperature of the high temperature heat source needs to be raised to increase the efficiency of the heat engine to 80% if the low temperature heat source remains unchanged;
(3) how much the temperature of the low-temperature heat source should be reduced to increase the efficiency of the heat engine to 80% if the high-temperature heat source remains unchanged.

8.16 (1) How much work is needed to extract 1000 J of heat from a heat source at 7℃ to a heat source at 27℃ with a Carnot cycle refrigerator? How about from -173℃ to 27℃?

(2) When a reversible Carnot machine is used as a heat engine and the temperature difference between two working heat sources is larger, it will be more beneficial to doing work. When such a machine is used as a refrigerator and the temperature difference between two heat sources is larger, is it more beneficial for refrigeration? Why?

Challenging Problems

8.17 The equation of state for 1 mole of van der Waals gas is

$$\left(p+\frac{a}{v^2}\right)(v-b)=RT$$

where b is a correction term for volume and a is a correction term for pressure.

(1) Above the critical temperature, compare the isotherms of the van der Waals gas and ideal gas.

(2) What are the characteristics of the isotherm of the van der Waals gas near the critical temperature?

8.18 The cycle performed by a diesel engine is approximated in Figure T8-9.

(1) Suppose the working medium is a diatomic ideal gas, $a \to b$ and $c \to d$ are adiabatic processes, $b \to c$ is an isobaric process, and $d \to a$ is an isovolumic process. The adiabatic compression ratio $k_1 = V_0/V_1 = 15$, and the adiabatic expansion ratio $k_2 = V_0/V_2 = 5$. Find the cycle efficiency and the work done on the environment in one cycle.

(2) Discuss the relationships of the cycle efficiency and the work done on the environment in one cycle with the compression and expansion ratios respectively.

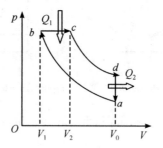

Figure T8-9 Figure of Problem 8.18

高等学校公共基础课系列教材

University Physics
(Ⅱ)

主　编　王　真　侯兆阳
副主编　沈　浩　令狐佳珺
　　　　王　欢　邹鹏飞
　　　　王学智

西安电子科技大学出版社

内 容 简 介

本书是为适应当前教学改革的需要，根据教育部高等学校非物理类专业物理基础课程教学指导分委员会制定的"非物理类理工学科大学物理课程教学基本要求"，结合编者多年的大学物理双语教学实践和教改经验编写而成的。

全书分为上、下两册，共14章。上册包括力学和热学；下册包括电磁学、波动光学和近代物理。本书除了介绍理工科普通物理教学大纲要求的基本内容外，还穿插介绍了物理学理论的发展历史和物理知识点在工程技术中的应用，并选编了将物理知识向当今科学前沿延伸的阅读材料，同时将课程思政元素融入物理知识的学习中。

本书配有一定数量的例题和习题，可作为各类高等院校非物理专业的理工科各专业以及经管类、文科相关专业的大学物理双语教材，也可作为对大学物理感兴趣的读者的自学参考书。

前言
PREFACE

随着全球化的不断加强,教育也在快速全球化。我国与不同国家的高校进行合作,建立中外合作办学机构,不但提升了我国教育的国际竞争力,也培养了大量高层次国际化人才。2017年全国教育工作会议工作报告中指出:"中外合作办学是实现不出国留学的重要手段,也是借鉴世界经验、引进优质资源的试验田。要下力气建设高水平示范性中外合作办学机构和项目,全面发挥中外合作办学辐射作用,深化对国内教育教学改革推动作用。"

大学物理(双语)是工程技术类各专业的一门重要的必修基础课;同时,它也是培养科学素质的一门重要课程。本课程的基本概念、基本理论和基本方法是构成科学素养的重要组成部分,对激发探索和创新精神、培养逻辑思维能力、提升整体科学素养、形成科学价值观起着其他课程无法代替的作用。

为了适应当前教育改革的需要,响应教育部"推进新工科建设与发展,开展新工科研究和实践"的号召,适应面向未来新技术和新产业发展的需要,在"长安大学教材建设项目"支持推动下,我们编写了本书。

本书的主要特色和创新点如下:

(1) 将基础物理知识与新技术应用有机融合。每项高新技术的产生和发展都与物理学的发展密不可分,本书努力将基础的物理知识与现代高新工程技术相结合。全书每章均选取一个合适的关键物理知识点,阐明它们在高新工程技术中的应用,实现理论和实际的有机融合。例如,在第1章质点运动学中,介绍完求解质点运动的两类常见问题这一知识点后,又分析了此计算方法在高速铁路建设所使用的北斗惯导小车中的应用;在第3章动量、能量和动量矩中,介绍完动量定理这一知识点后,又分析了其在火箭飞行控制中的应用。通过将物理知识与实际工程技术融合,将理论与实际应用紧密连接起来,解决学生关心的"学物理有什么用"的问题,激发学生的学习兴趣。

(2) 将基础物理知识向当今科学前沿延伸。在教育部高等学校非物理类专业物理基础课程教学指导分委员会制定的"非物理类理工学科大学物理课程教学基本要求"中提到,大学物理课程的基本内容是几十甚至几百年前就建立起来的理论体系,其中有些内容不免与现实脱节。为此,本书在每章主要内容的后面增加了延伸阅读,努力将本章的基础物理知识向当今科学前沿延伸。例如,第4章增加了"从猫下落翻身到运动生物学"阅读材料,第5章增加了"非线性振动"阅读材料。这样可以使学生尽早接触前沿科学技术的发展脉搏和现代物理的前沿课题,具有鲜明的时代特色。

(3) 将课程思政元素融入物理知识的学习中。为了响应教育部"推动课程思政全程融入课堂教学"的号召,本书精心选择和组织教材内容,将课程思政元素融入物理知识的学习中。本书在每章主要内容的后面都增加了"科学家简介"一栏,介绍与本章物理知识密切相关的国内外知名物理学家,介绍他们的学习环境、成长经历和学术成就。例如,在第1章中,通过介绍伽利略的生平,让学生感受物理学家追求真理、实事求是的科学态度,以及不向宗教迷信势力妥协的可贵品质；在第 3、4、6 章中，通过介绍钱学森、钱伟长、邓稼先等人的生平，让学生感受科学家的爱国主义情怀。同时，在每章中引入我国物理学理论的发展和技术应用成果，如我国自主研制的"北斗导航系统""光学干涉绝对重力仪"和"墨子号量子通信卫星"等，以增强学生的民族自豪感，激发学生的爱国情怀。

（4）将物理理论处理的问题实际化。本书在每章的课后习题中，除了常规的思考题和练习题外，还增加了一些接近实际情形、难度稍大的提升题，这类采用物理知识处理的问题与生活中的实际情况更接近，使物理理论指导工科实践的意义得到彰显，也有助于激发学生学习物理理论的兴趣。

本书由长安大学应用物理系教材编写组共同编写，王真、侯兆阳担任主编，沈浩、令狐佳珺、王欢、邹鹏飞、王学智担任副主编。王真编写了第 5、6、8、10 章，侯兆阳编写了第 11 章，沈浩编写了第 1、2、9 章，令狐佳珺编写了第 4、7、14 章，王欢编写了第 12 章，邹鹏飞编写了第 8 章，王学智编写了第 3、13 章。王真和侯兆阳对全书进行了校对和审定。长安大学应用物理系在线课程建设组提供了讲课视频资料。在本书的编写过程中，西安电子科技大学出版社刘小莉等编辑给予了大力协助，在此表示诚挚的谢意。

由于编者水平有限，书中难免存在不足之处，敬请广大读者批评指正。

编　者

2024 年 8 月

目 录
CONTENTS

Part 3　Electromagnetism

Chapter 9　Electrostatic Field ⋯ 2
- 9.1　Electric charge and Coulomb's law ⋯ 2
  - 9.1.1　Charge ⋯ 2
  - 9.1.2　Coulomb's law ⋯ 3
- 9.2　Electric field and electric field strength ⋯ 5
  - 9.2.1　Electric field ⋯ 5
  - 9.2.2　Electric field strength ⋯ 6
  - 9.2.3　Calculation of electric field strength ⋯ 6
  - 9.2.4　Electric dipole ⋯ 9
- 9.3　Gauss's law ⋯ 13
  - 9.3.1　Electric field line ⋯ 14
  - 9.3.2　Electric flux ⋯ 15
  - 9.3.3　Gauss's law ⋯ 16
  - 9.3.4　Application of Gauss's law ⋯ 19
- 9.4　Electric potential energy and electric potential ⋯ 23
  - 9.4.1　Work done by electrostatic force ⋯ 24
  - 9.4.2　Circulation of an electrostatic field ⋯ 25
  - 9.4.3　Electric potential ⋯ 25
  - 9.4.4　Calculation of electric potential ⋯ 28
- 9.5　Differential relationship between electric field strength and electric potential ⋯ 32
  - 9.5.1　Equipotential surface ⋯ 32
  - 9.5.2　Differential relationship between electric field strength and electric potential ⋯ 33
- 9.6　Conductor and dielectric in an electrostatic field ⋯ 36
  - 9.6.1　Electrostatic equilibrium of a conductor in an electrostatic field ⋯ 36
  - 9.6.2　Electrostatic shielding ⋯ 38
  - 9.6.3　Dielectric polarization ⋯ 41
  - 9.6.4　Polarization intensity ⋯ 42
  - 9.6.5　Gauss's law in a dielectric ⋯ 43
- 9.7　Capacitor ⋯ 45
  - 9.7.1　Capacitance of an isolated conductor ⋯ 45
  - 9.7.2　Capacitance of a capacitor ⋯ 46
  - 9.7.3　Calculation of the capacitance of a capacitor ⋯ 46

9.7.4　Capacitors in parallel and series ⋯⋯⋯⋯⋯⋯⋯⋯⋯⋯⋯⋯⋯⋯⋯⋯⋯⋯⋯⋯⋯⋯ 49
9.7.5　Application of capacitors in engineering technology ⋯⋯⋯⋯⋯⋯⋯⋯⋯⋯⋯ 50
9.8　Electrostatic energy ⋯⋯⋯⋯⋯⋯⋯⋯⋯⋯⋯⋯⋯⋯⋯⋯⋯⋯⋯⋯⋯⋯⋯⋯⋯⋯⋯⋯⋯⋯⋯ 52
9.8.1　Electrostatic energy of a charged system ⋯⋯⋯⋯⋯⋯⋯⋯⋯⋯⋯⋯⋯⋯⋯⋯ 52
9.8.2　Electrostatic energy of an electric field ⋯⋯⋯⋯⋯⋯⋯⋯⋯⋯⋯⋯⋯⋯⋯⋯⋯⋯ 53
Scientist Profile ⋯⋯⋯⋯⋯⋯⋯⋯⋯⋯⋯⋯⋯⋯⋯⋯⋯⋯⋯⋯⋯⋯⋯⋯⋯⋯⋯⋯⋯⋯⋯⋯⋯⋯⋯⋯⋯ 54
Extended Reading ⋯⋯⋯⋯⋯⋯⋯⋯⋯⋯⋯⋯⋯⋯⋯⋯⋯⋯⋯⋯⋯⋯⋯⋯⋯⋯⋯⋯⋯⋯⋯⋯⋯⋯⋯⋯ 55
Discussion Problems ⋯⋯⋯⋯⋯⋯⋯⋯⋯⋯⋯⋯⋯⋯⋯⋯⋯⋯⋯⋯⋯⋯⋯⋯⋯⋯⋯⋯⋯⋯⋯⋯⋯⋯ 60
Problems ⋯⋯⋯ 61
Challenging Problems ⋯⋯⋯⋯⋯⋯⋯⋯⋯⋯⋯⋯⋯⋯⋯⋯⋯⋯⋯⋯⋯⋯⋯⋯⋯⋯⋯⋯⋯⋯⋯⋯⋯⋯ 63

Chapter 10　Magnetic Field ⋯⋯⋯⋯⋯⋯⋯⋯⋯⋯⋯⋯⋯⋯⋯⋯⋯⋯⋯⋯⋯⋯⋯⋯⋯⋯⋯⋯⋯ 64

10.1　Current intensity and current density ⋯⋯⋯⋯⋯⋯⋯⋯⋯⋯⋯⋯⋯⋯⋯⋯⋯⋯⋯⋯⋯ 65
10.1.1　Current intensity ⋯⋯⋯⋯⋯⋯⋯⋯⋯⋯⋯⋯⋯⋯⋯⋯⋯⋯⋯⋯⋯⋯⋯⋯⋯⋯⋯⋯⋯ 65
10.1.2　Current density ⋯⋯⋯⋯⋯⋯⋯⋯⋯⋯⋯⋯⋯⋯⋯⋯⋯⋯⋯⋯⋯⋯⋯⋯⋯⋯⋯⋯⋯⋯ 65
10.2　Magnetic induction intensity ⋯⋯⋯⋯⋯⋯⋯⋯⋯⋯⋯⋯⋯⋯⋯⋯⋯⋯⋯⋯⋯⋯⋯⋯⋯⋯⋯ 67
10.2.1　Magnetic field ⋯⋯⋯⋯⋯⋯⋯⋯⋯⋯⋯⋯⋯⋯⋯⋯⋯⋯⋯⋯⋯⋯⋯⋯⋯⋯⋯⋯⋯⋯⋯ 67
10.2.2　Magnetic induction intensity ⋯⋯⋯⋯⋯⋯⋯⋯⋯⋯⋯⋯⋯⋯⋯⋯⋯⋯⋯⋯⋯⋯⋯ 67
10.2.3　The Biot-Savart law ⋯⋯⋯⋯⋯⋯⋯⋯⋯⋯⋯⋯⋯⋯⋯⋯⋯⋯⋯⋯⋯⋯⋯⋯⋯⋯⋯⋯ 68
10.2.4　Magnetic field generated by moving charges ⋯⋯⋯⋯⋯⋯⋯⋯⋯⋯⋯⋯⋯⋯ 71
10.3　Gauss's law for a magnetic field ⋯⋯⋯⋯⋯⋯⋯⋯⋯⋯⋯⋯⋯⋯⋯⋯⋯⋯⋯⋯⋯⋯⋯⋯ 73
10.3.1　Magnetic induction line ⋯⋯⋯⋯⋯⋯⋯⋯⋯⋯⋯⋯⋯⋯⋯⋯⋯⋯⋯⋯⋯⋯⋯⋯⋯⋯ 73
10.3.2　Magnetic flux ⋯⋯⋯⋯⋯⋯⋯⋯⋯⋯⋯⋯⋯⋯⋯⋯⋯⋯⋯⋯⋯⋯⋯⋯⋯⋯⋯⋯⋯⋯⋯⋯ 73
10.3.3　Gauss's law for a magnetic field ⋯⋯⋯⋯⋯⋯⋯⋯⋯⋯⋯⋯⋯⋯⋯⋯⋯⋯⋯⋯⋯ 74
10.3.4　The Ampère circuital theorem ⋯⋯⋯⋯⋯⋯⋯⋯⋯⋯⋯⋯⋯⋯⋯⋯⋯⋯⋯⋯⋯⋯⋯ 74
10.4　Effect of magnetic fields on current-carrying wires ⋯⋯⋯⋯⋯⋯⋯⋯⋯⋯⋯⋯⋯⋯ 79
10.4.1　The Ampère force ⋯⋯⋯⋯⋯⋯⋯⋯⋯⋯⋯⋯⋯⋯⋯⋯⋯⋯⋯⋯⋯⋯⋯⋯⋯⋯⋯⋯⋯ 79
10.4.2　Effect of uniform magnetic fields on current-carrying coils ⋯⋯⋯⋯⋯⋯⋯ 81
10.4.3　Work done by magnetic force ⋯⋯⋯⋯⋯⋯⋯⋯⋯⋯⋯⋯⋯⋯⋯⋯⋯⋯⋯⋯⋯⋯⋯ 83
10.5　Effect of magnetic fields on moving charges ⋯⋯⋯⋯⋯⋯⋯⋯⋯⋯⋯⋯⋯⋯⋯⋯⋯⋯ 85
10.5.1　The Lorentz force ⋯⋯⋯⋯⋯⋯⋯⋯⋯⋯⋯⋯⋯⋯⋯⋯⋯⋯⋯⋯⋯⋯⋯⋯⋯⋯⋯⋯⋯ 85
10.5.2　Movement of charged particles in a uniform magnetic field ⋯⋯⋯⋯⋯⋯ 86
10.5.3　The Hall effect ⋯⋯⋯⋯⋯⋯⋯⋯⋯⋯⋯⋯⋯⋯⋯⋯⋯⋯⋯⋯⋯⋯⋯⋯⋯⋯⋯⋯⋯⋯⋯ 87
10.5.4　Application of the Hall effect in engineering technology ⋯⋯⋯⋯⋯⋯⋯⋯ 89
10.6　Magnetism of matter ⋯⋯⋯⋯⋯⋯⋯⋯⋯⋯⋯⋯⋯⋯⋯⋯⋯⋯⋯⋯⋯⋯⋯⋯⋯⋯⋯⋯⋯⋯⋯ 92
10.6.1　Classification of magnetic material ⋯⋯⋯⋯⋯⋯⋯⋯⋯⋯⋯⋯⋯⋯⋯⋯⋯⋯⋯⋯ 92
10.6.2　Microscopic mechanism of paramagnetism and diamagnetism ⋯⋯⋯⋯⋯ 93
10.6.3　Magnetization and magnetization current ⋯⋯⋯⋯⋯⋯⋯⋯⋯⋯⋯⋯⋯⋯⋯⋯⋯ 95
10.6.4　The Ampère circuital theorem in magnetic material ⋯⋯⋯⋯⋯⋯⋯⋯⋯⋯ 96
10.7　Ferromagnetic material ⋯⋯⋯⋯⋯⋯⋯⋯⋯⋯⋯⋯⋯⋯⋯⋯⋯⋯⋯⋯⋯⋯⋯⋯⋯⋯⋯⋯⋯⋯ 98
10.7.1　Magnetization law of ferromagnetic material ⋯⋯⋯⋯⋯⋯⋯⋯⋯⋯⋯⋯⋯⋯ 98
10.7.2　Classification of ferromagnetic material ⋯⋯⋯⋯⋯⋯⋯⋯⋯⋯⋯⋯⋯⋯⋯⋯⋯ 99
10.7.3　Magnetic domain ⋯⋯⋯⋯⋯⋯⋯⋯⋯⋯⋯⋯⋯⋯⋯⋯⋯⋯⋯⋯⋯⋯⋯⋯⋯⋯⋯⋯⋯ 100

Scientist Profile ·········· 101
Extended Reading ·········· 102
Discussion Problems ·········· 108
Problems ·········· 108
Challenging Problems ·········· 110

Chapter 11 Electromagnetic Induction ·········· 111
11.1 Electromotive force ·········· 111
11.2 Faraday's law of induction ·········· 113
    11.2.1 Electromagnetism induction phenomenon ·········· 113
    11.2.2 Faraday's law of electromagnetic induction ·········· 114
11.3 Motional EMF and induced EMF ·········· 116
    11.3.1 Motional EMF ·········· 117
    11.3.2 Induced EMF ·········· 119
    11.3.3 Application of induced electric field in science and technology ·········· 120
11.4 Self induction and mutual induction ·········· 122
    11.4.1 Self induction ·········· 122
    11.4.2 Mutual induction ·········· 125
11.5 Self-induced magnetic energy and magnetic field energy ·········· 128
    11.5.1 Self-induced magnetic energy ·········· 128
    11.5.2 Magnetic field energy ·········· 128
11.6 Introduction to Maxwell's electromagnetic theory ·········· 131
    11.6.1 Displacement current and the Maxwell-Ampère theorem ·········· 131
    11.6.2 Maxwell's equations of an electromagnetic field ·········· 135
Scientist Profile ·········· 136
Extended Reading ·········· 137
Discussion Problems ·········· 140
Problems ·········· 141
Challenging Problems ·········· 144

Part 4 Wave Optics

Chapter 12 Wave Optics ·········· 146
12.1 Electromagnetic theory of light ·········· 146
12.2 Coherent light ·········· 147
    12.2.1 Luminous mechanism of a light source ·········· 147
    12.2.2 Coherent light ·········· 149
    12.2.3 Method of obtaining coherent light ·········· 151
12.3 Optical path length and optical path difference ·········· 152
    12.3.1 Optical path length ·········· 152
    12.3.2 Optical path difference ·········· 152
12.4 Young's double-slit interference ·········· 154

12.4.1	Young's double-slit interference	154
12.4.2	The Lloyd mirror interference experiment	159
12.5	Thin film interference	160
12.5.1	Thin film interference	160
12.5.2	Wedge interference	164
12.5.3	Newton's ring	168
12.5.4	The Michelson interferometer	171
12.6	Diffraction of light	175
12.6.1	Diffraction of light waves	175
12.6.2	Classification of diffraction of light waves	175
12.6.3	The Huygens-Fresnel principle	176
12.7	The Fraunhofer single slit diffraction	177
12.8	Diffraction grating	181
12.8.1	Grating diffraction	182
12.8.2	Law of grating diffraction	183
12.8.3	Grating spectrum	186
12.8.4	X-ray diffraction	188
12.9	The Fraunhofer diffraction with circular aperture	190
12.9.1	The Fraunhofer diffraction with circular aperture	190
12.9.2	Resolution of an optical instrument	191
12.10	Polarization of light	193
12.10.1	Polarization of light	194
12.10.2	Polarization generation and examination of a polarizer	196
12.10.3	Malus's law	198
12.10.4	Polarization of reflection and refraction	200
12.11	Birefringence of light	202
12.11.1	Birefringence phenomenon of a crystal	202
12.11.2	Application of polarized light in science and technology	205
Scientist Profile		207
Extended Reading		209
Discussion Problems		213
Problems		213
Challenging Problems		216

Part 5 Modern Physics

Chapter 13 Fundamentals of the Special Theory of Relativity — 218

13.1	Space-time view of classical mechanics	218
13.1.1	Principle of mechanical relativity	218
13.1.2	The Galileo transformation	219
13.1.3	Classical space-time view	220
13.2	Basic principles of the special theory of relativity	221

    13.2.1   Background of the special theory of relativity ·· 221
    13.2.2   Basic assumptions of the special theory of relativity ·· 223
    13.2.3   The Lorentz transformation ·· 224
  13.3 Space-time view of the special theory of relativity ··· 228
    13.3.1   Relativity of simultaneity ··· 228
    13.3.2   Time dilation effect ··· 229
    13.3.3   Length contraction effect ·· 230
  13.4 Dynamic basis of the special theory of relativity ··· 232
    13.4.1   Relativistic mass-velocity relationship ·· 232
    13.4.2   Basic equation of relativistic dynamics ··· 233
    13.4.3   Mass-energy relationship ··· 233
    13.4.4   Momentum-energy relationship ·· 234
    13.4.5   Application of relativistic mass-energy relationship in nuclear energy ············· 235
Scientist Profile ·· 239
Extended Reading ·· 241
Discussion Problems ·· 245
Problems ·· 245
Challenging Problems ·· 246

Chapter 14   Fundamentals of Quantum Physics ·· 248

  14.1 Planck's energy quantum hypothesis ··· 249
    14.1.1   Thermal radiation ··· 249
    14.1.2   Experimental law of black body radiation ··· 250
    14.1.3   Planck's quantum hypothesis ··· 252
  14.2 Photoelectric effect ··· 254
    14.2.1   Experimental law of the photoelectric effect ·· 254
    14.2.2   Difficulties of classical electromagnetic theory ··· 256
    14.2.3   Einstein's photon theory ·· 257
    14.2.4   Wave-particle duality of light ·· 258
    14.2.5   Application of the photoelectric effect ·· 259
  14.3 The Compton effect ··· 260
    14.3.1   Experimental law of the Compton effect ··· 260
    14.3.2   Theoretical explanation ·· 261
  14.4 Bohr's theory of the hydrogen atom ·· 264
    14.4.1   Experimental law of the hydrogen atom spectrum ··· 264
    14.4.2   Bohr's theory of the hydrogen atom ·· 265
  14.5 Wave-particle duality of a physical particle ·· 268
    14.5.1   The de Broglie hypothesis ··· 269
    14.5.2   Experimental verification of a matter wave ·· 269
    14.5.3   Statistical interpretation of a matter wave ·· 271
  14.6 Uncertainty relation ·· 272
  14.7 Wave function and the Schrödinger equation ··· 275
    14.7.1   Wave function ·· 275
    14.7.2   The Schrödinger equation ·· 277

14.7.3 Application of the Schrödinger equation ... 279
14.8 Application of quantum mechanics to the hydrogen atom ... 281
    14.8.1 The Schrödinger equation for the hydrogen atom ... 281
    14.8.2 Electron spin ... 283
14.9 Electron shell structure of an atom ... 285
Scientist Profile ... 286
Extended Reading ... 287
Discussion Problems ... 291
Problems ... 292
Challenging Problems ... 293

Part 3 Electromagnetism

Electromagnetism is a discipline that studies the laws of electromagnetic motion and its applications. Electromagnetic motion is one of the four basic interactions in nature, and the electromagnetic field is an important part of the natural world.

Humans understood the electromagnetic phenomena in a very early age, but it was not until 1819 that Oster discovered the magnetic effect of current, and Ampère discovered the effect of magnets on current in 1820. At this time, people began to realize the relationship between electricity and magnetism. In 1831, Faraday established the law of electromagnetic induction, and first put forward the concept of electromagnetic field. On the basis of the predecessors'work, Maxwell established a complete electromagnetic field theory in 1865. The study of electromagnetism is of epoch-making significance to the development of human civilization. The production and utilization of electric energy developed on the basis of electromagnetism has led to a new technological revolution and brought mankind into the electrification age. In the middle of the 20th century, microelectronics technology and electronic computers developed on the basis of electromagnetism brought humanity into the information age. Besides, electromagnetism is also an indispensable theoretical basis for human beings to deeply understand the natural world. From the extension of the subject system, electromagnetism is a basic theory for studying electrical engineering, radio electronics, remote and automatic control, and communication engineering.

According to the nature of the content, electromagnetism mainly includes two parts: field and circuit. University physics focuses on the study of field. The field is different from physical particles, and the concept and description method are entirely new for beginners. It should be emphasized that flux and circulation are two critical quantities and basic tools to study the vector field. Grasping this idea and method will help you clarify the framework of electromagnetic field theory, which is beneficial to the study of electromagnetism.

This part first studies the properties and laws of electrostatic fields, then discusses constant magnetic fields, and finally focuses on electromagnetic induction and basic knowledge of electromagnetic fields.

Chapter 9

Electrostatic Field

An electric field exists around any electric charge (static or moving), the electric field excited by a stationary electric charge relative to the observer is called an electrostatic field, and the force exerted by an electrostatic field on other electric charges is called electrostatic force. The spatial distribution of the electrostatic field does not change with time, that is, the electrostatic field is a constant field independent of time.

This chapter not only studies the properties of the electrostatic field in a vacuum, but also introduces the basic characteristics of conductors and dielectrics in the electrostatic field and their influence on the electric field. In addition, the electrostatic field energy, which can represent the character of the electric field is also introduced.

9.1 Electric charge and Coulomb's law

9.1.1 Charge

There are two types of charges in nature: positive charge and negative charge. According to the theory of modern physics about the structure of matter, any matter is made up of positively charged nuclei and negatively charged extranuclear electrons. In the normal state, the positive and negative charges inside an object are equal in magnitude, and the macroscopic object is electrically neutral. When the electrically neutral state is destroyed by some action (friction, photoelectric action, etc.), the object will obtain electrons and carry negative charges. Otherwise, the object will carry positive charges if it loses electrons.

The amount of charge an object carries is called the electric quantity, which is usually represented by the symbol Q or q. The unit of electric quantity in the International System of Units is the Coulomb, and the symbol is C.

In 1897, the British physicist Thomson discovered the electron. The electron is the particle with the smallest negative charge, and the modern measurement value of its electric quantity is $e = 1.602 \times 10^{-19}$ C. In 1913, the American physicist Millikan conducted the famous "oil drop experiment" and determined the electric quantity of fog droplets. A large number of experimental data have confirmed that the electric quantity of each oil droplet is always an integer multiple of e, that is, each droplet has several electrons. Rutherford discovered the proton in 1919 and determined that it is a particle with an electric quantity of $+e$. In nature, charge always appears in integer multiples of a basic unit, and this characteristic of charge is called the quantum property of charge. In 1964, the American physicist Gellmann first predicted that elementary particles were composed of several kinds of quarks or antiquarks, and each quark or antiquark might possess the electric quantity of $\pm\frac{1}{3}e$ or $\pm\frac{2}{3}e$. But so far no quark has been discovered in isolation experimentally. Charge can only transfer from one place to another, but can neither be created nor destroyed. This conclusion is called the **conservation of charge**. Experiments show that the electric quantity has nothing to do with the reference frame. In other words, it has relativistic invariance.

 ## 9.1.2 Coulomb's law

1. Point Charge

The interactive electric force exists between charged bodies, which is related to the shape, size, charge distribution, relative position, and surrounding medium of the charged body. When the dimension of the charged body has little influence on the discussed question, its shape and size can be completely ignored and then the charged body can be regarded as a point with charges, which is called a point charge. The point charge is an ideal physical model that is relevant to the problem we study.

2. Coulomb's Law

In 1785, Coulomb designed an accurate torsion balance, as shown in Figure 9.1.1, to measure the relationship between the interaction force of charges and their distance. Then he proposed Coulomb's law as follows:

The magnitude of the interaction force between two point charges at rest in a vacuum is proportional to the product of their electric quantities and inversely proportional to the square of their distance. The direction of the force is along their connecting line. When two point charges have like charges, there is a repulsive force between them. When they have unlike charges, there is

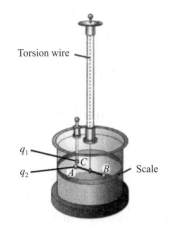

Figure 9.1.1 Coulomb's torsion balance

an attractive force between them.

As shown in Figure 9.1.2, q_1 and q_2 represent the amount of charge carried by two point charges, respectively. r_0 represents the unit vector from the point charge q_1 to the point charge q_2, and r represents the distance between the two charges. Therefore, the force on q_1 by q_2 is

Figure 9.1.2 Coulomb's Law

$$F_{12} = k\frac{q_1 q_2}{r^2} r_0 \qquad (9.1.1)$$

where $k = \dfrac{1}{4\pi\varepsilon_0}$ and ε_0 is called the dielectric constant of vacuum, or vacuum permittivity. It is a basic physical constant of electromagnetism, and its value is $\varepsilon_0 \approx 8.85 \times 10^{-12}\ \text{C} \cdot \text{N}^{-1} \cdot \text{m}^{-2}$.

If q_1 and q_2 have the same sign, the product of the two is positive, and the directions of $\boldsymbol{F}_{12}$ and $\boldsymbol{r}_0$ are the same. Thus, the force is repulsive. If q_1 and q_2 have different signs, the product of the two is negative, and $\boldsymbol{F}_{12}$ and $\boldsymbol{r}_0$ are in opposite directions, so the force is attractive.

Similarly, the force on q_1 by q_2 is $\boldsymbol{F}_{21}$:

$$\boldsymbol{F}_{21} = k\frac{q_1 q_2}{r^2}\boldsymbol{r}_0$$

Obviously,

$$\boldsymbol{F}_{12} = -\boldsymbol{F}_{21}$$

If the positive direction of the unit vector $\boldsymbol{r}_0$ is specified to be from force applying charge to force receiving charge, the electrostatic force $\boldsymbol{F}$ on the charges can be expressed as

$$\boldsymbol{F} = \frac{1}{4\pi\varepsilon_0}\frac{q_1 q_2}{r^2}\boldsymbol{r}_0 \qquad (9.1.2)$$

Coulomb's law suggests that the force between two charges is a centripetal force. Experiments show that the law is extremely accurate when the distance between two point charges is in the range from 10^{-14} m to 10^7 m.

Example 9.1.1 In a hydrogen atom, the sizes of the electron and nucleus are much smaller than the distance between them. Thus, they can both be regarded as point charges. It is already known that the distance between the electron and the nucleus is $r = 5.29 \times 10^{-11}$ m, the electric quantity of the electron is $-e$ and the mass of electron is $m = 9.11 \times 10^{-31}$ kg. The electric quantity of the hydrogen nucleus (proton) is e, and the mass is $M = 1.67 \times 10^{-27}$ kg. Compare the magnitude of the electrostatic attraction $\boldsymbol{F}_e$ and the gravitational force $\boldsymbol{F}_m$ between them.

Solution: According to Coulomb's law, the magnitude of electrostatic attraction between two particles is

$$F_e = \frac{1}{4\pi\varepsilon_0}\frac{e^2}{r^2} = 9.0 \times 10^9 \times \frac{(1.6 \times 10^{-19})^2}{(5.29 \times 10^{-11})^2} = 8.2 \times 10^{-8}\ \text{N}$$

According to Newton's law of universal gravitation, the magnitude of the gravitational force between them is

$$F_m = G \frac{Mm}{r^2} = 6.67 \times 10^{-11} \times \frac{9.11 \times 10^{-31} \times 1.67 \times 10^{-27}}{(5.29 \times 10^{-11})^2} = 3.6 \times 10^{-47} \text{N}$$

Thus, there is

$$\frac{F_e}{F_m} = 2.28 \times 10^{39}$$

It can be seen that the electrostatic force between electron and nucleus is far greater than the gravitational force between them. So, when discussing the interaction between electrons and nuclei, the gravitational force can be ignored.

9.2 Electric field and electric field strength

9.2.1 Electric field

Force is the interaction between objects and cannot exist without matter. How does the electrostatic force between charged bodies transmit? There are two opinions in history. One is action at a distance, and it is believed that the transmission does not need a medium or time. The other is short-range action and it is believed that the interaction between objects requires a medium and time. This medium was thought to be aether until Faraday and Maxwell established the electromagnetic theory of short-range action and proved it experimentally.

The development of modern physics has proved that the view of action at short range in the field theory is correct, while the view of action at a distance is wrong. Besides, aether does not exist, which reflects historical limitations of human beings in understanding objective things. The interaction between charged bodies is transmitted through the electric field. In a vacuum, a charge generates an electric field or excites an electric field around itself. The electric field has a strong effect on the charges in the field. The force produced by the electric field on a charge is determined only by the electric field where the charge is located, and has nothing to do with the electric field in other places. This interaction can be expressed as

Charge⇔Electric field⇔Charge

The electric field is a form of matter. Both theory and experiment show that fields have energy, momentum, and mass. The charge that generates the electric field is called the source charge. When the source charge is stationary and the electric quantity does not change with time, it generates an **electrostatic field**.

9.2.2 Electric field strength

Since there is an electrostatic field around a stationary charge, there should be a method to determine the magnitude and direction of the electrostatic field. Thus, the concept of electric field strength is proposed.

To determine the electric field at a certain point in space, the test charge q_0 can be placed at this point. For the test charge q_0, it is required to meet the following two conditions: ① Its size should be small enough so that it has a definite position in space and can be regarded as a point charge. Its position can be represented by a point (x, y, z). ② Its electric quantity should also be positive and small enough so that it will not affect the initial field after being introduced into the electric field. If there is a force on the test charge q_0, it means that there is an electric field at that point.

Experiments show that at a given point of an electrostatic field, when the magnitude of the test charge q_0 changes, the electric field force F applied to it changes, while the ratio F/q_0 remains unchanged but changes in various positions in the electric field. Thus, the ratio F/q_0 has nothing to do with the test charge q_0 and only varies with the field point. It can be used to reflect the strength of the field at each point in the electric field, which is called the electric field strength of the point, or the field strength for short. Its definition formula is expressed as

$$E = \frac{F}{q_0} \qquad (9.2.1)$$

The electric field strength E is a vector describing the property of the electric field itself. Its magnitude is equal to the magnitude of the electric field force on the unit test charge at this point. The direction of E is defined as the direction of the electric field force on the positive test charge at this point. The SI unit of electric field strength is $N \cdot C^{-1}$.

For different positions in a field, E is generally different. Thus, E is the vector function of the position in space $E(x, y, z)$. Combined with Formula (9.2.1), the electric field force $F(x, y, z)$ of the charge q at this point is

$$F(x, y, z) = qE(x, y, z) \qquad (9.2.2)$$

where $E(x, y, z)$ is the net electric field strength at this point generated by all charges except the test charge q. Obviously, when $q > 0$, F and E have the same signs, indicating that the electric field force F and the field strength E have the same direction. When $q < 0$, F and E have different signs, that is, the electric field force F and the field strength E have opposite directions.

9.2.3 Calculation of electric field strength

This section introduces the solution methods for the electric field strength generated by several different types of charge distributions.

1. Electric field produced by a point charge

Suppose there is a point charge q in a vacuum, as shown in Figure 9.2.1, now let's find the field strength at an arbitrary point P. Assuming that a test charge q_0 is placed at point P, according to Coulomb's law, the electric field force on the test charge q_0 is

$$F = \frac{1}{4\pi\varepsilon_0} \frac{qq_0}{r^2} \boldsymbol{r}_0$$

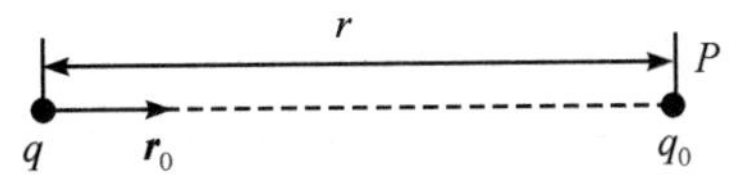

Figure 9.2.1 Electric field produced by a point charge

where $\boldsymbol{r}_0$ represents the unit vector from the point charge q to point P. From the definition of electric field strength, we have

$$\boldsymbol{E} = \frac{\boldsymbol{F}}{q_0} = \frac{1}{4\pi\varepsilon_0} \frac{q}{r^2} \boldsymbol{r}_0 \tag{9.2.3}$$

It can be seen that the magnitude of the electric field strength is proportional to the electric quantity of the field source charge and inversely proportional to the square of the distance from the field point to the field source charge. When $q > 0$, the direction of $\boldsymbol{E}$ deviates from the field source charge; when $q < 0$, it points to the field source charge, as shown in Figure 9.2.2.

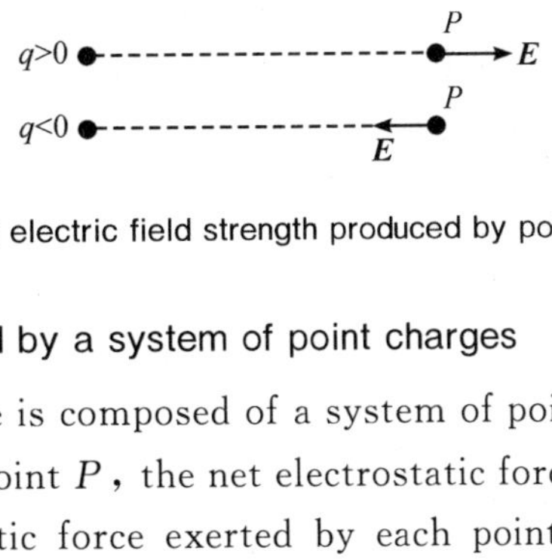

Figure 9.2.2 Comparison of electric field strength produced by point charges with different signs

2. Electric field produced by a system of point charges

If the field source charge is composed of a system of point charges $q_1, q_2, \cdots, q_n$ and a test charge q_0 is placed at point P, the net electrostatic force exerted on q_0 is equal to the vector sum of the electrostatic force exerted by each point charge ($\boldsymbol{F}_1, \boldsymbol{F}_2, \cdots, \boldsymbol{F}_n$), that is

$$\boldsymbol{F} = \boldsymbol{F}_1 + \boldsymbol{F}_2 \cdots + \boldsymbol{F}_n = \sum_{i=1}^{n} \boldsymbol{F}_i$$

Dividing both sides by q, the following expression could be derived:

$$\frac{\boldsymbol{F}}{q_0} = \frac{\boldsymbol{F}_1}{q_0} + \frac{\boldsymbol{F}_2}{q_0} + \frac{\boldsymbol{F}_3}{q_0} + \cdots + \frac{\boldsymbol{F}_n}{q_0}$$

According to the definition of electric field strength, each item on the right side is the field strength generated by each point charge alone, and the left side is the total field strength. So

$$\boldsymbol{E} = \boldsymbol{E}_1 + \boldsymbol{E}_2 \cdots + \boldsymbol{E}_n = \sum_{i=1}^{n} \boldsymbol{E}_i \tag{9.2.4}$$

That is to say, the total field strength of a certain point in the electric field is equal to the vector sum of the field strengths excited at this point when each point charge exists alone. This is called the **superposition principle of electric field strengths**. According to the formula of field strength produced by a point charge and the superposition principle of field strength, the electric field generated by any charged body can be solved.

3. Electric field produced by a continuous charge distribution

In practice charges of charged bodies are usually distributed continuously and can be regarded as a continuous distribution in a certain volume, a certain area, or a certain curve, which is called volume, surface and line distribution, respectively. Accordingly, the volume density ρ, surface density σ, and line density λ of the charge can be introduced:

$$\rho = \lim_{\Delta V \to 0} \frac{\Delta q}{\Delta V} = \frac{dq}{dV}$$

$$\sigma = \lim_{\Delta S \to 0} \frac{\Delta q}{\Delta S} = \frac{dq}{dS}$$

$$\lambda = \lim_{\Delta l \to 0} \frac{\Delta q}{\Delta l} = \frac{dq}{dl}$$

where ΔV, ΔS and Δl are the volume element, surface element and line element obtained by dividing the charged body, respectively. When ΔV, ΔS and Δl are infinitely small, the corresponding charge element dq can be regarded as a point charge. In this way, the whole charged body can be regarded as a point charge system composed of infinite point charges, so the field strength of the electric field of any charged body can be calculated by using the principle of superposition of field strengths.

Assuming that the field strength generated by any charge element dq at point P is $d\boldsymbol{E}$, according to Equation (9.2.3), we have

$$d\boldsymbol{E} = \frac{1}{4\pi\varepsilon_0} \frac{dq}{r^2} \boldsymbol{r}_0$$

where r is the distance from the charge element dq to the field point P, and $\boldsymbol{r}_0$ is the unit vector pointing to point P from dq. According to the concept of charge distribution, the charge element dq can be written as

$$dq = \begin{cases} \lambda dl & \text{(Line distribution)} \\ \sigma dS & \text{(Surface distribution)} \\ \rho dV & \text{(Volume distribution)} \end{cases}$$

The total field strength at point P can be obtained by superposition of the field strength vectors generated by all charge elements at point P. For continuously distributed charges, the superposition should be replaced by an integral, which could be expressed as

$$\boldsymbol{E} = \int d\boldsymbol{E} = \int \frac{1}{4\pi\varepsilon_0} \cdot \frac{dq}{r^2} \boldsymbol{r}_0 \qquad (9.2.5)$$

The above expression is a vector integral. In the actual calculation, $d\boldsymbol{E}$ is generally decomposed in the direction of the selected coordinate axis, and then the components are integrated respectively, that is,

$$E_x = \int dE_x, \quad E_y = \int dE_y, \quad E_z = \int dE_z$$

Based on the relation $\boldsymbol{E} = E_x \boldsymbol{i} + E_y \boldsymbol{j} + E_z \boldsymbol{k}$, the total electric field strength $\boldsymbol{E}$ can be solved.

9.2.4 Electric dipole

As shown in Figure 9.2.3, the distance between two charges $+q$ and $-q$ is l. If l is much smaller than the distance r from their center to the field point, the pair of point charges constitutes an **electric dipole**. The line connecting the two point charges is called the axis of the electric dipole, the vector $\boldsymbol{p} = q\boldsymbol{l}$ is called the electric moment of the electric dipole, and the direction of $\boldsymbol{l}$ is specified as from the negative charge to the positive charge.

As shown in Figure 9.2.3, in order to calculate the electric field strength at a point on the vertical line of the electric dipole in a vacuum, the midpoint of the electric dipole axis is set as the coordinate origin O, and the vertical line is marked as the OY axis. The distance between any point P on the vertical line and the coordinate origin O is r. If point P is on the vertical line, the distances between point P and $+q$ or $-q$ are r_+ or r_-, respectively, and $r_+ = r_-$. The field strengths generated by $+q$ and $-q$ at point P are

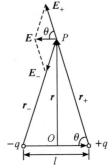

Figure 9.2.3 Electric dipole

$$E_+ = \frac{q}{4\pi\varepsilon_0 r_+^2} = \frac{q}{4\pi\varepsilon_0 (r^2 + l^2/4)}$$

$$E_- = \frac{q}{4\pi\varepsilon_0 r_-^2} = \frac{q}{4\pi\varepsilon_0 (r^2 + l^2/4)}$$

The magnitudes of E_+ and E_- at point P are equal, but their directions are different. According to the principle of symmetry, the magnitude of the field strength $\boldsymbol{E}$ at point P is $E = 2E_+ \cos\theta$.

Substitute $\cos\theta = \dfrac{l/2}{\sqrt{r^2 + l^2/4}}$ into the above expression, and there is

$$E = \frac{ql}{4\pi\varepsilon_0 (r^2 + l^2/4)^{3/2}}$$

Since the direction of $\boldsymbol{p} = q\boldsymbol{l}$ is opposite to $\boldsymbol{E}$, the electric field strength at point P should be

$$\boldsymbol{E} = -\frac{\boldsymbol{p}}{4\pi\varepsilon_0 (r^2 + l^2/4)^{3/2}}$$

If $r \gg l$, then

$$\boldsymbol{E} = -\frac{\boldsymbol{p}}{4\pi\varepsilon_0 r^3} \qquad (9.2.6)$$

Equation (9.2.6) shows that the magnitude of the electric field strength **E** at any point on the vertical line of the electric dipole is proportional to the electric moment **p**, and the direction of the electric field strength is opposite to that of the electric moment.

In Figure 9.2.4, if an electric dipole is at a uniform electric field **E** and the distance between its positive charge and negative charge is r_0, the electric field forces exerted on positive and negative charges are respectively $\boldsymbol{F}_+ = q\boldsymbol{E}$ and $\boldsymbol{F}_- = -q\boldsymbol{E}$, which are equal in magnitude and opposite in direction. Thus, the vector sum of $\boldsymbol{F}_+$ and $\boldsymbol{F}_-$ is **0**, but their lines of action are not on the same straight line. These two forces are called force couples. They have the same moment direction relative to the midpoint O, and both of their force arms are $\frac{1}{2}r_0\sin\theta$, where θ is the angle between **p** and **E**. So the total moment (also called the couple moment) is

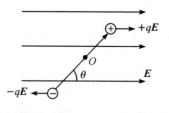

Figure 9.2.4 Electric dipole subjected to force in a uniform external electric field

$$M = F_+ \cdot \frac{1}{2}r_0\sin\theta + F_- \cdot \frac{1}{2}r_0\sin\theta = qr_0 E\sin\theta = pE\sin\theta \qquad (9.2.7)$$

It can be represented in vector form as

$$\boldsymbol{M} = \boldsymbol{p} \times \boldsymbol{E} \qquad (9.2.8)$$

Equation (9.2.8) shows that the electric dipole rotates under the action of the resultant moment **M** in the uniform electric field **E**. When $\theta = \frac{\pi}{2}$, the moment reaches the maximum. When $\theta = 0$, the moment is equal to zero. When the direction of **p** is parallel with the direction of **E**, it means that the dipole axis is consistent with the direction of the external field. That is to say, when the dipole is introduced into the electric field, no matter what the direction of its electric dipole moment is, it will eventually point to the direction of the electric field under the action of the electric field. Thus, the external electric field has an orientation effect on the dipole.

Example 9.2.1 As shown in Figure 9.2.5, charges are uniformly distributed on the thin rod, and the line density of charges is λ. Point P is located outside the rod, and the distance from point P to the rod is a. The angle between the lines connecting the two ends of the rod to point P and the rod is θ_1 and θ_2, respectively. Find the magnitude and direction of the electric field strength excited by the uniformly charged thin rod at point P.

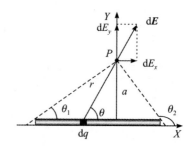

Figure 9.2.5 Figure of Example 9.2.1

Solution Set up the coordinate system as shown in the figure and take an arbitrary infinitesimal line element dx. Its electric charge is $dq = \lambda dx$. Suppose the distance between the element charge and point P is r, and the angle between r and X axis is θ. Thus, the direction of the elemental field strength $d\boldsymbol{E}$ at point P is at an angle θ with the X axis (as shown in the figure), and the magnitude of $d\boldsymbol{E}$ is $dE = \dfrac{1}{4\pi\varepsilon_0} \cdot \dfrac{\lambda dx}{r^2}$.

Since the direction of the field strength generated by each charge element on the rod at point P is different, $d\boldsymbol{E}$ should be decomposed in the selected coordinate system before integration:

$$dE_x = dE\cos\theta = \frac{\lambda dx}{4\pi\varepsilon_0 r^2}\cos\theta, \quad dE_y = dE\sin\theta = \frac{\lambda dx}{4\pi\varepsilon_0 r^2}\sin\theta$$

According to the geometric relationship in the figure, there is

$$\cot(\pi - \theta) = \frac{x}{a}$$

It can be rewritten as

$$x = a\cot(\pi - \theta) = -a\cot\theta$$

Differentiate both sides, and it can be obtained that

$$dx = -a \cdot \left(-\frac{1}{\sin^2\theta}\right) d\theta = \frac{a}{\sin^2\theta} d\theta$$

Since

$$r^2 = x^2 + a^2 = a^2(1 + \cot^2\theta) = \frac{a^2}{\sin^2\theta}$$

Substitute the above relationships into dE_x and dE_y, and there are

$$dE_x = \frac{\lambda dx}{4\pi\varepsilon_0 r^2}\cos\theta = \frac{\lambda \cos\theta}{4\pi\varepsilon_0 a} d\theta$$

$$dE_y = \frac{\lambda dx}{4\pi\varepsilon_0 r^2}\sin\theta = \frac{\lambda \sin\theta}{4\pi\varepsilon_0 a} d\theta$$

Calculate the definite integral from θ_1 and θ_2, and we obtain

$$E_x = \int dE_x = \int_{\theta_1}^{\theta_2} \frac{\lambda}{4\pi\varepsilon_0 a}\cos\theta d\theta = \frac{\lambda}{4\pi\varepsilon_0 a}(\sin\theta_2 - \sin\theta_1)$$

$$E_y = \int dE_y = \int_{\theta_1}^{\theta_2} \frac{\lambda}{4\pi\varepsilon_0 a}\sin\theta d\theta = \frac{\lambda}{4\pi\varepsilon_0 a}(\cos\theta_1 - \cos\theta_2)$$

Then

$$\boldsymbol{E} = E_x \boldsymbol{i} + E_y \boldsymbol{j} = \frac{\lambda}{4\pi\varepsilon_0 a}\left[(\sin\theta_2 - \sin\theta_1)\boldsymbol{i} + (\cos\theta_1 - \cos\theta_2)\boldsymbol{j}\right]$$

If $L \gg a$, the rod can be regarded as a charged infinitely long rod, and the relationships could be obtained: $\theta_1 \to 0$, $\theta_2 \to \pi$. Then the above expression can be rewritten as

$$\boldsymbol{E} = \frac{\lambda}{2\pi\varepsilon_0 a}\boldsymbol{j}$$

Obviously, the electric field excited by the infinitely long, uniformly charged thin rod has an **axisymmetric distribution characteristic**. The electric field around the rod is perpendicular to the rod and radiates around it. The magnitude of **E** is inversely proportional to the distance from the rod, and is the same for points that are equidistant from the rod. This result can be used directly in subsequent calculations.

Example 9.2.2 The radius of a thin ring is R, and it is uniformly charged with electric quantity of q. Find the electric field strength at point P, which is located on the central axis of the ring and a distance of x from the center point O.

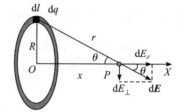

Figure 9.2.6　Figure of Example 9.2.2

Solution Establish a coordinate system as shown in Figure 9.2.6, and suppose that the distance between any point P on the axis and the coordinate origin O is x. Take an arbitrary line element dl on the ring, and its electric quantity is dq. Assume that the distance between the charge element dq and point P is r, the field strength generated by dq at point P is $d\mathbf{E}$, and the parallel and perpendicular components of $d\mathbf{E}$ are $dE_{/\!/}$ and $dE_{\perp}$, respectively. By using the symmetry principle, the vertical components cancel each other, so the electric field strength at point P is the sum of the parallel components.

$$E = \int dE_{/\!/} = \int \frac{dq}{4\pi\varepsilon_0 r^2} \cos\theta$$

where θ is the angle between the vector of $d\mathbf{E}$ and the X axis; then

$$E = \int \frac{dq}{4\pi\varepsilon_0 r^2} \cos\theta = \frac{\cos\theta}{4\pi\varepsilon_0 r^2} \int dq = \frac{q\cos\theta}{4\pi\varepsilon_0 r^2}$$

Since $\cos\theta = \dfrac{x}{r}$ and $r^2 = x^2 + R^2$, it can be rewritten as

$$E = \frac{\lambda x}{4\pi\varepsilon_0 (x^2+R^2)^{3/2}} \oint dl = \frac{\lambda x \cdot 2\pi R}{4\pi\varepsilon_0 (x^2+R^2)^{3/2}} = \frac{qx}{4\pi\varepsilon_0 (x^2+R^2)^{3/2}} = \frac{qx}{4\pi\varepsilon_0 (R^2+x^2)^{3/2}}$$

The direction of **E** is the same as the positive direction of the X axis.

Notes:

(1) If $x \gg R$, we have $(R^2+x^2)^{3/2} \approx x^3$ and then $E \approx \dfrac{q}{4\pi\varepsilon_0 x^2}$, indicating that the electric field far from the center of the ring is equivalent to the electric field generated by a point charge q.

(2) If $x \ll R$, we have $(R^2+x^2)^{3/2} \approx R^3$ and then $E \approx \dfrac{qx}{4\pi\varepsilon_0 R^3}$, indicating that the magnitude of the electric field on the axis near the center of the circle is proportional to x.

Example 9.2.3 Suppose the radius of a disk is R, charges are continuously and uniformly distributed on it, and the surface density of charges is σ. Find the field strength at any point P on the center axis of the disk and at a distance x from the center O.

Solution Establish a coordinate system as shown in Figure 9.2.7, and let the distance between any point P on the axis and the coordinate origin O be x. Take a thin ring of radius r and width dr on the disk as a surface element. The area of the ring is $2\pi r dr$, and the electric quantity is $dq = \sigma 2\pi r dr$. From Example 9.2.2, the field strength at point P produced by this thin ring charge is

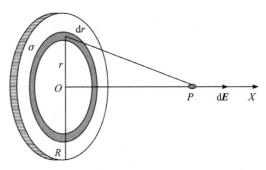

Figure 9.2.7 Figure of Example 9.2.3

$$dE = \frac{dq x}{4\pi\varepsilon_0 (r^2 + x^2)^{3/2}} = \frac{\sigma 2\pi r dr x}{4\pi\varepsilon_0 (r^2 + x^2)^{3/2}}$$

The direction of $d\boldsymbol{E}$ points to the positive X axis. Since the direction of the electric field $d\boldsymbol{E}$ of each ring that constitutes the circular surface is the same, the total field strength at point P is the integral of the field strength of each ring at point P, namely,

$$E = \int dE = \int_0^R \frac{\sigma 2\pi r dr x}{4\pi\varepsilon_0 (r^2 + x^2)^{3/2}} = \frac{\sigma}{2\varepsilon_0}\left[1 - \frac{x}{(R^2 + x^2)^{1/2}}\right]$$

Notes:

(1) If $x \ll R$, there is $\dfrac{x}{(R^2 + x^2)^{1/2}} = \dfrac{1}{\left(\dfrac{R^2}{x^2} + 1\right)^{1/2}} \approx 0$ and then $E \approx \dfrac{\sigma}{2\varepsilon_0}$. The charged disk can be regarded as an infinite charged plane whose electric field is uniform.

(2) If $x \gg R$, there is $\dfrac{x}{(R^2 + x^2)^{1/2}} = \dfrac{1}{\left(\dfrac{R^2}{x^2} + 1\right)^{1/2}} \approx 1 - \dfrac{R^2}{2x^2}$ and then $E \approx \dfrac{R^2 \sigma}{4\varepsilon_0 x^2} = \dfrac{\pi R^2 \sigma}{4\pi\varepsilon_0 x^2} = \dfrac{q}{4\pi\varepsilon_0 x^2}$.

This result shows that the electric field far from the charged circular surface is equivalent to the electric field of a point charge.

9.3 Gauss's law

This section introduces the electric field lines that describe the electric field by means of diagrams, and then leads to the concept of electric flux. From this, Gauss's law (general relationship between electric field and field source charge) is obtained.

9.3.1 Electric field line

1. Diagram of electric field lines

It is known that there is an electric field around a charged body. For a certain charge distribution, there is a certain electric field distribution in the corresponding space. The field strength E is a function of position, and the functional expression of E can be obtained by calculation. Although the functional expression of the field strength E accurately describes the field source distribution, it is not intuitional enough. In order to present an intuitional and comprehensive image of the field intensity distribution at each point in an electric field and the spatial distribution of the electric field, the electric field line is introduced.

Electric field lines are used for the spatial overview of electric field strength distribution. The following rules should be followed when drawing electric field lines: ① The tangent direction of any point on the curve is consistent with the direction of the field strength E at that point. For any point on the curve, the tangent can have two opposite directions. For correct selection, an arrow is generally marked on the curve, which points to the direction of the electric field line; ② At any point in the electric field, the number of electric field lines passing through a unit area perpendicular to the direction of the field strength is equal to the magnitude of the field strength at that point, i.e., $E = \dfrac{\Delta N}{\Delta S}$ called density of electric field lines, where ΔS is the surface element perpendicular to the direction of the field strength, and ΔN is the number of electric field lines passing through ΔS.

Figure 9.3.1 shows the electric field lines of several common electric fields.

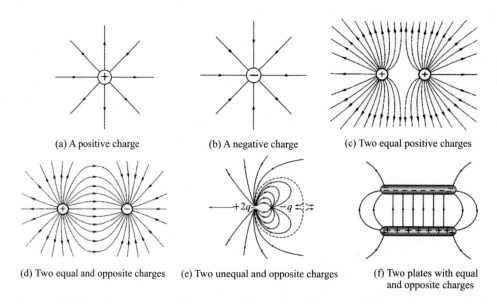

(a) A positive charge (b) A negative charge (c) Two equal positive charges

(d) Two equal and opposite charges (e) Two unequal and opposite charges (f) Two plates with equal and opposite charges

Figure 9.3.1 Electric field lines of several common electric fields

The electric field line pattern can be demonstrated experimentally. The method is to sprinkle acicular single crystal quinine or gypsum powder on a glass plate or float them on insulating oil, and place the plate or oil in an electric field. Then arrangement along the electric field lines will appear, as shown in Figure 9.3.2.

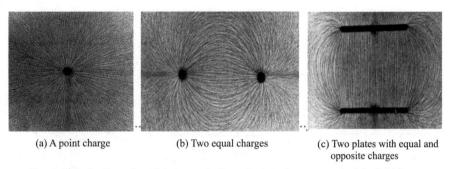

(a) A point charge (b) Two equal charges (c) Two plates with equal and opposite charges

Figure 9.3.2 Experimental demonstration of several common electric field lines

2. Properties of electric field lines

From the above electric field line diagrams, some basic properties of electric field lines can be summarized:

(1) The electric field line always starts from a positive charge or infinity, and ends at a negative charge or infinity;

(2) The electric field line is a curve that never closes, indicating that the electrostatic field is irrotational;

(3) Electric field lines in the same electric field do not cross each other.

The electric field lines are artificially drawn for the purpose of visually and intuitively depicting the field strength of the electric field space, and are not real. The most common use of electric field lines is to give an overview of electric field distribution in space.

9.3.2 Electric flux

The number of electric field lines passing through a certain surface in an electric field is called the electric field strength flux, or electric flux, which is represented by the symbol Φ_e.

In a uniform electric field E, as shown in Figure 9.3.3(a), the electric flux passing through the plane S perpendicular to the E direction is

$$\Phi_e = ES$$

If the angle between the normal n of the plane S and E is θ, as shown in Figure 9.3.3(b), the plane S is projected in the direction perpendicular to the field strength, and the electric flux passing through the plane S is

$$\Phi_e = ES\cos\theta = \boldsymbol{E} \cdot \boldsymbol{S}$$

When calculating the electric flux passing through any surface **S** in a non-uniform electric field, divide the surface into an infinite number of surface elements d**S**, as shown in Figure 9.3.3(c). The electric flux passing through the surface element d**S** is $d\Phi_e = \mathbf{E} \cdot d\mathbf{S}$. Then the electric flux through the surface **S** is

$$\Phi_e = \int_S \mathbf{E} \cdot d\mathbf{S}$$

If the surface **S** is a closed surface, the above formula can be written as

$$\Phi_e = \oint_S \mathbf{E} \cdot d\mathbf{S} \tag{9.3.1}$$

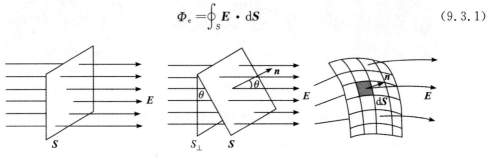

Figure 9.3.3 Electric flux

For a closed surface, it is defined that its normal direction points to the outside of the surface. Therefore, when the electric field lines cross from the inside, the electric flux is positive. When the electric field lines cross from the outside, the electric flux is negative. The electric flux Φ_e of the entire closed surface is equal to the difference between the number of electric field lines leaving and entering the closed surface, which is the total number of net electric field lines that pass through the closed surface.

9.3.3 Gauss's law

Gauss, a German mathematician and physicist, studied electric flux and deduced an important theorem satisfied by the electrostatic field, namely Gauss's law. Gauss's law is a basic principle of the electrostatic field, which gives the relationship between the electric flux passing through any closed surface and the electric charge enclosed inside the closed surface. It profoundly reflects the internal relationship between the electric field and the field source.

Gauss's law states that the net electric flux Φ_e through any closed (Gaussian) surface **S** is equal to the net charge inside the surface $\sum q_{int}$ divided by ε_0, as the following formula shows:

$$\Phi_e = \oint_S \mathbf{E} \cdot d\mathbf{S} = \frac{1}{\varepsilon_0} \sum q_{int} \tag{9.3.2}$$

Gauss's law of electrostatic field can be derived from Coulomb's law and the principle of superposition of electric fields. In the following we will prove Gauss's law of electrostatic field step by step, from special to general.

1. Electric flux through a spherical surface centered on a point charge

In the electric field of a point charge q, a spherical surface with radius r is drawn with q as the center, as shown in Figure 9.3.4. The electric flux passing through this spherical surface is $\Phi_e = \oint_S \boldsymbol{E} \cdot \mathrm{d}\boldsymbol{S}$. Since the direction of electric field anywhere on the sphere is consistent with the normal direction of the surface element,

$$\boldsymbol{E} \cdot \mathrm{d}\boldsymbol{S} = E\,\mathrm{d}S\cos 0 = E\,\mathrm{d}S$$

And the magnitude of $\boldsymbol{E}$ is equal everywhere on the same spherical surface. Thus there is

$$\Phi_e = \oint_S \boldsymbol{E} \cdot \mathrm{d}\boldsymbol{S} = \oint_S \frac{q}{4\pi\varepsilon_0 r^2}\mathrm{d}S = \frac{q}{4\pi\varepsilon_0 r^2}\oint_S \mathrm{d}S = \frac{q}{\varepsilon_0}$$

The result shows that the electric flux passing through the closed spherical surface is proportional to the electric quantity of the charges enclosed by the spherical surface and has nothing to do with the radius of the sphere.

2. Electric flux through an arbitrary closed surface surrounding a point charge

For an arbitrary surface S' surrounding a point charge q, as shown in Figure 9.3.4, a spherical surface S with q as the center could be drawn outside the surface S'. Since the electric field lines emitted from q will not be interrupted, the number of electric field lines passing through the S' surface is equal to the number of electric field lines passing through the S surface. The electric flux through any closed surface is the same. Then, there is

$$\Phi_e = \oint_{S'} \boldsymbol{E} \cdot \mathrm{d}\boldsymbol{S} = \frac{q}{\varepsilon_0}$$

In conclusion, no matter what the shape or size of a closed surface is, as long as it covers a point charge, the electric flux passing through the closed surface is equal to $\frac{q}{\varepsilon_0}$.

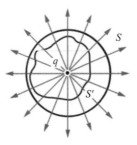

Figure 9.3.4 Electric flux through an arbitrary closed surface surrounding a point charge

3. Electric flux through any closed surface that does not enclose a point charge

A point charge q is outside a closed surface, as shown in Figure 9.3.5. From the continuity of the electric field lines, it can be known that each electric field line penetrates from one place and must go out from another place on the surface. The numbers of incoming and outgoing electric field lines are the same. In other words, the electric flux through the closed surface S is zero. Since the point charge is outside the closed surface, the point charge inside the closed surface is zero. Thus, the electric flux is.

$$\Phi_e = \oint_{S'} \boldsymbol{E} \cdot \mathrm{d}\boldsymbol{S} = \frac{q}{\varepsilon_0} = 0$$

where q represents the electric charge inside the closed surface and equals zero here.

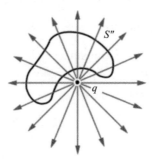

Figure 9.3.5 Electric flux through any closed surface that does not enclose a point charge

4. Electric flux through a closed surface in an arbitrary electrostatic field

First, let's consider the electric flux through an arbitrary closed surface in the electrostatic field of a point charge system. As shown in Figure 9.3.6, there is a point charge system composed of m point charges in space, and an arbitrary closed surface contains n of them: $q_1, q_2, \cdots, q_n$ ($n < m$), which means that there are still some point charges outside the closed surface. According to the principle of superposition of field strengths, the field strength at any point in space is equal to the vector sum of the excited field strengths when each point charge exists alone. The total field strength at any surface element on the surface is the vector sum of the field strengths excited by all point charges when existing alone at that location.

$$\boldsymbol{E} = (\boldsymbol{E}_1 + \boldsymbol{E}_2 + \cdots + \boldsymbol{E}_n) + (\boldsymbol{E}_{n+1} + \boldsymbol{E}_{n+2} + \cdots + \boldsymbol{E}_m)$$

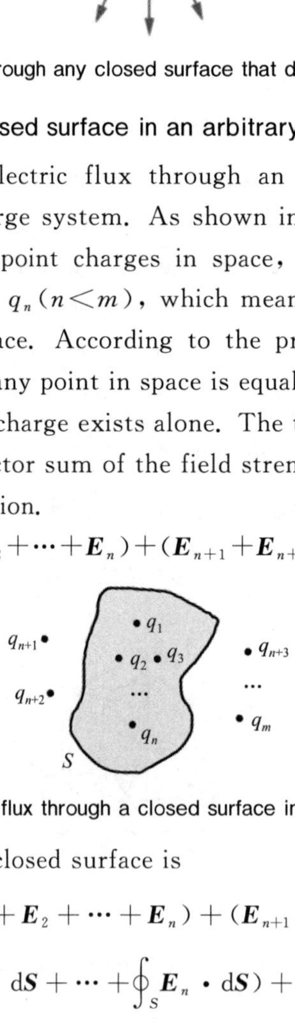

Figure 9.3.6 Electric flux through a closed surface in a general electric field

The total flux through the closed surface is

$$\Phi_e = \oint_S \boldsymbol{E} \cdot \mathrm{d}\boldsymbol{S} = \oint_S [(\boldsymbol{E}_1 + \boldsymbol{E}_2 + \cdots + \boldsymbol{E}_n) + (\boldsymbol{E}_{n+1} + \boldsymbol{E}_{n+2} + \cdots + \boldsymbol{E}_m)] \cdot \mathrm{d}\boldsymbol{S}$$

$$= (\oint_S \boldsymbol{E}_1 \cdot \mathrm{d}\boldsymbol{S} + \oint_S \boldsymbol{E}_2 \cdot \mathrm{d}\boldsymbol{S} + \cdots + \oint_S \boldsymbol{E}_n \cdot \mathrm{d}\boldsymbol{S}) +$$

$$\left(\oint_S \boldsymbol{E}_{n+1} \cdot \mathrm{d}\boldsymbol{S} + \oint_S \boldsymbol{E}_{n+2} \cdot \mathrm{d}\boldsymbol{S} + \cdots + \oint_S \boldsymbol{E}_m \cdot \mathrm{d}\boldsymbol{S} \right)$$

$$= \left(\frac{q_1}{\varepsilon_0} + \frac{q_2}{\varepsilon_0} + \cdots + \frac{q_n}{\varepsilon_0} \right) + (0 + 0 + \cdots + 0)$$

$$= \frac{1}{\varepsilon_0} \sum_{i=1}^{n} q_i$$

$$\Phi_e = \frac{1}{\varepsilon_0} \sum_{i=1}^{n} q_i \qquad (9.3.3)$$

Since any charged body can be regarded as a system composed of many point charges, Equation (9.3.3) can be extended to the electric field in the space generated by any charged body.

Notes:
(1) Gauss's law holds for any closed surface S.

(2) $\boldsymbol{E}$ of dS on the Gaussian surface is the total field strength generated by all charges inside and outside the Gaussian surface, and q is just an algebraic sum of the charges inside the Gaussian surface.

(3) The direction of the outer normal of dS on the Gaussian surface is defined as the positive direction of the surface element.

(4) If the electric flux passing through the Gaussian surface is zero ($\Phi_e = 0$), it doesn't mean that the electric field strength on the Gaussian surface is zero ($\boldsymbol{E} = 0$).

(5) If the Gaussian surface is surrounded by positive charges, there is $\Phi_e > 0$. There must be electric field lines starting from the positive charges and going out the Gaussian surface, indicating that the positive charge is the source of the electric field. On the contrary, if the Gaussian surface is surrounded by negative charges, there is $\Phi_e < 0$. There must be electric field lines entering the Gaussian surface and ending at the negative charges, indicating that the negative charge is the end point of the electric field. Thus, the electric field lines always start from positive charges and end at negative charges.

9.3.4 Application of Gauss's law

Gauss's law describes the relationship between the electric flux through a closed surface and the charge in the closed surface, and has general applicability. In practice, it usually applies to the following two situations:

(1) If the distribution of the electric field intensity $\boldsymbol{E}$ on the Gaussian surface is known, the electric flux through the Gaussian surface can be calculated. Then the charges enclosed in the Gaussian surface $\sum q_{\text{int}}$ can be obtained according to Gauss's law.

(2) If the charge distribution is known, the electric field strength on the Gaussian surface can be obtained by solving the surface equation. This method will encounter mathematical difficulties for any distribution of electric quantity, and other additional conditions must be given. However, when the charge distribution has a certain symmetry, the distribution of the electric field on the Gaussian surface also has a certain symmetry. It is possible to use Gauss's law to solve the electric field distribution easily.

The following example illustrates the method of applying Gauss's law to solve the electric field strength.

Example 9.3.1 Find the distribution of field strength inside and outside a uniformly charged spherical surface. The electric quantity of the sphere is known to be q and the radius of the sphere is R.

Solution Since the charge distribution has spherical symmetry, the generated field strength also has spherical symmetry. So the field strength at each point on the sphere and at the same distance from the center of the sphere O is equal in magnitude, and the direction is radial along the radius.

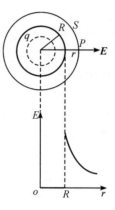

As shown in Figure 9.3.7, a concentric spherical Gaussian surface S with a radius r is constructed through any point P, the direction of the field strength at this point is consistent with the normal direction of the surface, and the electric flux passing through the Gaussian surface is

Figure 9.3.7 Figure of Example 9.3.1

$$\Phi_e = \oint_S \boldsymbol{E} \cdot \mathrm{d}\boldsymbol{S} = \oint_S E \mathrm{d}S = E \oint_S \mathrm{d}S = E 4\pi r^2$$

If point P is outside the charged spherical surface ($r > R$), the electric quantity surrounded by the Gaussian surface is q and according to Gauss's law, there is

$$\Phi_e = E 4\pi r^2 = \frac{q}{\varepsilon_0}$$

Then

$$E = \frac{q}{4\pi\varepsilon_0 r^2}$$

If point P is inside the charged spherical surface ($r < R$), the charge surrounded by the Gaussian surface is zero. According to Gauss's law,

$$\Phi_e = E 4\pi r^2 = \frac{q}{\varepsilon_0} = 0$$

Then

$$E = 0$$

Thus, the field strength outside the uniformly charged spherical surface is the same as the electric field generated by a point charge where the charges on the spherical surface are all concentrated at the center, but the field strength inside the sphere is zero, as shown in Figure 9.3.7.

Example 9.3.2 Find the electric field strength distribution of a uniformly charged sphere. The radius of the sphere is known to be R, and the electric quantity is q.

Solution Assuming that the charge density of the uniformly charged sphere is ρ. Since the sphere is uniformly charged, it can be divided into layers of uniformly charged spherical surfaces and the resulting electric field strength distribution has spherical symmetry.

For any point outside the sphere ($r \geqslant R$), the electric field strength generated by the charged sphere is the same as the electric field strength generated by the point charge formed by all the charges concentrated at the center of the sphere.

$$E = \frac{1}{4\pi\varepsilon_0} \cdot \frac{q}{r^2} \boldsymbol{r}_0$$

For any point P in the sphere ($r < R$), in order to calculate the electric field strength, a concentric spherical surface S with a radius r is drawn as a Gaussian surface through point P, as shown in Figure 9.3.8. The electric flux passing through the Gaussian surface S is

$$\Phi_e = \oint_S \boldsymbol{E} \cdot d\boldsymbol{S} = \oint_S E\,dS = E\oint_S dS = E 4\pi r^2$$

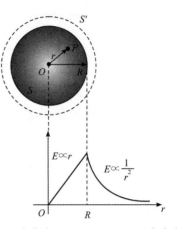

Figure 9.3.8 Figure of Example 9.3.2

The electric quantity enclosed in the Gaussian surface is

$$q' = \frac{4}{3}\pi r^3 \rho$$

According to Gauss's law, there is

$$E \cdot 4\pi r^2 = \frac{1}{\varepsilon_0} \frac{4}{3}\pi r^3 \rho$$

Then

$$E = \frac{\rho}{3\varepsilon_0} r$$

Example 9.3.3 Find the electric field distribution of an infinitely long, uniformly charged thin rod. The line charge density of the rod is known to be λ.

Solution Due to the axial symmetry charge distribution of the infinitely long charged straight line, the resulting electric field distribution is axisymmetric. The field strengths at all points equidistant from the straight line are equal in magnitude, and their directions are perpendicular to the axis outward.

Draw a coaxial closed cylindrical Gaussian surface S with a base radius r and a height l through any point P, as shown in Figure 9.3.9. The upper bottom surface, the lower bottom surface and the side surface are marked as S_1, S_2, and S_0, respectively.

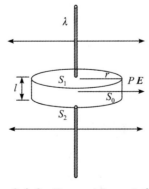

Figure 9.3.9 Figure of Example 9.3.3

The electric flux through the Gaussian surface is

$$\Phi_e = \oint_S \boldsymbol{E} \cdot d\boldsymbol{S} = \int_{S_1} \boldsymbol{E} \cdot d\boldsymbol{S} + \int_{S_2} \boldsymbol{E} \cdot d\boldsymbol{S} + \int_{S_0} \boldsymbol{E} \cdot d\boldsymbol{S}$$

Since the normal directions of the upper bottom surface and the lower bottom surface are perpendicular to the direction of the field strength, the electric flux passing through the upper and lower bottom surfaces is zero. Thus

$$\Phi_e = E \int_{S_0} dS = E 2\pi r l$$

The charge enclosed in the Gaussian surface is $q = \lambda l$. According to Gauss's law

$$\Phi_e = E 2\pi r l = \frac{\lambda l}{\varepsilon_0}$$

Then

$$E = \frac{\lambda}{2\pi\varepsilon_0 r}$$

For infinitely long, uniformly charged thin rods, it is much easier to solve the field strength by using Gauss's law than the direct integration method. This method can be extended to solving similar problems, such as the electric field distribution of a uniformly charged infinitely, long cylinder plane and the electric field of a uniformly charged, infinitely long cylinder.

Example 9.3.4 Find the electric field distribution in an infinite uniformly charged plane. The surface charge density is known to be $\sigma (\sigma > 0)$.

Solution Due to plane symmetry of charge distribution, the field strengths at all points equidistant from the plane are equal, and their directions are perpendicular to the plane. Draw a closed cylindrical Gaussian surface perpendicular to the plane and symmetrical about the plane through any point P outside the plane. The bottom area of the cylinder is ΔS and the upper bottom surface, the lower bottom surface and the side surface are marked as S_1, S_2, and S_3, respectively, as Figure 9.3.10 shows.

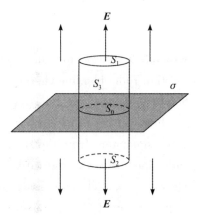

Figure 9.3.10 Figure 1 of Example 9.3.4

The electric flux through the Gaussian surface is

$$\Phi_e = \oint_S \boldsymbol{E} \cdot d\boldsymbol{S} = \int_{S_1} \boldsymbol{E} \cdot d\boldsymbol{S} + \int_{S_2} \boldsymbol{E} \cdot d\boldsymbol{S} + \int_{S_3} \boldsymbol{E} \cdot d\boldsymbol{S}$$

Since the normal direction of the side surface is perpendicular to the direction of the field strength, the electric flux passing through the side surface is zero. Thus,

$$\Phi_e = \int_{S_1} \boldsymbol{E} \cdot d\boldsymbol{S} + \int_{S_2} \boldsymbol{E} \cdot d\boldsymbol{S} = E S_1 + E S_2 = 2 E \Delta S$$

The electric quantity enclosed in the Gaussian surface is $q = \sigma \Delta S$. According to Gauss's law,

$$\Phi_e = 2E\Delta S = \frac{\sigma \Delta S}{\varepsilon_0}$$

Then
$$E = \frac{\sigma}{2\varepsilon_0}$$

Based on this result as well as the superposition principle for electric field, the electric field strength produced by two infinite uniformly charged plates can be directly calculated.

As shown in Figure 9.3.11, if the right direction is selected as the positive direction of the field strength, the total electric field strength is

(a) $\begin{cases} E_{\mathrm{I}} = -\dfrac{\sigma}{2\varepsilon_0} - \dfrac{\sigma}{2\varepsilon_0} = -\dfrac{\sigma}{\varepsilon_0} \\ E_{\mathrm{II}} = 0 \\ E_{\mathrm{III}} = \dfrac{\sigma}{2\varepsilon_0} + \dfrac{\sigma}{2\varepsilon_0} = \dfrac{\sigma}{\varepsilon_0} \end{cases}$
(b) $\begin{cases} E_{\mathrm{I}} = 0 \\ E_{\mathrm{II}} = \dfrac{\sigma}{2\varepsilon_0} + \dfrac{\sigma}{2\varepsilon_0} = \dfrac{\sigma}{\varepsilon_0} \\ E_{\mathrm{III}} = 0 \end{cases}$

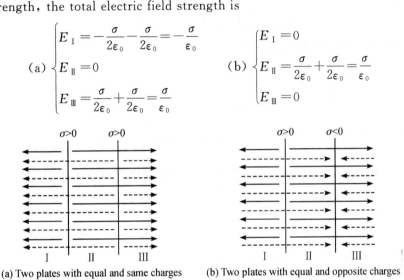

(a) Two plates with equal and same charges (b) Two plates with equal and opposite charges

Figure 9.3.11　Figure of Example 9.3.4

The above example shows that when the charged body's electric quantity has a certain symmetrical distribution, the electric field strength can be obtained using Gauss's law. The general steps are: ① Analyze the symmetry of the electric quantity distribution and the electric field; ② Draw a suitable Gaussian surface, on which the electric field is symmetric and whose magnitude is a constant; ③ Calculate the algebraic sum of electric quantity enclosed within the Gaussian surface and the electric flux through the Gaussian surface; ④ Calculate the electric filed strength using Gauss's law.

9.4　Electric potential energy and electric potential

As is known, when an electric charge moves in an electrostatic field, the electrostatic force does work on it. In this section, the work done by electrostatic forces will be discussed, and then the circuital theorem and electric potential of the electrostatic field will be deduced according to the characters of work.

9.4.1 Work done by electrostatic force

1. Work done by electrostatic force generated by a point charge

In the electric field generated by a point charge q, a test charge q_0 is moved from point a to point b through an arbitrary path, as shown in Figure 9.4.1. Take an infinitesimal displacement dl at any point on the route. It is known that the direction of the electric field is in the radial direction. In this displacement, the element work done by the electrostatic force is

$$dW = \mathbf{F} \cdot d\mathbf{l} = q_0 \mathbf{E} \cdot d\mathbf{l} = q_0 E \cos\theta \, dl$$

where θ is the angle between $\mathbf{E}$ and $d\mathbf{l}$.

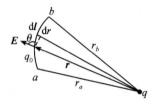

Figure 9.4.1 Work done by electrostatic force generated by a point charge

According to Coulomb's law

$$E = \frac{1}{4\pi\varepsilon_0} \cdot \frac{q}{r^2}$$

Substituting it in the above equation, we get

$$dW = \frac{q_0 q}{4\pi\varepsilon_0 r^2} \cos\theta \, dl$$

$dl \cos\theta = dr$, as shown in Figure 9.4.1. Substituting it in the above equation, we get

$$dW = \frac{q_0 q}{4\pi\varepsilon_0 r^2} dr$$

When the test charge is moved from a to b, the total work done by the electrostatic force should be the sum of all the elemental work.

$$W_{ab} = W = \int_a^b q_0 E \cos\theta \, dl = \int_{r_a}^{r_b} \frac{q_0 q}{4\pi\varepsilon_0 r^2} dr = \frac{q_0 q}{4\pi\varepsilon_0} \int_{r_a}^{r_b} \frac{1}{r^2} dr = \frac{q_0 q}{4\pi\varepsilon_0} \left(\frac{1}{r_a} - \frac{1}{r_b} \right) \quad (9.4.1)$$

where r_a and r_b represent the distance from the starting point and the end point of the path to the point charge, i.e., initial position and final position, respectively. This result indicates that in the electric field generated by a point charge q, when the initial and final positions are determined, no matter which path the test charge q_0 moves along, the electrostatic force does the same work on it. In short, the work done by the electrostatic force generated by a point charge depends only on the initial and final positions of the test charge and the electric quantities q and q_0, not on the path.

2. Work done by electrostatic force generated by a charged body

Any charged body can be divided into many charge elements, and each charge element can be regarded as a point charge, so the charged body can be considered as a point charge system. In the electric field generated by a point charge system of q_1, q_2, $\cdots$, q_n, if the test charge q_0 is moved from point a to point b, the work done by the electrostatic force is

$$W_{ab} = \int_a^b \boldsymbol{F} \cdot \mathrm{d}\boldsymbol{l} = \int_a^b q_0 \boldsymbol{E} \cdot \mathrm{d}\boldsymbol{l}$$

According to the principle of superposition of electric field strengths, the electric field strength $\boldsymbol{E}$ in the formula should be the vector sum of the electric filed strength generated by each point charge alone, namely,

$$\boldsymbol{E} = \boldsymbol{E}_1 + \boldsymbol{E}_2 + \cdots + \boldsymbol{E}_n$$

Thus,

$$W_{ab} = q_0 \int_a^b \boldsymbol{E}_1 \cdot \mathrm{d}\boldsymbol{l} + q_0 \int_a^b \boldsymbol{E}_2 \cdot \mathrm{d}\boldsymbol{l} + \cdots + q_0 \int_a^b \boldsymbol{E}_n \cdot \mathrm{d}\boldsymbol{l}$$

Each term on the right side of the equation represents the work done on q_0 by the electrostatic force generated by each point charge alone. Since they are all independent of the path, the total work W_{ab} done by the electrostatic force doesn't depend on the path. From this, it can be concluded that when the test charge moves in any given electrostatic field, the work done by the electrostatic force depends only on the electric quantity of the test charge and its initial and final positions, and not on the path. It indicates that the electrostatic force is a conservative force, and the corresponding electrostatic field is called a conservative field.

 9.4.2 Circulation of an electrostatic field

The electrostatic force is a conservative force; its work only depends on the initial and final positions, and doesn't depend on the path. Therefore, when q_0 moves along any closed path in the electrostatic field, the work done by the electrostatic force on it is equal to zero.

$$\oint q_0 \boldsymbol{E} \cdot \mathrm{d}\boldsymbol{l} = 0$$

Since $q_0 \neq 0$, it can be simplified to

$$\oint \boldsymbol{E} \cdot \mathrm{d}\boldsymbol{l} = 0 \tag{9.4.2}$$

This formula shows that in the electrostatic field, the integral of the electric field strength $\boldsymbol{E}$ along any closed path is zero, or the circulation of $\boldsymbol{E}$ is zero, which is called **the circuital theorem of the electrostatic field**. This theorem combined with Gauss's law, suggests that the electrostatic field is active, conservative and irrotational.

 9.4.3 Electric potential

The electrostatic field is a conservative field, for which the corresponding potential energy can be introduced. In this section, the electric potential energy, electric potential, and electric potential difference will be discussed.

1. Electric potential energy

Since the electrostatic field is a conservative force field, the corresponding potential energy can be introduced, which is similar to the universal gravitational field. In other words, the test charge q_0 has a certain potential energy at a certain position in the electrostatic field, which is called **electric potential energy**, and is represented by E_p.

When the test charge moves in the electrostatic field, the electrostatic force will work on it. And the potential energy of the test charge will change accordingly. According to the characteristic of conservative force doing work, the value of work done by the electrostatic force is equal to the negative value of the potential energy increment. Assuming that a and b are the initial and final positions of the movement, there is

$$-(E_{pb} - E_{pa}) = \int_a^b q_0 \boldsymbol{E} \cdot \mathrm{d}\boldsymbol{l} \tag{9.4.3}$$

where E_{pa} is the initial potential energy, and E_{pb} is the final potential energy. The unit of potential energy is the Joule (J).

It can be seen from Equation (9.4.3) that when the electrostatic force does positive work, $E_{pa} > E_{pb}$ and the potential energy decreases. When the electrostatic force does negative work, $E_{pa} < E_{pb}$ and the potential energy increases.

Electric potential energy is similar to gravitational potential energy and is a relative quantity. In order to determine the magnitude of the potential energy of a charge at a certain point in an electric field, a reference point with zero potential energy must be selected. The choice of this reference point can be arbitrary, mainly depending on the convenience of research. Usually, the zero potential energy reference point is selected at an "infinite distance" from the field source charge.

If the zero potential energy point is selected at the "infinite distance" from the field source charge $E_{p\infty} = 0$, then the potential energy of the test charge q_0 at point a in the electric field is

$$E_{pa} = \int_a^\infty q_0 \boldsymbol{E} \cdot \mathrm{d}\boldsymbol{l} \tag{9.4.4}$$

This expression shows that the potential energy of a charge at point a in the electric field is numerically equal to the work done by the electrostatic force in the process of moving the charge from point a to infinity.

It should be noted that the magnitude of the electric potential energy is relative and depends on the selection of the zero potential energy point. As is known, the gravitational potential energy is the energy belonging to the system composed of two objects, and the electric potential energy is the same, which belongs to the system of the electric field and charge placed in the electric field. The potential energy depends not only on the properties of the electric field itself, but also on the moved charge q_0, so it cannot be used as a physical quantity to describe the properties of the electric field itself.

2. Electric potential

It can be seen from Equation (9.4.4) that the potential energy of the charge q_0 at point a depends on the electric quantity of q_0, while the ratio $\dfrac{E_{pa}}{q_0}$ does not. It depends only on the spatial distribution of the electric field $\boldsymbol{E}$ and the corresponding position in the electric field. Therefore, this ratio $\dfrac{E_{pa}}{q_0}$ is a physical quantity that characterizes the properties of the electric field. This physical quantity is called electric potential, and U_a is usually used to represent the electric potential at point a. Then, there is

$$U_a = \frac{E_{pa}}{q_0} = \int_a^\infty \boldsymbol{E} \cdot \mathrm{d}\boldsymbol{l} \tag{9.4.5}$$

The integral path in the equation can be taken arbitrarily, as long as the initial position is at point a and the final position is at infinity. Therefore, the potential at a certain position in the electric field is numerically equal to the potential energy of a unit positive charge placed at that position, or numerically equal to the work done by the electrostatic force in moving a unit positive charge from this point to the zero potential point through any path.

Electric potential is a scalar quantity, and in the International System of Units, the unit of electric potential is the volt, represented by the symbol V.

Electric potential is a physical quantity that describes the properties of an electric field from the perspective that the electric field can do work or the electric field has energy, and it is generally a scalar function of spatial position. It should be noted that potential is also a relative quantity like potential energy, which depends on the selection of the zero potential point. It is easy to find that the selection of the zero potential point is consistent with that of the corresponding potential energy. Usually, the zero potential point is selected at an infinite distance from the field source charge. Then, the potential can be calculated. It should be noted that if the field source charge is distributed in an infinite range, the zero potential point cannot be selected at an infinite distance. Therefore, in general, the potential at a certain point U_p should be defined as

$$U_p = \int_p^{``0"} \boldsymbol{E} \cdot \mathrm{d}\boldsymbol{l} \tag{9.4.6}$$

The potential at any point p in the electric field is equal to the work done by the electrostatic force to move the unit positive charge from this point to the zero potential point via any path.

3. Potential difference

In an electrostatic field, the difference of potential between any two points a and b is called the potential difference between a and b, which is also known as voltage in a circuit, and is represented by the symbol ΔU_{ab}.

$$\Delta U_{ab} = U_a - U_b \tag{9.4.7}$$

According to Equation (9.4.5), we get

$$U_a - U_b = \int_a^\infty \boldsymbol{E} \cdot \mathrm{d}\boldsymbol{l} - \int_b^\infty \boldsymbol{E} \cdot \mathrm{d}\boldsymbol{l} = \int_a^\infty \boldsymbol{E} \cdot \mathrm{d}\boldsymbol{l} + \int_\infty^b \boldsymbol{E} \cdot \mathrm{d}\boldsymbol{l} = \int_a^b \boldsymbol{E} \cdot \mathrm{d}\boldsymbol{l}$$

Then

$$U_{ab} = \int_a^b \boldsymbol{E} \cdot \mathrm{d}\boldsymbol{l} \tag{9.4.8}$$

The potential at any position in the electric field is related to the selection of the reference point, while the potential difference between any two points does not, which is more important for studying the properties of the electric field.

If the potential difference between any two points a and b in the electric field is known, it is easy to calculate the work done by the electrostatic force to move any charge q from point a to point b:

$$W_{ab} = q\int_a^b \boldsymbol{E} \cdot \mathrm{d}\boldsymbol{l} = q(U_a - U_b) \tag{9.4.9}$$

Using this equation, potentials at any two points a and b in an electric field can be compared. If the electric field does positive work to move a unit positive charge from any point a to another point b, that is $\int_a^b \boldsymbol{E} \cdot \mathrm{d}\boldsymbol{l} > 0$, then $U_a > U_b$. In other words, the electric potential at point a is larger than that at point b. On the contrary, if the electric field does negative work, that is $\int_a^b \boldsymbol{E} \cdot \mathrm{d}\boldsymbol{l} < 0$, then $U_a < U_b$, which means that the potential at point a is smaller than that at point b.

9.4.4 Calculation of electric potential

The calculation of electric potential in the electrostatic field can be divided into two types. One is to determine electric potential according to the known field source charge distribution; the other is to determine electric potential according to field strength distribution.

1. Find the electric potential according to the known field source charge distribution

(1) A point charge q

In the electric field of the point charge q, if a position at infinity is taken as the reference point of zero electric potential and a straight line parallel to the electric field line is selected as the integration path, according to the definition of electric potential Equation (9.4.5), the potential of point p at a distance r from the point charge q is

$$U_p = \int_r^\infty \boldsymbol{E} \cdot \mathrm{d}\boldsymbol{l} = \int_r^\infty \frac{q}{4\pi\varepsilon_0} \cdot \frac{1}{r^2} \cos 0 \mathrm{d}l = \frac{q}{4\pi\varepsilon_0 r}$$

So

$$U_p = \frac{q}{4\pi\varepsilon_0 r} \tag{9.4.10}$$

This is the expression for the potential at any point in the electric field produced by a point charge q.

It can be seen from Equation (9.4.10) that if q is a positive charge, the potential is also positive. The farther away from the point charge q, the lower the potential, and it approaches zero at infinity. On the contrary, if q is a negative charge, the potential is also negative. The farther away from the point charge q, the higher the potential, and it approaches zero at infinity.

(2) A point charge system

Suppose there are n point charges q_1, q_2, $\cdots$, q_n and the electric fields generated by them are $\boldsymbol{E}_1$, $\boldsymbol{E}_2$, $\cdots$, $\boldsymbol{E}_n$, respectively, then the net electric field is $\boldsymbol{E}=\boldsymbol{E}_1+\boldsymbol{E}_2+\cdots+\boldsymbol{E}_n$. According to the definition of electric potential Equation (9.4.5), the electric potential at a certain position p in the electric field is

$$\begin{aligned} U_p &= \int_p^\infty \boldsymbol{E} \cdot \mathrm{d}\boldsymbol{l} = \int_p^\infty (\boldsymbol{E}_1 + \boldsymbol{E}_2 + \boldsymbol{E}_3 + \cdots + \boldsymbol{E}_n) \cdot \mathrm{d}\boldsymbol{l} \\ &= \int_p^\infty \boldsymbol{E}_1 \cdot \mathrm{d}\boldsymbol{l} + \int_p^\infty \boldsymbol{E}_2 \cdot \mathrm{d}\boldsymbol{l} + \cdots + \int_p^\infty \boldsymbol{E}_n \cdot \mathrm{d}\boldsymbol{l} \\ &= U_1 + U_2 + U_3 + \cdots + U_n = \sum_{i=1}^n U_i = \sum_{i=1}^n \frac{q_i}{4\pi\varepsilon_0 r_i} \end{aligned} \qquad (9.4.11)$$

where U_i is the potential generated by the ith point charge q_i at point p, and r_i is the distance from q_i to point p. The above equation shows that in the electric field of the point charge system, the potential at any point is equal to the algebraic sum of the potentials generated at that point when each point charge exists alone. This conclusion is called **the principle of superposition of electric potentials**.

(3) A continuous charged distribution

For a continuously distributed charged body, it can be divided into many infinitesimal charged volume elements. Any charge element $\mathrm{d}q$ can be regarded as a point charge, and the corresponding element potential generated at the field point p is

$$\mathrm{d}U = \frac{\mathrm{d}q}{4\pi\varepsilon_0 r}$$

So the total potential at this point generated by the whole charged body is

$$U = \int \mathrm{d}U = \int \frac{\mathrm{d}q}{4\pi\varepsilon_0 r} \qquad (9.4.12)$$

According to the different distribution of charges on the charged body, $\mathrm{d}q$ can be replaced by $\rho \mathrm{d}V$, $\sigma \mathrm{d}S$ and $\lambda \mathrm{d}l$. This is the integration method to calculate the electric potential. Since the electric potential is a scalar, its integration is much easier than the direct integration of electric field $\boldsymbol{E}$.

Example 9.4.1 Find the electric potential at any point on the axis passing through the center of the uniformly charged fine ring with radius R and electric quantity q.

Solution Establish a coordinate system as shown in Figure 9.4.2, and let the distance between any point P on the axis and the coordinate origin O be x. Take any line element $\mathrm{d}l$ on the ring, and the electric quantity of it is $\mathrm{d}q$.

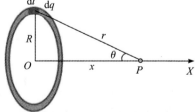

Figure 9.4.2 Figure of Example 9.4.1

Assuming that the distance from the element charge dq to point P is r and the potential generated by the element charge dq at point P is dU, it can be known from Equation (9.4.12) that the potential at the point P is

$$U_P = \int dU = \int \frac{dq}{4\pi\varepsilon_0 r} = \frac{q}{4\pi\varepsilon_0 (R^2+x^2)^{1/2}}$$

When $x=0$, the electric potential at the ring center O is

$$U_o = \frac{q}{4\pi\varepsilon_0 R}$$

When $x \gg R$, there is $U_P = \frac{q}{4\pi\varepsilon_0 x}$, which indicates that if a certain position on the axis of the ring is far enough from the center of the ring, its electric potential is equivalent to that generated by the point charge formed by concentrating all charges at the center of the ring.

Example 9.4.2 Find the electric potential at any point on the axis of a uniformly charged disk with a radius R and surface density σ.

Solution As shown in Figure 9.4.3, take any point P on the axis, and the distance between point P and the center O of the disk is x. The charged disk can be regarded as composed of numerous concentric rings. Take a thin ring with a radius of r and a width of dr, and take a very small fan-shaped surface element with an angle of $d\theta$ on the thin ring. The area of this fan-shaped surface element is $dS = r\,dr\,d\theta$ and its electric quantity is $dq = \sigma dS = \sigma r\,dr\,d\theta$. The charged surface element can be regarded as a point charge, and the potential generated at point P is

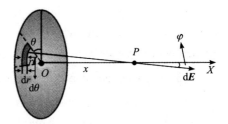

Figure 9.4.3 Figure of Example 9.4.2

$$dU = \frac{1}{4\pi\varepsilon_0} \cdot \frac{dq}{\sqrt{r^2+x^2}} = \frac{1}{4\pi\varepsilon_0} \cdot \frac{\sigma r\,dr\,d\theta}{\sqrt{r^2+x^2}}$$

The electric potential generated by the whole charged circular plate at point P is

$$U = \int dU = \frac{1}{4\pi\varepsilon_0} \int_0^R \int_0^{2\pi} \frac{\sigma r\,dr\,d\theta}{\sqrt{r^2+x^2}} = \frac{\sigma}{2\varepsilon_0}(\sqrt{R^2+x^2} - x)$$

When $x=0$, point P is at the center O of the disk and then $U = \frac{\sigma R}{2\varepsilon_0}$.

When $x \gg R$, there is

$$\sqrt{R^2+x^2} = (R^2+x^2)^{\frac{1}{2}} = x + \frac{R^2}{2x} + \cdots$$

and then

$$U = \frac{1}{4\pi\varepsilon_0} \cdot \frac{\pi R^2 \sigma}{x} = \frac{1}{4\pi\varepsilon_0} \cdot \frac{q}{x}$$

This result shows that when the distance between point P and the disk is much larger than the radius of the disk, the electric potential generated by the charged disk at point P

is the same as the electric potential generated when the charges of the disk are concentrated at the center of the disk.

In addition, this problem can also be solved by using the result of Example 9.4.1. The charged disk can be regarded as composed of numerous concentric rings. Take a thin ring with a radius of r and a width of dr, and the electric quantity of it is $dq=\sigma(2\pi r)dr$. According to the result of Example 9.4.1, there is

$$dU = \frac{1}{4\pi\varepsilon_0} \cdot \frac{dq}{\sqrt{r^2+x^2}} = \frac{\sigma r\, dr}{2\varepsilon_0 \sqrt{r^2+x^2}}$$

The electric potential generated by the whole charged circular plate at point P is

$$U = \int dU = \int_0^R \frac{\sigma r\, dr}{2\varepsilon_0 \sqrt{r^2+x^2}} = \frac{\sigma}{2\varepsilon_0}(\sqrt{R^2+x^2} - x)$$

which is the same as the above result.

2. Find the electric potential according to the known electric field

According to the definition of electric potential $U_p = \int_p^0 \mathbf{E} \cdot d\mathbf{l}$, if the distribution of the electric field from the field point P to the zero electric potential point is known, the electric potential at point P can be directly obtained by taking the integral.

Example 9.4.3 Find the electric potential at any point inside and outside a uniformly charged spherical surface with a radius R and an electric quantity q.

Solution As shown in Figure 9.4.4, the electric field distribution inside and outside the charged sphere is

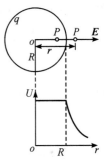

Figure 9.4.4 Figure of Example 9.4.3

$$\begin{cases} E=0 & (r<R) \\ E=\dfrac{q}{4\pi\varepsilon_0 r^2} & (r>R) \end{cases}$$

The direction of the electric field is in the radial direction, the distance between point P and the center of the sphere is r, the integral path is taken as an electric field line passing through point P, and infinity is taken as the zero potential energy point.

The electric potential at any point P outside the spherical surface is

$$U_P = \int_P^\infty \mathbf{E} \cdot d\mathbf{l} = \int_r^\infty \frac{q}{4\pi\varepsilon_0 r^2} dr = \frac{q}{4\pi\varepsilon_0 r}$$

The electric potential at any point P inside the spherical surface is

$$U_P = \int_P^\infty \mathbf{E} \cdot d\mathbf{l} = \int_r^R \mathbf{E} \cdot d\mathbf{r} + \int_R^\infty \mathbf{E} \cdot d\mathbf{r} = 0 + \int_R^\infty \frac{q}{4\pi\varepsilon_0 r^2} dr = \frac{q}{4\pi\varepsilon_0 R}$$

It can be seen that the electric potential at each point outside the sphere is the same as the electric potential generated by the point charge with all charges concentrated at the center of the sphere, and the electric potential at any point inside the sphere is a constant, which is the same as the electric potential of the sphere.

Example 9.4.4 Find the potential distribution in the electric field generated by an infinitely long straight uniform line with line charge density λ, as is shown in Figure 9.4.5.

Solution As shown in Figure 9.4.5, the electric field of any point P at a distance r from the long straight line is

$$E = \frac{\lambda}{2\pi\varepsilon_0 r}$$

The direction is radially outward.

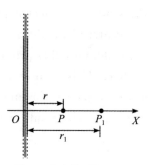

Figure 9.4.5 Figure of Example 9.4.4

If infinity is chosen as the zero potential point, then the potential at point P is

$$U_P = \int_P^\infty \boldsymbol{E} \cdot \mathrm{d}\boldsymbol{l} = \int_r^\infty \frac{\lambda}{2\pi\varepsilon_0 r} \mathrm{d}r = \frac{\lambda}{2\pi\varepsilon_0}(\ln\infty - \ln r)$$

Obviously the integral diverges and the result is infinite, which is obviously unreasonable. Therefore, the place at infinity can not be chosen as the zero potential point. In this question, point P_1 that is r_1 away from the straight line is chosen as the zero potential point and then the electric potential at point P, which is r away from the straight line, is

$$U_P = \int_P^{P_1} \boldsymbol{E} \cdot \mathrm{d}\boldsymbol{l} = \int_r^{r_1} \frac{\lambda}{2\pi\varepsilon_0 r} \mathrm{d}r = \frac{\lambda}{2\pi\varepsilon_0}\ln r_1 - \frac{\lambda}{2\pi\varepsilon_0}\ln r = \frac{\lambda}{2\pi\varepsilon_0}\ln\left(\frac{r_1}{r}\right)$$

According to the above equation, when $r_1 < r$, U_P is negative. When $r_1 > r$, U_P is positive.

9.5 Differential relationship between electric field strength and electric potential

9.5.1 Equipotential surface

Generally, the electric potential of each point in an electrostatic field changes. But there are many points in the electric field whose electric potentials are equal, and the surface formed by these equal electric potential points is called the **equipotential surface**.

In order to make an equipotential surface map show the magnitude of the electric field strength at each point in an electric field, the potential difference between any two adjacent equipotential surfaces is set to be equal. Figure 9.5.1 displays the equipotential surfaces and electric field line diagrams of several common electric fields. The dotted line in the figure represents the equipotential surface, and the solid line is the electric field line. It can be seen from the figure that the electric field line at any point in the electric field is orthogonal to the equipotential surface, and where the equipotential surfaces are denser, the field strength is larger; where the equipotential surfaces are sparser, the field strength is smaller.

Chapter 9 Electrostatic Field 33

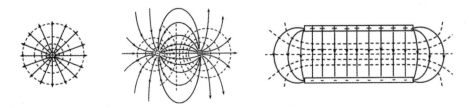

Figure 9.5.1 Equipotential surfaces and electric field line diagrams of several common electric fields

 9.5.2 Differential relationship between electric field strength and electric potential

The definition formula of electric potential $U_p = \int_p^0 \boldsymbol{E} \cdot \mathrm{d}\boldsymbol{l}$ reflects the integral relationship between electric potential and electric field strength in an electrostatic field, and electric potential distribution can be calculated by this equation after electric field distribution is obtained. However, in many practical problems, the electric potential distribution of the electrostatic field is often easy to obtain, and then the electric field distribution can be easily obtained according to the differential relationship between the electric field strength and the electric potential.

As shown in Figure 9.5.2, there are two closely spaced equipotential surfaces Ⅰ and Ⅱ in the electrostatic field, and their potentials are U and $U+\mathrm{d}U$, respectively, and $\mathrm{d}U>0$. Draw the normal n_0 of the equipotential surface Ⅰ through point a, and specify that n_0 points to the direction of potential increasing. Since the equipotential surfaces Ⅰ and Ⅱ are close together, it can be considered that their normals are in the same direction near point a, and the electric field strengths between the equipotential surfaces can also be regarded as a constant.

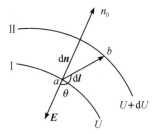

Figure 9.5.2 Differential relationship between electric field strength and electric potential

Assuming that the unit positive charge moves from point a on equipotential surface Ⅰ to point b on equipotential surface Ⅱ in the d$\boldsymbol{l}$ direction, the electric potential difference between points a and b can be obtained from Equation (9.4.8) as

$$U_\mathrm{I}-U_\mathrm{II}=-\mathrm{d}U=E\cos\theta\,\mathrm{d}l$$

where θ represents the angle between $\boldsymbol{E}$ and d$\boldsymbol{l}$, and $E\cos\theta=E_l$ is the component of $\boldsymbol{E}$ on

dl. Then there is
$$-dU = E_l \, dl$$
or
$$E_l = -\frac{dU}{dl} \tag{9.5.1}$$

The above equation shows that the component of the electric field E in the dl direction is equal to the negative value of the change rate of the electric potential in this direction.

Obviously, the value of $\frac{dU}{dl}$ will vary with the direction of dl. If dl is in the direction of the equipotential surface, then $\frac{dU}{dl} = 0$, which means that the component of the electric field strength in the direction of the equipotential surface is zero. If dl is in the normal direction of the equipotential surface, it could be written as dn, whose displacement is the smallest among all the displacements from equipotential surface I to equipotential surface II, as shown in Figure 9.5.2. Thus, the rate of change of the potential in the n_0 direction is the largest. Then Equation (9.5.1) can be written as

$$E_n = -\frac{dU}{dn} \tag{9.5.2}$$

where E_n is the component of the electric field strength in the normal direction. Since the equipotential surface is orthogonal to the electric field line everywhere, the component E_n is equal to the magnitude of E, so Equation (9.5.2) can be rewritten as

$$E = -\frac{dU}{dn} \tag{9.5.3}$$

When $\frac{dU}{dl} > 0$, $E < 0$, that is, the direction of E is always from high potential to low potential, which is opposite to the direction of n_0. Thus, the vector formula of Equation (9.5.3) is

$$\boldsymbol{E} = -\frac{dU}{dn}\boldsymbol{n}_0 \tag{9.5.4}$$

where $\boldsymbol{n}_0$ is the unit vector in the positive normal direction. Equation (9.5.4) shows that the electric field strength at any given point in the electric field is equal to the change rate of the potential in the normal direction of the equipotential surface. Equation (9.5.4) is the differential relationship between electric field strength and electric potential.

From the above analysis, it can be seen that at any point in the electric field, the change rate of the electric potential in different directions with distance is generally different, and the maximum value is called the electric potential gradient. The electric potential gradient is a vector. Its magnitude is equal to $\frac{dU}{dn}$ and its direction is the same as the direction of $\boldsymbol{n}_0$. It can be represented by the symbol grad or ∇, that is

$$\mathrm{grad}(U) = \nabla U = \frac{dU}{dn}\boldsymbol{n}_0$$

Thus, Equation (9.5.4) can be rewritten as
$$E = -\text{grad}(U) = -\nabla U \tag{9.5.5}$$

Equation (9.5.5) shows that the electric field strength at any point in the electric field is equal to the negative value of the potential gradient at that point, and the negative sign indicates that the electric field strength at this point is opposite to the potential gradient in direction, which suggests that the direction of electric field strength points to the direction of decreasing potential. In the Cartesian coordinate system, the potential U is a function of the coordinates, and the components of the electric field strength along the three coordinate axes can be obtained from Equation (9.5.1)

$$E_x = -\frac{\partial U}{\partial x}, \quad E_y = -\frac{\partial U}{\partial y}, \quad E_z = -\frac{\partial U}{\partial z} \tag{9.5.6}$$

By using these components, the vector description of electric field strength can be written as

$$\boldsymbol{E} = -\left(\frac{\partial U}{\partial x}\boldsymbol{i} + \frac{\partial U}{\partial y}\boldsymbol{j} + \frac{\partial U}{\partial z}\boldsymbol{k}\right) \tag{9.5.7}$$

From the differential relationship between electric field strength and electric potential, another unit of electric field strength can be obtained as volt · meter^{-1} (V · m^{-1}).

According to the differential relationship between electric field strength and electric potential, in order to find the electric field generated by a source charge, the electric potential distribution function can be obtained first. Then, the electric field strength can be calculated from the derivation of the electric potential function, which is an easier method to find the electric field distribution.

Example 9.5.1 Calculate the electric potential at any point P on the axis of a uniformly charged disk with a radius R and a surface density of charges σ ($\sigma > 0$), and use the differential relationship between electric field strength and electric potential to calculate the electric field strength at point P.

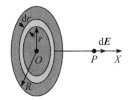

Figure 9.5.3 Figure of Example 9.5.1

Solution As shown in Figure 9.5.3, the distance between point P and the center O of the disk is set as x. Take a thin ring with a radius r and width dr on the disk, and the electric quantity of the thin ring is

$$dq = \sigma \cdot 2\pi r \, dr$$

According to the result of Example (9.4.1), the potential at point P is

$$dU = \frac{dq}{4\pi\varepsilon_0 \sqrt{r^2 + x^2}} = \frac{\sigma \cdot r \cdot dr}{2\varepsilon_0 \sqrt{r^2 + x^2}}$$

Then the electric potential generated by the whole charged disk at point P is

$$U = \int dU = \int_0^R \frac{\sigma r \, dr}{2\varepsilon_0 \sqrt{x^2 + r^2}} = \frac{\sigma}{2\varepsilon_0}(\sqrt{R^2 + x^2} - x)$$

It can be seen that the potential U at point P is a function of x. Using Equation (9.5.6), the component of field strength E at point P in the X axis direction can be obtained as

$$E_x = -\frac{\partial U}{\partial x} = -\frac{\partial}{\partial x}\left[\frac{\sigma}{2\varepsilon_0}(\sqrt{R^2+x^2}-x)\right] = \frac{\sigma}{2\varepsilon_0}\left(1-\frac{x}{\sqrt{R^2+x^2}}\right)$$

According to the symmetry of the charge distribution of the disk, it is obvious that

$$E_y = 0, \ E_z = 0$$

Then

$$\boldsymbol{E} = E_x \boldsymbol{i} = \frac{\sigma}{2\varepsilon_0}\left(1-\frac{x}{\sqrt{R^2+x^2}}\right)\boldsymbol{i}$$

9.6 Conductor and dielectric in an electrostatic field

9.6.1 Electrostatic equilibrium of a conductor in an electrostatic field

A conductor is an object that can conduct electricity. From a microscopic point of view, there are a lot of free charges in a conductor. When a conductor is uncharged or not affected by the external electric field, free charges in the conductor do random thermal motion. Positive and negative charges are evenly distributed, and the conductor does not exhibit electrical properties. If a conductor is placed in the electrostatic field with $\boldsymbol{E}_0$, free charges in the conductor will make a macroscopic directional movement under the action of the electric field force, as shown in Figure 9.6.1(a), causing the charges in the conductor to be redistributed. This phenomenon is called **electrostatic induction**. The charge generated by the conductor due to electrostatic induction is called **induced charge**. The induced charge will generate an additional electric field $\boldsymbol{E}'$, as shown in Figure 9.6.1(b). The direction of this electric field inside the conductor is opposite to the original electric field $\boldsymbol{E}_0$, thereby weakening the magnitude of the original electric field inside the conductor. With the further advancement of electrostatic induction, the number of induced charges continues to increase, so that the additional electric field increases. When the electric field strength of the total electric field in the conductor is $\boldsymbol{E} = \boldsymbol{E}_0 + \boldsymbol{E}' = 0$, the redistribution process of the free charge stops. There is no macroscopic directional movement of the charge inside and on the surface of the conductor, and the conductor reaches a state of **electrostatic equilibrium**.

When the conductor reaches an electrostatic equilibrium state, the following two conclusions can be drawn:

(1) **The electric field at any point inside the conductor is zero.** If the electric field inside the conductor is not zero, the electric field will continue to drive the free charge to move, i.e. the conductor is not in electrostatic equilibrium.

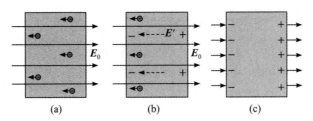

Figure 9.6.1 Electrostatic induction and electrostatic equilibrium of a conductor

(2) **The direction of the electric field strength near the surface of the conductor is perpendicular to the surface.** If the direction of the electric field near the surface of the conductor is not perpendicular to the surface, the tangential component of the electric field along the surface can also drive the directional motion of free charges, which is not an electrostatic equilibrium state either.

In addition, the following inferences can be drawn:

(1) The net charge everywhere inside the electrostatic equilibrium conductor is zero, and the charge carried by the conductor itself or its induced charge can only be distributed on the surface of the conductor. This conclusion can be proved by Gauss's law.

A closed surface S is arbitrarily made inside the conductor. According to Gauss's law $\oint_S \boldsymbol{E} \cdot \mathrm{d}\boldsymbol{S} = \dfrac{1}{\varepsilon_0} \sum q_{\text{int}}$, since the electric field inside the conductor is zero everywhere, the left side of the above equation is zero. The net charge surrounded by the surface S on the right side of the equation is also zero. Since the closed surface is arbitrarily made inside the conductor, it indicates that there is indeed no net charge in the conductor, and the charge can only be distributed on the surface of the conductor. It should be noticed that the conductor surface is not only the outer surface of the conductor, but also the inner surface, such as the inner surface of a conductor cavity, which may also have a charge distribution.

(2) The conductor is an equipotential body, and the surface of the conductor is equipotential. Since the electric field inside the conductor is zero everywhere, the electric potential difference between any two points a and b in the conductor is $U_{ab} = \int_a^b \boldsymbol{E} \cdot \mathrm{d}\boldsymbol{l} = 0$, so the conductor is an equipotential body, and its surface is an equipotential surface.

(3) The relationship between the magnitude of the electric field strength near the surface of the electrostatic equilibrium conductor and the charge density on the surface is

$$E = \dfrac{\sigma}{\varepsilon_0} \qquad (9.6.1)$$

which can be proved by Gauss's law.

As shown in Figure 9.6.2, a small cylindrical Gaussian surface S is drawn near the conductor surface. The upper base ΔS of the cylinder passes through point P and is parallel to the conductor surface, and the side surface of the cylinder is perpendicular to the conductor surface. According to Gauss's law, we have

$$\oint_S \boldsymbol{E} \cdot \mathrm{d}\boldsymbol{S} = E\Delta S = \frac{1}{\varepsilon_0}\sum q_{\text{int}} = \frac{\sigma \Delta S}{\varepsilon_0}$$

Since the electric field strength near the conductor surface is perpendicular to the surface and the side of the cylinder is parallel to the field strength $\boldsymbol{E}$, the flux through the side of the cylinder is zero. The bottom of the cylindrical surface is inside the conductor, where the electric field strength is identically equal to zero (i. e. $\boldsymbol{E}=0$), so the flux through the bottom is also zero. Since the upper bottom is perpendicular to electric field

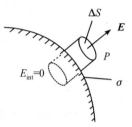

Figure 9.6.2 Electric field near the conductor surface

$\boldsymbol{E}$, which can be regarded as a uniform electric field in a small area ΔS, the flux through the upper bottom is $E\Delta S$. Since the electric charges can be regarded as uniformly distributed, the electric quantity enclosed by the cylindrical surface is $\sigma \Delta S$. Then, Equation (9.6.1) can be obtained.

It can be seen from Equation (9.6.1) that the distribution of charges on the conductor surface is related to the shape of the conductor itself and external conditions. For an isolated charged conductor, the charge density σ on the surface is closely related to the radius of curvature of the surface: Where the surface curvature is larger, σ is larger and where the curvature is smaller, σ is smaller, as shown in Figure 9.6.3.

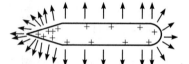

Figure 9.6.3 Electric field near the conductor surface

For a charged conductor with a tip, the charge surface density at the tip is relatively large, and the electric field strength near the conductor surface is also particularly large. When the electric field strength exceeds the breakdown field strength of the air, a discharge phenomenon in which the air is ionized will occur, and is called point discharge. The lightning rod is to use the principle of point discharge to prevent the damage of lightning strikes to buildings.

9.6.2 Electrostatic shielding

According to the principle that the electric field strength inside an electrostatic equilibrium conductor is zero, it is possible to use a cavity conductor to isolate the electric fields inside and outside the cavity, so that they do not affect each other. This effect is called **electrostatic shielding**.

1. Use a cavity conductor to shield the external electric field

As shown in Figure 9.6.4(a), a cavity conductor is placed in an electrostatic field, and the electric field strength inside the conductor is zero, so that the cavity conductor can be used to shield the external electric field. Objects inside the cavity are not affected by the external electric field.

2. Use a cavity conductor to shield the internal electric field

As shown in Figure 9.6.4(b), a point charge q is placed inside a cavity conductor, and the electric field strength outside the conductor is zero due to electrostatic equilibrium. According to Gauss's law, the inner surface of the conductor will induce equal and different charges $-q$. According to the law of conservation of charge, the outer surface will induce the same amount of the same charge q. If the outer surface of the cavity is connected to the ground, all the charges on the outer surface of the cavity will be introduced to the ground, and the electric field outside the cavity will also disappear. In this way, the charged body in the cavity will not have any influence on the outside of the cavity.

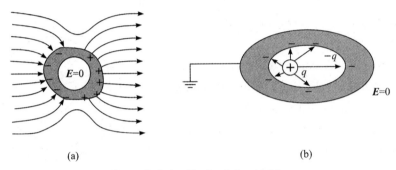

(a)   (b)

Figure 9.6.4 Electrostatic shielding

Electrostatic shielding has many applications in engineering technology. For example, precision devices can be covered with a metal shell or metal mesh to avoid the influence of external fields. When working with high voltage, the operator should wear a shielding suit made of wire mesh to prevent the electric field from harming the human body. The shielding suit may be charged and the electric potential may be high, but the electric field strength inside the shielding suit is zero, which ensures the safety of the operator.

Example 9.6.1 As shown in Figure 9.6.5, there is a large metal plate A with an area S and an electric quantity Q. Another large metal plate B is placed parallel next to it. This plate was originally uncharged. Ignoring the edge effect, find the charge distribution on plates A and B.

Solution Ignoring the edge effects, it can be considered that the charges on each surface are uniformly distributed, and the charge densities on the four surfaces are set to be σ_1, σ_2, σ_3 and σ_4, respectively, as shown in Figure 9.6.5. According to the law of conservation of charge,

$$\sigma_1 S + \sigma_2 S = Q \quad (1)$$
$$\sigma_3 S + \sigma_4 S = 0 \quad (2)$$

According to electrostatic equilibrium, the electric

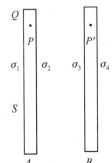

Figure 9.6.5 Figure of Example 9.6.1

field strength inside the conductor is zero everywhere. The electric field strength at any point P taken inside the conductor plate A is zero, so

$$E_P = \frac{\sigma_1}{2\varepsilon_0} - \frac{\sigma_2}{2\varepsilon_0} - \frac{\sigma_3}{2\varepsilon_0} - \frac{\sigma_4}{2\varepsilon_0} = 0 \tag{3}$$

Take any point P' in the conductor plate B, and the electric field strength at this point is zero, and then

$$E_{P'} = \frac{\sigma_1}{2\varepsilon_0} + \frac{\sigma_2}{2\varepsilon_0} + \frac{\sigma_3}{2\varepsilon_0} - \frac{\sigma_4}{2\varepsilon_0} = 0 \tag{4}$$

According to Equations (1)~(4),

$$\sigma_1 = \sigma_2 = -\sigma_3 = \sigma_4 = \frac{Q}{2S}$$

Example 9.6.2 As shown in Figure 9.6.6, there is a conductor sphere with a radius R_1 and an electric quantity Q_1. A conductor spherical shell with an inner radius R_2 and outer radius R_3 and an electric quantity Q_2 is placed concentrically outside the conductor sphere. If the conductor sphere is connected to the ground, find

(1) the electric quantity of the conductor sphere;

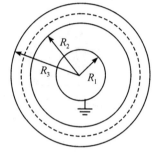

Figure 9.6.6   Figure of Example 9.6.2

(2) the electric potential of the conductor spherical shell.

Solution (1) Assume that after the conductor sphere is connected to the ground the electric quantity of the conductor sphere is q, the electric quantity of the inner surface of the conductor sphere is Q_{inner}, and the electric quantity of the outer surface of the conductor sphere is Q_{outer}.

For the conductor spherical shell, according to the law of conservation of charge, there is

$$Q_{inner} + Q_{outer} = Q_2 \tag{1}$$

Take a Gaussian surface as shown in Figure 9.6.6. When the conductor reaches electrostatic equilibrium, the electric field strength on the Gaussian surface is zero everywhere. According to Gauss's law, there is

$$\oint E \cdot dS = \frac{q + Q_{inner}}{\varepsilon_0} = 0 \tag{2}$$

According to Equations (1) and (2),

$$Q_{inner} = -q$$
$$Q_{outer} = Q_2 + q$$

By using Gauss's law, the distribution of electric field strength in space can be obtained:

When $r < R_1$, $E_1 = 0$;

When $R_1 < r < R_2$, $E_2 = \dfrac{q}{4\pi\varepsilon_0 r^2}$;

When $R_2 < r < R_3$, $E_3 = 0$;

When $r > R_3$, $E_4 = \dfrac{Q_2 + q}{4\pi\varepsilon_0 r^2}$.

The electric potential of the conductor sphere is

$$U = \int_{R_1}^{\infty} \boldsymbol{E} \cdot \mathrm{d}\boldsymbol{l} = \int_{R_1}^{R_2} \boldsymbol{E}_2 \cdot \mathrm{d}\boldsymbol{l} + \int_{R_2}^{R_3} \boldsymbol{E}_3 \cdot \mathrm{d}\boldsymbol{l} + \int_{R_3}^{\infty} \boldsymbol{E}_4 \cdot \mathrm{d}\boldsymbol{l}$$

$$= \int_{R_1}^{R_2} \dfrac{q}{4\pi\varepsilon_0 r^2}\mathrm{d}r + \int_{R_3}^{\infty} \dfrac{Q_2+q}{4\pi\varepsilon_0 r^2}\mathrm{d}r = \dfrac{q}{4\pi\varepsilon_0 R_1} - \dfrac{q}{4\pi\varepsilon_0 R_2} + \dfrac{Q_2+q}{4\pi\varepsilon_0 R_3}$$

Since the conductor sphere is connected to the ground, the potential of the conductor sphere is zero, that is,

$$\dfrac{q}{4\pi\varepsilon_0 R_1} - \dfrac{q}{4\pi\varepsilon_0 R_2} + \dfrac{Q_2+q}{4\pi\varepsilon_0 R_3} = 0$$

Then

$$q = \dfrac{Q_2 R_1 R_2}{R_1 R_3 - R_1 R_2 - R_2 R_3}$$

(2) The electric potential of the conductor spherical shell is

$$U = \int_{R_3}^{\infty} \boldsymbol{E}_4 \cdot \mathrm{d}\boldsymbol{l} = \int_{R_3}^{\infty} \dfrac{Q_2+q}{4\pi\varepsilon_0 r^2}\mathrm{d}r = \dfrac{Q_2+q}{4\pi\varepsilon_0 R_3} = \dfrac{Q_2}{4\pi\varepsilon_0 R_3} + \dfrac{Q_2 R_1 R_2}{4\pi\varepsilon_0 R_3(R_1 R_3 - R_1 R_2 - R_2 R_3)}$$

 ### 9.6.3 Dielectric polarization

A dielectric generally refers to a non-conductive insulator in which there are no charges that can move freely. However, under the action of an external electric field, the positive and negative charges in the dielectric can still perform microscopic relative motion, making the dielectric appear charged. This phenomenon of electrification of a dielectric under the action of an external electric field is called **dielectric polarization**. The electric charges that appear in the polarization of the dielectric are called **polarized charges**, also known as bound charges. These charges excite additional electric fields, weakening the external electric field.

Dielectric molecules consist of equal numbers of positive and negative charges, which can be treated equivalently as two point charges. If the positive and negative charge centers of the molecules do not coincide, such a pair of extremely close, equal and opposite charges form an electric dipole, and the dielectric formed by this molecule is called **polar molecular dielectric**, such as HCl, H_2O and CO. If the positive and negative charge centers of the molecule coincide, the electric dipole moment of the molecule is zero, and the dielectric formed by this molecule is called **non-polar molecular dielectric**, such as H_2, O_2, N_2 and CO_2.

When there is no external field, due to the thermal motion of molecules in the polar molecular dielectric, the molecular dipole moments are arranged randomly and cancel each other out, and the dielectric does not appear to be charged macroscopically. When there is an external field E_0, the dipole moment of the molecules will rotate to form orderly arrangement in the direction of the electric field, as shown in Figure 9.6.7(a), so that the dielectric is charged. This phenomenon is called **orientation polarization**.

The non-polar molecular dielectric is not charged when there is no external field; under the action of the external field, the positive and negative charge centers are subjected to force and relative displacement occurs, thereby forming an electric dipole moment. The directions of these electric dipole moments are all in the direction of the external field, as shown in Figure 9.6.7(b), so positive and negative polarized charges will appear on the surface of the dielectric. This polarization is called **displacement polarization**.

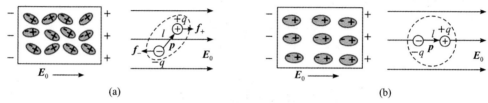

Figure 9.6.7 Polarization of dielectrics

Obviously, the microscopic mechanisms of displacement polarization and orientation polarization are different, but the result is the same. The molecular electric dipole moment vector sum in the medium is not zero, that is, the dielectric is polarized. Therefore, if the problem does not involve the mechanism of polarization, the macroscopic processing is not often distinguished.

 ### 9.6.4 Polarization intensity

In order to quantitatively describe the polarization degree of a dielectric, the physical quantity of the electric polarization intensity can be introduced.

When a dielectric is in a polarized state, in any tiny volume element ΔV of the dielectric, the vector sum of the electric dipole moments of all molecules is generally not zero, that is $\sum p_e \neq 0$. The vector sum of the molecular electric dipole moments in a unit volume is defined as the **polarization intensity vector**, referred to as the polarization, which is represented by the vector P

$$P = \frac{\sum p_e}{\Delta V} \tag{9.6.2}$$

P is a physical quantity that describes the polarization state of a dielectric (including polarization state and polarization direction). The SI unit of polarization intensity is Coulomb per square meter (C · m^{-2}).

In general, the polarization intensity of each point in a medium is not the same. If the magnitude and direction of the polarization intensity of each point in a dielectric are the same, the dielectric is called uniform polarization.

Experiments show that for most common isotropic linear dielectrics, the polarization intensity P at any point is proportional to the net electric field strength E at that point, and the direction is the same.

$$P = \chi_e \varepsilon_0 E \tag{9.6.3}$$

where χ_e is called the **polarizability of the dielectric**, which is a physical quantity describing the polarization properties of the dielectric and does not depend on the electric field strength E. A dielectric whose χ_e value is a constant at each point is called a homogeneous dielectric.

It should be noted that the electric field E in Equation (9.6.3) is the net electric field strength, which includes both the external electric field E_0 and the additional electric field generated by the polarized charge E', so

$$E = E_0 + E'$$

In general, the additional electric field E' generated by the polarized charge inside the dielectric always plays a role in weakening the original external electric field E_0 and the polarization of the dielectric. It's sometimes also called the depolarization field, and its magnitude depends on the geometry and polarizability of the dielectric χ_e.

The polarized charges of a non-uniform dielectric appear not only on the surface of the dielectric, but also inside it. The polarized charges of a uniform dielectric only appear on the surface of the dielectric, and the surface density of the polarized charge is

$$\sigma' = P \cdot e_n \tag{9.6.4}$$

where e_n is the unit vector in the normal direction of the dielectric surface.

Taking an arbitrary closed surface S inside the dielectric, it can be proved that the algebraic sum of polarized charges enclosed by S is

$$\sum q' = \oint_S P \cdot dS \tag{9.6.5}$$

This equation describes the general relationship between the polarization density P and the distribution of polarized charges, which indicates that the flux of the polarized P of any closed surface is equal to the negative value of the total polarized charges enclosed by the closed surface.

The above results show that the dielectric will be polarized under the action of the external electric field E_0. The distribution of the polarized charges σ' depending on the polarization density P and the shape of the dielectric determines the additional field E' and affects the net electric field strength E inside the dielectric, which in turn affects the polarization P. It can be seen that the physical quantities P, σ', E_0 and E are dependent on each other and restrict each other.

9.6.5 Gauss's law in a dielectric

As mentioned above, the physical quantities such as P, σ', E_0 and E in a dielectric are dependent on each other, which makes it difficult to find the electric field strength E. However, by using Gauss's law in the dielectrics, calculation can be simplified.

As is known, Gauss's law is based on Coulomb's law, and it is still valid in the dielectric. However, the charges enclosed on the Gaussian surface should be the sum of free charges q_0 and the polarized charges q', that is

$$\oint_S \boldsymbol{E} \cdot \mathrm{d}\boldsymbol{S} = \frac{1}{\varepsilon_0} \sum (q_0 + q') \tag{9.6.6}$$

Combining it with Equation (9.6.5), there is

$$\oint_S (\varepsilon_0 \boldsymbol{E} + \boldsymbol{P}) \cdot \mathrm{d}\boldsymbol{S} = \sum q_0 \tag{9.6.7}$$

Here, an auxiliary physical quantity, electric displacement denoted by $\boldsymbol{D}$, is introduced, so

$$\boldsymbol{D} = \varepsilon_0 \boldsymbol{E} + \boldsymbol{P} \tag{9.6.8}$$

Then, Equation (9.6.7) can be rewritten as

$$\oint_S \boldsymbol{D} \cdot \mathrm{d}\boldsymbol{S} = \sum q_0 \tag{9.6.9}$$

It indicates that the electric displacement flux through a closed Gaussian surface is equal to the algebraic sum of the free charges enclosed inside the closed surface, which is the Gauss's law in dielectrics.

Substituting Equation (9.6.3) into Equation (9.6.8), we have

$$\boldsymbol{D} = \varepsilon_0 \boldsymbol{E} + \boldsymbol{P} = \varepsilon_0 \boldsymbol{E} + \chi_e \varepsilon_0 \boldsymbol{E} = (1 + \chi_e) \varepsilon_0 \boldsymbol{E} \tag{9.6.10}$$

Assuming that

$$\varepsilon_r = 1 + \chi_e$$

$$\varepsilon = \varepsilon_0 \varepsilon_r$$

where ε is the permittivity of the dielectric (or dielectric constant), and ε_r is the relative permittivity of the dielectric (or relative dielectric constant), which is a dimensionless quantity. Equation (9.6.10) can be rewritten as

$$\boldsymbol{D} = \varepsilon \boldsymbol{E} = \varepsilon_0 \varepsilon_r \boldsymbol{E} \tag{9.6.11}$$

It describes the relationship between the electrical displacement $\boldsymbol{D}$ and the electric field strength $\boldsymbol{E}$ in an isotropic linear dielectric. The SI unit of electrical displacement $\boldsymbol{D}$ is $C \cdot m^{-2}$ (Coulombs per square meter).

According to Gauss's law of the dielectric, if the distribution of free charges is known, $\boldsymbol{D}$ can be obtained using Equation (9.6.9), and the electric field strength $\boldsymbol{E}$ in a uniform isotropic dielectric can be further calculated using Equation (9.6.11). The calculation of electric field strength $\boldsymbol{E}$ is simplified since the polarized charges are not involved in the calculation.

Example 9.6.3 A uniformly charged spherical surface with a radius R, electric quantity q_0, and charge surface density σ_0 is placed in a homogeneous infinite dielectric. The relative permittivity of the dielectric is ε_r. Find the distribution of electric field strength in the dielectric and the surface density of polarized charges on the surface of the dielectric σ'.

Solution Due to the spherical symmetry of the charge distribution on the metal spherical surface, the distribution of the electric field has spherical symmetry. As shown in Figure 9.6.8, a closed spherical surface S of radius r and concentric with the metal sphere is drawn through point P. According to Gauss's law in dielectrics, there is

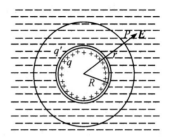

Figure 9.6.8 Figure of Example 9.6.3

$$\oint_S \boldsymbol{D} \cdot \mathrm{d}\boldsymbol{S} = q_0$$

Due to the principle of symmetry, the magnitude of $\boldsymbol{D}$ of each point on the spherical surface S is the same and there is

$$D 4\pi r^2 = q_0$$

Then

$$D = \frac{q_0}{4\pi r^2}$$

The electric field strength $\boldsymbol{E}$ in the dielectric is

$$E = \frac{D}{\varepsilon_0 \varepsilon_r} = \frac{q_0}{4\pi \varepsilon_0 \varepsilon_r r^2}$$

Since $\boldsymbol{D} = \varepsilon_0 \boldsymbol{E} + \boldsymbol{P}$, the magnitude of polarization intensity $\boldsymbol{P}$ in the dielectric is

$$P = \frac{(\varepsilon_r - 1) q_0}{4\pi \varepsilon_r r^2}$$

The direction of $\boldsymbol{P}$ is opposite to the direction of the outer normal of the inner surface of the dielectric, so the surface density of polarized charges σ' on the inner surface of the dielectric is

$$\sigma' = P_n = -\frac{(\varepsilon_r - 1) q_0}{4\pi \varepsilon_r R^2} = -\frac{\varepsilon_r - 1}{\varepsilon_r} \frac{q_0}{4\pi R^2} = -\sigma_0 \frac{\varepsilon_r - 1}{\varepsilon_r}$$

The above result shows that in the dielectric, the polarized charge on the surface is of the opposite sign to the free charge on the conductor spherical surface, and $|\sigma'| < |\sigma_0|$.

9.7 Capacitor

Capacitance is an important characteristic of a conductor or conductor group. Capacitors are a specific set of conductors that are widely used in practice.

9.7.1 Capacitance of an isolated conductor

The electric potential of an isolated conductor in a vacuum depends on its electric quantity and shape. The electric potential of an isolated spherical conductor of radius R and electric quantity q in a vacuum is

$$U=\frac{q}{4\pi\varepsilon_0 R}$$

It can be seen from the above equation that when the potential is a constant, the larger the radius of the sphere is, the more charges it carries. But the ratio of its electric quantity to the potential is a constant and only depends on the shape of the conductor. Thus, the concept of capacitance is introduced.

The ratio of the electric quantity of an isolated conductor to its electric potential is called the **capacitance of the isolated conductor**, denoted by C, that is,

$$C=\frac{q}{U} \tag{9.7.1}$$

The SI unit of capacitance is the farad, and the symbol is F. In practice, the common capacitance units are the microfarad (μF) and picofarad (pF), and their relationship with farad is 1 F$=10^6$ μF$=10^{12}$ pF.

9.7.2 Capacitance of a capacitor

When there are other conductors near a charged conductor, the electric potential of the conductor depends not only on the electric quantity it carries, but also on the shapes and positions of other conductors. In general, an electrical component composed of two conductors that are insulated from each other and close to each other is called a **capacitor**, and the two conductors that make up the capacitor are called the plates of the capacitor. When a capacitor is charged, the two polar plates are often charged with equal and opposite charges. The capacitance of a capacitor is defined as the ratio of the electric quantity q ($q>0$) on one plate to the electric potential difference ΔU between the two plates, that is,

$$C=\frac{q}{\Delta U} \tag{9.7.2}$$

The capacitance of a capacitor depends on the size, shape, and relative position of the plates and the permittivity of the dielectric between the plates, and it doesn't depend on whether the capacitor is charged or not.

9.7.3 Calculation of the capacitance of a capacitor

There are many kinds of capacitors used in production and scientific research, and their shapes are different, but their basic structures are the same. According to adjustability, they can be divided into adjustable capacitors, fine-tuning capacitors, and fixed capacitors, etc. According to the dielectric, they can be divided into air capacitors, mica capacitors, ceramic capacitors, and paper capacitors, etc. According to the shape, they can be divided into parallel plate capacitors, cylindrical capacitors, and spherical capacitors, etc.

The capacitance of a capacitor with a special shape can be calculated by the following steps. First, assume that the two plates of the capacitor are charged with equal and opposite charges. Second, calculate the electric field strength and the electric potential difference between the two plates. Finally, calculate the capacitance by using Equation (9.7.2).

1. Parallel plate capacitor

A parallel plate capacitor consists of two parallel metal plates of equal size that are placed close to each other. As shown in Figure 9.7.1, the area of the plate is S, the distance between the inner surfaces is d, and the two plates are filled with a uniform dielectric with a relative permittivity of ε_r. When the two plates of the capacitor are charged with q and $-q$, respectively, since d is much smaller than the linearity of the plates, the edge effect can be ignored. Consider the plate as an infinitely charged conductor plate, and the electric field between the two plates can be regarded as a uniform electric field. Assuming that the surface charge density on the inner surface of the polar plates are $\pm\sigma$, according to Gauss's law in dielectrics, the electric field strength between the two polar plates is

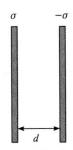

Figure 9.7.1 Parallel plate capacitor

$$E = \frac{\sigma_0}{\varepsilon_0 \varepsilon_r}$$

The electric potential difference between the two plates is

$$U = Ed = \frac{\sigma_0 d}{\varepsilon_0 \varepsilon_r}$$

According to the definition of a capacitor's capacitance Equation (9.7.2), the capacitance of a parallel plate capacitor is

$$C = \frac{q}{U} = \frac{\varepsilon_0 \varepsilon_r S}{d} \qquad (9.7.3)$$

2. Cylindrical capacitor

A cylindrical capacitor is a conductor set consisting of two coaxial cylindrical conductor cylinders. The conductors are filled with a uniform dielectric with a relative permittivity ε_r, and the radii of the inner and outer cylindrical conductors are R_1 and R_2, respectively. When the length $L \gg (R_2 - R_1)$, as shown in Figure 9.7.2, the edge effects at both ends can be ignored, and it can be treated as an infinitely long cylinder.

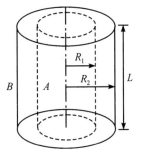

Figure 9.7.2 Cylindrical capacitor

Assuming that the two opposite surfaces of the two polar plates have electric quantities of q and $-q$, respectively, then the absolute value of the electric quantity per unit length of the cylinder is $\lambda = \dfrac{q}{L}$. According to Gauss's law in dielectrics, the electric field strength at a distance r from the axis is

$$E = \dfrac{\lambda}{2\pi\varepsilon_0\varepsilon_r r}$$

The electric potential difference between the two plates is

$$U = \int_{R_1}^{R_2} \boldsymbol{E} \cdot \mathrm{d}\boldsymbol{r} = \int_{R_1}^{R_2} \dfrac{\lambda}{2\pi\varepsilon_0\varepsilon_r r}\mathrm{d}r = \dfrac{\lambda}{2\pi\varepsilon_0\varepsilon_r}\ln\dfrac{R_2}{R_1} = \dfrac{q}{2\pi\varepsilon_0\varepsilon_r L}\ln\dfrac{R_2}{R_1}$$

According to the definition of a capacitor's capacitance Equation (9.7.2), the capacitance of a cylindrical capacitor is

$$C = \dfrac{q}{\Delta U} = \dfrac{2\pi\varepsilon_0\varepsilon_r L}{\ln\dfrac{R_2}{R_1}} \tag{9.7.4}$$

3. Spherical capacitor

A spherical capacitor is composed of two concentric conductor spherical shells, as shown in Figure 9.7.3. The radii of the inner and outer spherical shells are R_1 and R_2, respectively, and the space between them is filled with a uniform dielectric with a relative permittivity of ε_r. Assuming that the two opposite surfaces of the two polar plates have electric quantities of q and $-q$, respectively, according to Gauss's law in

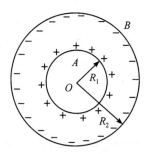

Figure 9.7.3 Spherical capacitor

dielectrics, the electric field at a distance r from the center of the sphere is

$$E = \dfrac{q}{4\pi\varepsilon_0\varepsilon_r r^2}$$

The electric potential difference between the two plates is

$$U = \int_{R_1}^{R_2} \boldsymbol{E} \cdot \mathrm{d}\boldsymbol{r} = \int_{R_1}^{R_2} \dfrac{q}{4\pi\varepsilon_0\varepsilon_r r^2}\mathrm{d}r = \dfrac{q}{4\pi\varepsilon_0\varepsilon_r}\left(\dfrac{1}{R_1} - \dfrac{1}{R_2}\right)$$

According to the definition of a capacitor's capacitance Equation (9.7.2), the capacitance of a spherical capacitor is

$$C = \dfrac{q}{\Delta U} = \dfrac{4\pi\varepsilon_0\varepsilon_r R_1 R_2}{R_2 - R_1} \tag{9.7.5}$$

It can be seen from the calculation results of the above three capacitors that the capacitance of the capacitor doesn't depend on the charge on the plate, but is related to the geometry of the capacitor. The smaller the distance between the two plates is, the greater the capacitance is. However, too small spacing can cause the capacitor to break down. For a rated voltage, the smaller the distance between the two plates, the stronger the electric field strength in the dielectric. When the electric field strength exceeds a certain limit (the breakdown field strength), bound charges in molecules can become free charges under the action of the strong electric field. The dielectric will lose its insulating properties and be converted into conductors, and the capacitor will be destroyed.

9.7.4 Capacitors in parallel and series

In practical applications, if the capacitance or withstand voltage of one capacitor does not meet requirements, several capacitors can be connected to form a combined capacitor. There are two basic ways of connection: parallel and series.

1. Capacitors in parallel

Figure 9.7.4 shows the parallel connection of n capacitors with a voltage U. The electric potential difference between the two plates of each capacitor is equal, and then we have

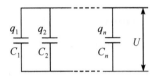

Figure 9.7.4 Capacitors in parallel

$$\Delta U = \Delta U_1 = \Delta U_2 = \cdots = \Delta U_n$$

The total electric quantity carried by the parallel combination is the sum of the electric quantities of all capacitors.

$$q = q_1 + q_2 + \cdots + q_n$$

So the equivalent capacitance of the parallel combination is

$$C = \frac{q}{U} = \frac{q_1}{\Delta U_1} + \frac{q_2}{\Delta U_2} + \cdots + \frac{q_n}{\Delta U_n} = \frac{q_1}{\Delta U} + \frac{q_2}{\Delta U} + \cdots + \frac{q_n}{\Delta U}$$

Since $\dfrac{q_i}{\Delta U} = C_i$ is the capacitance of each capacitor, there is

$$C = C_1 + C_2 + \cdots + C_n \tag{9.7.6}$$

The equivalent capacitance of a parallel combination is equal to the sum of the capacitances of all capacitors.

2. Capacitors in series

Figure 9.7.5 shows the series connection of n capacitors. Due to electrostatic induction, each capacitor is charged with equal and opposite charges $+q$ and $-q$, which is also the electric quantity carried by the series combination

$$q = q_1 = q_2 = \cdots = q_n$$

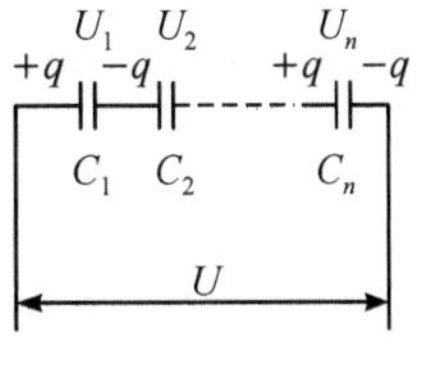

Figure 9.7.5 Capacitors in series

The total voltage across the series combination is the sum of the voltages across all individual capacitors

$$\Delta U = \Delta U_1 + \Delta U_2 + \cdots + \Delta U_n$$

The reciprocal of the equivalent capacitance is

$$\frac{1}{C} = \frac{\Delta U}{q} = \frac{\Delta U_1}{q_1} + \frac{\Delta U_2}{q_2} + \cdots + \frac{\Delta U_n}{q_n} = \frac{\Delta U_1}{q} + \frac{\Delta U_2}{q} + \cdots + \frac{\Delta U_n}{q}$$

Since $\dfrac{q_i}{\Delta U} = C_i$ is the capacitance of each capacitor, there is

$$\frac{1}{C} = \frac{1}{C_1} + \frac{1}{C_2} + \cdots + \frac{1}{C_n} \tag{9.7.7}$$

This formula indicates that the reciprocal of the capacitance of the series combination is equal to the sum of the reciprocals of all individual capacitors' capacitances.

Example 9.7.1 As shown in Figure 9.7.6, the distance between the two plates of a parallel plate capacitor is d, and the area of the parallel plate is S. A conductor plate of thickness t is inserted parallel between the plates. Ignoring the edge effect, find the capacitance of the capacitor.

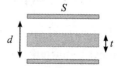

Figure 9.7.6 Figure of Example 9.7.1

Solution Assuming that the electric quantities of the two plates of the parallel plate capacitor are $+Q$ and $-Q$, then the surface density of the electric charge on the plate is $\sigma = \dfrac{Q}{S}$. The electric field strength between the two plates is

$$E = \frac{\sigma}{\varepsilon_0} = \frac{Q}{\varepsilon_0 S}$$

When a conductor plate with a thickness of t is inserted and the electrostatic equilibrium is reached, the electric field strength inside the conductor plate is zero ($E=0$). Then, the electric potential difference between the two polar plates is

$$\Delta U = E(d-t) = \frac{Q}{\varepsilon_0 S}(d-t)$$

Thus, the capacitance of the capacitor is $C = \dfrac{Q}{\Delta U} = \dfrac{\varepsilon_0 S}{d-t}$.

9.7.5 Application of capacitors in engineering technology

Due to the charging and discharging characteristics of capacitors, DC current cannot "pass through" them while AC current can "pass through" them; the charging or discharging process of a capacitor involves charge accumulation or release, so the voltage across it will not change suddenly. Because of the above characteristics, capacitors are widely used in coupling, bypassing, filtering, tuning, control and energy storage.

In recent years, with the increasing demand for high-power energy storage devices in electric vehicles, household appliances, and aerospace equipment, supercapacitors have come into being in place of traditional capacitors. Supercapacitors combine the fast large-current charging and discharging characteristics of traditional capacitors with the energy storage characteristics of batteries. Their energy density is thousands of times higher than those of traditional electrolytic capacitors, reaching the order of 1000 W·kg^{-1}, while the leakage current is thousands of times smaller. They can release hundreds to thousands of ampères of current instantly and large current discharging and even a short circuit will not have any effect on it; supercapacitors can be charged and discharged more than 100,000 times without any maintenance or care and with a lifespan of more than ten years.

Supercapacitors are capacitors that use the double layer principle. As shown in Figure 9.7.7, when an external voltage is applied to the two plates of a supercapacitor, the positive electrode of the plate stores positive charges and the negative plate stores negative charges, just like a normal capacitor. Under the action of the electric field generated by the charges on the two plates of the supercapacitor, opposite charges are formed on the interface between the electrolyte and the electrode to balance the internal electric field of the electrolyte. The positive and negative charges are arranged in opposite positions on the contact surface between two different phases and with an extremely short gap. This charge distribution layer is called the Helmholtz double layer with very large capacitance.

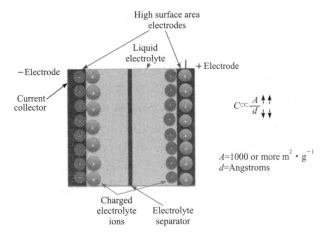

Figure 9.7.7　Double layer electric supercapacitor

In 2019, Shanghai's first batch of 10 fast-charging, high-energy smart supercapacitor buses were put into operation on Route 26; At the same time, the world's first DC super-fast smart charging bus terminal was built at the Bund in Shanghai. This kind of bus, which is equipped with China's independently developed supercapacitor fast-charging technology and a domestically produced intelligent flexible charging bow has a range of 10 kilometers after being charged for 3 minutes and can travel up to 30 kilometers when fully charged without the need for recharging during the journey, as Figure 9.7.8 shows.

Figure. 9.7.8　Bus with a supercapacitor and charging bow

9.8 Electrostatic energy

9.8.1 Electrostatic energy of a charged system

The charging process of objects can be regarded as the process of charge transfer and charge accumulation. In this process, the charges accumulated later move relative to the charges accumulated earlier. Since there is an interactive electric field force between the charges, the external force must overcome the electric field force to do work when moving these charges. According to the law of energy conversion and conservation, work done by an external force on a charged system is equal to the increase in the electrostatic energy of the charged system. Assume that charges in a charged system can be divided into an infinite number of small parts and these parts are initially dispersed at an infinite distance from each other, then the electrostatic energy in this state is usually specified to be zero. Therefore, the electrostatic energy of a charged system in any state is equal to the total work done to resist the electrostatic force when each part of the charges is gathered from an infinitely dispersed state into an existing charged system.

The electrostatic energy of a capacitor is discussed below by taking the charging process of a parallel plate capacitor as an example.

The capacitor does not store electric energy when it is not charged. During the charging process, the external force must overcome the electrostatic force to do work, and the positive charge is transported from the negatively charged negative plate to the positively charged positive plate. The work done by the external force is equal to electrostatic energy stored in the capacitor.

As shown in Figure 9.8.1, the parallel plate capacitor is in the charging process. At a certain time, the electric potential difference between the two plates is ΔU. If the positive charge of dq is moved from the negative plate to the positive plate, the work done by the external force to overcome the electric field force is

$$dW = \Delta U dq = \frac{q}{C} dq$$

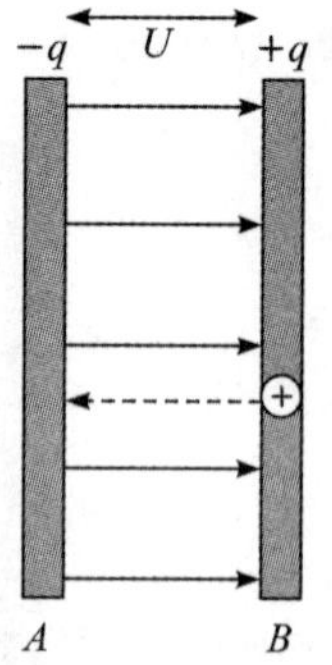

Figure 9.8.1 Electrostatic energy of a capacitor

When the two plates of the capacitor are charged with $\pm Q$, respectively, the work done by the external force is

$$W = \int dW = \int_0^Q \frac{q}{C} dq = \frac{Q^2}{2C} = \frac{1}{2} Q \Delta U = \frac{1}{2} C \Delta U^2 \qquad (9.8.1)$$

This is the electrostatic energy stored by the capacitor. It can be seen from Equation

(9.8.1) that when the electric potential difference is constant, the larger the capacitance of the capacitor, the more energy it can store. In this sense, the capacitance is a symbol of the capacity of the capacitor to store electrical energy.

9.8.2 Electrostatic energy of an electric field

From the above discussion, it can be seen that all charged systems have a certain amount of electrostatic energy. Then is this electrostatic energy concentrated on the charge or localized in the electric field? This question cannot be answered in electrostatics, because in the electrostatic field, the electric field is always accompanied by electric charges. We will know in later chapters that altering electric and magnetic fields, when propagating in space at a certain speed, form electromagnetic waves, which can carry energy and travel far away from the excitation field source. A large number of experimental facts show that electric energy is localized in the electric field, so the energy of a charged system is essentially the energy of the electric field established by the system.

Here, we will discuss the electrostatic energy of the electric field by taking a parallel plate capacitor as an example. For a parallel plate capacitor with a plate area S and plate spacing d, the volume of the space occupied by the electric field is Sd. According to Equation (9.8.1), the electrostatic energy stored in the capacitor is

$$W = \frac{1}{2}C\Delta U^2 = \frac{1}{2}\frac{\varepsilon S}{d}(Ed)^2 = \frac{1}{2}\varepsilon E^2 Sd = \frac{1}{2}\varepsilon E^2 \Delta V$$

The electrostatic energy per unit volume in the electric field, i.e., the electrostatic energy density is

$$w = \frac{W}{V} = \frac{1}{2}\varepsilon E^2 = \frac{1}{2}DE \qquad (9.8.2)$$

It can be proved that this equation for electrostatic energy density is valid for any electric field. For an arbitrary electric field, after choosing a volume element dV, the electrostatic energy density in dV can be regarded as a constant, and the energy stored in it can be obtained by taking an integral, that is

$$W = \int_V w\,dV = \int_V \frac{1}{2}\varepsilon E^2\,dV = \int_V \frac{1}{2}DE\,dV \qquad (9.8.3)$$

In anisotropic dielectrics, where D and E have different directions, a more general expression is

$$W = \int_V \frac{1}{2}\boldsymbol{D} \cdot \boldsymbol{E}\,dV \qquad (9.8.4)$$

Example 9.8.1 As shown in Figure 9.8.2, the radii of the inner and outer spherical shells of a spherical capacitor are R_1 and R_2, respectively. The space between the two spherical shells is filled with a dielectric whose relative permittivity is ε_r. Find the energy stored in this capacitor with an electric quantity Q.

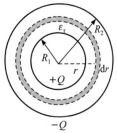

Figure 9.8.2 Figure of Example 9.8.1

Solution According to Gauss's law in dielectrics, there is

$$\oint_S \boldsymbol{D} \cdot d\boldsymbol{S} = \sum q_0$$

Then, the magnitude of the electrical displacement $\boldsymbol{D}$ between the spherical shells is

$$D = \frac{Q}{4\pi r^2}$$

Since $\boldsymbol{D} = \varepsilon_0 \varepsilon_r \boldsymbol{E} = \varepsilon \boldsymbol{E}$, there is

$$E = \frac{Q}{4\pi \varepsilon r^2}$$

The energy density of the electric field is

$$w = \frac{1}{2}\varepsilon E^2 = \frac{Q^2}{32\pi^2 \varepsilon r^4}$$

Take an element spherical shell with a radius r and thickness dr, and its volume is $dV = 4\pi r^2 dr$. The electrostatic energy in this volume element is

$$dW = w dV = \frac{Q^2}{32\pi^2 \varepsilon r^4} 4\pi r^2 dr = \frac{Q^2}{8\pi \varepsilon r^2} dr$$

The total energy stored in the electric field is

$$W = \int_{R_1}^{R_2} \frac{Q^2}{8\pi \varepsilon r^2} dr = \frac{Q^2}{8\pi \varepsilon} \left(\frac{1}{R_1} - \frac{1}{R_2} \right)$$

Scientist Profile

Yu Min (1926 - 2019), a native of Ninghe County, Hebei Province, was a famous nuclear physicist, a pioneer of the "two bombs and one satellite" project and a winner of the National Highest Science and Technology Award.

Yu Min graduated from the Department of Physics of Peking University in 1949. He served as vice president of the 9th Institute of the Second Ministry of Machine Building and deputy director of the Science and Technology Committee of the Ministry of Nuclear Industry. He was elected academician of the Chinese Academy of Sciences in 1980.

Yu Min was one of the outstanding leading figures in China's nuclear weapon theoretical research and defense high-tech development. In the 1960s, Yu Min began to devote himself to China's nuclear weapon cause and led and participated in the theoretical research and design of nuclear weapons. He organized and led a research team that discovered the key to achieving self-sustaining thermonuclear combustion of the hydrogen bomb, made a breakthrough in the hydrogen bomb technical approach, proposed a complete hydrogen bomb physical design plan including principles, materials and configurations and completed the theoretical design of the nuclear device, i. e. , China's first generation of nuclear weapon design scheme. As the main person in charge of the gaseous detonator bomb, the key to the miniaturization of the hydrogen bomb, Yu Min presided over the research and solved a series of key problems, including the compression of fission materials, neutron injection and its proliferation law, deuterium and tritium ignition and combustion law, the influence of light and heavy medium mixing on fusion and the high-energy neutron fission feedback law. He proposed specific measures to increase the design margin of the two key links. The successful development of the gas detonator laid a reliable foundation for the development of the second generation of nuclear weapons in China.

In the nuclear weapon development strategy, he and Deng Jiaxian proposed "accelerating the nuclear testing process". The proposal planned the deployment of China's nuclear tests in advance, enabling the Party Central Committee to make a decisive decision that won precious 10 years of nuclear testing time for China. It played an extremely important forward-looking role in improving China's nuclear weapon level, promoting the equipment of troops with nuclear weapons and forming combat effectiveness. In response to the nuclear test ban, the idea of weapon research such as precision laboratory experiments was proposed. Later, this proposal was adopted and evolved into the four pillars of the development of China's nuclear weapon industry and it is still the guiding ideology for the development of China's nuclear weapon industry. Since the 1970s, Yu Min played an important role in advocating and promoting national defense high-tech projects, especially the inertial confinement nuclear fusion research. He was a pioneer in theoretical research in the fields of inertial confinement fusion and X-ray laser in China.

Yu Min was a scientist who was loyal to the motherland, selfless, well-versed in both arts and sciences and possessed profound humanistic qualities. He made indelible historical contributions to China's nuclear weapon cause.

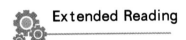

Extended Reading

Triboelectric nanogenerator

Triboelectricity is one of the most common phenomena in nature, occurring while combing hair, dressing, walking and driving. However, triboelectricity is difficult to collect and utilize, so it is often overlooked. In 2012, a research team led by Professor Wang Zhonglin of Georgia Institute of Technology in the United States developed a transparent flexible triboelectric nanogenerator, which successfully converted friction into usable electricity with the help of flexible polymer materials. The invention of the triboelectric nanogenerator is a milestone discovery in the fields of mechanical energy generation and self-driving system and it provides a new model for effectively collecting mechanical energy.

Triboelectrification refers to the process in which two different objects carry equal but opposite charges during the process of contact and separation. The triboelectric nanogenerator uses the principles of triboelectrification and electrostatic induction to promptly conduct charges carried by an object during friction to an external circuit, thereby converting mechanical energy into electrical energy. The triboelectric nanogenerator is mainly composed of two parts: the friction generation layer and the electrostatic induction layer. The friction generation layer is composed of two different dielectric layers with very different electronegativity (suppose the thicknesses are d_1 and d_2 and the dielectric constants are ε_1 and ε_2, respectively), which is mainly used for triboelectrification. The electrostatic induction layer is the electrode on the back of the dielectric, which is mainly used for electrostatic induction. When two dielectrics rub against each other under the driving force of external mechanical energy, charge transfer will occur at the interface where the two dielectrics contact due to the triboelectric effect. When the two are separated, the electrostatic field created by the friction charges will drive the electrons in the external circuit to flow due to the electrostatic induction effect. As the two continue to rub and separate, the charge density on the surface will gradually reach saturation.

Since the birth of the triboelectric nanogenerator, its output performance has reached a new height under the promotion of many scholars around the world. It is reported that under normal conditions, the surface charge density of the material can be increased to 1020 $\mu C \cdot m^{-2}$, the power generation capacity can be as high as 20 $W \cdot m^{-2}$, and the instantaneous conversion efficiency is nearly 50%. In addition, triboelectric nanogenerators have been shown to be used to collect a variety of mechanical energy in surrounding environment, such as human running, blinking, mechanical instrument vibrations, sound waves and ultrasound. Triboelectric nanogenerators can also convert mechanical signals into electrical signals, so sensors based on this are a popular application direction and related touch panels and electronic skins have been frequently reported.

Although triboelectric nanogenerators have many application scenarios and cover a wide range of fields, their basic working modes are mainly divided into the following four types.

The first is the vertical contact-separation mode. As shown in Figure ER 1(a), when two membrane materials with different electron binding abilities come into physical contact, they carry different charges on their surfaces. When the two membranes are separated, a potential difference is generated on the back electrodes of the two membranes, thus forming a current. The subsequent contact and separation of the two membranes also generates a potential difference, inducing the generation of current. This mode is called the vertical contact-separation mode. In order to realize the reciprocating motion in the mechanical structure, various structures have been developed, including arched, spring-supported, Z-shaped, and cantilever beam. This type of triboelectric generator is suitable for linear reciprocating motion and instantaneous linear impact. The electrical energy comes from the kinetic energy during periodic contact and separation and the output is also alternating current. This mode has the characteristics of high output and long membrane life and is a relatively commonly used working mode. However, the separation movement will form gap voids with volume changes, which brings certain challenges to the packaging of the generator. The vertical contact-separation mode has been widely used to collect kinetic energy such as human motion and mechanical vibration. Sensors developed based on it can better detect magnetic signals, pressure signals, acceleration signals, vibration signals, mercury vapor content, catechin content, sound wave information, etc.

The second is the horizontal sliding mode. As shown in Figure ER 1(b), the structure is similar to the vertical contact-separation mode except that the movement direction of the two membranes in contact and separation changes by 90 degrees. When the two membranes overlap, no external potential difference is generated due to the mutual compensation of positive and negative charges. Once relative sliding occurs between the membranes, the positive and negative charges that are unable to contact each other will form their own potentials, thereby generating a potential difference between the two back electrodes and inducing the generation of current. Similarly, the reciprocating overlapping and offsetting of the two membranes will form an electric current in the external circuit. Compared with the vertical contact-separation mode, the sliding direction within the surface is more flexible and changeable and is more suitable for collecting energy such as planar motion and turntable cylindrical motion. More importantly, there is a good linear relationship between the number of transferred charges and the amount of displacement and the fabricated sensor can quantitatively analyze external signals. However, long-term friction between the two membranes will wear the microscopic morphology of the membrane surface and result in a decrease in the membrane's microscopic surface area and charge surface density, thereby affecting the output performance. In addition, friction may also cause debris, further affecting the output performance of the generator. Generators based on this model have been successfully applied to collect energy in the form of wind energy, hydraulic kinetic energy, etc., and displacement sensors and speed sensors have also been developed.

The third is the single-electrode mode. In this mode a triboelectric nanogenerator has only one output electrode. Other materials produce friction with this electrode (or the materials on this electrode). After each of them carries different charges, they periodically contact and separate from this electrode or slide in and out like in an in-plane sliding mode, thereby forming a potential difference between the earth (zero potential) and the electrode and driving the flow of electrons between them to cause an output current, as shown in Figure ER 1(c). Because there is no need to connect electrodes to the moving object, the requirements for the moving object are lowered. Therefore, this model has good practicality and has been successfully applied to collect energy from air flow, rotating tires, falling liquid droplets, and flipping notepads. It can also be used to detect displacement signals, touch sensing, trajectory route, speed angle sensing, acceleration sensing, pressure sensing, identity authentication and many other signals. For lack of the

Coulomb force of charges on the other electrode (or its membrane) in the single-electrode mode, the charge transfer capability decreases, and thus the output is also greatly reduced. Generators that are mainly used for power generation usually avoid using this mode.

The fourth is the independent layer mode. This mode came into being with the characteristics of no electrode constraint in the single-electrode mode and high output in the dual-electrode mode. The two electrodes do not need to move in the same plane and the friction layer material can slide back and forth freely on it. As shown in Figure ER 1(d), the electrode covered by the friction layer obtains a lower potential (high potential when the friction layer is positively charged), and the larger the coverage area, the lower the potential. When the coverage areas are not equal, a potential difference will result between the two electrodes, thereby promoting the movement of electrons between the electrodes until the potential difference is offset. However, once the friction layer slides, causing the friction layer coverage area of the two electrodes to change, the potential difference between the two electrodes also changes accordingly, thereby inducing the generation of current between the electrodes. It can be seen from this that if the friction layer moves so that the coverage area of one electrode increases while the coverage area of the other electrode decreases, then the charge transfer capability will be greatly improved, resulting in better generator output performance. The triboelectric generator based on this model has the advantages of high output and flexibility, but it requires careful design of the electrodes and friction layer, so there is a lot of room for improvement in design. Successful applications of electrode patterns include radial electrodes, fork electrodes, chessboard electrodes, honeycomb three electrodes, etc. They each have their own unique advantages. In particular, generators based on radial electrodes can efficiently collect energy from rotating objects and are often used to make ultra-high-output nanogenerators, which can generate enough electricity to charge mobile phones and light up household LED lights.

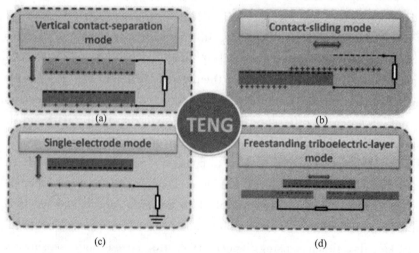

Figure ER 1   Four modes of a triboelectric nanogenerator

The materials that make up a triboelectric nanogenerator are divided into friction materials and electrode materials. Electrode materials are usually metal foils, metal particles, etc., and there are also studies using other conductive materials such as indium tin oxide conductive glass and graphene. The selection range of friction materials is very wide. Whether they are metal, polymer, oxide, or even human hair and skin, almost all materials we know have the triboelectric effect, which greatly expands the application range of the triboelectric nanogenerator. Generally speaking, when selecting materials for a triboelectric nanogenerator, the first thing to consider is the material's ability to gain or lose electrons. As early as 1957, Wilcke published the first friction sequence of different materials, which arranged the relative order of materials according to the properties of the surface electrons that are easy to lose and the electrons that are easy to gain when the materials come into contact. The farther apart two materials are in the sequence, the more charges they will have when they come into contact. Therefore, materials at the two ends of the friction sequence are used more frequently, such as polytetrafluoroethylene and polydimethylsiloxane at the negative charge end, and polyamide and polyethylene terephthalate at the positive charge end.

The triboelectric effect depends not only on the type of a material but also on the surface morphology and structure of the material. Therefore, the purpose of increasing the surface charge density and friction contact area can be achieved by designing the proper material morphology and structure. With the development of micro-nano technology, different micro-nano structures are introduced into the structural design of friction materials to achieve the purpose of improving output power. As shown in Figure ER 2, growing nanowires or nanorod arrays on the surface of the friction material can greatly increase the specific surface area of the material and thus increase the magnitude of the friction current.

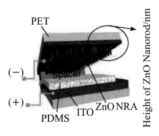

Figure ER 2　Triboelectric nanogenerator with zinc oxide nanorod arrays growing on the surface of the friction material

Since Faraday discovered the phenomenon of electromagnetic induction in 1831, the electromagnetic generator had been the main equipment for converting mechanical energy into electrical energy, providing electrical energy to the world. After hundreds of years of development and improvement, the electromagnetic generator can now convert high-speed mechanical energy with high quality, high efficiency and high output. However, it has strict requirements for the working environment and cannot work in extremely harsh environments such as humidity. Most importantly, its power source mainly comes from steam turbines, diesel engines or fuel combustion. In the face of decreasing non-renewable resources and increasingly serious environmental problems, its shortcomings are obvious. Compared with it, a triboelectric nanogenerator has incomparable advantages. It can collect a wide range of energy, clean and pollution-free, such as human mechanical energy, wind energy and water energy. As long as the energy causes friction, it can be collected. The preparation process is simple and the preparation price is low, which is conducive to large-scale production. Since the triboelectric nanogenerator relies on the coupling effect of triboelectrification and electrostatic induction, it is easy to control its generation of electrical energy and realize intelligence.

The triboelectric nanogenerator, which was born in 2012, has received widespread attention and developed rapidly since its launch. Now, more than 3,000 workers from more than 400 grassroots units in more than 40 countries are engaged in it. Judging from current research results, breakthroughs may be first achieved in three areas. First, it can be integrated into a self-driving system and take a place in the IoT sensor system. Secondly, if used as an active sensor, it also has potential application value in human-computer interaction, robotics, artificial intelligence and security systems, and wireless communication in these fields can be achieved in the future. Furthermore, due to its low cost, it can be integrated and laid on a large scale to form a large-scale array for collecting wave and wind energy, which can be used in the sea to collect blue energy. Finally, the triboelectric generator can provide a high-voltage source for specific high-voltage application scenarios, such as plasma excitation and field emission and there are more possibilities to be explored in the future.

Discussion Problems

9.1 What is the difference and correlation between $E=\dfrac{F}{q_0}$ and $E=\dfrac{1}{4\pi\varepsilon_0}\dfrac{q}{r^2}r_0$? What is the requirement on q_0 in the former formula?

9.2 If three equal charges are placed on the three vertices of an equilateral triangle, and a spherical surface is made with the center of the triangle as the center, can the field strength generated by them be calculated using Gauss's law? Does Gauss's law hold for this spherical surface?

9.3 If the electric flux Φ_e passing through a closed surface S is zero, is it true that
(1) the field strength at each point on the surface S is equal to zero?
(2) there is no charge surrounded by the surface S?
(3) the net charge enclosed by the surface S is zero?

9.4 There are two identical plates placed parallel to each other in a vacuum. The distance between them is d, the areas of the plates are S, and the charges are $+q$ and $-q$, respectively. Which of the following conclusions is correct and why?

(1) According to Coulomb's law, the force acting on the two plates is $f = \dfrac{q^2}{4\pi\varepsilon_0 d^2}$.

(2) Since $f = qE$, $E = \dfrac{\sigma}{\varepsilon_0}$ and $\sigma = \dfrac{q}{S}$, there is $f = \dfrac{q^2}{\varepsilon_0 S}$.

(3) The electric field produced by one plate at the other plate is $E = \dfrac{\sigma}{2\varepsilon_0}$. Thus $f = qE = \dfrac{q^2}{2\varepsilon_0 S}$.

Problems

9.5 As shown in Figure T9 - 1, two point charges with the same electric quantity $q = 10^{-10}$ C are placed at (0 m, 0.1 m) and (0 m, −0.1 m) in a plane rectangular coordinate system, respectively. Find

(1) the magnitude and direction of the electric force acting on a point charge at (0.2 m, 0 m) with an electric quantity of $Q = 10^{-8}$ C;

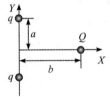

Figure T9 - 1 Figure of Problem 9.5

(2) the position of the point charge Q when the force is at its maximum.

9.6 Assuming that there are three point charges in space, their positions are at (0 m, 0 m), (3 m, 0 m) and (0 m, 4 m) and their electric quantities are 5×10^{-8} C, 4×10^{-8} C and -6×10^{-8} C, respectively. Find the total electric flux through a spherical surface of radius 5 m centered at (0, 0).

9.7 A straight wire AB of length l is uniformly charged with a line charge density λ, as shown in Figure T9 - 2. Find

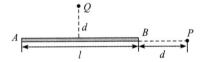

Figure T9 - 2 Figure of Problem 9.7

(1) the electric field strength and potential at point P, which is on the extension of the wire and has a distance of d away from one end of the wire (point B);

(2) the electric field strength and potential at point Q, which is on the vertical bisector of the wire and is at a distance d from the midpoint of the wire.

9.8 As shown in Figure T9 - 3, a very thin non-conductive plastic rod is bent into an arc with a radius of 50 cm, and the distance between its two ends is 2 cm. Positive charges of 3.12×10^{-9} C are uniformly distributed on the thin rod. Find the magnitude and direction of the electric field at the center of the circle and its electric potential.

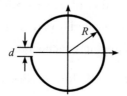

Figure T9-3   Figure of Problem 9.8

9.9   The volume charge density of a sphere of radius R is $\rho=kr$, where r is the radial distance from the sphere center and k ($k>0$) is a constant. Find the electric field and potential distribution in space, and plot the relationship curve between E and r.

9.10   As shown in Figure T9-4, there are two concentric spherical surfaces with radii of 10 cm and 20 cm, respectively. Positive charges with a charge density of $\rho=5.29\times 10^{-10}$ C·m^{-3} are evenly distributed in the space between them. Find the electric field and potential at a distance of 5 cm, 10 cm and 50 cm from the center of the spherical surface, respectively.

9.11   As shown in Figure T9-5, an infinite flat plate with a thickness of d is uniformly charged, and the volume charge density is ρ. Find the electric field

(1) at the center of the thin layer;

(2) in the plate at a distance r from the surface of the thin layer;

(3) outside the thin layer.

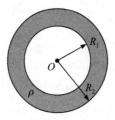

Figure T9-4   Figure of Problem 9.10   Figure T9-5   Figure of Problem 9.11

9.12   Four point charges q_1, q_2, q_3 and q_4 with the same electric quantity of 4×10^{-9} C are placed on the four vertices of a square, respectively. The distance between each vertex and the center point O of the square is 5 cm. Find

(1) the electric field and potential at point O;

(2) the work done by the electric field when moving a test charge ($q_0=4\times 10^{-9}$ C) from infinity to point O;

(3) the change in the potential energy of q_0 in the process described in Question (2).

9.13   As shown in Figure T9-6, the side length of the cube is $d=10$ cm. The components of the electric field are $E_x=bx^{1/2}$, and $E_y=E_z=0$, where $b=800$ N·C^{-1}·m$^{-1/2}$. Find

(1) the total electric flux through the surface of the cube;

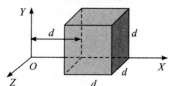

Figure T9-6   Figure of Problem 9.13

(2) the net charge in the cube.

9.14 As shown in Figure T9-7, the conductor sphere of radius $R_1 = 2.0$ cm has a concentric conductor shell. The inner and outer radii of the shell are $R_2 = 4.0$ cm and $R_3 = 5.0$ cm, respectively. The sphere has an electric quantity of $Q = 3.0 \times 10^{-8}$ C. Find

(1) the energy stored in the electric field;

(2) the energy stored in the electric field when the conductor shell is grounded;

(3) the capacitance of the capacitor.

9.15 As shown in Figure T9-8, a conductor sphere of radius R is originally uncharged, it is placed in the electric field of the point charge $+q$, and the distance between the center of the sphere and the point charge is d. Find

(1) the electric potential of the conductor sphere;

(2) the electric quantity of induced charges on the sphere when the conductor sphere is grounded.

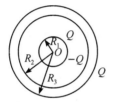
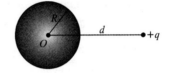

Figure T9-7 Figure of Problem 9.14   Figure T9-8 Figure of Problem 9.15

Challenging Problems

9.16 (1) Find the potential and electric field of an infinitely long and uniformly charged planar thin strip with a width of $2L$ and a surface charge density of σ ($\sigma > 0$). What are the characteristics of the potential and electric field strength surfaces?

(2) If there is another infinitely long thin strip with the same size, equal and opposite charges, which is placed parallel at a distance d, find the electric force acting on per unit length between the two thin strips.

9.17 A uniformly charged ring has a radius a and electric quantity Q ($Q > 0$).

(1) Find the electric potential and electric field in the plane of the ring. How do the electric potential and electric field strength change with position?

(2) Find the electric potential and electric field around the ring. What are the characteristics of the electric potential surface of the ring? What are the characteristics of equipotential lines and electric field lines?

Chapter 10

Magnetic Field

In the field of magnetism, the ancient Chinese people made great contributions. As early as the Spring and Autumn Period, with the development of metallurgy and the application of iron, people learned about the lodestone. There were descriptions and records about magnets in *Guiguzi*(《鬼谷子》), *Lvshi Chunqiu*(《吕氏春秋》) and other ancient books. After the Han Dynasty, more and more works recorded the phenomenon of magnets attracting iron. *Si Nanshao*(司南勺) was described in *Lunheng*(《论衡》). It is recognized as the earliest magnetic navigation device. In the 11th century, Chinese scientist Shen Kuo(沈括) invented the compass and discovered the geomagnetic declination. At the beginning of the 12th century, there was a clear record of the compass being used in navigation in our country.

For a long time in history, people thought that magnetism and electricity were two distinct phenomena, and their study had been developing independent of each other. Until 1820, the Danish scientist H. C. Oersted first discovered that the magnetic needle located near a current-carrying wire would be deflected. Later, A. M. Ampère and other scientists found that the current-carrying wire near a magnet was also affected, and interactive charged particles between two current-carrying wires would be deflected near the magnet. In 1822, Ampère put forward a hypothesis about the nature of magnetism of matter. He believed that the source of all magnetic phenomena is current, that is, the movement of electric charges and molecular current exists in the molecules of any object and is equivalent to elementary magnets, resulting in magnetic effect. His hypothesis is consistent with the modern theory of electrical structure of matter. In addition to the movement of electrons in a molecule around the nucleus, electrons themselves also have spin movements. These movements of electrons in the molecule are equivalent to the loop current. From this it can be seen that current is the root of all magnetic phenomena. This chapter will mainly discuss the properties of a stable magnetic field which does not change with time.

Chapter 10   Magnetic Field   **65**

10.1   Current intensity and current density

10.1.1   Current intensity

The directional motion of a large number of charged particles forms an electric current. Charged particles can be electrons, positive or negative ions, and positively charged "holes" in semiconductors, etc. These charged particles are collectively referred to as carriers.

The strength of current is described in terms of current intensity, which is defined as the electric quantity that passes through a particular section of a conductor per unit time. If the electric quantity passing through a section S of the conductor in time dt is dq, the current intensity passing through that section is

$$I = \frac{dq}{dt} \tag{10.1.1}$$

The current intensity is a scalar quantity, and it is customary to specify the direction of motion of the positive charge as the direction of the current. In the SI system of units, the unit of current strength is the ampère and the symbol is A.

$$1 \text{ A} = 1 \text{ C} \cdot \text{s}^{-1}$$

10.1.2   Current density

The common current is the current flowing along a wire. In practical problems, current often flows in a wire of uneven thickness or in a large conductor, such as the current in the earth, the current in the electrolyte of the electrolytic tank and the current through discharging gas. In this case, the magnitude and direction of the current in different parts of a conductor are different, resulting in a certain distribution of current. The concept of current density is introduced to describe the directional motion of charges in the conductor.

The current density j is defined as follows: At any point in a conductor, the direction of j is the same as the direction of the current at that point, and the magnitude of j is equal to the current per unit cross section area. In SI units, the unit of current density is A $\cdot$ m^{-2}. As shown in Figure 10.1.1, choose a point in the conductor and take an area element $dS_\perp$ perpendicular to the current direction, its normal direction n_0 taken as the direction of the current at this point. If the current through the area element is dI, then the current density vector at this point is

$$j = \frac{\mathrm{d}I}{\mathrm{d}S_\perp} n_0 \qquad (10.1.2)$$

The current density can accurately describe the magnitude and direction of the current at each point in a conductor. What is called current distribution actually refers to the distribution of the current density j, and the strength and direction of current in a strict sense refer to the magnitude and direction of the current density.

In bulk conductors, the current density may vary from place to place and may also vary over time. In this chapter we only discuss the case in which the current density does not vary with time in the conductor, that is, constant current.

An area element $\mathrm{d}S$ is taken at any point in the carrier conductor. As shown in Figure 10.1.1, the normal vector of $\mathrm{d}S$ is at an angle θ to the direction of the current there. The projected area of the surface element $\mathrm{d}S$ in the direction perpendicular to the current is $\mathrm{d}S_\perp = \mathrm{d}S\cos\theta$. The current intensity passing through the area element is $\mathrm{d}I$, and then the magnitude of current density at this point is

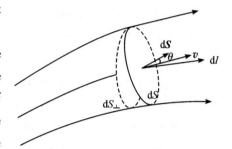

Figure 10.1.1 Current density

$$j = \frac{\mathrm{d}I}{\mathrm{d}S\cos\theta} \qquad (10.1.3)$$

The above equation can be written as

$$\mathrm{d}I = j\,\mathrm{d}S\cos\theta = \boldsymbol{j} \cdot \mathrm{d}\boldsymbol{S} \qquad (10.1.4)$$

Equation (10.1.4) states that the current intensity $\mathrm{d}I$ on a cross section is equal to the flux of the current density j through that cross section. The current intensity I through any cross section S in a conductor can be expressed as

$$I = \int_S \boldsymbol{j} \cdot \mathrm{d}\boldsymbol{S} \qquad (10.1.5)$$

Assuming that there is only one type of carrier in a conductor; in the absence of an external electric field, these carriers move randomly with an average velocity of zero, and no current is generated. In an applied electric field, carriers in a conductor will have an average orientation velocity v, forming a current, and this average orientation velocity is called the drift velocity. If the electric quantity carried by each carrier is q and the number density of carriers is n, then the current passing through the area element $\mathrm{d}S$ per unit time is

$$\mathrm{d}I = (nq\boldsymbol{v})\,\mathrm{d}\boldsymbol{S} \qquad (10.1.6)$$

Comparing Equation (10.1.4) with Equation (10.1.6), the current density can be written as

$$\boldsymbol{j} = nq\boldsymbol{v} \qquad (10.1.7)$$

For positive carriers ($q>0$), j is in the same direction as v. For negative charge carriers ($q<0$), j is in the opposite direction to v.

10.2 Magnetic induction intensity

10.2.1 Magnetic field

Both modern scientific theories and experiments have confirmed that interaction between static charges is transmitted through an electric field, and interaction between moving charges, magnets or currents is also transmitted through a field, which is called magnetic field. A magnetic field is a special form of matter that exists in the space around a moving electric charge (or current). The force of the magnetic field acting on the moving electric charge is called the magnetic force. Interactions between moving charges, between currents, and between currents or moving charges and magnets can all be seen as the result of the magnetic field excited by either party exerting a force on the other.

10.2.2 Magnetic induction intensity

In electrostatics, to quantitatively describe the distribution of an electric field, we use the effect of the electric field on the test charge to define the electric field strength. A method similar to the study of electrostatic fields is used to define the magnetic induction intensity B from the effect of the magnetic field on moving charges. A test charge with an electric quantity q and moving at a velocity v is introduced into a magnetic field. The experiment found that the force of the magnetic field on the moving test charge showed the following laws. ① When the charge q passes through different field points, the force on the charge is different. Even with the same field point, when the direction of the velocity is different, the force on the charge is different. But, there are alway a direction in which the force acting on the moving charge is zero. Then this direction can be defined as the direction of the magnetic direction. ② When the charge q is not parallel to the magnetic field, the direction of the magnetic force F on it is always perpendicular to the plane formed by the charge velocity and the magnetic field direction. ③ When the direction of the charge velocity is perpendicular to the direction of the magnetic field, the magnetic field force on the charge is the maximum. And the magnitude of the maximum magnetic force F_m is proportional to the magnitude of the charge q and the speed of the charge, but the ratio F_m/qv has nothing to do with qv of the moving charge. Thus, the magnitude of the magnetic induction intensity B is defined as

$$B = \frac{F_m}{qv} \tag{10.2.1}$$

Since the direction of the maximum magnetic force F_m is always perpendicular to the plane composed of **B** and **v**, as shown in Figure 10.2.1, the direction of the magnetic induction intensity **B** can be determined according to the maximum magnetic field force F_m and the moving speed of the positive charge. The direction of **B** is specified as the direction of $F_m \times q v$. This is consistent with the magnetic field direction determined by the N pole of a small magnetic needle.

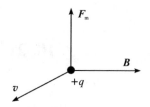

Figure 10.2.1 Relationship of **B**, **v** and F_m in the magnetic field

In the International System of Units, the unit of the magnetic induction intensity **B** is the Tesla (T). According to Equation (10.2.1), 1 Tesla $= 1 \text{ N} \cdot \text{s} \cdot \text{C}^{-1} \cdot \text{m}^{-1} = 1 \text{ N} \cdot \text{A}^{-1} \cdot \text{m}^{-1}$.

The magnetic induction value on the earth's surface is about 0.3×10^{-4} T (equator) to 0.6×10^{-4} T (poles). The magnetic induction intensity of a general permanent magnet is about 10^{-2} T, a large electromagnet can generate a magnetic field of 2 T, and a magnet made of superconducting material can generate a magnetic field of 10^2 T.

10.2.3 The Biot-Savart law

In 1820, the French physicists J. B. Biot and F. Savart did a lot of experimental research on the magnetic fields excited by different shapes of current-carrying wires. According to the analysis of the experimental results, the law of the magnetic field generated by the current element was obtained. The French mathematician and physicist P. S. Laplace summarized the results of Biot and Savart into a mathematical formula which is called the Biot-Savart law, expressed as follows: In a vacuum, the magnitude of the magnetic induction d**B** generated by any current element $I\mathrm{d}\boldsymbol{l}$ at a given point P is proportional to the magnitude of the current element, proportional to the sine of the angle θ between the current element and the vector pointing to point P from the current element, but inversely proportional to the square of the distance r from the current element to point P. The direction of d**B** is perpendicular to the plane formed by $I\mathrm{d}\boldsymbol{l}$ and r, and satisfies the right-hand rule, as shown in Figure 10.2.2. The vector form is

$$\mathrm{d}\boldsymbol{B} = \frac{\mu_0}{4\pi} \frac{I\mathrm{d}\boldsymbol{l} \times \boldsymbol{r}_0}{r^2} \qquad (10.2.2)$$

where $\boldsymbol{r}_0$ is the unit vector pointing from the current element $I\mathrm{d}\boldsymbol{l}$ to the field point P, and

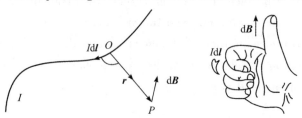

Figure 10.2.2 Magnetic induction intensity of a current element

$\mu_0 = 4\pi \times 10^{-7} \text{T} \cdot \text{m} \cdot \text{A}^{-1}$ is magnetic permeability of vacuum.

According to the principle of superposition of field strengths, the magnetic field generated by the entire current wire at point P is

$$\boldsymbol{B} = \int_l \mathrm{d}\boldsymbol{B} = \int_l \frac{\mu_0}{4\pi} \frac{I \mathrm{d}\boldsymbol{l} \times \boldsymbol{r}_0}{r^2} \tag{10.2.3}$$

When the direction of the magnetic induction intensity generated by each current element is different, an appropriate coordinate system must be selected. Project $\mathrm{d}\boldsymbol{B}$ in the direction of the coordinate axis, then integrate the coordinate component formula, and finally express the total magnetic induction intensity vector. It should be pointed out that the Biot-Savart law is based on the analysis of a large number of experimental facts and cannot be directly verified by experiments. However, the results calculated by the law are in good agreement with the experiments, which indirectly verifies the correctness of the Biot-Savart law. Now use the Biot-Savart law and the superposition principle to calculate the magnetic induction intensity of the magnetic field generated by some special current-carrying wires.

Example 10.2.1 It is known that there is a constant current I in the current-carrying straight wire, and the length of the wire is L. The angles between the two ends of the wire and the lines connecting point P are θ_1 and θ_2 respectively. The distance between point P and the wire is a. As shown in Figure 10.2.3, find the magnetic induction intensity generated by the straight wire at point P.

Solution As shown in the figure, select the foot point from point P to the wire as the coordinate origin, and establish a Cartesian coordinate system OXY. Take the current element $I\mathrm{d}\boldsymbol{l}$, and the corresponding position vector is $\boldsymbol{r}$ and the unit vector is $\boldsymbol{r}_0$. According to the Biot-Savart law, the magnetic field generated by this current element at point P is

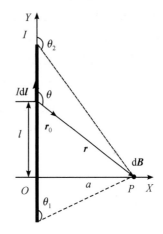

Figure 10.2.3 Figure of Example 10.2.1

$$\mathrm{d}\boldsymbol{B} = \frac{\mu_0}{4\pi} \frac{I \mathrm{d}\boldsymbol{l} \times \boldsymbol{r}_0}{r^2}$$

Its magnitude is

$$\mathrm{d}B = \frac{\mu_0}{4\pi} \frac{I \mathrm{d}l \, \sin\theta}{r^2}$$

By the right-hand spiral rule, the direction of $\mathrm{d}\boldsymbol{B}$ is perpendicular to the paper and facing inward, and the magnetic induction intensity generated by every current element on the wire at point P is in this direction. By the right triangle relationship

$$l = -a \cot\theta$$

Differentiating both sides of the equation can get $\mathrm{d}l = a \csc^2\theta \mathrm{d}\theta$, and $r^2 = a^2 + l^2 = a^2 \csc^2\theta$. According to the superposition principle, the magnetic induction intensity generated by the entire direct current at point P is

$$B = \int_{\theta_1}^{\theta_2} \frac{\mu_0}{4\pi} \frac{I dl \sin\theta}{r^2} = \int_{\theta_1}^{\theta_2} \frac{\mu_0 I}{4\pi a} \sin\theta \, d\theta = \frac{\mu_0 I}{4\pi a}(\cos\theta_1 - \cos\theta_2)$$

The direction is perpendicular to the paper and facing inward.

When the length of the wire is much greater than the distance a from point P to the straight wire, the wire can be regarded as "infinitely long". This means $\theta_1 \approx 0$ and $\theta_2 \approx \frac{\pi}{2}$, so $B = \frac{\mu_0 I}{2\pi a}$.

Example 10.2.2 As shown in Figure 10.2.4, a current-carrying ring of radius R carries a constant current I. Find the magnetic induction intensity at point P at a distance x from the center of the ring and on the axis of the ring.

Solution As shown in the figure, the OX axis is established in the axial direction, and a current element $I dl$ is arbitrarily selected on the ring. The position vector of point P relative to the

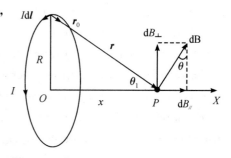

Figure 10.2.4 Figure of Example 10.2.2

current element is r, and the corresponding unit vector is r_0. According to the Biot-Savart law, the magnetic induction generated by the current element is

$$d\boldsymbol{B} = \frac{\mu_0}{4\pi} \frac{I d\boldsymbol{l} \times \boldsymbol{r}_0}{r^2}$$

Its magnitude is

$$dB = \frac{\mu_0}{4\pi} \frac{I dl}{r^2}$$

The direction of $d\boldsymbol{B}$ is perpendicular to the plane determined by $I d\boldsymbol{l}$ and $\boldsymbol{r}_0$ and facing upwards. As shown in the figure, if the current element is different, the direction of $d\boldsymbol{B}$ is different at point P. However, the direction in which each current element $d\boldsymbol{B}$ on the current-carrying ring is excited at point P is distributed on a conical surface with OP as the axis and P as the vertex. According to the principle of vector superposition, the direction of the magnetic induction intensity at point P is in the positive direction of the OX axis. Decompose $d\boldsymbol{B}$ into a vertical axial component $dB_\perp$ and an axial component $dB_\parallel$, and

$$dB_\perp = dB \cos\theta$$
$$dB_\parallel = dB \sin\theta$$

From the symmetry, $B_\perp = \int dB_\perp = 0$, so

$$B_\parallel = \int_L dB_\parallel = \int_L \frac{\mu_0 I}{4\pi r^2} \sin\theta \, dl = \frac{\mu_0 I R^2}{2\sqrt{(R^2 + x^2)^3}}$$

Thus,

$$\boldsymbol{B} = B_{\parallel}\boldsymbol{i} = \frac{\mu_0 I R^2}{2\sqrt{(R^2+x^2)^3}} \boldsymbol{i}$$

Discussion: (1) When $x=0$, the magnetic induction intensity generated by the circular current at the center of the circle is $B_0 = \frac{\mu_0 I}{2R}$. The magnetic induction intensity generated at the center of a circular arc current with a central angle α is $B = \frac{\mu_0 I \alpha}{4\pi R}$.

(2) When $x \gg R$, the magnetic induction intensity generated by the ring on the axis and away from the center of the circle is about $B \approx \frac{\mu_0 I R^2}{2x^3} = \frac{\mu_0 I S}{2\pi x^3}$.

Generally, for a planar current-carrying coil, the magnetic moment of the current-carrying coil is defined as

$$\boldsymbol{P}_m = I\boldsymbol{S} \qquad (10.2.4)$$

where $\boldsymbol{S} = S\boldsymbol{n}_0$ and $\boldsymbol{n}_0$ is the normal vector of the plane coil. $\boldsymbol{n}_0$ and the current I have a right helical relationship, as shown in Figure 10.2.5. Magnetic moment is an important physical quantity that is often used in the study of the magnetism of matter, as well as in the physics of molecules, atoms and nuclei.

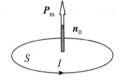

Figure 10.2.5 Magnetic moment of a current-carrying coil

Considering that the direction of the magnetic moment is the same as the direction of $\boldsymbol{B}$, the magnetic induction intensity on the axis and away from the center of the circle is

$$\boldsymbol{B} = \frac{\mu_0 \boldsymbol{P}_m}{2\pi x^3}$$

 10.2.4 Magnetic field generated by moving charges

Current in a conductor is the directional motion of a large number of charged particles, so the magnetic field generated by the current is actually the macroscopic representation of the magnetic field generated by moving charges. The following is an expression for the magnetic field generated by moving charges, derived from the Biot-Savart law.

As shown in Figure 10.2.6, suppose the cross-sectional area of the conductor is S, the number of carriers per unit volume is n, and the electric quantity of each carrier is q, moving in the current direction at the average velocity $\boldsymbol{v}$. Then the electric quantity passing through the cross section S in the unit time is $Q = I = qnvS$. The number of carriers in the current element $I d\boldsymbol{l}$ is $dN = nSdl$. According to the Biot-Savart law, the magnetic field generated by the current element $I d\boldsymbol{l}$ at point P is

$$d\boldsymbol{B} = \frac{\mu_0}{4\pi} \frac{I d\boldsymbol{l} \times \boldsymbol{r}_0}{r^2} = \frac{\mu_0}{4\pi} \frac{qnvS d\boldsymbol{l} \times \boldsymbol{r}_0}{r^2} = \frac{\mu_0}{4\pi} \frac{dN}{r^2} q\boldsymbol{v} \times \boldsymbol{r}_0$$

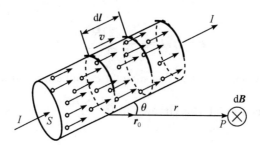

Figure 10.2.6 Magnetic field generated by moving charges

The magnetic field is generated jointly by the dN carriers in $I\mathrm{d}\boldsymbol{l}$. Then the magnetic induction intensity produced by a single moving charge at point P is

$$\boldsymbol{B} = \frac{\mathrm{d}\boldsymbol{B}}{\mathrm{d}N} = \frac{\mu_0}{4\pi} \frac{q\boldsymbol{v} \times \boldsymbol{r}_0}{r^2} \qquad (10.2.5)$$

where $\boldsymbol{r}_0$ is the unit vector directed at P of a moving charge, and r is the distance between the moving charge and point P.

Example 10.2.3 As shown in Figure 10.2.7, the charges q are uniformly distributed on a thin plastic disk of radius R. If the disk rotates at an angular velocity ω perpendicular to the central axis of the disk surface, try to find the magnetic induction at the center of the disk and the magnetic moment of the disk.

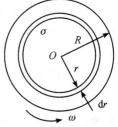

Figure 10.2.7 Figure of Example 10.2.3

Solution Think of the disk as an infinite number of rings. Take a ring with a radius r and width $\mathrm{d}r$. Its electric quantity is $\mathrm{d}q = \sigma 2\pi r \mathrm{d}r$, where $\sigma = \dfrac{q}{\pi R^2}$ is charge density. When the disk rotates at an angular velocity ω, the ring rotates to form a circular current, and the magnitude of the current is

$$\mathrm{d}I = \frac{\mathrm{d}q}{2\pi/\omega} = \frac{\omega}{2\pi}\mathrm{d}q$$

The magnetic induction generated by the circular current at the center of the disk is

$$\mathrm{d}B = \frac{\mu_0 \mathrm{d}I}{2r} = \frac{1}{2}\mu_0 \sigma \omega \mathrm{d}r$$

The disk rotates around the central axis perpendicular to the surface of the disk, and the magnetic induction at the center of the disk is

$$B = \int \mathrm{d}B = \int_0^R \frac{1}{2}\mu_0 \sigma \omega \mathrm{d}r = \frac{\mu_0 q \omega}{2\pi R}$$

A ring of radius r and width $\mathrm{d}r$ forms a circular current when it rotates at an angular velocity ω, and its magnetic moment is

$$\mathrm{d}P_\mathrm{m} = S\mathrm{d}I = \pi r^2 \mathrm{d}I = \frac{q\omega}{R^2}r^3\mathrm{d}r$$

So the magnetic moment of the disk is

$$P_m = \int dP_m = \int_0^R \frac{q\omega}{R^2} r^3 dr = \frac{R^2 q\omega}{4}$$

10.3　Gauss's law for a magnetic field

10.3.1　Magnetic induction line

Just as the electrostatic field is depicted with electric field lines, the magnetic field can also be depicted visually with magnetic induction lines. As with electric field lines, such regulations are made: ① The direction of **B** at any point is consistent with the tangent direction at this point; ② The strength of the magnetic field can be represented by the density of the magnetic induction line $\frac{dN}{dS_\perp} = B$.

Figure 10.3.1 shows schematic diagrams of several magnetic induction lines drawn according to experiment. It can be seen that magnetic induction lines have the following properties: ① Any two magnetic induction lines cannot intersect in space; ② Magnetic induction lines are closed curves; ③ The rotation direction of magnetic induction lines and the current direction satisfy the right-hand law. These characteristics of magnetic induction lines are very different from those of electric field lines of electrostatic fields.

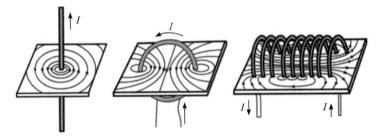

Figure 10.3.1　Magnetic induction lines of magnetic fields around several currents

10.3.2　Magnetic flux

Similar to the electric flux in the electrostatic field, the concept of magnetic flux can also be introduced in the magnetic field. The number of magnetic induction lines passing through a surface in the magnetic field is called the magnetic flux passing through the surface, and is represented by the symbol Φ_m.

In a non-uniform magnetic field, to calculate the magnetic flux passing through any surface, it is necessary to use the calculus method. As shown in Figure 10.3.2, the surface S is divided so that each area element d**S** can be regarded as a plane. The corresponding magnetic induction intensity can be regarded as uniform, and the magnetic flux passing through the area element d**S** is

$$d\Phi_m = \mathbf{B} \cdot d\mathbf{S} \quad (10.3.1)$$

The magnetic flux over the entire surface is

$$\Phi_m = \int_s d\Phi_m = \int_s \mathbf{B} \cdot d\mathbf{S} \quad (10.3.2)$$

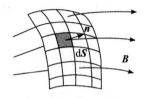

Figure 10.3.2 Magnetic flux

In SI units, the unit of magnetic flux is the Weber (Wb) and $1 \text{ Wb} = 1 \text{ T} \cdot \text{m}^2$.

For a closed surface S, we specify the positive direction is along the normal from the inside to the outside. In this way, when the magnetic induction line penetrates out of the curved surface, Φ_m is positive; and when it penetrates into the curved surface, Φ_m is negative. So the total magnetic flux over the closed surface is

$$\Phi_m = \oint_s d\Phi_m \quad (10.3.3)$$

It is equal to the number of magnetic induction lines penetrating out of the closed surface S minus the number of magnetic induction lines penetrating into the surface S.

10.3.3 Gauss's law for a magnetic field

Since the magnetic induction line generated by the current-carrying wire is a closed curve with no head or tail, the magnetic induction line that penetrates from one place of a closed surface must go out from another place, so the magnetic flux passing through any closed surface is always equal to zero, that is,

$$\oint_s \mathbf{B} \cdot d\mathbf{S} = 0 \quad (10.3.4)$$

Equation (10.3.4) is called Gauss's law for magnetic fields. It states that magnetic fields are passive fields.

In contrast to Gauss's law for electrostatic fields, that for magnetic fields actually states that it is impossible to have a monopole magnetic charge, or that a magnetic monopole does not exist.

10.3.4 The Ampère circuital theorem

In an electrostatic field, the integral of the electric field strength along any closed path L (circulation) is equal to zero, $\oint_L \mathbf{E} \cdot d\mathbf{l} = 0$, which reflects the important property that the electrostatic field is a conservative field. Then in a constant magnetic field, what is the integral of the magnetic induction $\mathbf{B}$ along any closed path L, $\oint_L \mathbf{B} \cdot d\mathbf{l}$? Take the magnetic field generated by a long straight current-carrying wire as an example to find the result.

As shown in Figure 10.3.3(a), in the plane perpendicular to the long straight current-carrying wire, make an arbitrary loop L around the current-carrying wire, and take any line element dl on L. When the winding direction of the closed loop L and the current direction satisfy the right-hand rule, the angle between dl and $\boldsymbol{B}$ is θ. The magnitude of magnetic induction at dl is $B=\dfrac{\mu_0 I}{2\pi r}$. The direction of $\boldsymbol{B}$ is along the tangent to the circle of radius r and determined by the right-hand rule. dφ is the central angle of dl to point O.

$$\oint_L \boldsymbol{B} \cdot \mathrm{d}\boldsymbol{l} = \oint_L B\cos\theta\, \mathrm{d}l = \oint_L \frac{\mu_0 I}{2\pi r} r\, \mathrm{d}\varphi = \mu_0 I$$

If the loop L turns in the opposite direction, then in the same way we can obtain

$$\oint_L \boldsymbol{B} \cdot \mathrm{d}\boldsymbol{l} = -\mu_0 I$$

If the loop L does not surround the current, as shown in 10.3.3(b), the above integral is equal to zero, that is

$$\oint_L \boldsymbol{B} \cdot \mathrm{d}\boldsymbol{l} = 0$$

If the loop is not in the plane perpendicular to the current, it can be decomposed into two parts: the loop in the plane and the loop perpendicular to the plane. For the part perpendicular to the plane, $\boldsymbol{B} \cdot \mathrm{d}\boldsymbol{l} = 0$, so it is only necessary to consider the loop in the plane.

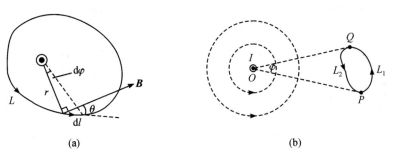

Figure 10.3.3 The Ampère circuital theorem

Although the above discussion is aimed at long straight current-carrying wire, the conclusions can be used in other shapes of current-carrying wires. For the magnetic field generated by any steady current, the closed loop L is not necessarily a plane curve, and there can be many currents passing through the closed loop, all of which have the same characteristics as we discussed above. The relation of this general regularity is called Ampère's circuital theorem and can be expressed as follows:

The line integral of the magnetic induction along any closed path L in a vacuum (also called a loop current) is equal to μ_0 times the algebraic sum of all currents enclosed by this loop L. Its mathematical expression is

$$\oint_L \boldsymbol{B} \cdot \mathrm{d}\boldsymbol{l} = \mu_0 \sum I_{\text{int}} \tag{10.3.5}$$

In the equation, when the direction of the current passing through the loop L and the detouring direction of the loop L obey the right-hand rule, the current I is positive.

Otherwise, I is negative. If the current I is not surrounded by the loop L, I does not contribute to the circulating current. For the convenience of description, the closed integral loop L in the above formula is called the Ampère loop.

It should be pointed out that the left of Equation (10.3.5) represents the magnetic induction intensity on the loop L, which is jointly excited by all currents in space. The right of Equation (10.3.5) determines the circulating current along the loop L, which is only related to the current enclosed by the loop. Ampère's circuital theorem states that the magnetic field is a rotating field, and the magnetic induction lines in the magnetic field are closed.

The magnetic field distribution of some current-carrying wires with certain symmetry can be easily calculated by applying Ampère's circuital theorem.

Example 10.3.1 It is known that an infinitely long, uniform current-carrying cylinder of radius R has a constant current I flowing through it, and the current is uniformly distributed on the cross section. Find the magnetic field distribution inside and outside the cylinder.

Solution As shown in Figure 10.3.4, the magnetic field generated by an infinitely long, uniform current-carrying cylinder has symmetry, and its magnetic induction lines are a series of concentric circles centered on the axis of the cylinder on a plane perpendicular to the cylinder. The magnitude of **B** at

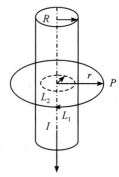

Figure 10.3.4 Figure of Example 10.3.1

each point on the same circumference is equal. The direction of the magnetic induction intensity **B** and the current direction satisfy the right-hand rule.

Now calculate the magnetic induction at any point P outside the cylinder. A circle of radius r passing through point P is taken as a closed loop, and the direction of the loop and the current direction satisfy the right-hand rule; then

$$\oint_{L_1} \boldsymbol{B} \cdot \mathrm{d}\boldsymbol{l} = \oint_{L_1} B \mathrm{d}l = B \oint_{L_1} \mathrm{d}l = B 2\pi r$$

Because

$$\mu_0 \sum I_{\text{int}} = \mu_0 I$$

according to Ampère's circuital theorem, the magnetic induction at point P outside the current-carrying cylinder is

$$B = \frac{\mu_0 I}{2\pi r}$$

If point P is inside the cylinder, the reference loop L_2 shown in the figure is established and then

$$\oint_L \boldsymbol{B} \cdot \mathrm{d}\boldsymbol{l} = \oint_L B \mathrm{d}l = B \oint_L \mathrm{d}l = B 2\pi r$$

Because

$$\mu_0 \sum I_{\text{int}} = \mu_0 \frac{Ir^2}{R^2}$$

according to Ampère's circuital theorem, the magnetic induction at point P inside the current-carrying cylinder is

$$B = \frac{\mu_0 Ir}{2\pi R^2}$$

Combining the two results, the distribution of B in space is

$$B = \begin{cases} \dfrac{\mu_0 Ir}{2\pi R^2} & (r<R) \\ \dfrac{\mu_0 I}{2\pi r} & (r>R) \end{cases}$$

It can be seen that inside the cylinder, the magnitude of the magnetic induction intensity B is proportional to the distance r from the axis while outside the cylinder, the magnitude of the magnetic induction intensity B is inversely proportional to the distance r from the axis.

Example 10.3.2 Find the magnetic field inside an infinitely long, current-carrying, straight, densely-wound solenoid. It is known that a long straight solenoid with a radius of R carries a constant current I, and the number of wire turns per unit length is n. Find the magnetic field distribution inside the long straight solenoid.

Solution It is known from experiments that the external magnetic field of the long straight and densely wound solenoid is very close to the tube wall, which can be approximately regarded as $\boldsymbol{B} = 0$. Inside the solenoid, it can be regarded as a uniform magnetic field, and the direction conforms to the right-hand rule, as shown in Figure 10.3.5. Make a closed rectangular loop L_{abcda}. According to Ampère's circuital theorem

$$\oint_L \boldsymbol{B} \cdot d\boldsymbol{l} = \int_{ab} \boldsymbol{B} \cdot d\boldsymbol{l} + \int_{bc} \boldsymbol{B} \cdot d\boldsymbol{l} + \int_{cd} \boldsymbol{B} \cdot d\boldsymbol{l} + \int_{da} \boldsymbol{B} \cdot d\boldsymbol{l}$$

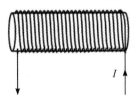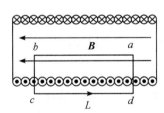

Figure 10.3.5　Figure of Example 10.3.2

The magnitude of $\boldsymbol{B}$ at each point on the ab segment is equal, the direction of $\boldsymbol{B}$ is the same as that of the loop, and then $\int_{ab} \boldsymbol{B} \cdot d\boldsymbol{l} = B\overline{ab}$. On the bc and da segments, the direction of $\boldsymbol{B}$ is perpendicular to the loop direction everywhere, and then $\int_{bc} \boldsymbol{B} \cdot d\boldsymbol{l} = 0$, $\int_{da} \boldsymbol{B} \cdot d\boldsymbol{l} = 0$. On the cd segment, $\boldsymbol{B}$ is zero everywhere, and then $\int_{cd} \boldsymbol{B} \cdot d\boldsymbol{l} = 0$.

Thus, $\oint_L \boldsymbol{B} \cdot d\boldsymbol{l} = B\overline{ab}$ and the algebraic sum of the currents enclosed by loop L is $nI\overline{ab}$.

From the Ampère circuital theorem,
$$B\overline{ab} = \mu_0 nI\overline{ab}$$
$$B = \mu_0 nI$$

From the randomness of the position of $\overline{ab}$, it can be known that the magnitude of **B** at each point in the long straight solenoid is $\mu_0 nI$, and the direction is parallel to the axis. In the laboratory, a current-carrying long straight solenoid is often used to obtain a uniform magnetic field.

Example 10.3.3 Suppose an infinitely large conductor thin plate is placed perpendicular to the surface of the paper, there is a current passing through it in a direction perpendicular to the surface of the paper, and the surface current density is j. Find the magnetic field distribution of the plate current.

Solution The magnetic field of an infinitely large plate current can be regarded as the magnetic field produced by countless parallel currents. Therefore, the magnetic field distribution in space has symmetry. Now calculate the magnetic induction intensity at any point in space. Suppose point P is in the upper half space. Taking the symmetrical long straight current elements $\mathrm{d}x_1$ and $\mathrm{d}x_2$ as shown in Figure 10.3.6, it can be seen that the combined magnetic field direction is horizontally to the left. Due to the symmetry, the magnetic field direction of the entire plane current at point P should point to the left, while the magnetic field direction should point to the right at point P below the plane. Because of the plane symmetry, the magnitude of **B** should be the same for all points that are equidistant from the plane. For points P and P' above and below the plane, although the directions of the magnetic fields are opposite, as long as they are at the same distance from the plane, the magnitudes of the magnetic induction should be the same.

Make an loop as shown in Figure 10.3.6, with both sides of bc and da equally divided by the plane where the current resides. ab and cd are parallel to the current plane. According to Ampère's circuital theorem, we have

$$\oint_L \boldsymbol{B} \cdot \mathrm{d}\boldsymbol{l} = \int_{ab} \boldsymbol{B} \cdot \mathrm{d}\boldsymbol{l} + \int_{bc} \boldsymbol{B} \cdot \mathrm{d}\boldsymbol{l} + \int_{cd} \boldsymbol{B} \cdot \mathrm{d}\boldsymbol{l} + \int_{da} \boldsymbol{B} \cdot \mathrm{d}\boldsymbol{l} = \mu_0 \overline{ab} j$$

The magnetic induction intensity of each point on the segment ab is equal, and the direction is the same as the loop direction. Then

$$\int_{ab} \boldsymbol{B} \cdot \mathrm{d}\boldsymbol{l} = B\overline{ab}$$

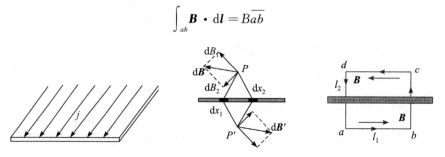

Figure 10.3.6 Figure of Example 10.3.3

On the sections of bc and da, the direction is perpendicular to the direction of the loop, and then

$$\int_{bc} \boldsymbol{B} \cdot \mathrm{d}\boldsymbol{l} = \int_{da} \boldsymbol{B} \cdot \mathrm{d}\boldsymbol{l} = 0$$

On the segment cd, the magnetic induction intensity of each point is equal, and the direction is consistent with the loop direction, which is $B=B'$. Therefore,

$$\int_{cd} \boldsymbol{B} \cdot \mathrm{d}\boldsymbol{l} = \int_{cd} \boldsymbol{B}' \cdot \mathrm{d}\boldsymbol{l} = B'\overline{cd} = B\overline{ab}$$

$$\oint_L \boldsymbol{B} \cdot \mathrm{d}\boldsymbol{l} = 2B\overline{ab} = \mu_0 \overline{ab} j$$

So

$$B = \frac{1}{2}\mu_0 j$$

It can be seen that the magnetic fields on both sides of the infinitely large uniform plane current are uniform magnetic fields but in opposite directions.

10.4 Effect of magnetic fields on current-carrying wires

10.4.1 The Ampère force

In 1820, Ampère summed up the expression of the force on the current element in a magnetic field based on a large number of experimental results, that is

$$\mathrm{d}\boldsymbol{F} = I\,\mathrm{d}\boldsymbol{l} \times \boldsymbol{B} \qquad (10.4.1)$$

This is called Ampère's law and this force is called the Ampère force, also known as the magnetic field force. $\boldsymbol{B}$ in the formula is the magnetic induction intensity at the position of the current element $I\mathrm{d}\boldsymbol{l}$. According to the principle of superposition of forces, the ampère force on a current-carrying wire is

$$\boldsymbol{F} = \int_l \mathrm{d}\boldsymbol{F} = \int_l I\,\mathrm{d}\boldsymbol{l} \times \boldsymbol{B} \qquad (10.4.2)$$

which is a vector integration. If the direction of the Ampère force on each current element on the wire is different, $\mathrm{d}\boldsymbol{F}$ must be decomposed in the selected coordinate system first, then integrated, and finally the resultant force can be obtained.

Since individual current element cannot be obtained, Ampère's law cannot be directly proved experimentally. But by using Equation (10.4.2), we can calculate the Ampère force on the current-carrying wires of various shapes in the magnetic field, and the results are consistent with the experiment.

Example 10.4.1 As shown in Figure 10.4.1, a semicircular wire with a radius R carries a current I. The wire is placed in a uniform magnetic field and the magnetic induction intensity is **B**. The direction of the magnetic field is perpendicular to the plane of the wire. Find the Ampère force of the magnetic field acting on the wire.

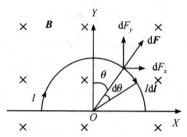

Figure 10.4.1   Figure of Example 10.4.1

Solution   The coordinate system OXY is established as shown in the figure, and a current element $I\mathrm{d}l$ is arbitrarily selected on the semicircular wire. The magnitude of the Ampère force is

$$\mathrm{d}F = IB\mathrm{d}l$$

The direction of $\mathrm{d}\boldsymbol{F}$ is perpendicular to the direction of $I\mathrm{d}\boldsymbol{l}$ and radially outward. Since the direction of the Ampère force on each current element on the wire is different, $\mathrm{d}\boldsymbol{F}$ is decomposed along the coordinate axis

$$\mathrm{d}F_x = \mathrm{d}F\sin\theta, \quad \mathrm{d}F_y = \mathrm{d}F\cos\theta$$

From the symmetry we have

$$F_x = 0$$

Because $\mathrm{d}l = R\mathrm{d}\theta$,

$$F_y = \int \mathrm{d}F_y = 2\int_0^{\frac{\pi}{2}} BIR\cos\theta\,\mathrm{d}\theta = 2BIR$$

The Ampère force acting on the semicircular wire is

$$\boldsymbol{F} = 2BIR\boldsymbol{j}$$

Thus, in a uniform magnetic field, the magnetic field force on a current-carrying wire of any shape is equivalent to the force on a straight current from the start point to the end point of the wire in the magnetic field.

Example 10.4.2   As shown in Figure 10.4.2, next to an infinitely long straight wire carrying current I_1 is placed a square coil of side length b, carrying current I_2. Find the Ampère force on the coil.

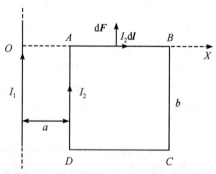

Figure 10.4.2   Figure of Example 10.4.2

Solution As shown in the figure, establish the OX axis, and the magnitude of the magnetic induction generated by the long straight wire I_1 is

$$B = \frac{\mu_0 I_1}{2\pi l}$$

and the direction is perpendicular to the paper and facing inward.

For segment AB, take the current element $I_2 d\boldsymbol{l}$, and the Ampère force acting on $I_2 d\boldsymbol{l}$ is

$$dF = I_2 dl B = \frac{\mu_0 I_1 I_2}{2\pi l} dl$$

and the direction of $d\boldsymbol{F}$ is perpendicular to I_2 and facing upward. Because all the current elements on segment AB are in the same direction, the magnitude of the Ampère force on segment AB is

$$F = \int_l dF = \int_a^{a+b} \frac{\mu_0 I_1 I_2}{2\pi l} dl = \frac{\mu_0 I_1 I_2}{2\pi} \ln \frac{a+b}{a}$$

The direction is perpendicular to I_2 and facing upward.

Similarly, the force acting on segment CD can be calculated as

$$F = \frac{\mu_0 I_1 I_2}{2\pi} \ln \frac{a+b}{a}$$

and the direction is perpendicular to I_2 and facing downward.

For segment BC, the magnitude of the magnetic field is $B = \frac{\mu_0 I_1}{2\pi(a+b)}$ at the position of segment BC, so the magnitude of the force is

$$F = I_2 B b = \frac{\mu_0 I_1 I_2 b}{2\pi(a+b)}$$

and the direction is horizontally to the right.

Similarly, the force acting on segment DA can be calculated as

$$F = \frac{\mu_0 I_1 I_2 b}{2\pi a}$$

and the direction is horizontally to the left.

Therefore, the Ampère force acting on the coil is

$$F = \frac{\mu_0 I_1 I_2 b}{2\pi} \left(\frac{1}{a} - \frac{1}{a+b} \right)$$

and the direction is horizontally to the left.

10.4.2 Effect of uniform magnetic fields on current-carrying coils

A current-carrying coil is subjected to a magnetic torque in an external magnetic field. Under the action of the magnetic torque, the coil will deflect, which is the basic principle of making electric motors and various electromagnetic instruments. The effect of a uniform magnetic field on a planar current-carrying coil is discussed below.

As shown in Figure 10.4.3, in a uniform magnetic field with a magnetic induction intensity of **B** is placed a rigid rectangular plane current-carrying coil *abcd*. The current intensity is I, the coil area is S, and the angle between the coil plane and the direction of the magnetic field is θ. The angle between the coil magnetic moment $\boldsymbol{P}_m$ and the magnetic field is $\varphi = \dfrac{\pi}{2} - \theta$.

According to the analysis of the Ampère force on the current-carrying wire in the magnetic field, it can be seen that the Ampère forces on the wires *ab* and *cd* are

$$F_1 = F'_1 = IB\overline{ab}\sin\theta$$

F_1 and F'_1 are equal in magnitude and opposite in direction, acting on the same straight line, and the resultant force is zero. Similarly, the Ampère forces on wires *bc* and *da* are

$$F_2 = F'_2 = IB\overline{bc}$$

F_2 and F'_2 are also equal in magnitude and opposite in direction, but not on the same line. As shown in Figure 10.4.3(b), a moment will be generated on the coil, causing the coil to rotate around the axis OO'. The magnitude of the moment is

$$M = F_2 \overline{ab} \sin\varphi = IBS\sin\varphi$$

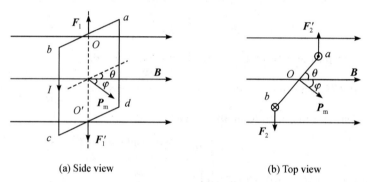

(a) Side view (b) Top view

Figure 10.4.3 Moment of a planar current-carrying coil in a uniform magnetic field

As mentioned before, the magnetic moment of the planar coil is $\boldsymbol{P}_m = I\boldsymbol{S} = IS\boldsymbol{n}$, and then, $M = P_m B \sin\varphi$. According to the vector relationship, we have

$$\boldsymbol{M} = \boldsymbol{P}_m \times \boldsymbol{B} \tag{10.4.3}$$

The direction of the moment $\boldsymbol{M}$ is consistent with the direction of $\boldsymbol{P}_m \times \boldsymbol{B}$. Equation (10.4.3) is the moment of the current-carrying coil in the uniform magnetic field. Although this formula is obtained from the special case of a rectangular coil, it can be proved that it is applicable to planar current-carrying coils of any shape. Even, for the movement of a charged particle along a closed loop and the magnetic moment caused by a charged particle's spin, the magnetic moment in the magnetic field can be described by Equation (10.4.3).

Let us discuss a few special cases:

(1) When $\varphi = \dfrac{\pi}{2}$, the coil plane is parallel to **B**, and $\boldsymbol{P}_m$ is perpendicular to **B**. The magnetic moment on the coil is the maximum, $M = BIS$, and has a tendency to reduce φ.

(2) When $\varphi=0$, the plane of the coil is perpendicular to $\boldsymbol{B}$, $\boldsymbol{P}_\mathrm{m}$ and $\boldsymbol{B}$ are in the same direction, the magnetic moment on the coil is zero, and the coil is in a stable equilibrium state.

(3) When $\varphi=\pi$, the plane of the coil is perpendicular to $\boldsymbol{B}$, but the direction of $\boldsymbol{P}_\mathrm{m}$ is opposite to that of $\boldsymbol{B}$, and the magnetic moment on the coil is also zero. At this time, the coil is in an unstable equilibrium position, which means that once the external disturbance makes the coil slightly deviate from this unstable equilibrium position, the magnetic moment of the magnetic field on the coil will make the coil continue to deviate until $\boldsymbol{P}_\mathrm{m}$ turns to the direction of $\boldsymbol{B}$.

From the above discussion, it can be seen that, in a uniform magnetic field, the rigid plane current-carrying coil is only subjected to the action of the magnetic moment, so only rotation occurs and the translation of the entire coil does not occur.

10.4.3 Work done by magnetic force

When a current-carrying wire or coil moves in a magnetic field, the Ampère force will do work. Now, the general expression for the work done by the Ampère force can be obtained from two special cases.

1. Work done by the magnetic force when a current-carrying wire moves in a magnetic field

As shown in Figure 10.4.4, in a uniform magnetic field with a magnetic induction intensity $\boldsymbol{B}$, a wire of length l and two parallel guide rails form a current-carrying closed loop $abcd$, the current intensity I remains unchanged, and the wire can slide along the guide rails. According to Ampère's law, the Ampère force on the wire ab is

$$F = BIl$$

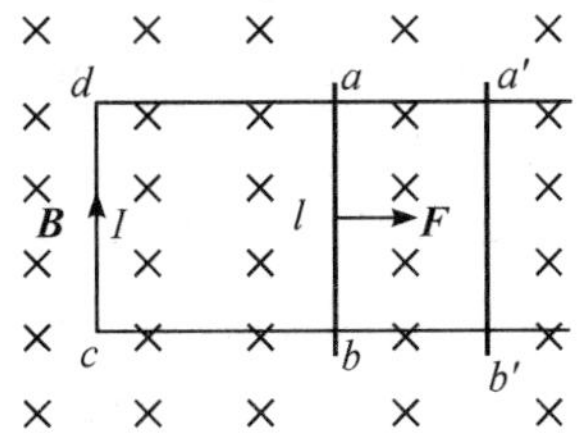

Figure 10.4.4 Work done by magnetic force

The direction is shown in Figure 10.4.4. In moving the wire from position ab to $a'b'$ in the direction of the force, the work done by the Ampère force is

$$A = F\overline{aa'} = BIl\overline{aa'} = BI\Delta S = I\Delta\Phi_\mathrm{m} \qquad (10.4.4)$$

The above expression shows that when the current-carrying wire moves in the magnetic field, if the current in the loop is constant, the work done by the Ampère force is equal to the current intensity multiplied by the increment of the magnetic flux in the area enclosed by the loop.

2. Work done by the magnetic moment when a current-carrying coil rotates in a uniform magnetic field

As shown in Figure 10.4.5, in a uniform magnetic field with a magnetic induction intensity of **B**, there is a planar coil with an area of S and a constant current intensity of I. When the coil rotates, the magnetic moment on the current-carrying coil is

$$M = P_m \times B$$

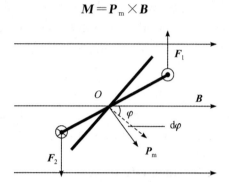

Figure 10.4.5 Work done by magnetic moment

The direction of the moment is perpendicular to the paper, and the coil deflects. If the current in the coil remains unchanged, the element work done by the magnetic moment for a small angle $d\varphi$ is

$$dA = -Md\varphi = -BIS \sin\varphi d\varphi = Id(BS \cos\varphi)$$

The negative sign indicates that when the magnetic moment does positive work, the angle φ decreases and $d\varphi$ is a negative value. When the coil goes from φ_1 to φ_2, the total work done by the magnetic moment is

$$A = \int_{\varphi_1}^{\varphi_2} Id(BS \cos\varphi) = I(BS \cos\varphi_2 - BS \cos\varphi_1) = I\Delta\Phi_m \quad (10.4.5)$$

The above equation shows that the work done by the magnetic moment on the current-carrying coil is also equal to the current intensity in the loop multiplied by the increase in the magnetic flux in the area enclosed by the loop. This result is the same as Equation (10.4.4), which is a general expression for the work done by the magnetic force.

Example 10.4.3 A semicircular closed coil carrying current I and with a radius R, is placed in a uniform external magnetic field **B** whose direction is parallel to the plane of the coil, as shown in Figure 10.4.6. Find

(1) the magnitude and direction of the moment that the coil is subjected to at this time;

(2) the work done by the magnetic moment when the coil plane turns to a position perpendicular to the magnetic field.

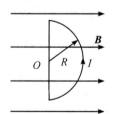

Figure 10.4.6 Figure of Example 10.4.3

Solution (1) The magnetic moment of the coil

$$P_m = IS = ISn = \frac{1}{2}I\pi R^2 n$$

In the position shown in the figure, the direction of the magnetic moment of the coil is perpendicular to the surface of the paper and facing outward, and the included angle with **B** is $\pi/2$.

According to $\boldsymbol{M} = \boldsymbol{P}_m \times \boldsymbol{B}$, the magnitude of the magnetic moment on the coil is

$$M = P_m B = \frac{1}{2} I B \pi R^2$$

The direction of the magnetic moment is perpendicular to **B** and upward.

(2) At the initial position, the plane of the coil is parallel to **B**, then the normal vector of the coil is perpendicular to **B**, and the magnetic flux passing through the coil is zero $\Phi_{m_1} = 0$. Under the action of the magnetic moment, the coil rotates by $\pi/2$, the normal vector of the coil is consistent with **B**, and the magnetic flux passing through the coil is $\Phi_{m_2} = \boldsymbol{B} \cdot \boldsymbol{S} = B \frac{1}{2} \pi R^2$. According to the work expression of the magnetic moment Equation (10.4.5),

$$A = I \Delta \Phi_m = I(\Phi_{m_2} - \Phi_{m_1}) = I\left(B \frac{1}{2} \pi R^2 - 0\right) = \frac{1}{2} I B \pi R^2$$

It can also be calculated by integration

$$A = \int_{\frac{\pi}{2}}^{0} -M \mathrm{d}\theta = \int_{\frac{\pi}{2}}^{0} -P_m B \sin\theta \, \mathrm{d}\theta = \frac{1}{2} I B \pi R^2$$

10.5 Effect of magnetic fields on moving charges

10.5.1 The Lorentz force

When a charged particle moves in a magnetic field, it will be affected by the force of the magnetic field, which is called the Lorentz force. In terms of the microscopic nature of the Ampère force that the current-carrying wire is subjected to in the magnetic field, it should be attributed to the Lorentz force.

Experiments have shown that the relationship between the Lorentz force $\boldsymbol{f}_m$ of a moving charged particle q with velocity $\boldsymbol{v}$ in a magnetic field and the magnetic induction intensity **B** is

$$\boldsymbol{f}_m = q\boldsymbol{v} \times \boldsymbol{B} \tag{10.5.1}$$

The magnitude of the Lorentz force is

$$f_m = qvB\sin\theta$$

where θ is the angle between **v** and **B**.

The direction of $\boldsymbol{f}_m$ is perpendicular to the plane formed by **v** and **B**, and is determined by the right-hand rule. It should be noted that the direction of $\boldsymbol{f}_m$ is related to the positivity and negativity of the charged particles. According to this, C.D. Anderson discovered the positron in 1932 and won the 1936 Nobel Prize in Physics.

Since the Lorentz force is perpendicular to the plane determined by the velocity and the magnetic field, it can only change the direction of motion of the charge and does not do work on the charge. The magnitude of the force changes with the angle between v and B, which makes the moving charges present various forms of motion in the magnetic field, and has practical applications in electronic control, magnetic focusing, etc.

10.5.2 Movement of charged particles in a uniform magnetic field

Assuming that there is a uniform magnetic field in space and the magnetic induction intensity is B. A charged particle with an electric quantity q and a mass m, regardless of gravity, enters the magnetic field at an initial velocity v and moves in the magnetic field. We shall analyze it in the following three situations.

1. **The initial velocity is parallel to the magnetic induction intensity B**

If the initial velocity v is parallel to the magnetic induction intensity B, the Lorentz force acting on the charged particle is equal to zero, and the charged particle is not affected by the magnetic field and moves in a straight line at a uniform velocity v after entering the magnetic field.

2. **The initial velocity is perpendicular to the magnetic induction intensity B**

If the initial velocity v is perpendicular to the magnetic induction intensity B, as shown in figure 10.5.1, the magnitude of the Lorentz force f_m on the particle is $f_m = qvB$. The direction is perpendicular to v and B, so the magnitude of the particle velocity does not change, and only the direction changes. The charged particle will make a uniform circular motion, and the Lorentz force acts as a centripetal force, so there is

$$qvB = m\frac{v^2}{R}$$

or

$$R = \frac{mv}{qB}$$

where R is the radius of the circular orbit of the particle.

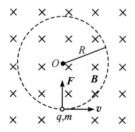

Figure 10.5.1 Motion caused by the Lorentz force in a uniform magnetic field

It can be seen that for a certain charged particle (that is, a constant value of $\frac{q}{m}$), the orbital radius is proportional to the speed of the charged particle and inversely proportional to the magnitude of the magnetic induction. The smaller the speed, the smaller the Lorentz force and the smaller the orbit radius.

The time required for a charged particle to rotate once around a circular orbit (period) is

$$T = \frac{2\pi R}{v} = 2\pi \frac{m}{qB} \tag{10.5.2}$$

It can be seen that the period has nothing to do with the speed of the charged particle, and this feature is the theoretical basis of magnetic focusing and the cyclotron to be introduced later.

3. The initial velocity and the magnetic induction intensity B form an angle θ

If the initial velocity v forms an angle θ with the magnetic induction intensity **B**, as shown in Figure 10.5.2, the velocity can be decomposed into the component parallel to **B**, $v_{/\!/} = v\cos\theta$ and the component perpendicular to **B**, $v_\perp = v\sin\theta$. Due to the action of the magnetic field, the charged particle not only moves in a circular motion at a uniform speed $v_\perp$ in the plane perpendicular to the magnetic field, but also moves in a straight line at a uniform speed $v_{/\!/}$ in the direction parallel to **B**, so the combined motion of the charged particle is an isometric spiral motion, and the trajectory is a helix. The radius of the helix is

$$R = \frac{mv_\perp}{qB} \tag{10.5.3}$$

The pitch is

$$h = v_{/\!/} T = v_{/\!/} \frac{2\pi R}{v_\perp} = 2\pi \frac{mv_{/\!/}}{qB} \tag{10.5.4}$$

where T is the period. The pitch h is only related to the velocity component parallel to **B** and has nothing to do with the velocity component perpendicular to **B**.

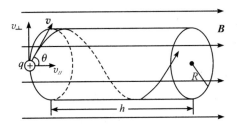

Figure 10.5.2 Spiral motion of a charged particle in a uniform magnetic field

10.5.3 The Hall effect

As shown in Figure 10.5.3, when a thin conductor plate is placed in a magnetic field **B** and a current I passes through the conductor in the direction perpendicular to **B**, a potential difference U_H will be generated on the upper and the lower sides of the conductor plate. This phenomenon called the Hall effect was first discovered by the young American physicist Hall in 1879, and the corresponding potential difference U_H is called the Hall voltage.

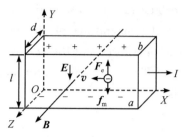

Figure 10.5.3 The Hall effect

The generation of the Hall effect can be explained by the Lorentz force. As shown in Figure 10.5.3, the carrier is assumed to be negative, moving to the left, and the direction is opposite to the current direction. At the beginning, the carrier is subjected to the downward Lorentz force f_m. The negative charges are shifted downwards, accumulating on the surface a, and an equal number of positive charges accumulate on the surface b, resulting in an electric field directed from b to a. Subsequently, the carriers in the conductor will be subject to the combined action of the upward electric field force F_e and the Lorentz force f_m until the accumulation of charges reaches a dynamic equilibrium. Then F_e and f_m reach a balance when the voltage between surfaces a and b is the Hall voltage U_H.

Let the electric quantity of the carrier be q, and the average velocity of the directional motion be v. When the forces are balanced

$$qvB = qE$$

Suppose that the width of the metal sheet is l and a uniform electric field is between surfaces a and b, then the Hall voltage is

$$U_H = El$$

If the number of carriers per unit volume is n, according to the definition of current intensity $I = nqvS$, where $S = ld$ is the cross-sectional area of the sheet, the Hall voltage can be expressed as

$$U_H = \frac{1}{qn} \frac{IB}{d} = R_H \frac{IB}{d} \tag{10.5.5}$$

where $R_H = \dfrac{1}{nq}$ is called the Hall coefficient, which is related to material of the Hall element.

If the carrier is a positive hole charge, the direction of the Lorentz force on the charge is downward, positive charges are accumulated on the lower surface, and negative charges are accumulated on the upper surface. The Hall effect provides an important method for the study of semiconductors. The conductivity type of a semiconductor can be judged by the positivity or negativity of U_H, and the concentration n of the carrier can also be measured.

About 100 years after the discovery of the Hall effect, the German physicist Klitzing discovered the quantum Hall effect while studying semiconductors at extremely low temperatures and in strong magnetic fields, which is one of the amazing advances in contemporary condensed matter physics. He won the Nobel Prize in Physics in 1985.

Afterwards, the Chinese-American physicist Cui Qi and American physicists Laughlin and Sturmer discovered the fractional quantum Hall effect in a stronger magnetic field and received the Nobel Prize in Physics in 1998. In 2013, the team of the Chinese physicist Xue Qikun first observed the quantum anomalous Hall effect from experiment. This discovery is a major breakthrough in related fields and an important scientific discovery in the field of world basic research.

10.5.4 Application of the Hall effect in engineering technology

In general, the Hall coefficients of metals and electrolytes are small, and the Hall effect is not significant while the Hall coefficients of semiconductor materials are large, and the Hall effect is obvious. Since the 1960s, with the rapid development of semiconductor materials and technology, it has been found that the Hall elements made of semiconductor materials are sensitive to magnetic fields, simple in structure, small in size, wide in frequency response, large in output voltage variation and long in service life. Therefore, the Hall effect is widely used in electromagnetic measurement, non-electrical measurement, automatic control, computing and communication devices.

(1) Measurement of semiconductor properties. The Hall effect can be used to measure the properties of semiconductor materials and high temperature superconductors. Suppose the direction of the current in the conductor is shown in Figure 10.5.3. The carrier is negative, the moving direction is opposite to the current direction, and the Lorentz force is downward. Therefore, the upper interface of the conductor has positive charges and the lower interface negative. If the carrier is positive, the upper interface of the conductor has negative charges and the lower interface positive. It can be seen that the sign of the carrier can be determined by the sign of the Hall voltage between the upper and lower interfaces. In this way, it is possible to determine whether a semiconductor is P-type or N-type. If the carrier is known, the concentration of the negative carrier in the conductor can be calculated by measuring the Hall coefficient R_H, and then the effect on the carrier concentration of objective factors can be obtained. For example, the LakeShore Hall effect measurement system, shown in Figure 10.5.4, can be used to measure the resistance,

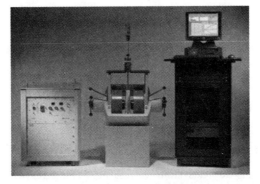

Figure 10.5.4 The Hall effect measurement system

resistivity, the Hall coefficient, the Hall mobility, carrier density and electronic properties of an sample, which is an essential tool for understanding and studying electrical properties of semiconductor devices and materials.

(2) Measurement of the magnetic field. The Hall effect can be used to make the Gaussimeter which can measure magnetic induction accurately. The probe of the Gaussimeter is a Hall element, inside which is a semiconductor sheet. According to Equation (10.5.5), the Hall voltage, U_H, can be measured by a millivolt meter, and the Hall coefficient R_H and current I can also be measured by corresponding instruments so that the magnetic induction intensity, B, can be conveniently calculated. The dial of the Gaussimeter is marked with the magnetic induction intensity. As long as the Gaussimeter is inserted into the magnetic field to be measured, B can be read directly, which is very convenient. For a high measurement accuracy requirement, such as better than $\pm 0.5\%$, the Gallium Arsenide Hall element is usually used, which has high sensitivity. For a low measurement accuracy requirement, such as less than $\pm 0.5\%$, silicon and germanium Hall elements can be selected.

(3) Magnetohydrodynamic power generation. In addition to the Hall effect in solids, the Hall effect also occurs in conductive fluids. The basic principle of Magnetohydrodynamic power generation technology is to use the Hall effect of the plasma such that under the action of the transverse magnetic field, the positive and negative charged particles of the plasma passing through the magnetic field are separated and accumulated on the two pole plates to form a power supply electromotive force, as shown in Figure 10.5.5. This new high-efficiency power generation method converts gas into a plasma stream to generate electrical energy through the heat generated by the combustion of fuel. It is not necessary to first convert heat energy released by the combustion of fuel into mechanical energy to drive the generator wheel to rotate and then convert mechanical energy into electrical energy, as in thermal power generation. This not only improves the utilization efficiency of thermal energy, but also meets the requirements for environmental protection.

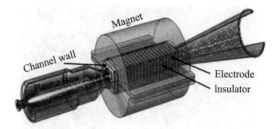

Figure 10 5.5 Schematic diagram of the Magnetohydrodynamic generator

(4) Electromagnetic non-destructive testing. The testing is based on the high permeability of ferromagnetic materials. Defects are detected by measuring the

permeability changes caused by defects in ferromagnetic materials. Ferromagnetic material is magnetized under the action of an external magnetic field. When there are no defects in the material, most of the magnetic lines of force pass through it, and the magnetic lines of force are uniformly distributed inside it. When there is a defect, the magnetic permeability at the defect in the material is much smaller than that of the ferromagnetic material itself, causing the magnetic field lines to bend, and some of the magnetic field lines leak out of the surface of the material. The hall element is used to detect the signal change of the leakage magnetic field **B**. The existence of defects can be effectively detected. The electromagnetic non-destructive testing method based on the Hall effect is safe and reliable, and can realize detection without being influenced by speed. Therefore, it is used in equipment fault diagnosis and material defect detection, such as steel wire rope flaw detection in lifting, transportation, hoisting and bearing equipment and non-destructive testing of pipeline cracks.

(5) The Hall sensor. The Hall elements, Hall integrated circuits, and Hall components based on the Hall effect are generally called the Hall effect magneto-sensitive sensors, or the Hall sensors for short. A variety of electrical and non-electrical linear sensors can be made by using the linear relationship between the Hall voltage and applied magnetic field. For example, when a certain current is controlled, the AC and DC magnetic induction intensity and magnetic field intensity can be measured; when the proportional relationship between current and voltage is controlled, so that the output Hall voltage is proportional to the voltage or current, a power measurement sensor can be made. When the magnetic field intensity is fixed, it can be used to measure AC and DC current and voltage. Using this principle, various non-electrical quantities such as force, displacement, differential pressure, angle, vibration, rotational speed and acceleration. can be accurately measured. The Hall sensors are widely used in daily life and industrial production. In daily life, for example, the commutation structure of the tape recorder uses the Hall sensor to detect the end of the tape and fulfill the automatic commutation function. The motor in the washing machine relies mainly on the detection of the Hall sensor to control the speed and steering of the motor to realize forward and reverse rotation as well as high and low speed rotation function. The Hall switch sensors are used in the temperature control of rice cookers and gas stoves and defrosting of refrigerators. In industrial production, the Hall-type car igniters are different from traditional igniters, with the advantages of high ignition energy, reliable high-speed ignition, and low failure rate. The Hall-effect speed and mileage testers can accurately measure the speed and mileage of cars.

10.6 Magnetism of matter

10.6.1 Classification of magnetic material

There are various substances in the actual magnetic field, and these substances are in a special state due to the action of the magnetic field, that is, the magnetized state. The magnetized matter in turn has an effect on the magnetic field.

Experiments show that the influence of different magnetic material on the magnetic field is very different. A magnetic medium (magnetic material) is placed in an external magnetic field with a magnetic induction intensity $\boldsymbol{B}_0$. The magnetic induction intensity of the additional magnetic field generated by the magnetic medium is $\boldsymbol{B}'$. Then the total magnetic induction intensity in the magnetic medium is the vector sum of $\boldsymbol{B}_0$ and $\boldsymbol{B}'$, that is,

$$\boldsymbol{B} = \boldsymbol{B}_0 + \boldsymbol{B}' \qquad (10.6.1)$$

For different magnetic material, the magnitude and direction of $\boldsymbol{B}'$ are different. The relative permeability μ_r of the magnetic medium is used to describe the influence of different magnetic mediums on the original external magnetic field after magnetization, and is defined as

$$\mu_r = \frac{B}{B_0} \qquad (10.6.2)$$

According to the magnitude of permeability, magnetic media can be divided into three types:

(1) Paramagnetic ($\mu_r > 1$), such as platinum, manganese, chromium, oxygen and nitrogen. In the external magnetic field, the additional magnetic induction intensity $\boldsymbol{B}'$ and $\boldsymbol{B}_0$ are in the same direction, so the magnitude of the total magnetic induction intensity is $B > B_0$.

(2) Diamagnetic ($\mu_r < 1$), such as sulfur, copper, bismuth, hydrogen and lead. In the external magnetic field, its additional magnetic induction $\boldsymbol{B}'$ is opposite to the direction of $\boldsymbol{B}_0$, so the magnitude of the total magnetic induction is $B < B_0$.

(3) Ferromagnetic ($\mu_r \gg 1$), such as iron, cobalt and nickel. In the external magnetic field, its additional magnetic induction intensity $\boldsymbol{B}'$ is in the same direction as $\boldsymbol{B}_0$, and $\boldsymbol{B}' \gg \boldsymbol{B}_0$, so the magnitude of the total magnetic induction intensity $B \gg B_0$.

The relative permeability μ_r of paramagnetic and diamagnetic substances is only slightly greater or less than 1, and is constant. They have little effect on the magnetic field and belong to weak magnetic materials while ferromagnetic materials have a great influence on the magnetic field and belong to strong magnetic materials.

10.6.2 Microscopic mechanism of paramagnetism and diamagnetism

In a molecule of matter, every electron is involved in two kinds of motion at the same time, namely the motion around the nucleus and the spin motion. The vector sum of these magnetic moments of all electrons in a molecule is called the intrinsic magnetic moment of the molecule, denoted by the symbol $\boldsymbol{P}_m$. The intrinsic magnetic moment of this molecule can be equivalently represented by a ring current, called molecular current.

1. Magnetization mechanism of paramagnetic materials

A paramagnetic material is a type of magnetic medium in which the intrinsic magnetic moment of the molecule is not zero $\boldsymbol{P}_m \neq 0$. In the absence of an external magnetic field, due to the random thermal motion of the molecules, the distribution of the intrinsic magnetic moment of each molecule is chaotic, as shown in Figure 10.6.1, and the material cannot show magnetism as a whole.

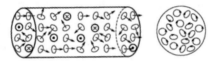

(a) Without an external magnetic field

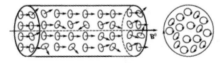

(b) With an external magnetic field

Figure 10.6.1 Magnetization mechanism of paramagnetic material

When there is an external magnetic field, $\boldsymbol{P}_m$ turns to the direction of the external magnetic field, and forms an orderly arrangement under the action of the magnetic moment, which is called the turning magnetization process. At the same time, an additional magnetic field $\boldsymbol{B}'$ is generated which is consistent with the direction of the external magnetic field. Therefore, the intensity of magnetic induction in paramagnetic material is $\boldsymbol{B} = \boldsymbol{B}_0 + \boldsymbol{B}' > \boldsymbol{B}_0$ and $\mu_r > 1$.

2. Magnetization mechanism of diamagnetic materials

For diamagnetic material, the intrinsic magnetic moment of each molecule is zero $\boldsymbol{P}_m = 0$, which does not exhibit magnetism in the absence of an external magnetic field. When there is an external magnetic field, although there is no turning magnetization for $\boldsymbol{P}_m = 0$, since the orbital magnetic moment and spin magnetic moment of each electron in the molecule are not zero, the orbital magnetic moment will be affected by the external magnetic field to produce an additional magnetic moment $\Delta \boldsymbol{P}_m$ in the opposite direction of the magnetic field. Therefore, an additional magnetic field $\boldsymbol{B}'$ opposite to the external magnetic field is generated, so the magnetic induction intensity in the diamagnetic material $B=B_0+B'<B_0$ and $\mu_r<1$.

Suppose the electrons in an atom make a circular motion around the nucleus at a velocity v under the action of the Coulomb force, as shown in Figure 10.6.2(a). If the direction of the external magnetic field B is consistent with that of the electron's orbital magnetic moment, the Lorentz force on the electron will be outward along the orbital radius, which will reduce the centripetal force of the electron. In order to keep the electron's orbital radius unchanged, the speed of the electron must be reduced. Since the magnitude of the electron's magnetic moment is proportional to the speed of motion, it decreases with the decrease of the speed, which can be equivalent to an additional magnetic moment ΔP_m in the opposite direction to B. If the direction of the external magnetic field B is opposite to that of the electron's orbital magnetic moment, as shown in Figure 10.6.2(b), the same analysis as above can be made to draw the conclusion that the additional magnetic moment ΔP_m is opposite in direction to B. Therefore, whether the direction of the magnetic moment of the electron orbit is the same as or opposite to the direction of B, an additional magnetic moment ΔP_m opposite to B in direction can be generated, and as a result, an additional magnetic field B' opposite in direction to B will be generated, so that the internal magnetic induction decreases, that is $B=B_0-B'$.

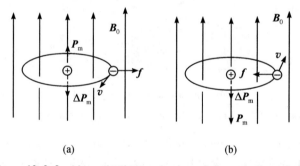

Figure 10.6.2 Magnetization mechanism of diamagnetic material

The diamagnetic effect is not unique to the anti-magnetic material but exists in all materials. Electrons in the molecules of any material are moving in a circular orbit around the core. With the presence of an external magnetic field, the additional magnetic moment opposite to the external magnetic field can be generated. However, the diamagnetic effect in paramagnetic material can be negligible compared with the paramagnetic effect, so its magnetization in the external magnetic field mainly depends on its paramagnetic effect, hence paramagnetism.

Whether it is paramagnetic or diamagnetic material, the additional magnetic field B' generated in the presence of an external magnetic field is much smaller than the original magnetic field B_0, hence weak magnetism.

10.6.3 Magnetization and magnetization current

Before a magnetic material is magnetized, the total magnetic moment of the molecules is zero. But after magnetization, the total magnetic moment of the molecules in the medium will not be zero, and the higher the magnetization of the material, the greater the total magnetic moment. Obviously, the magnetization degree of a magnetic material can be described by the vector sum of the molecular magnetic moments in the unit volume of the material, which can be defined as the magnetization intensity and represented by **M**. If the total magnetic moment of molecules in the volume element ΔV near a point in the magnetic medium is $\sum \boldsymbol{P}_m$, then the magnetization **M** at this point is

$$\boldsymbol{M} = \frac{\sum \boldsymbol{P}_m}{\Delta V} \qquad (10.6.3)$$

Whether it is paramagnetic or diamagnetic material, an equivalent current I_S will be generated on the surface of the magnetic material after magnetization, which is called magnetizing current. From a macroscopic point of view, the additional magnetic field $\boldsymbol{B}'$ in the magnetic material can be generated by this layer of magnetizing current I_S.

Similar to the relationship between polarization and polarization charge when a dielectric is polarized, the relationship between magnetization and magnetizing current of a magnetic material can be given by the example of a solenoid.

If there is a long solenoid carrying current I, and the tube is filled with isotropic paramagnetic material, a uniform magnetic field will be generated inside the solenoid, so that the material in the tube is uniformly magnetized. At this time, the magnetic moment of each molecule in the medium will be arranged in the direction of the magnetic field, as shown in figure 10.6.3. At any point in the material, the circular currents of adjacent molecules are always opposite in pairs and cancel each other out, but at the edge of the cross section, the circular currents cannot be canceled out. These circular currents are connected end to end with each other, resulting in a horizontal circular current at the edge of the section. Each section of the solenoid has a corresponding circular current. It is equivalent to a surface current flowing along a cylindrical surface on the surface of the material. The equivalent current in the material is the magnetizing current I_S which is in

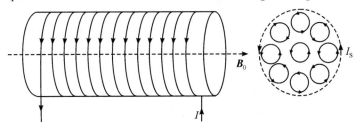

Figure 10.6.3 Magnetizing current appearing on a uniform paramagnetic surface in a long straight solenoid

the opposite direction to the conduction current I.

Obviously, both the magnetization intensity and the magnetizing current describe the degree of magnetization of the material and it can be proved that the relationship between the two is

$$\oint_L \boldsymbol{M} \cdot \mathrm{d}\boldsymbol{l} = \sum I_S \tag{10.6.4}$$

That is, the line integral of the magnetization $\boldsymbol{M}$ along any closed path L is equal to the algebraic sum of the magnetizing currents enclosed by the closed path.

10.6.4 The Ampère circuital theorem in magnetic material

From the above discussion, it can be seen that magnetization of the magnetic material in the solenoid is related to the equivalent magnetizing current I_S in the solenoid. Therefore, when there is a magnetic material, the magnetic field will be jointly generated by the conduction current I and the magnetizing current I_S. Then, the Ampère circuital theorem is written as

$$\oint_L \boldsymbol{B} \cdot \mathrm{d}\boldsymbol{l} = \mu_0 \left(\sum I_{\mathrm{int}} + I_S \right) \tag{10.6.5}$$

Because I_S cannot be known in advance, it is difficult to directly find the distribution of the magnetic field by the above equation. To solve this problem, the method of introducing auxiliary vectors can be adopted. Use Equation (10.6.4) to rewrite Equation (10.6.5) as

$$\oint_L \boldsymbol{B} \cdot \mathrm{d}\boldsymbol{l} = \mu_0 \left(\sum I_{\mathrm{int}} + \oint_L \boldsymbol{M} \cdot \mathrm{d}\boldsymbol{l} \right)$$

that is

$$\oint_L \left(\frac{\boldsymbol{B}}{\mu_0} - \boldsymbol{M} \right) \cdot \mathrm{d}\boldsymbol{l} = \sum I_{\mathrm{int}} \tag{10.6.6}$$

Introduce an auxiliary physical quantity to describe the magnetic field, the magnetic field strength vector $\boldsymbol{H}$, which is defined as

$$\boldsymbol{H} = \frac{\boldsymbol{B}}{\mu_0} - \boldsymbol{M} \tag{10.6.7}$$

Thus

$$\oint_L \boldsymbol{H} \cdot \mathrm{d}\boldsymbol{l} = \sum I_{\mathrm{int}} \tag{10.6.8}$$

The above expression shows that in a steady magnetic field, the line integral of the magnetic field strength vector $\boldsymbol{H}$ along any closed path is equal to the algebraic sum of the conduction currents enclosed in the loop, regardless of the magnetizing current. This is Ampère's circuital theorem in the presence of a magnetic material. Although the above equation is derived from the special case of the current-carrying solenoid, it can be applied in a general case.

When there is no magnetic material, the magnetization $\boldsymbol{M} = 0$, and Equation (10.6.8) is reduced to the form of Equation (10. 3. 5). Obviously, Equation (10. 6. 8) is a more general form of Ampère's circuital theorem for a steady magnetic field.

Experiments have shown that for an isotropic homogeneous magnetic material, the magnetization **M** at any point in the material is proportional to the magnetic field strength **H** at that point, and the proportionality coefficient χ_m is a constant, called the magnetic susceptibility of the magnetic material, that is

$$\boldsymbol{M} = \chi_m \boldsymbol{H} \qquad (10.6.9)$$

Substituting Equation (10. 6. 9) into Equation (10. 6. 7), we can obtain

$$\boldsymbol{B} = \mu_0 (\boldsymbol{H} + \boldsymbol{M}) = \mu_0 (1 + \chi_m) \boldsymbol{H} \qquad (10.6.10)$$

Generally

$$\mu_r = 1 + \chi_m$$

$$\mu = \mu_0 \mu_r$$

Thus, the relationship between **B** and **H** can be written as

$$\boldsymbol{B} = \mu \boldsymbol{H} \qquad (10.6.11)$$

where μ_r is called relative permeability of magnetic material, a dimensionless quantity and μ is called permeability of magnetic material.

The magnetic field distribution in the presence of a magnetic material can be easily solved by using Ampère's circuital theorem. When the magnetic field distribution has special symmetry, the distribution of **H** can be obtained by Equation (10. 6. 8) according to the distribution of conduction current, and then the distribution of **B** can be obtained by Equation (10. 6. 11).

Example 10. 6. 1 As shown in Figure 10. 6. 4, a current I flows uniformly in an infinitely long cylindrical conductor with a radius of R_1. There is a coaxial cylindrical surface with a radius of R_2 outside it, and the space between the two cylindrical surfaces is filled with the homogeneous magnetic material with relative magnetic permeability μ_r. The current I flows back along the outer cylindrical surface. Find the magnetic field distribution in the space.

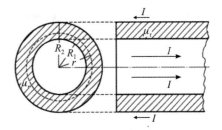

Figure 10.6.4 Figure of Example 10.6.1

Solution Since the conduction current I and the distribution of the magnetic material are axisymmetric, the magnetic field distribution is also axisymmetric. If a circle with a point on the axis as its center and of an arbitrary radius r is drawn in a plane perpendicular to the axis, the magnetic field strength **H** and magnetic induction strength **B** of each point on the circumference are equal, and the directions are both in the tangent direction of the circumference. Therefore, such a circle can be regarded as an integral loop L. According to the Ampère circuital theorem

$$\oint_L \boldsymbol{H} \cdot \mathrm{d}\boldsymbol{l} = H 2\pi r = \sum I_{\text{int}}$$

When $0 \leqslant r < R_1$

$$H_1 2\pi r = \frac{I}{\pi R_1^2} \pi r^2$$

Then

$$H_1 = \frac{Ir}{2\pi R_1^2}$$

$$B_1 = \mu_1 H_1 = \frac{\mu_0 Ir}{2\pi R_1^2}$$

When $R_1 < r < R_2$

$$H_2 2\pi r = I$$

Then

$$H_2 = \frac{I}{2\pi r}$$

$$B_2 = \mu_2 H_2 = \frac{\mu_0 \mu_r I}{2\pi r}$$

When $r > R_2$

$$H_2 2\pi r = 0$$

Then

$$H_3 = 0$$
$$B_3 = 0$$

10.7 Ferromagnetic material

With an external magnetic field, the additional magnetic field $\boldsymbol{B}'$ generated by ferromagnetic material is much larger than the external magnetic field $\boldsymbol{B}_0$, and its relative magnetic permeability μ_r changes with the external magnetic field and is not constant. This kind of magnetic material is called ferromagnetic material, such as iron, nickel, cobalt, and their metal compounds. The characteristics of ferromagnetic material cannot be explained by the magnetization theory of general magnetic material.

10.7.1 Magnetization law of ferromagnetic material

In the experiment, a solenoid is filled with the ferromagnetic material to be measured. With the current from small to large, the change curve of $B \sim H$ is measured, which is called $B \sim H$ magnetization curve, as shown in Figure 10.7.1. At the beginning, $B=0$, $H=0$, and the magnetic material is in an unmagnetized state. From point O to P, B increases with the increase of H, but the rate of increase slows down. After point P, B reaches a saturation state and no longer increases with the increase of H. The magnetic induction intensity B_s corresponding to point P is called the saturation magnetic induction intensity. The $O \sim P$ segment curve is called the initial magnetization curve. Then the external magnetic field H is dropped to zero, and B slowly decreases, that is, the $P \sim R$ segment. When $H = 0$, the corresponding B_r is called the remanence. To reduce the magnetic induction intensity of the material to 0, a magnetic field in the opposite direction must be added. H increases in the opposite direction, B decreases rapidly until $B = 0$, and then demagnetization is completed, which is the $R \sim C$ segment. The corresponding H_c at this point is called the coercive force. As H in the opposite direction continues increasing, the ferromagnetic material is reversely magnetized, and reaches the reverse saturation point P', which is the $C \sim P'$ segment. After that, the reverse H is reduced to 0, and then in the $P' \sim P$ positive direction H increases until it reaches the positive saturation point P. The resulting closed curve is called a hysteresis loop. The phenomenon that the change of B lags behind that of H is called hysteresis. Corresponding to the same H value, the multi-value of B is related to the process, which is enough to show the complexity of ferromagnetic magnetization.

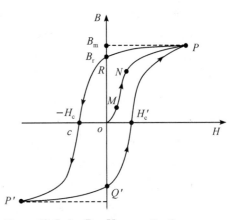

Figure 10.7.1 $B \sim H$ magnetization curve

Experiments show that the repeated magnetization of ferromagnetic material in the alternating magnetic field is accompanied by energy loss, which is called hysteresis loss. Both theory and practice have proved that the hysteresis loss is proportional to the area enclosed by the hysteresis loop.

10.7.2 Classification of ferromagnetic material

The hysteresis loops of different ferromagnetic materials are very different. According to the value of the coercive force, ferromagnetic materials can be divided into the following categories:

(1) Soft magnetic materials. The coercive force of pure iron, silicon steel, permalloy and other materials is small ($H_c < 10^2$ A · m^{-1}). The hysteresis loop is narrow and long and the surrounding area is small, as shown in Figure 10.7.2 (a), and this kind of material is called soft magnetic material. Soft magnetic materials have small hysteresis loss and are easy to magnetize or demagnetize. They can be used to make relays, transformers, electromagnets, motors, and magnetic cores for various high-frequency electromagnetic components.

(2) Hard magnetic materials. Carbon steel, tungsten steel, AlNiCo alloy and other materials have large coercivity ($H_c > 10^2$ A · m^{-1}), so the area surrounded by the hysteresis loop is large, as shown in Figure 10.7.2 (b), and these materials are called hard

magnetic materials. The hysteresis loss of hard magnetic materials is large, the remanence B_r is large, and they are not easy to demagnetize. They are suitable for making permanent magnets, such as permanent magnets used in magnetoelectric meters, speakers, and earphones.

(3) Rectangular magnetic materials. The hysteresis loop of material such as manganese magnesium ferrite and lithium manganese ferrite is close to a rectangle, as shown in Figure 10.7.2 (c), and they are called rectangular magnetic materials. The remanence B_r of the rectangular magnetic material is close to the saturation value B_s, and the high remanence ratio B_r/B_s and the low coercivity H_c are the remarkable characteristics of the rectangular magnetic material. According to this feature, rectangular magnetic materials can be used to make digital magnetic recording devices, storage elements in computers, etc.

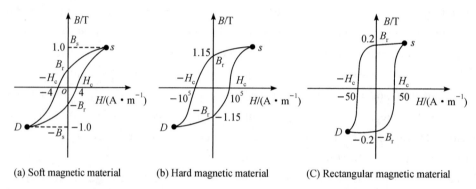

(a) Soft magnetic material   (b) Hard magnetic material   (C) Rectangular magnetic material

Figure 10.7.2　Hysteresis loops of various ferromagnetic materials

10.7.3　Magnetic domain

Recent studies have shown that magnetism of ferromagnetic materials is mainly derived from the spin magnetic moment of the electron. In the absence of an external magnetic field, the spin magnetic moments of electrons in a ferromagnetic material can be "spontaneously" arranged in a small range to form small "spontaneous magnetization regions", which are called magnetic domains. The linearity of the magnetic domain is on the order of millimeters and consists of about 10^{17} to 10^{21} molecules. According to the theory of quantum mechanics, there is an "exchange" between electrons, which makes the electron spins have lower energy when they are arranged in parallel, and the exchange is a pure quantum effect.

Usually in an unmagnetized ferromagnetic material, the spontaneous magnetization directions in various magnetic domains are different and in disorder, leading to non-magnetism on a macroscopic scale, as shown in Figure 10.7.3 (a). After an external magnetic field is applied, the magnetic moments of various magnetic domains in the ferromagnetic material tend to be arranged in the direction of the external magnetic field, as shown in Figure 10.7.3 (b). When the external magnetic field is enhanced to a certain extent, the magnetization directions of all magnetic domains in the ferromagnetic material are arranged in the direction of the external magnetic field. At this time, the magnetization of the ferromagnetic material reaches a saturation state, and because the magnetic moments in all magnetic domains are arranged in the external field, the ferromagnetic material shows strong magnetic properties.

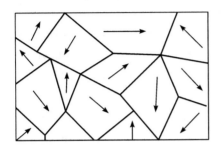
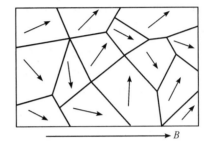

(a) Without an external magnetic field   (b) With an external magenetic field

Figure 10.7.3 Magnetic domains of ferromagnetic materials

When the ferromagnet is subjected to strong vibration, or under the influence of violent motion at high temperature, the magnetic domain will collapse, and a series of ferromagnetic properties associated with the magnetic domain will disappear. When the temperature is higher than a specific critical temperature, ferromagnetic material will lose ferromagnetism and become paramagnetic. Curie firstly discovered the phenomenon and this critical temperature is called Curie temperature or the Curie point of the ferromagnetic material. The Curie temperatures of iron, cobalt, and nickel are 1040 K, 1388 K, and 631 K, respectively.

Scientist Profile

Michael Faraday (1791 – 1867) was a British physicist, chemist, and well-known self-taught scientist. Because of his great contributions to electromagnetism, he is called "Father of Electricity" and "Father of Alternating Current".

Faraday was born in a poor blacksmith family in Surrey in 1791. Due to the poverty of his family, he only studied for two years in a primary school and had no formal education when he was a child. In 1803, forced to make a living, he took to the streets as a newsboy, and later became an apprentice in the home of a bookseller and stapler. During his apprenticeship in the bookstore, Faraday had a strong desire for knowledge, read all kinds of books eagerly, and learned a lot of basic knowledge in natural science. Faraday did not miss any opportunity to learn and joined the youth science organization—City of

London Philosophical Society. Through all this, he initially mastered the basic knowledge of physics, chemistry, astronomy, geology, meteorology, etc., which laid a good foundation for future research work. Faraday's eagerness to learn moved a patron of the bookstore. With his help, Faraday had a chance to listen to the famous chemist David's speeches. He recorded and organized all the speeches, and went back to discuss and study them with his friends seriously. Afterwards, he sent the organized speech records to David, and attached a letter to express his willingness to devote himself to the scientific career. In 1811, he became David's experimental assistant and began his scientific career.

From 1815, Faraday began to carry out independent research work under the guidance of David in the Royal Institute of Britain and achieved certain results. He and Stoddart pioneered the metallographic analysis method, obtained hexachloroethane and tetrachloroethylene by substitution reaction, and found the method for liquefaction of chlorine gas and other gases. In 1821, Faraday invented the electric motor, which was the first device in the world to use electric current to move objects. Although this device is very simple compared with modern technology, it has created the history of the development of electric motors.

In 1831, Faraday discovered the phenomenon of electromagnetic induction in an experiment, which became one of the greatest contributions in Faraday's life. Faraday's discovery cleared a stumbling block in the search for the nature of electromagnetism, opening a new avenue for generating large amounts of current except batteries. According to this experiment, Faraday invented the disc generator in the same year, which was Faraday's second major electrical invention. Although the structure of this disc generator was simple, it was the first generator created by human beings, and the generator that produces electricity in the modern world started from it.

In 1837, Faraday introduced the concepts of electric and magnetic fields, pointing out that there were fields around electricity and magnetism, which broke the traditional concept of "action at a distance" in Newtonian mechanics. In 1838, he proposed a new concept of electric field lines to explain electrical and magnetic phenomena, which was a major breakthrough in physical theory. In 1852, he introduced the concept of magnetic field lines, which laid the foundation for the establishment of classical electromagnetic theory.

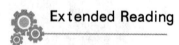
Extended Reading

Maglev train

The traditional wheel-rail traffic relies on the friction between the wheel and the track to provide the traction force required by the vehicle, and the traction force is restricted by the maximum kinetic friction coefficient. In the empirical formula, the kinetic friction coefficient is inversely proportional to speed. The higher the speed, the smaller the friction coefficient between the wheel and the rail and the smaller the traction force. This feature limits the traction capacity of the system and affects the acceleration performance and climbing ability of the train. As a new type of rail transportation, maglev transportation uses electromagnetic force to realize the non-contact support between the vehicle and the track, and realizes the traction of the vehicle through the linear motor. Compared with the traditional wheel-rail traffic, it avoids the mechanical contact between the vehicle and the track, and has the advantages of high speed, great climbing ability, small turning radius and low noise.

Magnetic levitation is the use of magnetic force to make objects in a frictionless, non-contact suspension state. Magnetic levitation looks simple, but the realization of specific magnetic levitation characteristics has gone through a long time. As early as 1842, the British physicist Enhuw proposed the concept of magnetic levitation, but he also pointed out that a ferromagnet could not be maintained in a free and stable suspension state in all six degrees of freedom by means of permanent magnets alone. In order to achieve stable magnetic levitation, the magnitude of the magnetic field force must be continuously adjusted according to the suspended state of the object, that is, a controllable electromagnet could be used to achieve this, and at least one degree of freedom of the suspended rotor must be actively controlled. In 1938, Kempel used an inductive sensor and a tube amplifier to make a controllable electromagnet, and successfully achieved stable magnetic levitation for an object weighing 2100 N, which was the prototype of a maglev train. In the 1960s, with the rapid development of modern control theory and electronic technology, the originally huge control equipment became very light, which provided the possibility of realizing the maglev train technology. In 1969, Maffei from the German Traction Locomotive Company developed a small maglev train system model, later named TR01, and it reached a speed of 165 km $\cdot$ h^{-1} on a 1 km track, which was a milestone in the development of maglev trains. After 1970, the economic strength of industrialized countries in the world was increased. In order to improve the transportation capacity to meet the needs of their economic development, developed countries such as Germany, Japan, the United States, Canada, France, and the United Kingdom successively started to plan the maglev transportation system. However the United States and the Soviet Union abandoned their research programs in the 1970s and 1980s, respectively. The United Kingdom began to study maglev trains in 1973, but was one of the first countries to put maglev trains into commercial operation. Currently in the race to manufacture maglev trains, the most mature research is in Japan and Germany. In 1999, the superconducting maglev train developed by Japan reached a speed of 550 km $\cdot$ h^{-1} on the experimental line. In 1977, Germany developed normal-conducting and super-conducting test trains respectively, but after analysis and comparison, they concentrated on developing only normal-conducting maglev trains. At present, Germany's technology in this research is very mature. The Transrapid system has become the first maglev high-speed railway system in the world to enter the mature stage of technical application. the TR08 model of the system is applied on the Shanghai high-speed maglev demonstration line, and the TR09 maglev train is currently being developed. China's research on maglev trains started relatively late. In 1989, the National University of Defense Technology developed China's first maglev test prototype. In 1994, Southwest Jiaotong University built our first maglev railway test line, and at the same time carried out a manned test of a maglev train with a speed of 30 km $\cdot$ h^{-1}, which indicated that China mastered the technology of manufacturing maglev trains. In 2016, the Changsha maglev express train with a speed of 100 km $\cdot$ h^{-1}, led by CRRC Zhuzhou Machinery Co., Ltd., was put into operation. In 2018, China's first commercial maglev version 2.0 train rolled off the production line at CRRC Zhuzhou Electric Locomotive Co., Ltd. The train's design speed increased to 160 km $\cdot$ h^{-1}, and it was composed of three sections and could carry a maximum of 500 passengers. In May 2019, China's 600 km $\cdot$ h^{-1} high-speed maglev test prototype rolled off the assembly line in Qingdao, marking a major breakthrough in China's high-speed maglev technology.

The basic principle of a maglev train is shown in Figure ER 1(a). Some magnets are installed on the floor of the train. When the train moves, these magnets pass over the metal guide rails. Due to the change of the surrounding magnetic field, electromagnetic induction occurs and then induced current is formed in the guide rail, thus generating magnetic fields, just as numbers of magnets are formed in the guide rail, and the polarity of these magnets is opposite to that of the magnets in the train, thereby producing an upward repulsive force. When the upward thrust is strong enough, the train can leave the ground and float above the metal rails. Although magnets of the same polarity repel each other, they can produce buoyancy for suspension; But magnets of the opposite polarity attract, and maglev trains use this principle to move. As shown in Figure ER 1(b), when the guide rail under the train generates buoyancy due to the electronic motion, the lines of the guide rails on both sides begin to energize, generating another set of magnetic fields

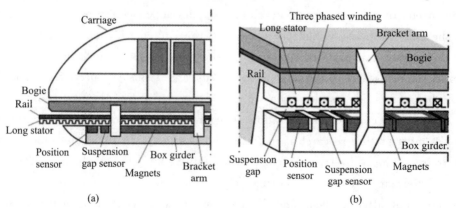

Figure ER 1 Schematic diagram of the maglev train principle

slightly ahead of the train. After a special arrangement, the magnetic south pole of the guide rail will be close to the magnetic north pole of the train. Because of this suction, the train can move forward. By adjusting the current through both sides of the guide rail, the attractive magnetic force can be made to act just in front of the train. Maglev trains can resist the earth's gravity, suspended over the track. According to different working principles, there are normal-conducting electromagnetic attraction suspension and superconducting repulsion suspension.

Normal-conducting electromagnetic attraction suspension: Using the normal-conducting magnets (levitation electromagnets) installed on the bogies on both sides of the vehicle and the magnets laid on the line guide rails, the attraction generated under the action of the magnetic field makes the vehicle float. The gap between the vehicle and the rail surface is inversely proportional to the magnitude of the attraction force. In order to ensure the reliability of the suspension, the smooth operation of the train and a higher power of the linear motor, it is necessary to accurately control the current in the electromagnet for the magnetic field to maintain a stable strength and suspension force and for the train body and guide rail to maintain a gap of about 10 mm. The feedback control of the system is usually carried out by using an air gap sensor for measuring the gap. This suspension method does not require special ground support devices and auxiliary ground wheels, and the requirements for the control system can also be slightly lower. The speed of this maglev train is usually in the range of 300 – 500 km $\cdot$ h^{-1}, which is suitable for intercity and suburban transportation. The HSST system in Japan, the TR ultra-high speed maglev train in Germany and the medium-low speed maglev train in China all use this technology.

Superconducting repulsion suspension: In this form, superconducting magnets are installed at the bottom of the vehicle, and a series of aluminum loop coils are laid on both sides of the track. When the train is running, a strong magnetic field is generated after the coil (superconducting magnet) on the train is energized, and the ground coil cuts the magnetic field line to generate an induced current in the ground coil. The magnetic field produced by the induced current is opposite to the magnetic field of the superconducting magnets on the vehicle, and the two magnetic fields produce a repulsive force. When the repulsive force is greater than the weight of the vehicle, the vehicle floats. Since the resistance of superconducting magnets is zero, almost no energy is lost during operation, and the magnetic field strength is large. When the vehicle is displaced downward, the distance between the superconducting magnet and the suspension coil decreases and the current increases, so that the suspension force increases, and the vehicle automatically returns to the original suspension position. This gap is related to the value of the speed. Generally, the train body can be suspended when it reaches 100 km $\cdot$ h^{-1}. Therefore, auxiliary mechanical support devices, such as auxiliary support wheels and corresponding spring supports, must be installed on the vehicle to ensure the safe and reliable landing of the train. The maximum operating speed of this maglev train can reach 1000 km · h^{-1}. Of course, its construction technology and cost are much higher than those of the normal-conducting attraction maglev train. Japan's MLU series high-speed maglev trains use this technology.

It has been more than 80 years since the emergence of maglev train technology, but it has not been widely used due to the high cost, high power consumption and large radiation. The Shanghai maglev demonstration line is about 30 kilometers long, with a total investment of 8.9 billion yuan, and the average cost per kilometer is about 300 million yuan, which is 2 to 3 times that of high-speed rail. The maglev high-speed rail of the Central Shinkansen currently under construction in Japan, with a total length of 286 kilometers, is expected to cost about 525 billion yuan after completion, with an average cost of about 1.8 billion yuan per kilometer. Faced with such high construction costs and long cost recovery cycles, it is difficult to achieve profitability. Since the magnetic levitation system must be supplemented by electromagnetic force to complete the suspension, guidance and driving, the safety of the train in the case of power failure is still an unsolved problem. In addition, the impact of strong magnetic fields on the human body and the environment also needs further research.

Maglev trains have not been widely used so far. The world's first maglev line was the Birmingham International Airport Line in the United Kingdom. It was completed and used in 1984 and had a total length of 600 meters. Later, it was abandoned due to reliability problems. Germany built the 32-kilometer-long maglev test line in Emsland in 1984, but it was shut down in 2006 due to a serious derailment accident, which seriously affected the promotion of maglev train technology in Germany. The manned test of Japan's maglev train was successful in 1982, but the actual operation of the maglev high-speed rail line planned by Japan was not approved due to the high cost of construction. In 2013, Japan once again launched the Central Shinkansen project connecting Tokyo to Nagoya, aiming to open in 2027. At present, there are only four maglev train lines in actual commercial operation in the world, and all of them are in China, Japan and South Korea. They are Shanghai maglev demonstration line, Japan's Aichi eastern hilly line, South Korea's Incheon airport maglev line, and Changsha maglev fast line. Among the four maglev lines, the Shanghai maglev demonstration line is the only high-speed maglev line in commercial operation in the world, as shown in Figure ER 2. The other three are medium- or low-speed maglev lines, running at about 100 kilometers per hour or below 100 kilometers per hour.

In China, Japan, Germany and other typical countries, the high-speed maglev railway is basically mature in technology, but there is a lack of successful attempts in commercial application. The low-temperature superconducting materials used in Japan have high environmental requirements, huge costs in engineering applications, and certain safety hazards. For this reason, high-temperature superconducting materials are being actively

Figure ER 2   Shanghai maglev train

tested in order to obtain greater convenience in engineering applications. On the whole, the future high-speed maglev railway will develop in the direction of practical technology and low cost. The low-vacuum pipeline maglev railway has the advantages of higher speed, more energy saving, environmental protection, all-weather transportation and so on. It has the technical and economic characteristics of new transportation mode and has a good prospect. At present, Virgin Hyperloop One, HTT, SpaceX and other companies in the United States are actively carrying out research on low-vacuum pipeline maglev railway related technologies, and at the same time rapidly deploying them around the world. China is also speeding up research on low-vacuum pipeline maglev railway. As shown in Figure ER 3, CRRC, Southwest Jiaotong University, China Aerospace Science and Industry Corporation, China Academy of Railway Sciences Group Co., LTD and other units are actively exploring the related technology, and Shenzhen, Guizhou and other places are actively introducing this technology. Although the low-vacuum pipeline maglev railway is still far from commercial application, its unique speed advantage and the characteristic of being unaffected by the external environment are unmatched by other ground transportation methods.

Figure ER 3   Experimental platform of China's first vacuum pipeline superhigh-speed maglev train prototype

Discussion Problems

10.1 A current source Idl is placed at a certain point in a magnetic field. When it is placed along the X axis, it is not subjected to force. If it is turned to the positive direction of the Y axis, the force it receives is in the negative direction of the Z axis. Find the direction of the magnetic induction intensity B at this point.

10.2 (1) In a space without current, if the magnetic induction lines are parallel straight lines, will the magnitude of the magnetic induction intensity B vary along the magnetic induction line and in the direction perpendicular to it (that is, is the magnetic field uniform)?

(2) If there is current, is the above conclusion correct?

10.3 Can Ampère's circuital law be used to find the magnetic induction of a current-carrying wire of a finite length? Why?

10.4 If two coils with the same area and current are placed in a uniform magnetic field, one being a triangle and the other a rectangle, are the maximum magnetic moments on the two coils equal? Are the resultant magnetic forces equal?

10.5 Someone says that B and H of paramagnetic substances are in the same direction, while B and H of diamagnetic substances are in opposite directions. Do you think it is correct? Why?

Problems

10.6 As shown in Figure T10 - 1, a single layer of wire is evenly and closely wound on the whole wooden ring with rectangular cross section and there are N turns in total. Find the distribution of magnetic induction intensity in the ring and the magnetic flux on the cross section after the current I flows in the wire.

10.7 In an infinitely long semi-cylindrical metal sheet with a radius 1.0 cm, a current of 5.0 A passes from bottom to top, and the current distribution is uniform, as shown in Figure T10 - 2. Find the magnetic induction at any point on the axis of the cylinder.

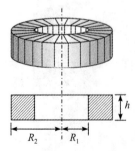

Figure T10 - 1 Figure of Problem 10.6

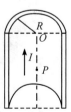

Figure T10 - 2 Figure of Problem 10.7

10.8 As shown in Figure T10‑3, AB and CD are long straight wires, $\overset{\frown}{BC}$ is an arc-shaped wire with the center at the point O, and its radius is R. If a current I flows, find the magnetic induction intensity at point O.

10.9 As shown in Figure T10‑4, in an infinitely long current-carrying plate with a width of a, passes a uniform current I along its length. Find the magnetic induction intensity at point P that is coplanar with the plate and is b away from one side of the plate.

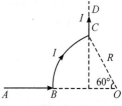

Figure T10‑3 Figure of Problem 10.8

10.10 Inside the long straight cylindrical conductor of radius R, a long straight cylindrical cavity of radius r is dug parallel to the axis, the distance between the two axes is a, $a > r$, and the cross section is shown in Figure T10‑5. Now the current I flows along the conductor tube, the current is evenly distributed over the cross section of the tube, and the current direction is parallel to the axis of the tube. Find

(1) the magnitude of the magnetic induction intensity on the cylinder axis;

(2) the magnitude of the magnetic induction on the axis of the hollow part.

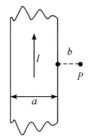 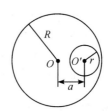

Figure T10‑4 Figure of Problem 10.9   Figure T10‑5 Figure of Problem 10.10

10.11 A long straight coaxial cable consists of a cylindrical wire and a coaxial cylindrical conductor shell, the dimensions of which are shown in Figure T10‑6. The current in the cable flows from the center wire and returns from the outer conductor cylinder shell. Assuming that the current is evenly distributed, and vacuum treatment can be performed between the inner cylinder and the outer cylinder. Find the distribution of the magnetic induction intensity.

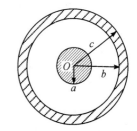

Figure T10‑6 Figure of Problem 10.11

10.12 As shown in Figure T10‑7, a current $I_1 = 20$ A passes through the long straight wire AB, and a current $I_2 = 10$ A passes through the rectangular coil $CDEF$. AB is coplanar with the coil, and both CD and EF are parallel to AB. $a = 9.0$ cm, $b = 20.0$ cm and $d = 1.0$ cm. Find

(1) the force applied on each side of the rectangular coil by the magnetic field of the wire AB;

(2) the resultant force and moment on the rectangular coil.

10.13 The equilateral triangle coil with side length $l=0.1$ m is placed in a uniform magnetic field with magnetic induction intensity $B=1$ T, and the coil plane is parallel to the direction of the magnetic field. As shown in Figure T10-8, in the coil passes the current $I=10$ A. Find

(1) the Ampère force on each side of the coil;

(2) the magnitude of the magnetic moment on the OO' axis;

(3) the work done by the magnetic force when turning from the position to the coil plane perpendicular to the magnetic field.

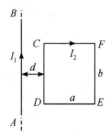
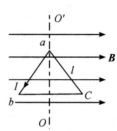

Figure T10-7 Figure of Problem 10.12   Figure T10-8 Figure of Problem 10.13

10.14 There are two infinitely long coaxial conductor cylindrical surfaces of radii r and R, through which current I in opposite directions passes respectively, and the two cylindrical surfaces are filled with a uniform magnetic medium with a relative permeability μ_r. Find

(1) the magnetic induction in the magnetic medium;

(2) the magnetic induction intensity outside the two cylindrical surfaces.

Challenging Problems

10.15 The length of a uniformly and densely wound solenoid is $2L$, the radius is a, there are n turns of coils wound on the unit length, and there is a current I.

(1) Find the magnetic field distribution on the axis of the solenoid. What are the characteristics of the field strength curve on the axis when the ratio of length to diameter is different?

(2) What are the characteristics of the magnetic induction line of the solenoid? What are the characteristics of the magnetic induction intensity distribution surface?

10.16 A non-uniform magnetic field $\boldsymbol{B}$ is in the Z direction, and its magnitude is proportional to Z: $\boldsymbol{B} = Kz\boldsymbol{k}$ (K is the proportionality constant). A charged particle with mass m and electric quantity q is injected into the magnetic field from the origin at an initial velocity $\boldsymbol{v}_0$, the initial velocity direction is in the OXZ plane, and the included angle with the Z direction is θ, as shown in Figure T10-9. What are the characteristics of the particle motion trajectory?

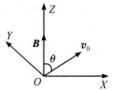

Figure T10-9 Figure of Problem 10.16

Chapter 11

Electromagnetic Induction

In 1820, Oersted discovered the magnetic effect of electric current, which led to extensive research on the electrical effect of magnetism. In 1831, Faraday, a British physicist, discovered electric currents could be induced by magnetic fields and summarized the law of electromagnetic induction. Then Maxwell proposed that the changing electric field could excite a magnetic field. In 1865, based on the work of Coulomb, Oersted, Ampère, Faraday et al., Maxwell proposed a mathematical theory that showed a close relationship between all electric and magnetic phenomena. This theory predicted the existence of the electromagnetic wave and calculated that it traveled at the speed of light. In 1887, Hertz first detected the electromagnetic wave in experiments. The establishment of electromagnetic theory laid a foundation for the development of modern communication theory.

This chapter focuses on the definition of electromotive force and Faraday's law of electromagnetic induction, discusses two forms of induced electromotive force: motional electromotive force and induced electromotive force, and introduces the magnetic field energy combined with self-inductance and mutual inductance and Maxwell's equations.

11.1 Electromotive force

As shown in Figure 11.1.1, when the loop is connected, the positive charges move from plate A with higher potential (positive electrode of the source) to plate B with lower potential (negative electrode of the source) through the external circuit under the electrostatic force, and they neutralize with the negative charges. As a result, the number of charges of the two plates decreases, and the potential difference between the two plates would decrease to zero. The function of the

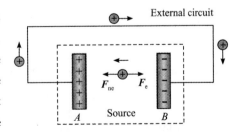

Figure 11.1.1 Non-electrostatic force

source is to transfer positive charges from the low potential plate B to the high potential plate A so that the potential difference between the electrodes maintains constant. This process couldn't be realized if there were only an electrostatic force $\boldsymbol{F}_e$ in the source, and a non-electrostatic force $\boldsymbol{F}_{ne}$ is required. The source could provide such non-electrostatic force.

There are many types of sources, such as a dry battery, accumulator, photocell, and generator. The property of non-electrostatic force is different in different sources. For example, the non-electrostatic force in a chemical battery comes from chemical action, while it comes from electromagnetic induction in an ordinary generator. From the standpoint of energy, the non-electrostatic force inside the source does work to overcome the electrostatic force during the transfer of charges, so that the electrostatic potential energy increases and the energy of some form (mechanical, chemical, thermal, and so on) is converted into electric energy.

The work done by the non-electrostatic force in different sources is different when it transfers a certain number of charges from lower to higher potential. To quantitatively describe the ability to convert energy, a new concept of **electromotive force** (EMF) $\mathscr{E}$ is proposed, and it is defined as the work W_{ne} done by a non-electrostatic force in transferring the unit positive charge from lower to higher potential. So we have

$$\mathscr{E} = \frac{W_{ne}}{q} \tag{11.1.1}$$

The SI unit of EMF is the same as that of electric potential, the volt (1 V = 1 J · C^{-1}).

Note that "electromotive force" is a poor term because EMF is not a force but an energy-per-unit-charge quantity, like electric potential. Moreover, the EMF and potential difference are two completely different physical quantities, although they have the same unit. Electromotive force is a physical quantity that describes the ability of non-electrostatic force to do work in a source. It depends on the property of source itself and has nothing to do with the external circuit.

In analogy with the electrostatic field, the non-electrostatic force is equivalent to the action of non-electrostatic field, represented by $\boldsymbol{E}_{ne}$. Then its action on a charge q is $\boldsymbol{F}_{ne} = q\boldsymbol{E}_{ne}$. Inside the source, the work done by the non-electrostatic force in transferring the charge q from lower to higher potential is

$$W_{ne} = \int_{(-)}^{(+)} q\boldsymbol{E}_{ne} \cdot d\boldsymbol{l} \tag{11.1.2}$$

By substituting Equation (11.1.2) into Equation (11.1.1), we can obtain

$$\mathscr{E} = \int_{(-)}^{(+)} \boldsymbol{E}_{ne} \cdot d\boldsymbol{l} \tag{11.1.3}$$

Then Formula (11.1.3) is the EMF expressed in the viewpoint of field.

EMF is a scalar, but the direction of the increase of the electric potential inside a source is usually specified as the direction of EMF, that is, from the lower to higher potential inside a source.

Since E_{ne} outside a source is zero, the EMF of a source can also be defined as the work done by the non-electrostatic force when the unit positive charge is transferred around a closed circuit, that is

$$E = \oint_L E_{ne} \cdot dl \qquad (11.1.4)$$

11.2 Faraday's law of induction

11.2.1 Electromagnetism induction phenomenon

During the 1830s, several pioneering experiments with magnetically induced EMF were carried out in England by Faraday and in USA by Henry. Figure 11.2.1 shows two examples. In Figure 11.2.1(a), a coil of wire is connected to a galvanometer (G). When the nearby magnet is stationary, the meter shows no current. But when we move the magnet toward or away from the coil, the meter shows current in the coil. In Figure 11.2.1(b), when coil A is powered on or powered off, a current is generated in coil B. After extensive experiments, Faraday summarized several situations that could generate induced current: a changing current, changing magnetic field, moving magnet, and moving conductor in a magnetic field. These experiments can be summarized into two situations: the closed coil remains stationary but the surrounding magnetic field changes; there is relative motion between the closed coil and the magnetic field. And no matter which method is used to generate current, it can be found that the magnetic flux passing through the closed coil has changed. Thus, the following conclusion can be obtained. When the magnetic flux passing through the area enclosed by a closed conductor coil changes, a current is generated in the coil. This phenomenon is called **electromagnetic induction**, and the current generated in the coil is called **induced current**. The presence of current in the coil indicates the presence of EMF. This EMF results from the change of

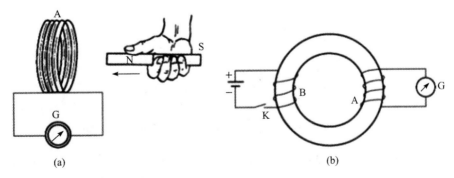

Figure 11.2.1 Experiment of electromagnetic induction

magnetic flux, so it is called **induced EMF**.

11.2.2 Faraday's law of electromagnetic induction

In 1831, Faraday summarized the law of electromagnetic induction through a large number of experiments, that is, **the magnitude of the induced EMF generated in a coil is proportional to the rate of change of the magnetic flux passing through the coil.** This is Faraday's law of electromagnetic induction, and it is expressed as

$$\mathscr{E}_i = -\frac{d\Phi_m}{dt} \tag{11.2.1}$$

The minus sign in the equation indicates the direction of the induced EMF. The method of determining the direction of induced EMF is as follows. First, the direction of a coil is selected, and then the normal direction of the area enclosed by the loop is determined according to the right-hand rule, as shown in Figure 11.2.2. The direction of the loop is specified to determine the positive or negative

Figure 11.2.2 Determining the normal direction of a loop

induced EMF, and the normal direction of the area enclosed by the loop is specified to determine the positive or negative magnetic flux. In this way, when the angle between the magnetic field and the normal direction of the area is less than 90°, the magnetic flux is positive. When the angle is greater than 90°, the magnetic flux is negative. Then, the direction of the induced EMF is determined according to the rate of the change of magnetic flux with time $\frac{d\Phi_m}{dt}$. If $\frac{d\Phi_m}{dt} > 0$, the induced electromotive force is negative, and the direction of the induced EMF is opposite to the direction of the selected loop. On the contrary, if $\frac{d\Phi_m}{dt} < 0$, the induced EMF is positive, and the direction of the induced EMF is the same as the direction of the selected loop. As shown in Figure 11.2.3, the change of magnetic flux in the loop is given, and then we can use the above method to determine the direction of the induced EMF in the loop.

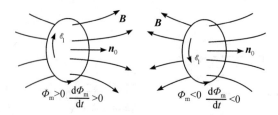

Figure 11.2.3 Determining the direction of induced EMF by Faraday's law

Shortly after Faraday's discovery, Lenz, a Russian physicist, proposed a law to determine the direction of induced current, known as Lenz's law. The statement is that **the direction of the induced current in a closed coil is such that the original magnetic field created**

by that current opposes the change in the magnetic flux. As shown in Figure 11.2.4 (a), when the N pole of the magnet moves towards a conducting coil, the magnetic flux passing through the coil increases. According to Lenz's law, the magnetic field generated by the induced current will hinder the increase of magnetic flux in the coil. Therefore, the direction of the magnetic field generated by the induced current is opposite to that of the original magnetic field. According to the right-hand rule, the direction of induced current can be determined, as shown in Figure 11.2.4 (a). In Figure 11.2.4 (b), when the N pole of the magnet moves away from the coil, the magnetic flux passing through the coil decreases. According to Lenz's law, the direction of the magnetic field generated by the induced current hinders the reduction of magnetic flux in the coil. Therefore, the direction of the induced current can be determined, as shown in Figure 11.2.4 (b).

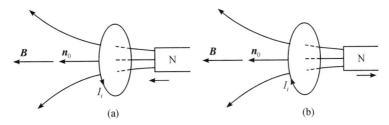

Figure 11.2.4 Determining the direction of induced current by Lenz's law

For a coil with N turns, the induced EMF in the coil is equal to the sum of the induced EMFs in all turns of the coil.

$$\mathcal{E}_i = -\frac{d\Phi_{m1}}{dt} - \frac{d\Phi_{m2}}{dt} - \cdots - \frac{d\Phi_{mN}}{dt} = -\frac{d}{dt}(\Phi_{m1}+\Phi_{m2}+\cdots+\Phi_{mN}) = -\frac{d\Psi_m}{dt} \quad (11.2.2)$$

where $\Psi_m = \Phi_{m1}+\Phi_{m2}+\cdots+\Phi_{mN}$ is the total magnetic flux passing through all coils, also known as the **magnetic flux linkage**.

If the resistance of a closed conductor coil is R, the induced current in the coil can be obtained from Ohm's law of the entire coil

$$I = \frac{\mathcal{E}_i}{R} = -\frac{1}{R}\frac{d\Phi_m}{dt}$$

If the magnetic flux passing through the coil at time t_1 is Φ_{m1} and the magnetic flux passing through the coil at time t_2 is Φ_{m2}, then the quantity of induced charges passing through the coil during time $\Delta t = t_2 - t_1$ is

$$q = \int_{t_1}^{t_2} I\,dt = -\frac{1}{R}\int_{\Phi_{m1}}^{\Phi_{m2}} d\Phi = \frac{1}{R}(\Phi_{m1} - \Phi_{m2}) \quad (11.2.3)$$

Equation (11.2.3) indicates that the induced electric quantity passing through the coil is only related to the change in magnetic flux. Therefore, if the induced electric quantity is measured, the change in magnetic flux can be calculated. The commonly used magnetic flux meter (also known as the Gaussimeter) for measuring magnetic strength is made based on this principle.

Example 11.2.1 A rectangular wire frame *abcd* with N turns is placed parallel to a long straight wire carrying a steady current I, as shown in Figure 11.2.5. The length and width of the rectangular wire frame are l_1 and l_2, respectively. When $t=0$, the distance between the *ad* side and the long straight wire is r. The rectangular wire frame moves away from the straight wire at a constant velocity v. Find the magnitude and direction of the induced EMF in the wire frame at any time t.

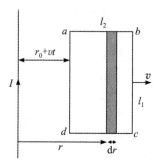

Figure 11.2.5　Figure of Example 11.2.1

Solution　Take the clockwise direction as the positive direction, and the normal direction of the area $\boldsymbol{n}_0$ enclosed by the frame is perpendicular to the paper and downwards. Take a rectangular area element at a distance r from the long straight wire, $d\boldsymbol{S} = l_1 dr \boldsymbol{n}_0$. The magnetic flux passing through the area element at time t is

$$d\Phi_m = \boldsymbol{B} \cdot d\boldsymbol{S} = \frac{\mu_0 I l_1}{2\pi r} dr$$

The magnetic flux passing through one coil at time t is

$$\Phi_m = \int_S \boldsymbol{B} \cdot d\boldsymbol{S} = \int_{r_0+vt}^{r_0+l_2+vt} \frac{\mu_0 I l_1}{2\pi r} dr = \frac{\mu_0 I l_1}{2\pi} \ln \frac{r_0 + l_2 + vt}{r_0 + vt}$$

The induced EMF in the N-turn coils is

$$\mathscr{E}_i = -N \frac{d\Phi_m}{dt} = \frac{N \mu_0 I l_1 l_2 v}{2\pi (r_0 + vt)(r_0 + l_2 + vt)}$$

$\mathscr{E}_i$ is positive, so the direction of the induced EMF is consistent with the selected direction of the rectangular wire frame, that is, it is clockwise.

11.3　Motional EMF and induced EMF

Faraday's law of electromagnetic induction states that when the magnetic flux passing through an area enclosed by a coil changes, an induced EMF is generated in the coil. There are various ways to cause a change of magnetic flux, but there are two basic methods. In one case, the distribution of magnetic field remains unchanged while the conductor coil or conductor moves in the magnetic field. The EMF generated by the motion of the conductor is called motional EMF. In the other case, the conductor coil remains stationary while the magnetic field changes with time. The EMF generated by the change of magnetic field is called induced EMF.

11.3.1 Motional EMF

As shown in Figure 11.3.1, a conductor rod with a length l slides at the velocity v along a U-shaped conducting rail. The area enclosed by the moving rod and rail is increasing as the rod moves. Thus the magnetic flux enclosed by the rod and rail is increasing. A motional EMF is induced according to Faraday's law of electromagnetic induction. The free electrons inside the conductor rod also move at the velocity v, and the electrons are subjected to the Lorentz force

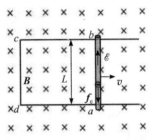

Figure 11.3.1 Motional EMF

$$f_e = (-e) v \times B \tag{11.3.1}$$

The direction of the Lorentz force points to a from b. Under the action of the Lorentz force, free electrons drift downwards, so that the negative charges accumulate at the end a of the conductor rod, and positive charges accumulate at the end b, which results in the formation of an electrostatic field inside the conductor rod. When the Lorentz force equals the electrostatic forces acting on the electrons, a stable potential difference is formed between the two ends of the conductor rod ab, and the potential at the end b is higher than that at the end a. The non-electrostatic force is the Lorentz force, thus the non-electrostatic field E_{ne} can be expressed as

$$E_{ne} = \frac{f_e}{-e} = v \times B$$

The work done in moving the unit positive charge from the end a to end b by this non-electrostatic field is just the motional EMF

$$\mathscr{E}_{ab} = \int_{-}^{+} E_{ne} \cdot dl = \int_{a}^{b} (v \times B) \cdot dl \tag{11.3.2}$$

Equation (11.3.2) is the general expression for motional EMF within a wire of any shape moving in a magnetic field. In Figure 11.3.1, v, B, and dl are mutually perpendicular to each other and Equation (11.3.2) becomes

$$\mathscr{E}_{ab} = \int_{a}^{b} (v \times B) \cdot dl = BLv \tag{11.3.3}$$

Example 11.3.1 Faraday devised the world's first generator in 1831, as shown in Figure 11.3.2(a). When a copper disk rotates between the poles of a permanent magnet, a motional EMF will be induced with the rim terminal positive and the axis terminal negative. As long as the rotation rate is kept constant, the EMF will be quite steady. Assume that the copper disk of radius R rotates at a constant angular speed ω in a uniform magnetic field **B**, calculate the induced EMF.

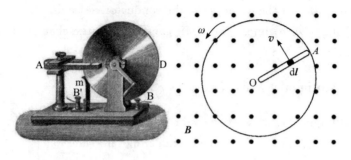

(a) First generator (b) A side view of the copper disk

Figure 11.3.2   Figure of Example 11.3.1

Solution   The induced EMF in the radial direction of the disk is equivalent to an EMF which a rotating conducting rod produces, as shown in Figure 11.3.2(b), and the rotating conducting rod is located along any radial line of the disk.

The EMF of a length element dl is

$$d\mathscr{E} = (\boldsymbol{v} \times \boldsymbol{B}) \cdot d\boldsymbol{l} = vB \, dl = B\omega l \, dl$$

The total EMF of the rod is the sum of EMFs of all length elements, that is

$$\mathscr{E} = \int d\mathscr{E} = \int_0^R B\omega l \, dl = \frac{1}{2} B\omega R^2$$

Example 11.3.2   A conductor rod AB with a length L is placed perpendicular to a long straight wire carrying a steady current I, as shown in Figure 11.3.3. The conductor rod moves at a constant velocity v parallel to the long straight wire. The distance between the near end of the rod and the wire is d. Find the EMF in the conductor rod AB and determine which end has the higher potential.

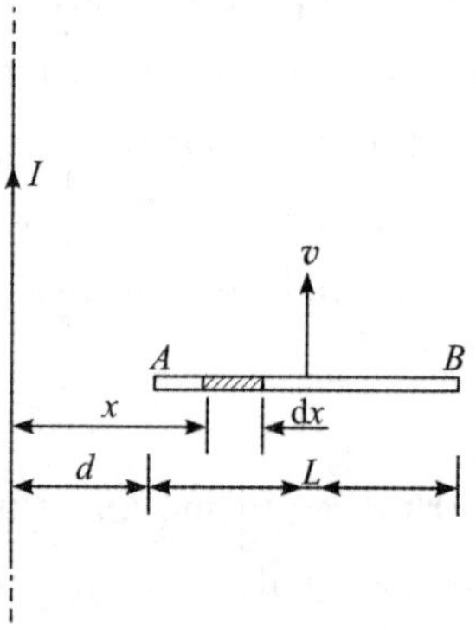

Solution   The magnetic field excited by the long straight wire is a non-uniform magnetic field, so the magnetic strength of the conductor rod AB varies at different positions. Take a length element dx on the rod AB at a distance of x from the long straight wire. The magnetic strength at the length element is

Figure 11.3.3   Figure of Example 11.3.2

$$B = \frac{\mu_0 I}{2\pi x}$$

The EMF of the length element dx is

$$d\mathscr{E} = (\boldsymbol{v} \times \boldsymbol{B}) \cdot dx \boldsymbol{i} = -vB \, dx = -\frac{\mu_0 Iv}{2\pi x} dx$$

The total EMF of the rod is the sum of all length element EMFs, that is

$$\mathscr{E} = \int d\mathscr{E} = \int_d^{d+L} -\frac{\mu_0 Iv}{2\pi x} dx = -\frac{\mu_0 Iv}{2\pi} \ln \frac{d+L}{d}$$

The direction of the motional EMF is from B to A, so the potential at the end A is higher than that at the end B.

11.3.2 Induced EMF

The Lorentz force can explain the origin of motional EMF. But when the conductor or circuit is stationary, the EMF generated by the change of magnetic field cannot be explained by the previous analysis. In 1861, Maxwell proposed the hypothesis of induced electric field, based on the analysis of electromagnetic induction phenomena. **A changing magnetic field produces an electric field with closed electric field lines, called an induced electric field or vortex electric field**, and its electric field strength is represented by E_v. The non-electrostatic force of the induced EMF is provided by the induced electric field. Later, a large number of experiments confirmed the correctness of Maxwell's hypothesis.

According to the definition of EMF and Faraday's law of electromagnetic induction, it can be obtained that

$$\mathscr{E} = \oint_L \boldsymbol{E}_v \cdot \mathrm{d}\boldsymbol{l} = -\frac{\mathrm{d}\Psi_m}{\mathrm{d}t} = -\frac{\mathrm{d}}{\mathrm{d}t}\int_S \boldsymbol{B} \cdot \mathrm{d}\boldsymbol{S} \tag{11.3.4}$$

When the circuit is fixed, the change in magnetic flux only comes from the change in magnetic field, and Equation (11.3.4) can be rewritten as

$$\mathscr{E} = \oint_L \boldsymbol{E}_v \cdot \mathrm{d}\boldsymbol{l} = -\int_S \frac{\partial \boldsymbol{B}}{\partial t} \cdot \mathrm{d}\boldsymbol{S} \tag{11.3.5}$$

Equation (11.3.5) states that in a changing magnetic field, the line integral of the induced electric field over any closed loop is equal to the rate of change in magnetic flux over the area enclosed by the closed loop. The above equation is one of the fundamental equations in electromagnetics, which provides a quantitative relationship between the changing magnetic field and the induced electric field it generates. The minus sign in the equation represents the left-hand relationship, as shown in Figure 11.3.4.

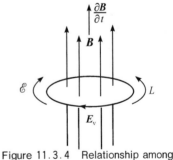

Figure 11.3.4 Relationship among $\frac{\partial \boldsymbol{B}}{\partial t}$, $\boldsymbol{E}_v$, L, $\mathscr{E}$ and $\boldsymbol{B}$

Just like the electrostatic field, the induced electric field is an objective substance in nature, and it can exert force on charges. The difference is that the induced electric field is generated by a changing magnetic field, and its electric field lines are a series of closed curves, so its line integral around a closed loop $\oint_L \boldsymbol{E}_v \cdot \mathrm{d}\boldsymbol{l}$ is not zero. The induced electric field is not a conservative field, while the electrostatic field is a conservative field.

Example 11.3.3 As shown in Figure 11.3.5, a long straight solenoid with a radius of R carries a variable current, and the magnetic field inside the solenoid is uniform. When the magnetic field increases at a constant rate of $\dfrac{\partial \boldsymbol{B}}{\partial t}$, find the induced electric field inside and outside the solenoid, and the induced EMF in the concentric circular conductor circuit with a radius r.

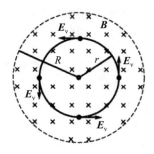

Figure 11.3.5 Figure of Example 11.3.3

Solution The change in current causes a change in the magnetic field inside the solenoid, and the changing magnetic field generates a rotating electric field inside and outside the solenoid. Due to the cylindrical symmetry of the magnetic field, the electric field lines of the induced electric field are a series of concentric circles, and the induced electric field has the same value at any points of a circle. Take any circle with a radius of r as the integral loop L, and its rotation direction is counterclockwise. Then we have

$$\mathscr{E} = \oint_L \boldsymbol{E}_v \cdot \mathrm{d}\boldsymbol{l} = \oint_L E_v \mathrm{d}l = E_v \oint_L \mathrm{d}l = E_v 2\pi r$$

Inside the solenoid,

$$E_v 2\pi r = \dfrac{\partial B}{\partial t} \pi r^2$$

$$E_v = \dfrac{r}{2} \dfrac{\partial B}{\partial t}$$

$$\mathscr{E} = \pi r^2 \dfrac{\partial B}{\partial t}$$

$\mathscr{E}$ is in the same direction as E_v, which is counterclockwise.

Outside the solenoid, there is no magnetic field, so

$$\mathscr{E} = E_v 2\pi r = \pi R^2 \dfrac{\partial B}{\partial t}$$

Thus,

$$E_v = \dfrac{R^2}{2r} \dfrac{\partial B}{\partial t}$$

The directions of $\mathscr{E}$ and E_v both are counterclockwise.

11.3.3 Application of induced electric field in science and technology

1. Betatron

A betatron uses the induced electric field to accelerate electrons. The first betatron was built in 1940, and its device is shown in Figure 11.3.6. It has a circular vacuum chamber between two poles of an electromagnet. The electromagnet is excited by alternating current, and an alternating magnetic field is generated between the two poles. This alternating magnetic field generates an induced electric field in the vacuum chamber, and its electric field lines are a series of concentric circles around the magnetic field lines.

When the electrons go into the annular vacuum chamber along the tangent of the induced electric field, they will be accelerated by the induced electric field in the circular vacuum chamber. At the same time, the electrons are also subjected to the Lorentz force of the magnetic field, which makes them move on circular orbits. This is the working principle of a betatron.

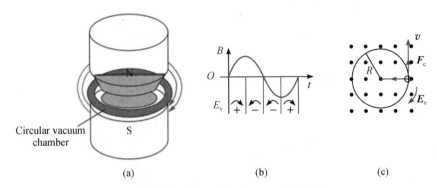

Figure 11.3.6 Schematic diagram of a betatron

2. Eddy current

When a conductor moves in a magnetic field or is in a changing magnetic field, a vortex-like induced current will appear in the conductor. These currents form closed circuits inside the conductor, which are called **eddy currents** or **eddy**. Figure 11.3.7 (a) shows an iron core coil with alternating current, which generates eddy currents when exposed to an alternating magnetic field. Due to the extremely small resistance of metals, extremely strong eddy currents can be generated, releasing a large amount of Joule heat inside the iron core. This is the principle of induction heating.

There are many unique advantages of eddy current heating. This method heats different parts of an object simultaneously, rather than conducting heat layer by layer from the outside. It can be used to melt easily oxidizable or insoluble metals, like the high-frequency induction furnace commonly used in alloy melting. This method has the advantages of high heating speed, uniform temperature, easy control, and no material contamination.

The thermal effect generated by eddy current has some drawbacks though. For example, iron cores of transformers or other motors often generate useless heat due to eddy current, which not only consumes electrical energy, but also may make the transformer fail to work due to the heating of the iron core. To reduce loss of thermal energy, iron cores usually are composed of a stack of high resistivity silicon-steel sheets. And the insulation layer of the silicon-steel sheets is perpendicular to the direction of the eddy current, as shown in Figure 11.3.7 (b). This reduces the eddy current, restricts it to the thin sheet, and greatly reduces energy loss.

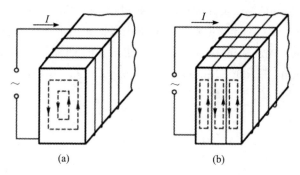

Figure 11.3.7 Eddy currents in transformer cores

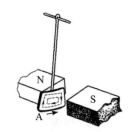

In addition to the thermal effect, eddy current also has magnetic effect. As shown in Figure 11.3.8, when there is no current in the electromagnet, the damping pendulum A made of copper plate will swing back and forth multiple times before stopping. If the electromagnetic iron works, the magnetic field generates eddy current in the swinging copper plate. As the induced eddy current always produces a retarding force when the copper plate enters or leaves the magnetic field, the swinging plate quickly comes to rest.

Figure 11.3.8 Electromagnetic damping pendulum

This phenomenon is called electromagnetic damping. The electromagnetic damper in electromagnetic instruments is made based on the eddy current's magnetic effect, which can quickly stabilize the instrument pointer at the position it should indicate.

11.4 Self induction and mutual induction

According to Faraday's law of electromagnetic induction, when the magnetic flux through a closed loop changes, there must be an induced electromotive force in the closed loop. In the previous section, we studied motional and induced electromotive force. In this section, we will apply Faraday's law of electromagnetic induction to practical circuits and discuss two kinds of electromagnetic induction phenomena, self induction and mutual induction, which are widely used in practice.

11.4.1 Self induction

If the current in a coil circuit changes, the magnetic flux of the magnetic field it stimulates through the coil itself also changes, thus causing the coil to generate an induced electromotive force. The electromagnetic induction phenomenon caused by the current change in the coil itself is called self induction phenomenon and the generated electromotive force is called self-induced electromotive force, donated as $\mathscr{E}_L$.

The phenomenon of self induction can be observed by the following experiment. As shown in Figure 11.4.1, S_1 and S_2 are two bulbs of the same size, and L is a self inductance coil. Adjust the rheostat R so that its resistance is equal to that of the coil L before the experiment. The moment the switch K is turned on, it can be observed that the bulb S_1 lights up immediately, while the bulb S_2 gradually lightens. After a while both bulbs reach the same brightness. This is because the change of current generates a self induced electromotive force in the coil in the S_2 branch as the switch K is turned on. According to Lenz's law, the induced electromotive force prevents the increase of current, so the increase of current in the S_2 branch is slower than that in the S_1 branch without the self inductance coil, thus the bulb S_2 lights up more slowly than S_1. When the switch K is turned off, one can see that the two bulbs are not immediately extinguished, but are brighter and then slowly dim. This is because the current in the coil is rapidly reduced when the power supply is cut off, resulting in a self-induced electromotive force in the coil. At this time, although the power supply has been disconnected, the coil L and the two bulbs form a closed loop, and the self-induced electromotive force causes an induced current in the loop.

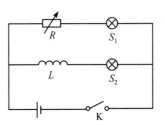

Figure 11.4.1 Self induction phenomenon

If the geometry of a loop remains the same, then according to the Biot-Savart law, the magnetic induction **B** at any point in space is proportional to the current in the loop. The flux linkage Ψ_m through the area enclosed by the loop is also proportional to I. Thus

$$\Psi_m = LI \qquad (11.4.1)$$

where L is the proportional coefficient, called the self induction coefficient in the loop, or self inductance. Experiments show that the self inductance has nothing to do with the current in the loop, but is determined by the size, geometry, number of turns of the loop and the permeability of the surrounding magnetic medium.

In the International System of Units, the unit of self inductance is the Henry (H), but millihenry (mH) and microhenry (μH) are commonly used in practical applications.

According to Faraday's law of electromagnetic induction, the self-induced electromotive force generated in a loop can be expressed as

$$\mathscr{E}_L = -\frac{d\Psi_m}{dt} = -L\frac{dI}{dt} \qquad (11.4.2)$$

When parameters of the coil itself are unchanged, and the surrounding medium is weak magnetic (no ferromagnetic material), the self inductance is a constant quantity independent of current. The minus sign in Equation (11.4.2) is a mathematical representation of Lenz's law, which shows that the self induced electromotive force will resist the change of current in the circuit. When the change of current $\frac{dI}{dt} > 0$ in the loop,

$\mathscr{E}_L < 0$, that is, the self-induced electromotive force is opposite to the direction of the original current. On the contrary, when $\dfrac{dI}{dt} < 0$ in the loop, $\mathscr{E}_L > 0$, that is, the self-induced electromotive force is in the same direction as the original current. Thus, the stronger the self inductance of the loop, the stronger the property of keeping the current in its loop constant. This property of the self inductance coefficient L is similar to the mass m in mechanics, so the self inductance L is often referred to as "electromagnetic inertia".

Self induction is widely used in electrical and radio technology. In a circuit, the self inductance coil facilitates DC and low frequency while blocking AC and high frequency, such as ballasts in electrotechnics and oscillating coils in radio technology. In addition, a self inductance coil and capacitor together form a filtering circuit, which can make AC signals at some frequencies pass smoothly and AC signals at other frequencies blocked, so as to achieve the purpose of filtering. The self inductance coil and capacitor can also be used to form a resonant circuit.

In some cases, self inductance can be very harmful. For example, winding coils of large motors, generators, etc. have a large self inductance. When the switch is on and off, the strong self inductance electromotive force may cause dielectric breakdown, so measures must be taken to ensure the safety of personnel and equipment.

The calculation of self inductance coefficient is generally complicated, and it is generally measured by an experimental method. For some simple loops with a regular shape, it can be calculated.

Example 11.4.1 There is a long tightly-wound straight solenoid, the length is l, the cross-sectional area is S, the total number of turns of the coil is N, and the permeability of the medium inside is μ. Try to find the self inductance of the solenoid.

Solution Ignoring the edge effect, the magnetic field in the solenoid can be regarded as approximately uniform, and the magnitude of its magnetic induction intensity $\boldsymbol{B}$ is

$$B = \mu \frac{N}{l} I$$

The direction of $\boldsymbol{B}$ can be viewed as parallel to the axis of the solenoid. Then, the flux linkage through the solenoid is

$$\Psi_m = N\Phi_m = NBS = \mu \frac{N^2}{l} IS$$

The self inductance coefficient can be obtained from $\Psi_m = LI$

$$L = \frac{\Psi_m}{I} = \mu \frac{N^2}{l} S$$

If the number of turns of the solenoid on the unit length is n, and the volume of the solenoid is V, then the self inductance coefficient of the solenoid is

$$L = \mu \frac{N^2}{l} S = \mu \frac{N^2}{l^2} lS = \mu n^2 V$$

Example 11.4.2 As shown in Figure 11.4.2, a long coaxial cable consists of an inner cylinder of radius R_1 and an outer cylinder of radius R_2, the space between which is filled with a magnetic medium of permeability μ. The currents passing through the two cylinder surfaces are equal in magnitude and opposite in direction. Find the self inductance coefficient per unit length of the cable.

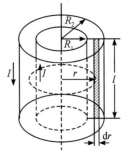

Figure 11.4.2 Figure of Example 11.4.2

Solution According to the Ampère circuital theorem, the magnetic field is limited to the space between two cylinder surfaces, where the magnetic induction intensity at the distance r from the axis is

$$B = \frac{\mu I}{2\pi r} \quad (R_1 < r < R_2)$$

Consider a part of the cable with length l, and the magnetic flux through the area of the shaded part is

$$d\Phi_m = Bl\, dr = \frac{\mu I l}{2\pi r} dr$$

The self inductance per unit length of the cable is

$$L = \frac{\Phi_m}{lI} = \frac{\mu}{2\pi} \ln \frac{R_2}{R_1}$$

 11.4.2 Mutual induction

For two adjacent current-carrying loops, the current change of one loop results in an induced electromotive force in the other loop, which is called mutual inductance phenomenon, and the induced electromotive force generated is called mutual inductance electromotive force. Shown in Figure 11.4.3 are two adjacent coils 1 and 2. When the current of coil 1 changes, according to the Biot-Savart law, the magnetic induction intensity generated by I_1 at any point in the surrounding space is proportional to I_1. Therefore, the magnetic flux of the magnetic field generated by coil 1 and through coil 2 must also be proportional to I_1, which is

$$\Phi_{m21} = M_{21} I_1$$

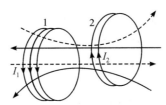

Figure 11.4.3 Mutual inductance phenomenon

Similarly, when the current of coil 2 changes, the magnetic flux of the magnetic field generated by coil 2 and through coil 1 is

$$\Phi_{m12} = M_{12} I_2$$

In the expressions, M_{21} and M_{12} are two proportional coefficients, which are only related to the shape and size of the two coil loops and the permeability of the surrounding magnetic medium. It can be proved that $M_{21} = M_{12}$, expressed both as M and called the mutual inductance coefficient of the coil.

According to Faraday's law of electromagnetic induction, under the condition that the shape, size, and relative position of the coil loop and the permeability of the surrounding magnetic medium are unchanged, the mutual inductance electromotive force generated by current I_1 in coil 2 is

$$\mathscr{E}_{21} = -M \frac{dI_1}{dt} \tag{11.4.3}$$

Similarly, the mutual inductance electromotive force generated by current I_2 in coil 1 is

$$\mathscr{E}_{12} = -M \frac{dI_2}{dt} \tag{11.4.4}$$

Thus, the mutual inductance electromotive force in one coil is proportional to the rate of current change in the other coil, and is also proportional to their mutual inductance coefficient. When the current changes, the greater the mutual inductance coefficient, the greater the mutual inductance electromotive force. The mutual inductance coefficient is a physical quantity characterizing the mutual inductance ability of two coils. In the SI units, the unit of mutual inductance is also the Henry.

Mutual inductance is widely used in radio technology and electromagnetic measurement. All kinds of power transformers, voltage transformers, current transformers and so on are manufactured using the principle of mutual inductance. However, mutual inductance between circuits will cause mutual interference, and the magnetic shielding method must be used to reduce such interference. Like the self inductance, the mutual inductance coefficient is usually measured experimentally and can only be calculated in the case of some simple loops.

Example 11.4.3 Transformers are made according to the principle of mutual inductance. Suppose that the primary coil and secondary coil of a transformer are two coaxial straight solenoids of the same length l and radius R, as Figure 11.4.4 shows. Their turns are respectively N_1 and N_2. The permeability of the magnetic medium in the solenoid is μ. Find

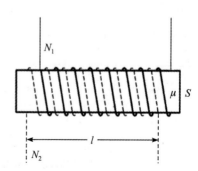

Figure 11.4.4 Figure of Example 11.4.3

(1) the mutual inductance coefficient of the two coils;

(2) the relationship between the self inductance coefficient and the mutual inductance coefficient of the two coils.

Solution (1) If there is a current I_1 in the primary coil 1, the magnetic field in the solenoid can be regarded as a uniform magnetic field, and the magnetic induction intensity $\boldsymbol{B}_1$ is

$$B_1 = \mu \frac{N_1}{l} I_1$$

The direction of $\boldsymbol{B}_1$ is parallel to the axis of the solenoid, and the flux linkage through the secondary coil is

$$\Psi_{m21} = N_2 B_1 S_2 = \mu \frac{N_1 N_2 \pi R^2}{l} I_1$$

According to the definition of mutual inductance,

$$M = \frac{\Psi_{m21}}{I_1} = \mu \frac{N_1 N_2 \pi R^2}{l}$$

(2) According to the calculated result in Example 11.4.1, the self inductance of a long straight solenoid is $L = \mu \frac{N^2 \pi R^2}{l}$. The self inductance coefficients of the primary and secondary coils are respectively,

$$L_1 = \mu \frac{N_1^2 \pi R^2}{l}, \quad L_2 = \mu \frac{N_2^2 \pi R^2}{l}$$

Thus,

$$M^2 = L_1 L_2$$
$$M = \sqrt{L_1 L_2}$$

In general, when two coils with mutual inductance coupling are connected in series, due to mutual inductance, the total self inductance coefficient can be expressed as $L = L_1 + L_2 \pm 2M$, where L_1, L_2, and M respectively represent the self inductance of the two self inductance coils and the mutual inductance of them. The sign depends on whether the series of the two self inductance coils is forward (the magnetic fields generated by the two coils are in the same direction) or reverse (the magnetic fields generated by the two coils are in the opposite direction). The mutual inductance coefficient of the two coils can be expressed by their respective self inductance coefficient and coupling coefficient, $M = k\sqrt{L_1 L_2}$, where k is the coupling coefficient of the two coils. Generally, $k \leqslant 1$. When there is no magnetic leakage, that is, the magnetic flux generated by one loop passes entirely through the other loop and vice versa, $k = 1$, in which case the two circuits are fully coupled. Two long straight tightly wound solenoids around the same cylinder, and two coils on an iron core, can be approximated as fully coupled.

11.5 Self-induced magnetic energy and magnetic field energy

Magnetic fields, like electric fields, have energy. The following is an analysis of the energy conversion relationship in the self inductance phenomenon.

11.5.1 Self-induced magnetic energy

In the circuit shown in Figure 11.5.1, when the switch S is on, the luminous and heating energy of the bulb H is provided by the power supply. When the power supply in the circuit is quickly disconnected, the bulb H will not go off immediately, but after a sudden flash. At this time, the magnetic energy stored in the coil is released by the self-induced

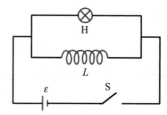

Figure 11.5.1   RL circuit diagram

electromotive force, and becomes the light energy and heat energy emitted by the bulb H in a very short time. The circuit shown in Figure 11.5.1 is taken as an example to derive the magnetic field energy formula. When the power supply is cut off, after a period of time, the current in the coil is reduced from I to zero. At this time, the self-induced electromotive force in the coil will prevent the reduction of the current, that is, the direction of the self-induced electromotive force is the same as the direction of the current. In time dt, the work done by the self-induced electromotive force is

$$dA = \mathscr{E}_L i\, dt = -Li\, di \tag{11.5.1}$$

The total work done by the self-induced electromotive force in this process is

$$A = \int_I^0 -Li\, di = \frac{1}{2}LI^2 \tag{11.5.2}$$

The work done by the self-induced electromotive force is equal to the magnetic energy stored in the self-induced coil. Obviously, the magnetic energy stored in a coil with a self inductance of L and a current of I is

$$W_m = \frac{1}{2}LI^2 \tag{11.5.3}$$

11.5.2 Magnetic field energy

Take a long straight solenoid as an example to understand magnetic field energy. When in a solenoid passes a current, the magnetic induction intensity in it is

$$B = \mu n I$$

The self inductance coefficient of the solenoid is

$$L = \mu n^2 V$$

where n is the number of turns per unit length of the solenoid and V is the volume of the solenoid. Substitute it into the self inductance magnetic energy formula, and the magnetic field energy it stores is

$$W_m = \frac{1}{2}LI^2 = \frac{1}{2}\mu n^2 V I^2 = \frac{1}{2}\frac{B^2}{\mu}V \tag{11.5.4}$$

For $H = \dfrac{B}{\mu} = nI$,

$$W_m = \frac{1}{2}BHV \tag{11.5.5}$$

Because the magnetic field inside the long straight solenoid is uniform, the magnetic field energy per unit volume is

$$w_m = \frac{W_m}{V} = \frac{1}{2}\frac{B^2}{\mu} = \frac{1}{2}BH \tag{11.5.6}$$

w_m is called the magnetic field energy density. Although Equation (11.5.6) is derived from a special case, it proves to be true in the case of non-uniform magnetic fields.

It can be seen that the magnetic field, like the electric field, is a form of matter and therefore has energy. Magnetic field energy and other forms of energy can be converted into each other, and electromagnetic induction is a specific form of energy conversion. Under normal circumstances, the magnetic energy density is a function of spatial position; for a non-uniform magnetic field, the space where the magnetic field is located can be divided into countless volume elements, and the magnetic energy in any volume element dV is

$$dW_m = w_m dV = \frac{1}{2}\frac{B^2}{\mu}dV$$

The magnetic energy in a finite volume V is

$$W_m = \int_V dW_m = \frac{1}{2}\int_V \frac{B^2}{\mu}dV \tag{11.5.7}$$

The magnetic field energy stored in a current-carrying coil can be expressed as follows

$$W_m = \frac{1}{2}LI^2 = \int_V \frac{1}{2}\frac{B^2}{\mu}dV \tag{11.5.8}$$

This provides another method to calculate self inductance L, namely the magnetic energy method to define self inductance

$$L = \frac{2W_m}{I^2} \tag{11.5.9}$$

Example 11.5.1 As shown in Figure 11.5.2, a long coaxial cable consists of an inner cylinder of radius R_1 and an outer cylinder of radius R_2. The space between R_1 and R_2 is filled with a magnetic medium with magnetic permeability μ. There are equal and opposite axial currents I in the inner and outer cylinder. The current is evenly distributed in the cylinder. Find the magnetic

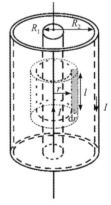

Figure 11.5.2 Figure of Example 11.5.1

energy stored in a cable with a length l.

Solution According to the Ampère circuital theorem, the magnetic field is limited to the space between two cylinder surfaces. In the space between R_1 and R_2, the magnetic induction intensity at the point r from the axis is

$$B=\frac{\mu I}{2\pi r} \quad (R_1<r<R_2)$$

The magnetic energy density here is

$$w_{m1}=\frac{B^2}{2\mu}=\frac{\mu I^2}{8\pi^2 r^2}$$

The magnetic energy density between two conductors is a function of r. Take the cylindrical shell volume dV of radius r, thickness dr and length l as the volume element, and $dV=2\pi rl\,dr$, where the magnetic energy is

$$dW_{m1}=w_{m1}dV=\frac{\mu I^2}{8\pi^2 r^2}2\pi rl\,dr=\frac{\mu I^2 l\,dr}{4\pi r}$$

Therefore, the total magnetic energy stored in a length l between the inner and outer current-carrying conductors is

$$W_{m1}=\int dW_{m1}=\int_{R_1}^{R_2}\frac{\mu I^2 l\,dr}{4\pi r}=\frac{\mu I^2 l}{4\pi}\ln\frac{R_2}{R_1}$$

Since the current is evenly distributed in the cross section of the inner cylinder, the magnitude of the magnetic induction intensity in the cylinder can be obtained from the Ampère circuital theorem

$$B=\frac{\mu_0 Ir}{2\pi R_1^2}$$

Since the permeability of a conductor is close to the permeability in a vacuum, it is taken as μ_0. Using the same method described above, we can find the magnetic energy stored in a cable of length l is

$$W_{m2}=\int_V\frac{B^2}{2\mu_0}dV=\int_0^{R_1}\frac{\mu_0 I^2 lr^3\,dr}{4\pi R_1^4}=\frac{\mu_0 I^2 l}{16\pi}$$

Therefore, the total magnetic energy stored in a coaxial cable with current I is

$$W_m=W_{m1}+W_{m2}=\frac{\mu I^2 l}{4\pi}\ln\frac{R_2}{R_1}+\frac{\mu_0 I^2 l}{16\pi}$$

Note that if W_m is known, then the self inductance coefficient can be found according to $W_m=\frac{1}{2}LI^2$. The self inductance coefficient of the coaxial cable is

$$L=\frac{\mu l}{2\pi}\ln\frac{R_2}{R_1}+\frac{\mu_0 l}{8\pi}$$

Select an appropriate cable size so that $\frac{\mu_0 l}{8\pi}$ is negligible. Or select a cable whose inner conductor is a hollow cylinder. Because the magnetic field in the cylinder is zero, the term does not exist, and then the self inductance of the unit length coaxial cable is

$$L = \frac{\mu}{2\pi} \ln \frac{R_2}{R_1}$$

11.6 Introduction to Maxwell's electromagnetic theory

In the previous chapters, we introduced the basic properties and basic laws of the electrostatic field and the constant magnetic field, which are static fields that do not change with time, but the most common case is the electromagnetic field that changes with time. Faraday's law of electromagnetic induction involves a changing magnetic field that can excite an electric field. Maxwell proposed the concept of a changing electric field to excite a magnetic field after studying the contradiction in the application of the Ampère circuital theorem to circuit currents that change with time. It further revealed the internal relation and dependency between electric field and magnetic field. On this basis, Maxwell summarized the laws of electromagnetic phenomena under special conditions into a complete system of universal electromagnetic field theory—Maxwell's equations. From the theory of electromagnetic field, Maxwell also predicted the existence of electromagnetic waves. In 1887, Hertz experimentally confirmed the existence of electromagnetic waves, which gave decisive support to Maxwell's theory of electromagnetic field.

11.6.1 Displacement current and the Maxwell-Ampère theorem

From previous study, we can see that the magnetic field of constant current satisfies the Ampère circuital theorem

$$\oint_L \boldsymbol{H} \cdot \mathrm{d}\boldsymbol{l} = I = \oint_S \boldsymbol{j} \cdot \mathrm{d}\boldsymbol{S}$$

This law states that the circulation of the magnetic field strength along any closed loop is equal to the algebraic sum of the conducting currents surrounding the closed loop. Now, does this law hold true in the case of unsteady currents? This can be illustrated by the following special case.

In the charging and discharging process of a capacitor, the conduction current is discontinuous for the entire circuit, and the current I in the circuit conductor is a non-constant current that changes with time. As shown in Figure 11.6.1, if a closed loop L is taken near plate A, two surfaces S_1 and S_2 can be made as the boundary of this loop, where S_1 intersects with the wire, and S_2 is between the two poles and does not intersect with the wire. S_1 and S_2 form a closed surface. If surface S_1 is taken as the basis for measuring whether there is a current through the area surrounded by L, then because it intersects with the wire, the current through the area surrounded by L (that is, the S_1

surface) is I_c, so by the Ampère circuital theorem

$$\oint_L \boldsymbol{H} \cdot \mathrm{d}\boldsymbol{l} = I_c$$

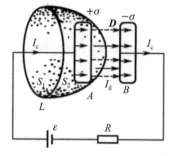

Figure 11.6.1 Displacement current

If surface S_2 is taken as the basis, then no current passes through S_2, so by the Ampère circuital theorem

$$\oint_L \boldsymbol{H} \cdot \mathrm{d}\boldsymbol{l} = 0$$

This shows that the Ampère circuital theorem is contradictory in the magnetic field of unstable constant current, and is not applicable in the case of unsteady current.

Maxwell believed that this contradiction arose because the circulation of magnetic field strength was thought to be determined by a single conducting current, but the conduction current was interrupted between the two plates of the capacitor. He noted that in the process of charging and discharging, although there was no conducting current between the two plates of the capacitor, there was an electric field; as a result, the free charge on the plate of the capacitor changed with time, and the electric field between the plates also changed with time.

Let the charge surface density on the capacitor plate A at any time be $+\sigma$ and the charge surface density on plate B be $-\sigma$. The charging current is I_c in the circuit with a plate area S. Then

$$I_c = \frac{\mathrm{d}q}{\mathrm{d}t} = \frac{\mathrm{d}(S\sigma)}{\mathrm{d}t} = S\frac{\mathrm{d}\sigma}{\mathrm{d}t}$$

The conduction current density is

$$j_c = \frac{\mathrm{d}\sigma}{\mathrm{d}t}$$

The magnitude of the electric displacement ($D=\sigma$) and the electric displacement flux ($\Phi_D = DS = \sigma S$) between the two plates also change with time. Substitute them into the above two equations respectively, and

$$I_c = \frac{\mathrm{d}q}{\mathrm{d}t} = \frac{\mathrm{d}\Phi_D}{\mathrm{d}t}$$

$$j_c = \frac{\mathrm{d}D}{\mathrm{d}t}$$

It can be seen that the rate of change of the electric displacement flux between the two plates with time is numerically equal to the charging current I_c in the circuit, and when the capacitor is charged, the direction of $\frac{dD}{dt}$ between the plates is also from the positive plate to the negative plate, which is in the same direction as the conduction current density in the circuit. Therefore, Maxwell assumed the changing electric field as current and introduced the concept of displacement current. The displacement current I_d passing through a section of an electric field is equal to the rate of change with time of the electric displacement flux passing through the section. The displacement current density j_d at a certain point in the electric field is equal to the change rate of the electric displacement with time at that point.

$$I_d = \frac{d\Phi_D}{dt} \tag{11.6.1a}$$

$$j_d = \frac{dD}{dt} \tag{11.6.1b}$$

Maxwell believed that displacement current and conduction current can both excite the magnetic field, and the magnetic field is exactly the same as the magnetic field excited by its equivalent conduction current. In this way, in the entire circuit, where the conduction current is interrupted, the displacement current takes over, and their values are equal and the directions are the same. For the general case, Maxwell believed that both conduction and displacement currents could exist. Thus, he generalized the concept of electric current, calling the sum of the two the full current I_s, expressed by

$$I_s = I_c + I_d \tag{11.6.2}$$

For any loop, the full current is continuous everywhere. By using the concept of full current, the Ampère circuital theorem can be naturally extended to the magnetic field of unstable constant current, which also solves the problem of current continuity in the charging and discharging process of the capacitor. The core of Maxwell's displacement current hypothesis is that a rotating magnetic field is excited by a changing electric field. Although displacement current has the name of "current", it is essentially a changing electric field. The correctness of Maxwell's displacement current hypothesis has been proved by many conclusions and experimental results derived from it.

According to the concept of full current, in general, the Ampère circuital theorem is modified as

$$\oint_L \mathbf{H} \cdot d\mathbf{l} = I_s = I_c + \frac{d\Phi_D}{dt} \tag{11.6.3a}$$

or

$$\oint_L \mathbf{H} \cdot d\mathbf{l} = \iint_S \left(j_c + \frac{\partial \mathbf{D}}{\partial t} \right) \cdot d\mathbf{S} \tag{11.6.3b}$$

The above formula shows that the circulation of the magnetic field strength $\mathbf{H}$ along any closed loop is equal to the full current passing through the surface surrounding the closed loop, which is the Maxwell-Ampère theorem. Maxwell creatively linked the changing electric field with the magnetic field. Not only can the conducting current excite the magnetic field in space, but also the changing electric field can excite the magnetic field in space, and both are rotating magnetic fields. This means that the displacement current is equivalent to the conduction current in terms of magnetic effect. However, no conductor is needed to form a displacement current, so it has no thermal effect.

Example 11.6.1 As shown in Figure 11.6.2, a disc-shaped parallel plate capacitor with a radius of R is placed in a vacuum and the edge effect is ignored. During charging, the electric field change rate between the plates is dE/dt. Find

(1) the displacement current between the plates;

(2) the magnetic inductance **B** at a point r from the axis of the circular plate.

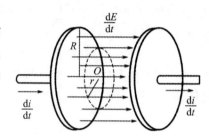

Figure 11.6.2　Figure of Example 11.6.1

Solution　(1) According to the definition of displacement current, I_d can be obtained

$$I_d = \frac{d\Phi_D}{dt} = \frac{d(\mathbf{D} \cdot \mathbf{S})}{dt} = \frac{d(\varepsilon_0 E \cdot \pi R^2)}{dt} = \pi R^2 \varepsilon_0 \frac{dE}{dt}$$

(2) Because the magnetic field distribution is symmetrical and unstable, the Ampère circuital theorem of full current should be used to solve the magnetic induction intensity.

When $r < R$,

$$\oint_{L_1} \mathbf{H}_1 \cdot d\mathbf{l} = \frac{I_d}{\pi R^2} \pi r^2$$

$$H_1 2\pi r = \frac{I_d}{\pi R^2} \pi r^2$$

$$H_1 = \frac{\varepsilon_0 r}{2} \frac{dE}{dt}$$

According to the relationship between magnetic induction intensity and magnetic field intensity, it can be obtained that

$$B_1 = \mu_0 H_1 = \frac{\mu_0 \varepsilon_0 r}{2} \frac{dE}{dt}$$

When $r > R$,

$$\oint_{L_2} \mathbf{H}_2 \cdot d\mathbf{l} = I_d$$

$$H_2 2\pi r = I_d$$

$$H_2 = \frac{\varepsilon_0 R^2}{2r} \frac{dE}{dt}$$

According to the relationship between magnetic induction intensity and magnetic field intensity, it can be obtained that

$$B_2 = \mu_0 H_2 = \frac{\mu_0 \varepsilon_0 R^2}{2r} \frac{dE}{dt}$$

11.6.2 Maxwell's equations of an electromagnetic field

Through the efforts of many physicists such as Coulomb and Ampère, the basic laws of an electrostatic field and constant magnetic field had been established. Taking into account the electric and magnetic fields that change with time, Maxwell proposed two basic hypotheses: the induced electric field and the displacement current, the former indicating that the changing magnetic field will excite the induced electric field, and the latter indicating that the changing electric field will excite the induced magnetic field. These two hypotheses reveal the intrinsic connection between electric and magnetic fields. A space with a changing electric field must have a changing magnetic field, and likewise a space with a changing magnetic field must have a changing electric field. That is, the changing electric field and the changing magnetic field are closely linked together, and they constitute a unified electromagnetic field as a whole. This is Maxwell's basic concept of electromagnetic fields. In 1865, Maxwell proposed four equations to describe the general law of electromagnetic fields, known as Maxwell's equations.

The integral forms of Maxwell's equations can be expressed as follows:

(1) The electric displacement flux through any closed surface is equal to the algebraic sum of the free charges enclosed by the surface.

$$\oint_S \boldsymbol{D} \cdot \mathrm{d}\boldsymbol{S} = \sum q_{\text{int}} \qquad (11.6.4)$$

(2) The line integral of the electric field strength along an arbitrary closed curve is equal to the negative value of the rate of change with time of the magnetic flux that passes through an arbitrary closed surface bounded by the curve, which for the first time relates the changing magnetic field to the electric field.

$$\oint_L \boldsymbol{E} \cdot \mathrm{d}\boldsymbol{l} = -\int_S \frac{\partial \boldsymbol{B}}{\partial t} \cdot \mathrm{d}\boldsymbol{S} \qquad (11.6.5)$$

(3) The magnetic flux through any closed surface is identical to zero.

$$\oint_S \boldsymbol{B} \cdot \mathrm{d}\boldsymbol{S} = 0 \qquad (11.6.6)$$

(4) The line integral of the magnetic field strength along any closed curve is equal to the total current passing through the surface bounded by the curve.

$$\oint_L \boldsymbol{H} \cdot \mathrm{d}\boldsymbol{l} = I_c + \int_S \frac{\partial \boldsymbol{D}}{\partial t} \cdot \mathrm{d}\boldsymbol{S} \qquad (11.6.7)$$

Maxwell's equations are a concise and perfect summary of the basic laws of electromagnetic fields. Summarized from macroscopic electromagnetic phenomena, Maxwell's electromagnetic field theory can be applied to various macroscopic electromagnetic phenomena, and it is also correct in the field of high speed. However, in the electromagnetic phenomena of microscopic processes such as molecules and atoms, Maxwell's theory is not fully applicable and the more general quantum electrodynamics is needed. Maxwell's theory can be regarded as the approximate law of quantum electrodynamics under some special conditions.

In the application of Maxwell's equations to solve practical problems, interaction between the electromagnetic field and matter is often involved, so it is necessary to consider the influence of the medium on the electromagnetic field.

$$\boldsymbol{D} = \varepsilon \boldsymbol{E}, \ \boldsymbol{B} = \mu \boldsymbol{H}$$

In non-uniform media, the boundary value relationship of the electromagnetic field quantity at the interface should also be considered, as well as the initial value conditions of $\boldsymbol{E}$ and $\boldsymbol{B}$ in the specific problem. By solving equations, we can find $\boldsymbol{E}(x, y, z)$ and $\boldsymbol{B}(x, y, z)$ at any time so as to determine the electromagnetic field at any point in space at any time. Maxwell's electromagnetic field theory had brought profound changes to production technology and human life since the end of last century and the beginning of this century. At the same time, the establishment of Maxwell's equations laid a solid foundation for the development of modern communication theory and electromagnetic wave theory of light.

Scientist Profile

James Clerk Maxwell (1831 – 1879), a British physicist and mathematician, was the founder of classical electrodynamics and one of the founders of statistical physics.

Maxwell was clever and premature. He submitted a scientific research paper to the Royal College of Edinburgh in 1846, entered the University of Edinburgh in 1847 to study mathematics and physics, transferred to the Department of Mathematics at Trinity College, Cambridge in 1850, and was employed in the newly created position of Cavendish Professor of experimental physics at Cambridge University in 1871, responsible for the preparation of the famous Cavendish Laboratory.

Maxwell's main contributions were the establishment of Maxwell's equations, the establishment of classical electrodynamics, the prediction of the existence of electromagnetic waves, and the formulation of the electromagnetic theory of light. Maxwell was the mastermind of electromagnetism. He was born when Faraday, the founder of electromagnetic theory, put forward the electromagnetic induction theorem. Later he formed a friendship with Faraday, and jointly built the scientific system of electromagnetic theory. In the history of physics, Newton's classical mechanics opened the door to the mechanical age, while Maxwell's theory of electromagnetism laid the cornerstone for the electrical age.

Maxwell also made important contributions to many other disciplines, including astronomy and thermodynamics. Maxwell realized that not all gas molecules move at the same speed; some move slowly, some move fast, and some move at very high speeds. Maxwell derived the formula for the percentage of molecules in a gas that are known to move at a certain speed, or the Maxwell distribution, which is one of the most widely used scientific formulas and plays an important role in many branches of physics.

Another important part of Maxwell's work was the founding of Cambridge's first physics laboratory, the famous Cavendish Laboratory. The laboratory has had an extremely important influence on the development of experimental physics as a whole, and many famous scientists have worked in the laboratory. The Cavendish Laboratory is even known as the "cradle of Nobel Prize winners in physics". As the first director of the laboratory, Maxwell made the inaugural address in 1871, wonderfully expounding the future teaching policy and research spirit of the laboratory, which was a significant speech in the history of science. Maxwell's profession was theoretical physics, but he was well aware that the era of experimental dominance had not yet passed. He criticized the traditional British "chalk" physics at that time, called for strengthening the research of experimental physics and its role in university education, and established the spirit of experimental science for later generations.

Maxwell's achievements in electromagnetism have been hailed as the second great unification of physics after Newton, and Maxwell was widely regarded as the most influential nineteenth century physicist in the twentieth century. In 1931, at a conference commemorating Maxwell's centenary, Einstein called his work "the most profound and fruitful work in physics since Newton."

Extended Reading

Electromagnetic theory and symmetry

Symmetry is an ancient concept produced in the process of observing and understanding nature, which gives people a kind of full, symmetrical, balanced and smooth aesthetic feeling. Symmetry is found in all aspects of nature, such as the left and right symmetry of the human body, mirror symmetry when looking in the mirror, and the center symmetry of the square. Symmetry is the embodiment of a certain essence and inherent law of the material world. As a natural science, physics has symmetry everywhere. Symmetry in physics contains two aspects of significance. On the one hand, it refers to a kind of symmetry pursued by physical theory itself, which is one of the three major standards of physical aesthetics (simplicity, symmetry and harmony); On the other hand, it means that physical laws are a reflection of the symmetry of nature.

One goal of science is to seek connections between different phenomena, and Newton unified heaven and earth through the correlation between a falling apple and the moon. Similarly, scientists in the 19th century discovered that both electricity and magnetism were generated by the presence of charged bodies, and that these two forces could be seen as two aspects of a single electromagnetic force between charged bodies. In 1820, Oersted discovered the magnetic effect of electric currents, and the connection between electrical and magnetic phenomena began to be recognized. After repeating Oersted's experiment, Faraday realized that it should also be possible to think in reverse, that magnetism also had an electrical effect. This is the result of symmetric thinking, which is actually phenomenological, that is, physicists just think that it should be this way, but whether it is this way and why it is this are not clear. Faraday finally turned this idea into reality after a long period of scientific exploration and discovered the basic law of electromagnetic induction.

In the 1860s, Maxwell thought deeply about both electricity and magnetism, and he carefully compared the three basic laws of electromagnetism that had been discovered at the time: Coulomb's law, which reflected the electric field of a charged body; Biot-Savart's law, which reflected the magnetic field excited by a fluid; and the law of electromagnetic induction for electromagnetic interactions. Like most theoretical physicists, Maxwell believed that a correct and fundamental description of the laws of nature should be harmonious and symmetrical, and that the electromagnetic law should discuss its two aspects—electrical and magnetic phenomena—in a symmetrical manner. In this regard, the three basic laws seemed to be missing something. The first and second laws reflected that electric fields were generated by charged bodies and magnetic fields were generated by moving charges. Faraday's law of electromagnetic induction, extended by Maxwell, suggested that electric fields could be generated in another way—by changing magnetic fields. In Maxwell's view, this was obviously not perfect enough, and there should be a fourth law, that is, a law reflecting a second way of generating magnetic fields, which should be symmetric with Faraday's law. Since a changing magnetic field could produce an electric field, a changing electric field should also produce a magnetic field.

In 1862, Maxwell published his paper on electromagnetism, *On the Lines of Force in Physics*. Thomson, the British physicist and discoverer of the electron, later recalled, "I still remember that paper very well. I was an eighteen-year-old student at the time, and I was so excited to read it! It was a very long article, and I copied it all down." This was an epoch-making paper, which was no longer a simple mathematical translation and extension of Faraday's point of view, but a major development. The decisive significance was that Maxwell proposed according to the symmetry of electricity and magnetism—since a changing magnetic field would cause an induced electric field, then a changing electric field would also cause an induced magnetic field. Before this, when people talked about electric currents producing magnetic fields, they always referred to conduction current, that is, the current formed by the movement of free electrons in a conductor. If the changing electric field produced a magnetic field, then the role of the changing electric field was equivalent to the role of conducting current, but it was not the real current formed by charge flow. Maxwell, through rigorous mathematical derivation, found the equation representing this current and called it the displacement current.

It was a major breakthrough in electromagnetism after Faraday's electromagnetic induction to introduce the concept of displacement current in theory. Based on this scientific hypothesis, Maxwell derived a set of highly abstract differential equations, which became known as Maxwell's equations. This set of equations developed Faraday's achievement in two ways. One was the displacement current, which showed that a changing magnetic field produced an electric field and vice versa, that is, a changing electric field also produced a magnetic field. The second was the induced electric field: Where there was a changing magnetic field, whether it was a conductor or a dielectric around it, there was an induced electric field. After Maxwell's creative summary, the laws of electromagnetic phenomena were finally revealed by him in mathematical form. Coulomb's law, Biot-Savart's law, electromagnetic induction law and displacement current hypothesis constituted the simple logic basis of Maxwell's equations.

In the history of physics, the pursuit of symmetry of physical laws and formulas often plays a positive role in the development of theories. For electromagnetism, the inverse proportion to the square in electrostatic force and magnetic force corresponds to that in gravity. Maxwell, out of the love of classical physicists for perfection, symmetry and harmony, added a displacement current to Ampère's law in accordance with the symmetry of electricity and magnetism without any experimental basis, so that the formula showed a beautiful image symmetry. Harmonious unity of physics, and description of the laws of nature with the most concise theory are the goal of physicists. Maxwell's equations are a very beautiful unity, essentially the unity of electric and magnetic fields. They reveal the beauty of symmetry in the conversion between electric and magnetic fields, and this beauty is fully expressed in modern mathematical form, worthy of the title of "poetic beautiful equations".

In the history of natural science, it is only when a certain science has reached its peak that it can be mathematically expressed in the form of laws. These laws can not only explain known physical phenomena, but also reveal certain things that have not yet been discovered. Just as Newton's law of gravitation foresaw Neptune, Maxwell's theory foresaw the existence of electromagnetic waves. By formalizing the four partial differentials of Maxwell's equations into two second-order partial differential equations, it is found that both the electric and magnetic fields satisfy the wave equation, that is, they are waves. Maxwell pointed out that since the alternating electric field would produce an alternating magnetic field, and the alternating magnetic field would produce an alternating electric field, then this alternating electromagnetic field would be propagated to space in the form of waves, forming electromagnetic waves. According to Maxwell's theory, the speed of electromagnetic wave propagation in a vacuum was

$$c = \frac{1}{\sqrt{\varepsilon_0 \mu_0}}$$

Within the experimental range of error, this constant c was equal to the measured speed of light. Maxwell did not regard this result as a coincidence, believing that there must be a physical mystery. He boldly predicted that light was also an electromagnetic wave, stating that the speed of shear waves in the hypothetical medium calculated from the Kohlerausch and Weber electromagnetism experiment was so consistent with the speed of light calculated from the Fizeall optical experiment that light was a shear wave, and that this was the cause of electrical and magnetic phenomena. From this, Maxwell proposed the electromagnetic theory of light.

Maxwell's electromagnetic field theory achieved a great unification of physics in the 1860s, that is, the unification of electricity, magnetism and light. Like Newton's theory of mechanics, the electromagnetic theory with Maxwell's equations at its core was one of the proudest achievements of classic physics, and the perfect unity of physical theories it revealed led to a rush for unity in physics, among which the most prominent figure was Einstein. Physicists found that Newton's laws remained unchanged after the Galilean transformation (the principle of relativity in mechanics), so Newton's laws were symmetric with respect to the Galilean transformation. However, Maxwell's equations were not symmetrical in the Galilean transformation, which was very frustrating to physicists. It was this problem that brought about a revolution in modern physics. Einstein believed in the symmetry of physical theory and thought that there should be a new transformation law that would keep Maxwell's equations unchanged after this transformation, and that the new transformation law should include the Galilean transformation. This new transformation was the Lorentz transformation, which was the space-time transformation followed by objects moving at high speed. Einstein derived the Lorentz transformation, and then Maxwell's equations, on the basis of his two basic principles of special relativity. In this way, a higher level symmetry of the spatiotemporal transformation of Maxwell's equations was achieved and at the same time a new theory of spatiotemporal transformation was born.

Discussion Problems

11.1 In the law of electromagnetic induction $\mathscr{E}_i = -\dfrac{d\Phi_m}{dt}$, what is the meaning of the minus sign? How to judge the direction of induced electromotive force according to the negative sign?

11.2 When a bar magnet is inserted into a copper ring along its axis, is there an induced current and an induced electric field in the copper ring? If a plastic ring is used instead of a copper ring, is there still an induced current and an induced electric field in the ring?

11.3 How to place two similar flat circular coils so that their mutual inductance coefficient is minimal? Let the distance between the two centers remain constant.

11.4 What is displacement current? How is it different from conducting current?

Problems

11.5 A circular loop with a radius of $r=10$ cm is placed in a uniform magnetic field of $B=0.8$ T. The plane of the loop is perpendicular to the magnetic field. The loop radius shrinks at a constant rate $dr/dt = 80$ cm·s^{-1}. Find the induced electromotive force in the loop.

11.6 As shown in Figure T11-1, a coil with a radius $r=1$ cm and $N=10$ turns is placed at the middle point A of the horseshoe magnet. And the coil plane's normal is parallel to the magnetic induction intensity at point A. Move this coil far enough and the total electric quantity flowing through the coil during this period is $Q = \pi \times 10^{-6}$. What is the magnetic induction at point A (The resistance of the coil is $R = 10$ Ω)?

11.7 A quarter-arc wire of radius R is located in a uniform magnetic field $\boldsymbol{B}$, and the line between the end a of the arc and the center of the circle O is perpendicular to the magnetic field. Let the arc ac rotate at an angular speed ω with ao as the axis. When ac turns to the position shown in Figure T11-2 (At this time the motion direction of point c is inward), find the induced electromotive force of the arc of the wire.

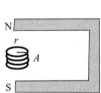

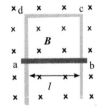

Figure T11-1 Figure of Problem 11.6 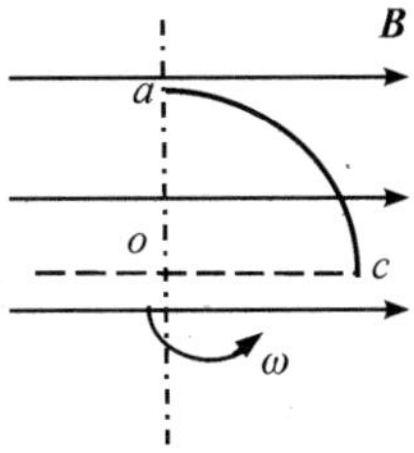   Figure T11-2 Figure of Problem 11.7

11.8 A long rectangular U-shaped guide rail of width l is placed vertically. The bare wire ab can slide frictionlessly along the metal guide rail (Resistance can be ignored), and the guide rail is located in the horizontal uniform magnetic field of magnetic induction intensity $\boldsymbol{B}$. As shown in Figure T11-3, let the mass of wire ab be m and its resistance in the circuit be R. $abcd$ forms a loop. When $t=0$, $v=0$. Try to find the relationship between the sliding speed of wire ab and time t.

Figure T11-3 Figure of Problem 11.8

11.9 Wire ab is l in length and rotates at a uniform angular speed ω around the vertical axis passing point O. $aO = \dfrac{l}{3}$. The magnetic induction intensity is parallel to the rotating axis, as shown in Figure T11-4. Find

(1) the potential difference between the two ends of ab;

(2) which point has the higher potential, a or b.

11.10 As shown in Figure T11-5, there is a rectangular coil in the plane of two parallel current-carrying infinitely long straight wires. The currents in the two wires are opposite in direction and equal in magnitude, and increase with a rate of change dI/dt. Find

(1) the magnetic flux passing through the coil at any time;

(2) the induced electromotive force in the coil.

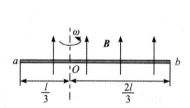
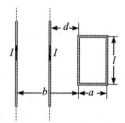

Figure T11-4 Figure of Problem 11.9   Figure T11-5 Figure of Problem 11.10

11.11 A uniform magnetic field with magnetic induction of B fills a cylindrical space with radius R. A metal rod is placed in the position shown in Figure T11-6 and the rod length is $2R$, half of which is in the magnetic field and the other half is out of the magnetic field. When $dB/dt > 0$, find the magnitude and direction of the induced electromotive force at both ends of the rod.

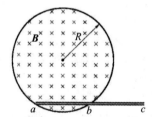

Figure T11-6 Figure of Problem 11.11

11.12 As shown in Figure T11-7, in a straight solenoid of radius R is a magnetic field and $dB/dt > 0$. There is an arbitrary closed wire $abca$, part of which is stretched straight into an ab string in the solenoid, and two points a and b are insulated from the solenoid. If $ab=R$, try to find the induced electromotive force in the closed wire.

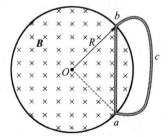

Figure T11-7 Figure of Problem 11.12

11.13 The spiral ring with a rectangular section has a total of N turns, and the dimensions are shown in Figure T11 - 8. The two rectangles at the bottom of the figure represent the cross section of the spiral ring. On the axis of the spiral ring there is another straight wire of infinite length. Find

(1) the self inductance coefficient of the spiral ring;

(2) the mutual inductance coefficient of the long straight wire and spiral ring.

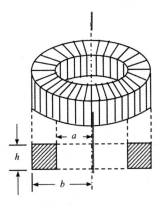

Figure T11 - 8 Figure of Problem 11.13

11.14 The core radius of the long straight coaxial cable is R_1, the outer cylinder radius is R_2, the space between the core wire and the cylinder is filled with magnetic medium, and the relative permeability is μ_r. Current I flows out of the core wire and back through the outer cylinder, and the current is evenly distributed on the cross section of the core wire. Find

(1) the magnetic energy stored per unit length of the cable;

(2) the self inductance coefficient per unit length of the cable.

11.15 A straight wire of infinite length and a square coil are placed as shown in Figure T11 - 9 (The wire and coil contact is insulated). Find the mutual inductance between the coil and wire.

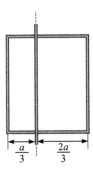

Figure T11 - 9 Figure of Problem 11.15

11.16 Two circular plates with a radius of $R = 0.10$ m constitute a parallel plate capacitor and are placed in a vacuum. Charge the capacitor so that the change rate of the electric field between the two plates is $dE/dt = 1.0 \times 10^{13}$ V $\cdot$ m^{-1} $\cdot$ s^{-1}. Find

(1) the displacement current between the plates;

(2) the magnetic induction intensity of the capacitor at a point $r = 9 \times 10^{-3}$ m from the center axis.

Challenging Problems

11.17   A and C are two coaxial rings with radii a and c, respectively, separated by d. If the magnetic field at the C ring is approximated as a uniform magnetic field, find the mutual inductance coefficient of the two coils. For different c/a values, observe the relationship between the mutual inductance coefficient and the distance between the two rings and compare it with the exact formula of mutual inductance

$$M = \frac{\mu_0 \sqrt{ac}}{k}[(2-k^3)E(k) - 2K(k)]$$

where K and E are the first kind of complete elliptic integral and the second kind of complete elliptic integral, respectively, k is the parameter and

$$k^2 = \frac{4ac}{d^2 + (a+c)^2}$$

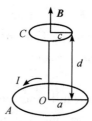

Figure T11-10   Figure of Problem 11.17

Part 4 Wave Optics

Optics is a fundamental subject in physics. Human beings have been studying light for at least two thousand years. Rectilinear propagation of light, reflection of light, and refraction of light were the earliest light phenomena observed and recognized by humans. Around the middle of the 17th century, the laws of reflection and refraction of light were established, laying the foundation of geometric optics. The 17th and 18th centuries were an important period in the development history of optics when scientists not only observed and studied light experimentally but also systematized and theorized existing optical knowledge. In the early 17th century, Lipsch, Galileo, Kepler and others invented telescopes for astronomical observation. In 1621, Fresnel discovered the law of refraction that light changed direction when passing through the interface of two media. Soon after, Descartes derived the law of refraction expressed by a sine function. The understanding of the nature of light has always been a focus of debate in the development of optics. As early as the 17th century, there were two different theories about the nature of light. One theory was the particle theory represented by Newton, which formulated that light was a collection of particles that propagated in space at a certain speed. The particle theory of light could explain phenomena such as rectilinear propagation of light, reflection of light and refraction of light. The other theory was the wave theory of light represented by Huygens, which stated that light was a wave that propagated in a medium. Due to Newton's authority at that time, the particle theory of light remained dominant until the 18th century. In addition, since people then were unable to accurately measure the speed of light in the air and water through experiments, it was impossible to judge the pros and cons of the two theories based on experimental facts.

In the early 19th century, with the development of science and experimental conditions, people discovered the interference, diffraction, polarization and other phenomena of light. These phenomena were incompatible with the particle theory, but the wave theory of light could successfully explain them. It was not until 1850, two hundred years after Newton proposed the particle theory, that Foucault and Fizeau respectively used experimental methods to clarify that the speed of light in water is less than that in the air. These became strong evidence for the wave theory of light, and this theory achieved a decisive victory.

In the mid-19th century, scientists represented by Maxwell found the connection between light and electromagnetic theory and believed that light was a kind of electromagnetic wave. From the late 19th century to the early 20th century, the study of optics went deep into the microscopic field of generation of light and interaction between light and matter. Through the study of a series of new phenomena such as black body radiation, the photoelectric effect and the Compton effect, people's understanding of the nature of light advanced a step further, and it was established that light was a particle stream composed of particles with a certain energy and momentum, which were called photons. The photons mentioned here are different from the particles in Newton's particle theory. Light has both particle and wave characteristics, that is, light has the wave-particle duality.

In the 1950s, new progress was made in optics with the discovery of laser. Many branches of optics emerged: fiber optic technology, laser holography, nonlinear optics, modern optics, and quantum optics.

Chapter 12

Wave Optics

This chapter mainly introduces wave optics, which uses the model of optical waves to describe the physical characteristics of light, and studies the phenomena that occur in the process of light propagation. In this way, it is more convenient to study the phenomena of light interference, diffraction and polarization. Through some classical experiments in optics, we will understand the methods for observing experimental phenomena and forming the theoretical systems of optics.

12.1 Electromagnetic theory of light

According to Maxwell's electromagnetic theory, an electromagnetic wave is formed by the periodic change of the electric field intensity vector E and magnetic field intensity vector H in space. Hertz firstly produced electromagnetic waves by means of electromagnetic oscillation and proved that their properties were exactly the same as those of light waves. Physicists have done a lot of experiments, which not only prove that light waves are electromagnetic waves, but also prove that X rays, γ-rays and so on are all electromagnetic waves. Since the element human eyes or photosensitive instruments are more sensitive to is E, the vibration vector of the light wave that we shall mention in the later discussion only refers to the electric vector, which is also called light vector. When vector E and vector H appear at the same time and position, their directions are perpendicular to each other, and both of them are perpendicular to the propagation direction of the light speed (u). These three vectors satisfy the right-hand spiral relationship, as shown in Figure 12.1.1, indicating that light is a transverse wave which will be discussed in detail in the polarization of light.

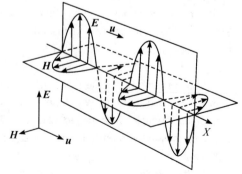

Figure 12.1.1   Schematic diagram of the light vector E

According to Maxwell's electromagnetic theory, the velocity of an electromagnetic wave in a vacuum is

$$u = \frac{1}{\sqrt{\epsilon_0 \mu_0}} \qquad (12.1.1)$$

Within the range of experimental error, u is equal to the propagation speed of light in a vacuum, which is approximately $3 \times 10^8 \, \text{m} \cdot \text{s}^{-1}$. The propagation speed of light waves in a medium is

$$u = \frac{c}{\sqrt{\epsilon_r \mu_r}} \qquad (12.1.2)$$

The ratio of the light speed in a vacuum to its speed in a medium is defined as the refractive index, that is

$$n = \frac{c}{u} \qquad (12.1.3)$$

By substituting Equation (12.1.3) into Equation (12.1.2), we can obtain

$$n = \sqrt{\epsilon_r \mu_r} \qquad (12.1.4)$$

In general, the refractive index n of most optically transparent substances is determined by the properties of the medium itself. Compared with the medium with a small refractive index, the medium with a large refractive index is called optically dense medium, and the opposite is called optically thin medium.

For a light wave in a vacuum with velocity c, frequency ν, and wavelength λ, there is

$$u = \nu \lambda_n \qquad (12.1.5)$$

The frequency of light is only related to the light source, and it is independent of the medium, so the propagating speed of light waves in a medium with a refractive index n is

$$\lambda_n = \frac{u}{\nu} = \frac{\lambda}{n} \qquad (12.1.6)$$

That is to say, as for light of the same frequency, its wavelength in a medium is smaller than that in a vacuum.

12.2 Coherent light

12.2.1 Luminous mechanism of a light source

Any object that emits light can be called a light source. Common light sources are the sun, fluorescent lamps, mercury lamps, etc. Light generally refers to visible light, that is, electromagnetic waves that can be seen by people. Its frequency is in the range from 3.9×10^{14} to 7.5×10^{14} Hz, corresponding to the wavelength between 760 nm and 400 nm in a vacuum. Different frequencies of visible light correspond to different colors from purple to red. According to the luminous mechanism, the light sources can be divided into the ordinary light source and laser light source. Ordinary light sources can be divided into the following types according to the excitation method: ① Electroluminescence caused by electric excitation, such as lightning, neon lights, and semiconductor emitting photodiodes; ② Photoluminescence caused by light excitation, such as the fluorescent lamp using ultraviolet light which originates from gas discharge in the lamp tube to excite the phosphor on the tube wall; ③ Chemical luminescence resulting from a chemical reaction, such as combustion, firefly glow, and phosphorescence emitted by the slow oxidation of phosphorus in the air; ④ Thermal radiation. Any object can radiate electromagnetic waves, and the object mainly radiates infrared light at low temperature, whereas it can radiate visible light, ultraviolet light and so on at high temperature.

Luminescence of an ordinary light source is formed by the spontaneous radiation of atoms (or molecules) in an excited state. According to modern physics theory, the energy of an atom (or molecule) can only be in a series of discrete energy levels. The state at the lowest energy level is called the ground state, and the state at any other higher energy is called the excited state. When the atom absorbs external energy in an excited state, the excited state is extremely unstable and will spontaneously return to a lower energy level. During this process, the atom emits light waves (electromagnetic waves). The transition of the atom from a high energy level to a low energy level, that is, the duration of each atom luminescence is very short (about 10^{-8} s). Therefore, each atom luminescence can only radiate a light wave of limited length and a certain frequency, known as the light wave train. Generally speaking, the excitation and radiation of numerous atoms in a light source are independent of each other, random and intermittent; Light emitted by different atoms at the same moment, or by the same atom level at different times possesses different frequency, initial phase and vibration direction, which are independent of each other. Hence, a beam of ordinary light is composed of a series of wave trains which have different frequency, independent vibration direction, and uncertain phase difference.

Light waves with a single wavelength are called monochromatic light. However, strict monochromatic light does not exist in reality, because light from the general light source is emitted by a large number of molecules or atoms at the same time, which contains a variety of different wavelength components, called polychromatic light. If the light wave contains a very narrow range of wavelength components, then this light is called quasi monochromatic light, which is commonly referred to as monochromatic light. The narrower the wavelength range, the better its monochromaticity. Spectrometers can be used to separate different wavelength components of light emitted by the light source from each other, and all the wavelength components constitute what is called the spectrum. The bright or dark lines corresponding to each wavelength component of the spectrum are called spectral lines and they all have a certain width, as shown in Figure 12.2.1. Each light source has its specific spectral structure, which can be used to analyze chemical elements, and to study the internal structure of atoms or molecules.

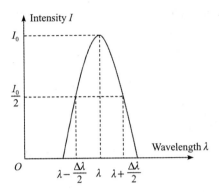

Figure 12.2.1 Spectral line and its width

 ## 12.2.2 Coherent light

In the chapter of mechanical waves, we discussed the superposition principle of waves. When two (or more) waves propagate in space, all points in space participate in the vibration caused by each wave at that point, and the vibrations at any point in the region where they meet is the synthesis of the vibrations generated at that point when each wave exists alone. The superposition principle of waves is also applicable to light waves. The superposition of light waves is the vibration superposition of E caused by two light waves at the same point.

Suppose that monochromatic light sources emit two harmonic light waves with the same frequency and vibrating in the same direction, as shown in Figure 12.2.2. When they propagate to any point P in space and meet in the same uniform medium, their vibration equations of the light vector are respectively

$$\begin{cases} E_1 = E_{10} \cos\left(2\pi\nu t - \dfrac{2\pi}{\lambda_n} r_1 + \varphi_{10}\right) \\ E_2 = E_{20} \cos\left(2\pi\nu t - \dfrac{2\pi}{\lambda_n} r_2 + \varphi_{20}\right) \end{cases} \quad (12.2.1)$$

Figure 12.2.2 Superposition of light waves

where ν is the frequency of the light wave, λ_n is the wavelength of the light wave in the medium, and φ_{10} and φ_{20} are the initial phases of the two light waves respectively.

According to the superposition principle, if E_1 and E_2 are in the same direction, the amplitude E of the resultant light at point P is

$$E_0 = \sqrt{E_{10}^2 + E_{20}^2 + 2E_{10}E_{20}\cos\Delta\varphi} \quad (12.2.2)$$

where $\Delta\varphi$ is the phase difference when the two light waves meet, that is

$$\Delta\varphi = (\varphi_{20} - \varphi_{10}) - \dfrac{2\pi}{\lambda_n}(r_2 - r_1) \quad (12.2.3)$$

Obviously, the amplitude of the combined vibration changes with time, and the actually observed light intensity is the average intensity over a longer period of time, so the average relative intensity of the combined vibration is the average value of E^2 with respect to time, that is

$$I = \overline{E^2} = E_{10}^2 + E_{20}^2 + 2E_{10}E_{20}\overline{\cos\Delta\varphi}$$
$$= I_1 + I_2 + 2\sqrt{I_1 I_2}\,\overline{\cos\Delta\varphi} \qquad (12.2.4)$$

where I_1 and I_2 are the light intensity generated at point P when the light source exists alone and do not change with time. Therefore, the combined light intensity I depends on the interference term $\sqrt{I_1 I_2}\,\overline{\cos\Delta\varphi}$, which will be discussed below.

1. Incoherent superposition

If two light waves are emitted by two independent ordinary light sources, due to the randomness of the light source, the initial phase difference of the two light waves will also change rapidly so that the phase difference $\Delta\varphi$ of the two light waves at point P can be an arbitrary value between 0 and 2π. Hence, in the observed time, the interference term is zero and by Equation (12.2.4)

$$I = \overline{E^2} = I_1 + I_2 \qquad (12.2.5)$$

That is to say, in a long time relative to the period of the light wave, the observed light intensity is the sum of light intensities when two light waves exist alone for a long time, and this superposition is incoherent superposition.

2. Coherent superposition

If the phase difference between two light sources is constant, $\varphi_{20} - \varphi_{10}$ is constant and does not change with time. At any point P in space, the change of the phase difference of two light waves is only related to position and does not change with time. Therefore, its average value with respect to time will not be zero, and the combined light intensity after superposition is

$$I = I_1 + I_2 + 2\sqrt{I_1 I_2}\,\cos\Delta\varphi \qquad (12.2.6)$$

It can be seen that for different points in space, due to different position, the phase difference $\Delta\varphi$ will take different values, causing the superposition of light waves in the encounter area to form stable, strong or weak distribution of light intensity, and this superposition is coherent superposition. The phenomenon that the light intensity changes periodically according to space due to the coherent superposition of light waves is called light interference, and the spatial distribution image is called interference pattern. Interference patterns are generally light and dark.

According to Equation (12.2.6), when $\Delta\varphi = \pm 2k\pi$ ($k = 0, 1, 2, \cdots$), that is, two light waves are in phase at the meeting position, there is

$$I_{\max} = I_1 + I_2 + 2\sqrt{I_1 I_2} \qquad (12.2.7)$$

At these positions, the light intensity reaches the maximum, called constructive interference (strengthening mode).

If $\Delta\varphi = \pm(2k+1)\pi$ ($k = 0, 1, 2, \cdots$), that is, two light waves are out of phase at the meeting positions, there is

$$I_{\min} = I_1 + I_2 - 2\sqrt{I_1 I_2} \qquad (12.2.8)$$

At these positions the light intensity reaches the minimum, called destructive interference (weakening mode).

When the phase difference $\Delta\varphi$ is any other value, the light intensity of coherent superposition is between the maximum and the minimum, which is the transition region of the light and dark pattern.

It can be found that the interference of light is essentially the redistribution of light intensity (the energy of light) in space. The spatial distribution of light intensity is determined by the phase difference, indicating the spatial phase distribution of the light waves involved in coherent superposition. That is, the interference pattern records the phase information, and this concept is the basis of information optics, as well as the basic principle of holography.

If the intensities of two light waves are equal, that is, $I_1 = I_2 = I_0$, by Equation (12.2.6),

$$I = 4I_0 \cos^2 \frac{\Delta\varphi}{2} \qquad (12.2.9)$$

At this point, the intensity is four times that of a single light wave; $I_{\min} = 0$ represents complete extinction. In this case, the interference pattern has the greatest contrast between light and dark.

From the above discussion, it can be seen that the requirements of interference when two light waves meet in space are: The vibration frequency of the light vector is the same, the vibration direction is the same, and the phase difference is constant. Light that satisfies these interference conditions is called coherent light, and the light source that can produce coherent light is called coherent light source. In an experiment, in order to obtain the interference pattern with clear contrast between strength and weakness, the amplitude difference of coherent light waves participating in superposition is also required not to be large.

12.2.3 Method of obtaining coherent light

A single-frequency laser source has good coherence, but in real life we can also observe the interference phenomenon of ordinary light. There are two specific methods to obtain coherent light from ordinary light: wavefront splitting method and amplitude splitting method. The wavefront splitting method is the coherence of the subwave generated from different parts of the same wavefront, such as the double-slit interference which will be discussed below; The amplitude splitting method uses the reflection and refraction of light on the surface of the transparent medium film to divide the same beam into two coherent beams with smaller amplitudes, such as the film interference, the Newton ring and the Michelson interferometer to be introduced later.

12.3 Optical path length and optical path difference

From the coherent superposition of light waves discussed in the previous section, it can be seen that the distribution of light intensity in the superposition region depends on the phase difference, and the phase difference is related to the propagation path of the two light waves. The concepts of optical path length and optical path difference will be introduced in order to calculate the phase difference of light waves propagating in different media.

12.3.1 Optical path length

The wavelength of a light wave indicates its spatial periodicity just like the wavelength of a mechanical wave. A light wave has a phase change of 2π whenever it travels a distance of one wavelength in its propagation path. Because the wavelength of the light wave is related to the refractive index of the medium, monochromatic light with the same frequency, propagating the same distance in different media, has different phase changes. For the same phase change, the distance of propagation in different media will also be different. If the distance the light wave travels in the medium is r and for the same phase change, the distance it travels in a vacuum is r_0, then there is

$$2\pi \frac{r}{\lambda_n} = 2\pi \frac{r_0}{\lambda} \tag{12.3.1}$$

Then, we can obtain

$$r_0 = \frac{\lambda}{\lambda_n} r = nr \tag{12.3.2}$$

The above expression shows that with the same change in phase, the propagating distance r of a light wave in a medium is equal to be the distance nr as the light wave propagates in a vacuum. We define the optical path length as the product of the distance traveled by a light wave in a medium and the refractive index of the medium. The optical path length is a reduced amount, and its physical meaning is that for the same phase change or time, the propagation distance of light in a medium is converted to the corresponding propagation distance of light in a vacuum. The purpose of such calculation is to facilitate the discussion of the propagation of light waves in different media. The optical path length of a light wave is equal to the distance only when it travels in a vacuum.

12.3.2 Optical path difference

The relationship between the phase difference and the optical path difference when two light waves meet in space is discussed below. As shown in Figure 12.3.1, the light waves emitted from the two coherent light sources (S_1 and S_2) are propagating through the two media with different refractive indexes (n_1 and n_2), and meet at point P. The phase difference $\Delta\varphi$ resulting from the different propagation paths is

$$\Delta\varphi = 2\pi\frac{r_2}{\lambda_2} - 2\pi\frac{r_1}{\lambda_1} = 2\pi\left(\frac{n_2 r_2}{\lambda} - \frac{n_1 r_1}{\lambda}\right) = \frac{2\pi}{\lambda}\delta \qquad (12.3.3)$$

where

$$\delta = n_2 r_2 - n_1 r_1 \qquad (12.3.4)$$

which represents the optical path difference between the two light waves when they reach point P, that is, the phase difference is the product of the optical path difference multiplied by $2\pi/\lambda$.

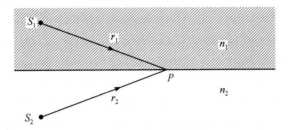

Figure 12.3.1 Light waves travel through different media

When coherent light interferes, the occurrence of interference strengthening or weakening depends on the optical path difference of the two coherent light, rather than the geometric distance difference. If two coherent light sources are obtained from the same wave source and the optical path difference satisfies the relation below

$$\delta = n_2 r_2 - n_1 r_1 = \pm k\lambda \quad (k=0, 1, 2, \cdots) \qquad (12.3.5)$$

the bright pattern (fringe) can be obtained due to the constructive interference. When the optical path difference satisfies the relation below

$$\delta = n_2 r_2 - n_1 r_1 = \pm(2k+1)\frac{\lambda}{2} \quad (k=0, 1, 2, \cdots) \qquad (12.3.6)$$

the dark pattern (fringe) can be obtained due to the destructive interference.

Points with an equal optical path difference can form the same order fringe in space, that is, fringes are the tracks of equal points, so the concepts of optical path length and optical path difference are very important. The correct calculation of the optical path difference is the key to discussing the coherent superposition of optical waves. When calculating the optical path difference, the following two points need to be noted.

1. Equivalent optical path length of a thin lens

In the optical system, a thin lens is generally placed on the optical path. As shown in Figure 12.3.2, when the light source (or object point) S is imaged by the thin lens, the image point is a bright spot, indicating that each light

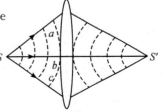

Figure 12.3.2 Equivalent optical path length of a thin lens

wave is in phase superposition. The optical path length between the object point and the image point is the same. Each light from the object point S to the image point S' has a different geometric path, and its path through the lens is also different. If the geometric path is longer, the path is shorter in the lens. However, the refractive index of the lens is greater than that of the air. As a result, each light has the same optical path length. It can be seen that the lens only changes the propagation direction of each ray, and does not produce additional optical path difference.

2. Half-wave loss

According to the electromagnetic theory of light, when a light wave travels from the optically thin medium to the optically dense medium with normal incidence (incidence angle of 0°) or grazing incidence (incidence angle of 90°), the phase different between the reflected light and incident light is π at the interface between the two media. This change is equivalent to a change in the reflected light path of an extra half wavelength, that is, an increase or loss of half a wavelength. This phenomenon is called half-wave loss. The optical path difference $\pm\lambda/2$ generated by the reflected light and the incident light at the incident point is also called the additional optical path difference, where λ is the wavelength in a vacuum, and the selection of the positive or negative sign has no effect on the calculation of the interference pattern distribution. Hence, the book defines a unified positive sign. If the incident light is reflected from the optically dense medium to the optically thin medium, there is no half-wave loss. In any case, there is no half-wave loss in refracted light. When calculating the optical path difference, it is necessary to pay attention to whether there is a half-wave loss.

12.4 Young's double-slit interference

As mentioned earlier, the wavefront splitting method is one of the methods to obtain coherent light. Since the vibration of each point on the same wavefront emanating from a light source has the same phase, two parts taken from the same wavefront can be used as coherent light sources. The coherent light waves in Young's double-slit and Lloyd mirror interference experiments will be introduced in this section, and they play an important role in the establishment of the wave theory.

12.4.1 Young's double-slit interference

In 1801, Thomas Young first used a single light source to obtain two coherent light waves, observed the interference phenomenon of light, and used the wave property of light to successfully explain the interference phenomenon of light, thus further confirming the wave theory of light. Thomas Young, a British physicist and one of the great founders of wave optics, made important contributions to optics, physiological optics, material mechanics and so on.

1. Experimental setup of Young's double-slit interference

As shown in Figure 12.4.1(a), the monochromatic light emitted by an ordinary monochromatic light source (such as a sodium light lamp) illuminates vertically the single slit S, which can be regarded as a monochromatic line light source. The outgoing light wave shines on two slits (S_1 and S_2), which are parallel to S and very close to each other, and the distances from S_1 and S_2 to S are equal. S_1 and S_2 can be regarded as two linear monochromatic wavelet sources derived from the same wavefront. They are coherent light sources. When the two wavelets emitted from S_1 and S_2 meet in space, interference will occur. A receiving screen parallel to the slit is placed behind S_1 and S_2. Linear interference fringes, alternating between light and dark and with equal spacing will appear on the screen, as shown in Figure 12.4.1(b). By using good coherence and high brightness of laser, we now can get a clear and bright interference fringe on the screen after directly illuminating the double holes with the laser beam.

(a) Setup of double-slit interference   (b) Fringe distribution of double-slit interference

Figure 12.4.1 Young's double-slit interference

2. Distribution of interference fringes

As shown in Figure 12.4.2, assuming the space between the two slits is d, and the distance between the two slits and the receiving screen is D. In general, the order of D is less than 10^{-3} m, while the order of d can reach 1 m. Let the midperpendicular line of the two slits intersect with the screen at point O, that is, $OS_1 = OS_2$. The distances from S_1 and S_2 to any point P on the receiving screen are r_1 and r_2, respectively. The distance from P to point O is x. $D \gg d$, and let the angle between the two light waves reaching any point P on the screen and the normal of the double-slit plane be θ.

Because the distances from the two coherent sources S_1 and S_2 to S are equal, the initial phase difference is zero, and the optical path difference between the two light waves emitted by the light sources is $\delta = r_2 - r_1$. It can be obtained from the geometric relationship that

$$r_2 - r_1 \approx d\sin\theta \approx d\tan\theta = d\frac{x}{D} \tag{12.4.1}$$

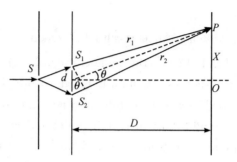

Figure 12.4.2 Schematic diagram of Young's double-slit interference

The optical path difference

$$\delta = r_2 - r_1 = \frac{d}{D}x \tag{12.4.2}$$

According to Equations (12.3.5) and (12.4.2),

$$\delta = \frac{d}{D}x = \pm k\lambda \tag{12.4.3}$$

The position of the bright fringe is

$$x = \pm k\frac{D\lambda}{d} \quad (k=0, 1, 2, \cdots) \tag{12.4.4}$$

If $k=0$, $x=0$. At the center of the receiving screen, the corresponding optical path difference is $\delta=0$, which is a zero order bright fringe. When $k=1, 2, \cdots$, the bright fringes are called the first order, the second order, $\cdots$ bright fringes respectively. The plus or minus sign indicate that the other bright fringes are symmetrical on both sides of the zero order bright fringe.

According to Equation (12.3.6), if

$$\delta = \frac{d}{D}x = \pm(k-1)\frac{\lambda}{2} \tag{12.4.5}$$

the position of the dark fringe can be obtained

$$x = \pm(2k-1)\frac{D}{d} \quad (k=1, 2, \cdots) \tag{12.4.6}$$

In this expression, the plus or minus sign indicate that the other dark fringes are symmetrical on both sides of the zero order bright fringe.

In Young's double-slit interference, if the width of the two slits and the intensity of the two beams are equal, the interference light intensity generated by the light waves passing through the double slits can be obtained from Equation (12.2.9)

$$I = 4I_0 \cos^2\frac{\Delta\varphi}{2} = 4I_0 \cos^2\frac{\delta\pi}{\lambda} \tag{12.4.7}$$

Figure 12.4.3 shows the relationship between the double-slit interference light intensity and the optical path difference, indicating the light intensities of the different order bright fringes are equal.

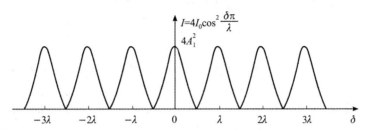

Figure 12.4.3 Light intensity distribution of Young's double-slit interference

The distance between centers of adjacent bright or dark fringes is called fringe spacing, which reflects the spatial periodicity of light intensity distribution of interference fringes. As can be obtained from Equations (12.4.4) and (12.4.6), the spacing between bright or dark fringes is

$$\Delta x = x_{k+1} - x_k = \frac{D\lambda}{d} \tag{12.4.8}$$

The above formula shows that the spacing of the fringes is independent of the order k, and the fringes of the double-slit interference are equally spaced. The distance between the centers of two dark fringes is also called the fringe width. The fringe width of the double-slit interference is equal to the fringe spacing. As can be seen from Figure 12.4.3, the light intensity of each point in the bright fringes is not equal but gradually transitions from the smallest to the largest, and finally drops the smallest.

As can be seen from Formula (12.4.8), the physical essence of the double-slit interference is to transform and amplify the wavelength (which reflects the longitudinal spatial periodicity of light waves and is difficult to directly observe) into an observable horizontal interference pattern by means of interference.

If the incident light is white, because the optical path difference of various colors is zero, the zero order bright fringe is still white. However, with a higher order, the positions of the bright fringes of different wavelengths are different, which are distributed from the short wavelength to long wavelength along the side near the zero order bright fringe and form a color stripe arranged from purple to red in turn. This phenomenon is called interference spectrum. Moreover, the k order fringe with the larger wavelength may overlap with the $k+1$ order fringe with the smaller wavelength, resulting in the blurred higher order fringe. Therefore, experiments are generally conducted using quasi-monochromatic light sources.

Example 12.4.1 In Young's double-slit interference experiment, the whole device is placed in the air, the monochromatic parallel light with a wavelength $\lambda = 500$ nm is vertically incident to the double slits with a spacing $d = 1 \times 10^{-4}$ m, and the distance from the screen to the double slits $D = 2$ m. Find the distance between centers of the two first order bright fringes on both sides of the central bright fringe.

Solution According to $x = \pm k \dfrac{D\lambda}{d}$, it can be deduced that

$$x_1 = \pm \dfrac{D\lambda}{d} = \pm \dfrac{2 \times 500 \times 10^{-9}}{1 \times 10^{-4}} = \pm 0.01 \text{ m}$$

Assuming $n=1$ and $k=1$, the distance between centers of the two first order bright fringes is

$$\Delta x = 2x_1 = \pm 0.02 \text{ m}$$

Example 12.4.2 The slit spacing of Young's double-slit interference experiment is set as d, and the whole device is immersed in the air. The visible light with a wavelength in the range of 400~750 nm shines vertically. Try to find the order of the visible spectrum that can be clearly observed.

Solution According to $\delta = r_2 - r_1 = \dfrac{d}{D}x = \pm k\lambda$ ($k = 0, 1, 2, \cdots$), the optical path difference of various colors is zero for the central bright fringe, and the overlapping of zero order bright fringe results in white, while other bright fringes of the same order and of various wavelengths are separated due to different wavelengths and different positions, and fringes of different orders may also overlap. What is called overlapping means that the fringes of different wavelengths and different orders are in the same position on the receiving screen, that is, the fringes of different wavelengths and different orders have the same optical path difference. Obviously, the light wave with the largest wavelength of a certain order (set as k) and the light wave with the smallest wavelength of a higher order $(k+1)$, denoted as $k l_{\text{red}} = (k+1) l_{\text{purple}}$, overlap first. By substituting the data, we get

$$k = \dfrac{\lambda_{\text{red}}}{\lambda_{\text{red}} - \lambda_{\text{purple}}} = \dfrac{400}{750 - 400} = 1.1$$

Because k can only be taken as integers, overlapping starts from the second order. We can only see a clear and complete $\pm$1st order light spectrum from purple to red. In the overlapping area, the observed color fringes near either side of the central bright fringe are formed by various colors of light, and the colors far from the central bright fringe are almost completely overlapping, which means the fringes are not visible.

Example 12.4.3 As shown in Figure 12.4.4, a double-slit interference device in the air is covered with a transparent film with a refractive index $n = 1.58$.

(1) If the wavelength of the incident monochromatic light is 550 nm and the thickness of the film is $e = 8.53 \times 10^{-6}$ m, which order of the original pattern will the zero order pattern move to?

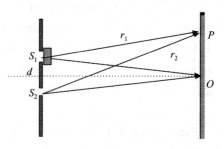

Figure 12.4.4 Figure of Example 12.4.3

(2) If the spacing d is 0.5 mm, the distance between the receiving screen and the double slits is 0.5 m, and the interference fringe on the screen moves by 10 mm after the

device is covered, find the thickness of the film.

Solution (1) When the device is not covered, point P is set as the k order bright fringe, and $r_2 - r_1 = k\lambda$; After covering the device, if point P becomes the zero order bright fringe, the optical path difference satisfies $r_2 - (r_1 - e + ne) = 0$, that is, $r_2 - r_1 = (n-1)e$.

From the above two equations, it is clear that

$$k = \frac{(n-1)e}{\lambda} = \frac{(1.58-1) \times 8.53 \times 10^{-6}}{550 \times 10^{-9}} \approx 9$$

So, the zero order fringe has been moved to the original order 9.

(2) According to what is given, the zero order bright fringe also moves up the same distance $x = 10$ mm, and the optical path difference from the two slits to point P before covering is

$$r_2 - r_1 = \frac{dx}{D}$$

After covering, point P is the zero order bright fringe, and then $r_2 - r_1 = (n-1)e$
So,

$$(n-1)e = \frac{dx}{D}$$

$$e = \frac{dx}{(n-1)D} = \frac{0.5 \times 10}{(1.58-1) \times 500} = 1.72 \times 10^{-2} \text{ mm}$$

It can be seen that the thickness and refractive index of the transparent film can also be measured by the variation of the double-slit interference fringe.

12.4.2 The Lloyd mirror interference experiment

In Young's double-slit experiment, only when the slits are very narrow will the light vibration at S_1 and S_2 have the same phase, but the light intensity through the slits is very weak. The interference fringe is often not bright or clear. In 1834, Lloyd improved Young's double-slit interference device and proposed a simpler device for generating double-beam interference. The proposed device not only produced clearer interference phenomena, but also proved that the reflected light had half-wave loss under certain conditions.

Figure 12.4.5 shows the Lloyd mirror interference experimental setup. The whole device is placed in the air, and one of the light waves emanating from the slit S_1 is directly transmitted to the receiving screen E, while the other beam is grazed to the flat mirror and then reflected to the receiving screen. The reflected light can be regarded as emitted by the virtual light source S_2, so S_1 and S_2 constitute a pair of coherent light sources, which meet in the shadow area and generate interference. There is a straight fringe alternating between light and dark, which is parallel to the slit.

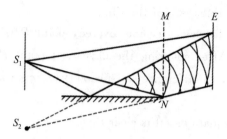

Figure 12.4.5 Setup of the Lloyd mirror interference

When the receiving screen is moved to the position N in contact with the edge of the mirror, the distances to the incident light and the reflected light are equal, and there should be bright fringes at N. However, the experimental result finds dark fringes, which indicates that the light wave directly incident on the screen and the light wave reflected by the mirror are in the opposite phase at N, that is, the phase difference is π. The same is true with the fringes at other positions, that is, the position of the bright fringes should be calculated according to Young's double-slit interference bright fringe formula Equation (12.4.4), but the dark fringes actually appear. The light directly incident on the screen is propagated in the air or uniform media, and there can be no phase change, so only the light reflected by the mirror has a phase mutation and there is a half-wave loss resulting in the additional optical path difference. This experimental result verifies the conclusion of half-wave loss in electromagnetic theory.

12.5 Thin film interference

The previous discussion was about the interference generated by the wavefront splitting method. In this section, we will study the interference generated by the amplitude splitting method. Thin film interference is one of the common amplitude splitting interference. What is called a film refers to a thin layer of medium film formed by a transparent medium, such as a soap film, oil film floating on a water surface and film layer plated on the surface of the optical instrument lens. When light shines on a transparent film, the reflected light (or transmitted light) generated by the upper and lower surfaces of the film is superimposed on each other. The interference is called thin film interference. For example, colorful stripes on soap bubbles, stripes on the oil film on a water surface, and colorful patterns on the wings of insects are the results of thin film interference.

12.5.1 Thin film interference

As shown in Figure 12.5.1, a transparent film with a thickness e and refractive index

n_2 is placed in media with refractive indexes n_1 and n_3 respectively. A light wave emitted by a monochromatic point light source shines on the upper surface of the film with the incident angle i, which is divided into two light waves at point A. One beam forms reflected light 1, and the other is transmitted onto the film to form refracted light. The refracted light is reflected at point B on the lower surface of the film to the upper surface C, and then refracted back to the original medium to form light wave 2. The two light waves 1 and 2 are derived from the same incident light wave, which are coherent light and parallel to each other. Their coherent region is located at infinity. With the lens L applied to converge light at point P on the receiving screen which is located in the focal plane of the lens, results a coherent superposition.

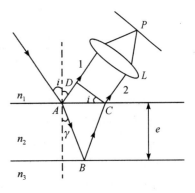

Figure 12.5.1 Thin film interference

Let $CD \perp AD$. Because the lens does not produce the additional optical path difference and the optical paths of CP and DP are equal, the optical path difference between the two light is generated before the wave surface DC. It is composed of two parts, one being the optical path difference caused by different transmission paths and the other being the additional optical path difference caused by half-wave loss which may exist when the light is reflected on the upper and lower surfaces of the film. The total optical path difference can be expressed as

$$\delta = n_2(AB+BC) - n_1 AD + \delta'$$

where δ' is the additional optical path difference in accordance with different refractive indexes, and the values are

$$\delta' = \begin{cases} 0 & (n_1 < n_2 < n_3 \text{ or } n_1 > n_2 > n_3) \\ \dfrac{\lambda}{2} & (n_1 < n_2 > n_3 \text{ or } n_1 > n_2 < n_3) \end{cases}$$

According to the geometry relation,

$$AB = BC = \frac{e}{\cos\gamma}$$

$$AD = AC \sin i = 2e \tan\gamma \sin i$$

So

$$\delta = 2n_2 \frac{e}{\cos\gamma} - 2n_1 e \tan\gamma \sin i + \delta' = \frac{2e}{\cos\gamma}(n_2 - n_1 \sin\gamma \sin i) + \delta'$$

According to the refractive law, $n_1 \sin i = n_2 \sin\gamma$, so

$$\delta = \frac{2n_2 e}{\cos\gamma}(1 - \sin^2\gamma) + \delta' = 2n_2 e \cos r + \delta' \qquad (12.5.1)$$

or

$$\delta = 2n_2 e \sqrt{1 - \sin^2\gamma} + \delta' = 2e\sqrt{n_2^2 - n_1^2 \sin^2 i} + \delta' \qquad (12.5.2)$$

Therefore, the position of the coherent fringe is

$$\delta = 2e\sqrt{n_2^2 - n_1^2 \sin^2 i} + \delta' = \begin{cases} k\lambda & (k=1, 2, 3, \cdots, \text{bright fringes}) \\ (2k+1)\dfrac{\lambda}{2} & (k=0, 1, 2, \cdots, \text{dark fringes}) \end{cases} \quad (12.5.3)$$

According to Equation (12.5.3), when n_1 and n_2 are constant, the optical path difference δ is determined by the film thickness e and incident angle i. Thin film interference can be divided into two cases:

(1) If e is constant, that is, the thickness of the medium film is equal, δ is only determined by the inclination i of the incident light. For the incident light with the same incident angle, its reflected light has the same optical path difference and corresponds to the same order interference fringes. This is called equal inclination interference and the interference fringes formed are called equal inclination interference fringes.

(2) If i is unchanged and the film thickness e changes, that is, parallel light shines onto the film with inhomogeneous thickness, δ is only related to the film thickness e. The optical path difference at the same film thickness is equal, which can form interference fringes of the same order. We call this interference equal thickness interference, and the interference fringes are called equal thickness interference fringes.

As shown in Figure 12.5.2, the light emitted by the monochromatic point light source S is incident on the surface of the film with the same angle, and the reflected light converges on the same circle on the receiving screen through the lens. When the angle of the incident light continuously changes, a series of light and dark concentric rings are formed.

Transmitted light also has the interference phenomenon, but the brightness is low and reverse from reflected light. That is, as for the same film thickness, if the reflected light interference is dark, the transmitted light interference is bright, and vice versa, which is also an inevitable consequence abiding by the law of conservation of energy.

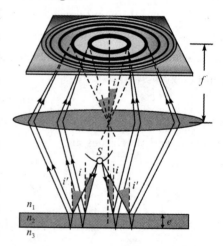

Figure 12.5.2 Equal inclination interference fringes of a point source

In modern optical instruments, in order to reduce the energy loss caused by the reflection of incident light on the surface of optical devices such as a lenses, a transparent film of uniform thickness is often plated on the surface (such as MgF_2), whose refractive index is between that of the air and that of glass. When the thickness of the film is appropriate, the reflected light of some wavelengths can be weakened due to interference, thereby increasing the light energy through the device. The film that can enhance the transmitted light is called an anti-reflection film.

As shown in Figure 12.5.3, a magnesium fluoride film with a thickness e and refractive index $n = 1.38$ is plated on the glass surface with a refractive index $n_g = 1.60$ and placed in the air. Therefore, the light reflected on the upper and lower surfaces of the magnesium fluoride film both has half-wave loss, and the additional optical path difference is not considered. As can be seen from Equation (12.5.3), when the light wave is close to the vertical incidence ($i = 90°$), the reflected light satisfies the weakening condition

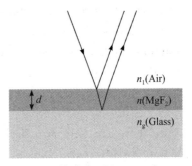

Figure 12.5.3 Diagram of an anti-reflection film

$$\delta = 2ne = (2k+1)\frac{\lambda}{2} \quad (k=0, 1, 2, \cdots)$$

The thickness of the film is

$$e = \frac{(2k+1)\lambda}{4n} \tag{12.5.4}$$

The minimum thickness of the anti-reflection film is (when $k=0$) $e = \dfrac{\lambda}{4n}$, which is called optical thickness.

According to Equation (12.5.4), a film of a certain thickness can only enhance the light of a specific wavelength and its similar wavelength. When the lens of a camera and vision aid is coated, the transmission of green light corresponding to the wavelength human eyes are most sensitive to often needs enhancing. Such a thickness will just make the reflection of blue and red light meet the conditions for interference enhancement, so the lens surface appears blue-purple or purple-red.

On the other hand, in some optical systems, some optical elements are required to have high reflectivity. For example, the reflector in a laser requires the reflectivity to monochromatic light of a certain frequency be more than 99%. In order to enhance the reflective energy, a transparent film with high reflectivity is often coated on the glass surface. By using the fact that the optical path difference between the reflected light on the upper surface and the lower surface meets the interference requirement, the intensity of reflected light can be enhanced. This film is called high-reflection film. Since the energy of reflected light accounts for about 5% of the incident light energy, in order to achieve the purpose of high reflectivity, the glass surface is often alternately plated with multi-layer dielectric films with different refractive indexes, generally 13 layers, even as many as 15 and 17 layers. Astronauts helmets and visors are plated with a multilayer film with high reflectivity to infrared light, in order to shield the extreme infrared radiation in space.

Example 12.5.1 A beam of parallel white light is vertically incident on a film of uniform thickness placed in the air. The refractive index of the film is $n = 1.4$. Two dark lines with wavelengths of 400 nm and 600 nm respectively appear in the reflected light. Find the thickness of the film.

Solution Because the reflected light has a half-wave loss, the optical path difference between the two light waves is

$$\delta = 2ne + \frac{\lambda}{2}$$

Since there are only two dark lines, the dark fringe order of the two wavelengths should only be different by one order, that is, $\lambda_1 = 400$ nm at the order k and $\lambda_2 = 600$ nm at the order $k-1$. So there are

$$2ne + \frac{\lambda_1}{2} = (2k+1)\frac{\lambda_1}{2}$$

$$2ne + \frac{\lambda_2}{2} = (2k-1)\frac{\lambda_2}{2}$$

It can be concluded that

$$2ne = k\lambda_1 \qquad (1)$$

$$2ne = (k-1)\lambda_2 \qquad (2)$$

From them, it is derived that

$$k\lambda_1 = (k-1)\lambda_2$$

So

$$k = \frac{\lambda_2}{\lambda_2 - \lambda_1} = \frac{600}{600 - 400} = 3$$

That is, the second order of 600 nm and the third order of 400 nm are observed, which means

$$e = \frac{k\lambda_1}{2n} = \frac{3 \times 400}{2 \times 1.4} = 428.6 \text{ nm}$$

Thin film interference is an important method for measuring and testing precision mechanical parts or optical components, which is widely used in modern science and technology. Two representative experiments based on equal thickness interference are introduced below.

12.5.2 Wedge interference

The wedge interference device commonly used in the laboratory is shown in Figure 12.5.4(a). Two glass plates are in contact at one end, and a thin sheet (or filament) with a thickness of h is inserted between the other ends, so that a wedge air layer with an angle of θ is formed between the two glass plates, hence an air wedge. The contact point of the two glasses is the edge of the wedge. If the space between the two glass plates is filled with a transparent medium of refractive index n, a wedge of one material is formed.

When the light emitted by the monochromatic light source S becomes parallel light after passing through the lens L, the light is reflected vertically ($i=0$) toward the wedge by the semi-reflector M with an inclination angle $\dfrac{\pi}{4}$. Because the wedge angle θ is very small, it can be approximated that the incident light is perpendicular to both the upper surface of the wedge film and its lower surface. At this time, the light formed by the reflection of the upper and lower surfaces of the wedge film is coherent light, so interference occurs. The distribution of interference fringes can be observed in the microscope, as shown in Figure 12.5.4(b). If the refractive indexes of the upper and lower glass are the same n_1, regardless of the relationship between n and n_1, the optical path difference between the two reflected light is

$$\delta = 2ne + \dfrac{\lambda}{2} \qquad (12.5.5)$$

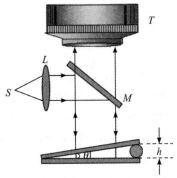

(a) Experimental setup of wedge interference   (b) Fringes of the wedge interference

Figure 12.5.4 Wedge interference

Wedge interference bright and dark fringe condition can be obtained as

$$\delta = 2ne + \dfrac{\lambda}{2} = \begin{cases} k\lambda & (k=1,\ 2,\ 3,\ \cdots,\ \text{bright fringes}) \\ (2k+1)\dfrac{\lambda}{2} & (k=0,\ 1,\ 2,\ \cdots,\ \text{dark fringes}) \end{cases} \qquad (12.5.6)$$

The above expression shows that for the edge of the wedge, there is $e=0$, so the edge of the wedge is zero order dark fringe. If the refractive indexes of the upper and lower glass are different, please discuss the bright and dark fringe conditions according to the relationship of the three refractive indexes.

From Equation (12.5.6), it can be obtained that the thickness difference Δe of the wedge film corresponding to two adjacent bright or dark fringes is

$$\Delta e = e_{k+1} - e_k = \dfrac{\lambda}{2n} = \dfrac{\lambda_n}{2} \qquad (12.5.7)$$

The thickness difference of the film corresponding to adjacent bright (dark) fringes is equal to half of the wavelength in the medium.

Let the angle of the wedge be θ, and as can be seen from the geometric relationship in Figure 12.5.5, the spacing l of adjacent bright fringes (or dark fringes) should satisfy the relation

$$\Delta e = l\sin\theta$$

Usually $\theta < 1°$, so

$$l = \frac{\Delta e}{\sin\theta} \approx \frac{\lambda}{2n\theta} = \frac{\lambda_n}{2\theta} \quad (12.5.8)$$

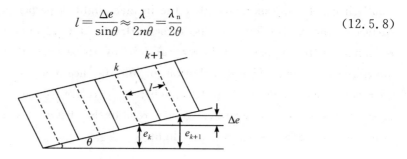

Figure 12.5.5 Fringe distribution of wedge interference

It can be concluded that the fringes are approximately equally spaced. For incident light of a certain wavelength, the fringe spacing is inversely proportional to the wedge angle and the medium refractive index n.

Each order of the interference bright or dark fringe corresponds to a certain film thickness e. The same thickness corresponds to the same order of the fringe, and the shape of the fringe depends on the trajectory of points with the same film thickness (isopach line). The isopach line of the wedge is parallel to the edge line, so the wedge interference fringes are straight light and dark fringes, with equal spacing and parallel to the edge. It should be emphasized that the film refers to the air film or medium film between the glass, rather than the glass, because the thickness of the glass is so large that an ordinary light source will not produce interference fringes.

As can be seen from Equation (12.5.8), if the wedge angle θ and the interference fringe spacing l are known, the wavelength of monochromatic light λ can be calculated; On the contrary, if the wavelength of monochromatic light λ and the distance between the interference fringes l are known, the wedge angle θ can be obtained, and then the linearity of the small object sandwiched between the two glass pieces can be obtained from the geometric relationship. This principle is commonly used in engineering technology to determine the diameter of a filament or the thickness of a film. For example, a very thin silicon dioxide film is often plated on the semiconductor material silicon (Si) sheet to protect semiconductor components. To measure the thickness of the silicon dioxide film, it can be made into a wedge shape, as shown in Figure 12.5.6. The number of interference fringes (or dark fringes) can be measured by the wedge interference device, and the thickness e of the silica film can be calculated by using the geometric relationship.

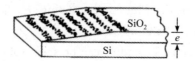

Figure 12.5.6 Fringe distribution of SiO$_2$ wedge interference

Wedge interference is also commonly used in detecting the flatness of a workpiece surface. If the upper and lower surfaces of the wedge are standard optical planes, the interference fringes are a series of parallel, equally spaced, bright and dark fringes. If one of the two pieces of glass is a standard optical plane and the other is a bumpy glass sheet or a polished metal surface, the interference fringe will no longer be straight, but uneven and irregular. As shown in Figure 12.5.7, according to the degree and direction of fringe bending, the concave-convex condition of the plate to be tested can also be judged. Because the thickness difference of the air film between adjacent bright (dark) fringes is λ, the precision of bump defect detected by this method can reach about 0.1 μm.

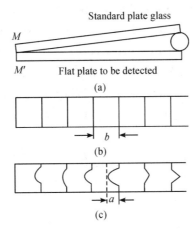

Figure 12.5.7 Inspection of tiny defects of a workpiece surface

Example 12.5.2 To measure the diameter D of a thin metal wire, an air wedge is formed by the method shown in Figure 12.5.8. Parallel monochromatic light is vertically incident to form an equal thickness interference fringe. D can be calculated by measuring the spacing of the interference bright fringes with a reading microscope. Let the wavelength of the incident light be 550 nm, the distance between the metal wire and the edge of the wedge be $l =$ 28.350 mm, and the distance from the 1st bright fringe to the 31st bright fringe be 4.328 mm. Find the diameter of the thin metal wire D.

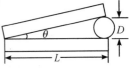

Figure 12.5.8 Figure of Example 12.5.2

Solution Because the wedge angle θ is small,

$$\sin\theta \approx \frac{D}{l}$$

The distance between two adjacent bright fringes satisfies the relation

$$a\sin\theta = \frac{\lambda}{2}$$

where a is the distance between two adjacent bright fringes. From what is given,

$$a = \frac{l}{30} = \frac{4.325}{30} \approx 0.144,17 \text{ mm}$$

Hence, the diameter of the thin metal wire

$$D = l\frac{\lambda}{2a} = 28.350 \times \frac{550 \times 10^{-6}}{2 \times 0.144\ 17} = 0.054,08 \text{ mm}$$

Example 12.5.3 Tiny defects of a workpiece surface can be checked by wedge interference. As shown in Figure 12.5.9, when monochromatic light of wavelength λ is incident vertically on the wedge, interference fringes are observed.

(1) Is the uneven position convex or concave?

(2) What is the convex or concave height?

Solution (1) For equal thickness interference, each point on the same order fringe has an equal air layer thickness. Since the interference fringes of the same order are curved into the direction of the edge, the equal air layer thickness corresponding to the fringes of the same order can only be ensured by being concave, so the uneven area is concave.

(2) As shown in Figure 12.5.9 (b), when the spacing of the interference fringes is b, the corresponding air layer thickness is $\lambda/2$. When the thickness interval is a and the corresponding air layer thickness is h, it is obtained by the similar triangle relationship that

$$\frac{\lambda/2}{h} = \frac{b}{a}$$

So,

$$h = \frac{a}{b}\frac{\lambda}{2}$$

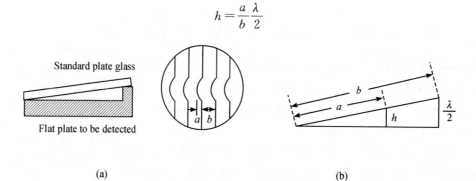

(a)   (b)

Figure 12.5.9   Figure of Example 12.5.3

12.5.3 Newton's ring

The "Newton ring" phenomenon was first discovered by Isaac Newton in 1675. The experimental device is shown in Figure 12.5.10. The surface of a plano-convex lens with a large curvature radius R is in contact with a flat glass, and a layer of plano-concave spherical air or other transparent medium film is formed between them. If monochromatic parallel light is incident on the film, a series of concentric ring patterns with alternating bright and dark fringes can be observed on the upper or lower surface of the spherical film. This kind of equal thickness interference fringe is called Newton's ring.

Suppose that the refractive index n_1 of the upper and lower glass is the same, and the refractive index of the film is n, the bright and dark interference condition for the Newton ring is

$$\delta = 2ne + \frac{\lambda}{2} = \begin{cases} k\lambda & (k=1, 2, 3, \cdots, \text{bright fringes}) \\ (2k+1)\frac{\lambda}{2} & (k=0, 1, 2, \cdots, \text{dark fringes}) \end{cases} \quad (12.5.9)$$

If the film thickness corresponding to a ring of radius r is e, it can be seen from the geometric relationship in Figure 12.5.10(a) that

$$(R-e)^2 + r^2 = R^2$$

Omitting the higher order small quantity e^2, it can be obtained that

$$e = \frac{r^2}{2R} \quad (12.5.10)$$

The optical path difference of the reflected wave is

$$\delta = 2ne + \frac{\lambda}{2} = \frac{nr^2}{R} + \frac{\lambda}{2} \quad (12.5.11)$$

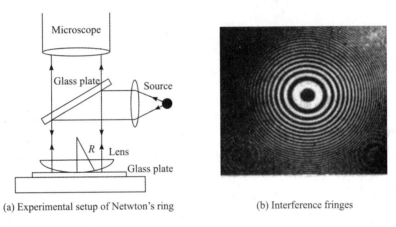

(a) Experimental setup of Netwton's ring (b) Interference fringes

Figure 12.5.10 Newton's ring

According to Equation (12.4.9), the radius of the bright and dark ring can be obtained

$$r = \begin{cases} \sqrt{\dfrac{\left(k - \dfrac{1}{2}\right)R\lambda}{n}} & (k=1, 2, 3, \cdots, \text{bright fringes}) \\ \sqrt{\dfrac{kR\lambda}{n}} & (k=0, 1, 2, \cdots, \text{dark fringes}) \end{cases} \quad (12.5.12)$$

The above expression shows that the center of Newton's ring is a zero order dark ring. The farther away from the center, the larger the optical path difference and the higher the order.

Because the thickness of the film is nonlinear, the radius of the bright or dark ring is proportional to the square root of the order k, so the fringe spacing is different. According to Equation (12.5.12), the difference between the radii of two adjacent dark rings is

$$\Delta r = r_{k+1} - r_k = (\sqrt{k+1} - \sqrt{k})\sqrt{\frac{R\lambda}{n}}$$

It can be seen that the higher the order of Newton's ring, the smaller the fringe spacing, which also shows that the distribution of the interference pattern becomes denser with the increase of the order. Therefore, Newton's ring is a series of concentric rings with alternating light and dark and uneven spacing.

In experiments, Newton's ring experiment is used to measure the curvature radius R of a lens. Because the center of the Newton's ring actually observed is not a dark point, but a dark spot of a certain size (due to influence factors such as the extrusion of actual instrument components). The appearance of the dark spot causes the fact that the radius of the interference ring is not easy to directly and accurately measure, nor is the order of the interference ring. In practical application, the diameter d_k of the k order dark ring and the diameter d_{k+m} of the $k+m$ order dark ring outward from it are often taken, and then by Equation (12.5.12)

$$r_{k+m}^2 - r_k^2 = \frac{(k+m)R\lambda - kR\lambda}{n}$$

$$R = \frac{r_{k+m}^2 - r_k^2}{m\lambda} n$$

The curvature radius of a plano-convex lens is

$$R = \frac{d_{k+m}^2 - d_k^2}{4m\lambda} n$$

Example 12.5.4 Figure 12.5.11 shows the experimental apparatus for measuring the refractive index of the oil film. Place a drop of oil on a flat glass sheet and slowly unfold the oil droplets into a spherical film. Under the vertical incidence of monochromatic light with a wavelength of 600 nm, the interference fringes formed by the reflected light of the oil film can be observed. We know the refractive index of the oil film $n_1 = 1.20$, and the refractive index of the glass $n_2 = 1.50$. When the highest point of the oil film center is $h = 875$ nm away from the upper surface of the glass sheet, what are the number of bright fringes and the thickness of the film at each bright fringe? Which state (bright or dark) is the center point in? How do the fringes change if the oil film unfolds?

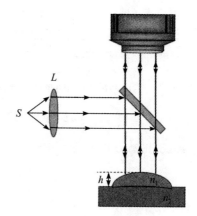

Figure 12.5.11　Figure of Example 12.5.4

Solution　The principle of this measuring device is similar to that of Newton's ring. The difference is when the light in the device reflects both on the upper and lower surfaces of the oil film, there is always a half-wave loss. Therefore, without considering the additional optical path difference, the condition for producing bright fringes is

$$\delta = 2hn = k\lambda$$

So,

$$h_k = \frac{k}{2n}\lambda$$

When $k=0$, $h_0=0$; When $k=1$, $h_1=250$ nm; When $k=2$, $h_2=500$ nm; When $k=3$, $h_3=750$ nm; When $k=4$, $h_4=1000$ nm.

Because the interference condition is the same where the oil film thickness is the same, the interference fringes observed from the reflected light are concentric rings with alternating light and dark. When $h=875$ nm, four bright fringes ($k=0, 1, 2, 3$) are observed. At the outer edge of the oil film $h=0$ is the zero order bright fringe center. When $h=875$ nm, the oil film center is between the bright fringe and the dark fringe in brightness.

Example 12.5.5 As shown in Figure 12.5.12, there is a small gap e_0 between the plano-convex lens of Newton's ring and the flat glass. The curvature radius of the plano-convex lens is known to be R. If the monochromatic light of wavelength λ is incident vertically, find the radius of each dark ring of Newton's ring formed by the reflected light.

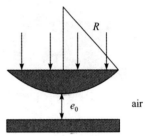

Figure 12.5.12   Figure of Example 12.5.5

Solution Let the radius of certain dark ring be r. If $e_0=0$, according to the geometry relationship,

$$e \approx \frac{r^2}{2R} \quad (1)$$

Consider e_0 and half-wave loss, and the optical path difference is

$$\delta = 2e + 2e_0 + \frac{\lambda}{2}$$

According to the interference weakening condition

$$2e + 2e_0 + \frac{\lambda}{2} = (2k+1)\frac{\lambda}{2} \quad (k=1, 2, 3, \cdots) \quad (2)$$

Combine the two equations, and

$$r = \sqrt{R(k\lambda - 2e_0)} \quad (k \text{ is an integer and } k \geqslant 2e_0/\lambda)$$

Obviously, the order of the central bright or dark fringe is determined by e_0. Only if $e_0=0$, will the center be a zero order dark fringe.

 12.5.4   The Michelson interferometer

The Michelson interferometer is a kind of precision instrument that generates double beam interference by the amplitude splitting method. It is designed by the American German physicists Michelson and Morey for the study of the speed of light. This interferometer has provided the experimental basis for the establishment of special relativity in the history of modern physics, and it can accurately measure length and small changes in length, as well as the refractive index of transparent materials. The construction photo of the Michelson interferometer is shown in Figure 12.5.13(a), and the photo of the Michelson interference fringe obtained with the laser light source is shown in Figure 12.5.13(b). In modern physics and modern metrology technology, a variety of special interferometers have been developed by using the principle of this instrument. Here, only the basic principle and simple application of the Michelson interferometer will be introduced.

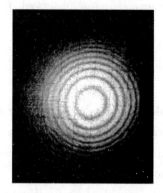

(a) Structure of the Michelson interferometer   (b) Interference fringes of the Michelson interferometer

Figure 12.5.13 The Michelson interferometer

The principle of the Michelson interferometer is shown in Figure 12.5.14. M_1 and M_2 are two finely polished flat mirrors mounted on two perpendicular arms, of which M_2 is fixed and M_1 is helically controlled which allows for small movements on the guide rail. G_1 and G_2 are two parallel glass sheets of the same material with uniform and equal thickness. The inclined angles between G_1 and G_2 and the arms are both 45°. A transparent thin silver layer is plated on a surface of G_1, which makes one half of light shining on the thin silver layer of G_1 reflected and the other half of light transmitted. G_1 is also called a splitter.

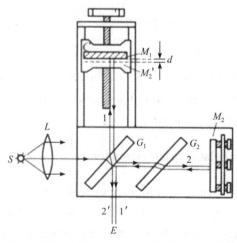

Figure 12.5.14 Schematic diagram of the Michelson interferometer

After the light emitted by the light source S is expanded by the lens L, it is directed towards the splitter plate. Part of the refracted light is reflected by the thin silver layer and then goes through G_1 again before going into M_1, denoted as light 1. Light 1 reflected by M_1 passes through G_1 for the third time, then propagates to position E and then reaches the observation screen. The other part of the beam, after splitting, goes through the thin silver layer, goes through G_2, is reflected by M_2, and then passes through G_2 again, marked as light 2. After going through G_1, light 2 is reflected to position E, and also collected to the observation screen. Obviously, Light 1 and 2 are two coherent light waves, and the interference fringe shown in Figure 12.5.13(b) can be seen at E. The function of the device G_2 is to make light 1 and 2 pass through the glass plate three times respectively, so as to avoid the large optical path difference caused by the different path experienced by the light. Therefore, G_2 is also called the compensation plate.

Assuming that the virtual image of M_2 formed by the silver layer is M_2', the light reflected from M_2 can be regarded as emanating from the virtual image, thus forming an equivalent "air film" between M_1 and M_2'. Interference from the two surfaces of M_1 and M_2', denoted as reflected light 1 and 2, can be treated as film interference. If M_1 and M_2 are not strictly perpendicular to each other, then the "air film" between M_1 and M_2 is a wedge, and the resulting interference fringes will be approximately parallel interference fringes with equal spacing. If M_1 and M_2 are strictly perpendicular to each other, and the "air film" between M_1 and M_2 is a film of air with uniform thickness, the interference fringes will be equally inclined interference fring rings.

According to the theory of thin film interference, when M_1 is adjusted to shift half a wavelength forward or backward (corresponding to the change in the thickness of the air film $\lambda/2$), one interference fringe can be observed to emerge from the center or disappear. Therefore, if the number of fringes that appear or disappear in the field of view is ΔN, the distance Δd traveled by M_1 is

$$\Delta d = \Delta N \frac{\lambda}{2} \tag{12.5.13}$$

The relationship between the number of moving fringes, the wavelength of monochromatic light and the tiny distance of movement can be established by Equation (12.5.13), which can be used to calculate the tiny change of length. The measurement accuracy can reach $\lambda/2 - \lambda/200$.

Michelson's interferometer was the first to measure the length of the international standard meter stick by terms of the wavelength of light, so that a permanent standard was established. In addition, Michelson used the interferometer to study the fine structure of spectra, advancing the development of atomic physics and metrology science. For this, Michelson was awarded the 1907 Nobel Prize in Physics. Later, people used the Michelson interferometer as a prototype to develop various forms of interferometers to determine the refractive index and impurity concentration of substances, as well as to check the quality of optical components. In 2015, the gravitational wave signal emitted by the merger of two black holes was measured by the Michelson interferometer based gravitational wave detector, whose distance was 1.3 billion light years from Earth, as shown in Figure 12.5.15.

Figure 12.5.15 Laser Interferometer Gravitational Wave Observatory, denoted as LIGO

Example 12.5.6 A 100 mm glass tube placed in one arm of the Michelson interferometer is filled with air at one atmosphere and illuminated with light with a wavelength of 585 nm. If the air in the glass tube is gradually pumped into a vacuum, it is found that 100 interference fringes have moved. Find the refractive index of the air.

Solution A glass tube is placed in one arm of the Michelson interferometer. Select the optical path length through the air in the glass tube as the reference object. Let the length of the glass tube be l and the refractive index of the air in the tube be n. Since the light beam passes twice in the glass tube, the optical path length of the light through the glass tube with air is $2nl$.

When the glass tube becomes a vacuum due to the air being pumped away, the optical path length in this condition will become $2l$. The optical path length difference in the tube before and after pumping is

$$2nl - 2l = 2(n-1)l$$

The optical path length of the other arm of the Michelson interferometer does not change in the experiment, so the change in the optical path difference between the two arms is only caused by the optical path length change in the glass tube. Every time the optical path length changes $\lambda/2$, a fringe will move through. The relation between the number of moving fringes and the change of the optical path length is

$$2(n-1)l = \Delta N \lambda$$

So, the refractive index of the air is

$$n = 1 + \frac{\Delta N \lambda}{2l}$$

By substituting the parameters of $\Delta N = 100$, $l = 100$ mm and $\lambda = 585$ nm, we can get

$$n = 1.000,292$$

Chapter 12 Wave Optics

12.6 Diffraction of light

12.6.1 Diffraction of light waves

When a wave meets an obstacle in propagation, it can go around the edge of the obstacle and reach the place that cannot be reached by linear propagation. This phenomenon of the wave deviating from the linear propagation is called the diffraction phenomenon of the wave. Diffraction, like interference, is also one of the main characteristics of waves. Light is a kind of electromagnetic wave, to which diffraction also happens. The light emitted by the light source S illuminates the adjustable slit K. When the slit width is relatively large, a uniform light streak will be displayed on the receiving screen E, as shown in Figure 12.6.1(a). If the width of the slit is gradually reduced, the light streak on the receiving screen is also reduced, which reflects the linear propagation of light. When the width of the slit becomes small and comparable to the wavelength of incident light, the light streak on the receiving screen no longer shrinks but becomes larger, which indicates that the light wave has been "curved" to the geometric shadow area of the slit. The brightness of the light streak also changes from the original uniform distribution to a series of bright and dark fringes, and the edge of each fringe loses its obvious boundary and becomes blurred, as shown in Figure 12.6.1(b). The light bypasses the edge of the obstacle and propagates into the geometric shadow, and the inhomogeneous distribution of light intensity appears on the receiving screen, a phenomenon called light diffraction.

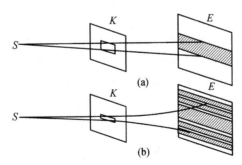

Figure 12.6.1 Diffraction of light waves

12.6.2 Classification of diffraction of light waves

To observe and measure the diffraction phenomenon of light in the laboratory, a light source, diffraction screen and observation screen are usually required. As shown in Figure

12.6.2, S is a monochromatic light source, K is a diffraction screen, which can be a slit or a small hole, etc., and E is an observation screen.

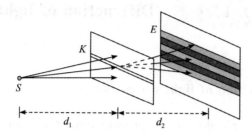

Figure 12.6.2 Experimental setup of light diffraction

According to the distance d_1 (between the light source and the diffraction screen) and the distance d_2 (between the diffraction screen and the observation screen), the diffraction phenomenon is divided into two categories: ① When at least one of d_1 and d_2 is at finite distance, such diffraction is called the Fresnel diffraction whose analysis method is more complex; ② When d_1 and d_2 both approach infinity, this type of diffraction is called the Fraunhofer diffraction. In reality, it is difficult to meet the requirement that d_1 and d_2 be both infinite. In the experiment, monochromatic parallel light is used to illuminate the diffraction screen, which is equivalent to d_1 being infinite. At the same time, a convergent convex lens is placed behind the diffraction screen, and the observation screen is placed on the focal plane of the lens, which is equivalent to d_2 being close to infinity. The Fraunhofer diffraction can be regarded as the diffraction of parallel light after going through the diffraction screen, and the diffraction fringe is observed at infinite distance, which makes the theoretical analysis simple.

12.6.3 The Huygens-Fresnel principle

When a wave propagates to any position in a medium, every point on the wavefront can be regarded as a new wavelet source, and the wavelet envelope at any time determines the new wavefront. Huygens' principle can explain that the direction of light propagation changes when it passes through a diffraction screen, but it cannot explain the location of diffraction fringes and the distribution of light intensity in detail.

Fresnel developed Huygens' principle by supplementing it with the quantitative expression describing the phase and amplitude of the wavelet and proposing the principle of wavelet coherent superposition, which is called the Huygens-Fresnel principle. It can be briefly described as follows: Each point on a wavefront can be regarded as a wavelet source that generates high-order wavelets. The wavelet emitted from each point on the same wavefront is a coherent wave. When these wavelets meet at a certain point in space, coherent superposition is generated.

According to the Huygens-Fresnel principle, if the wavefront S of the light wave at a certain time is shown in Figure 12.6.3, the optical vibration at any point P in space can be the combined vibration of the wavelets emitted by each surface element dS on the wavefront S after superposition at that point. If the optical vibration caused by the wavelet emitted by the surface element dS at point P is denoted as $d\boldsymbol{E}$, $d\boldsymbol{E}$ is proportional to dS and inversely proportional to the distance from point P to dS, which is also related to the inclination θ. The light vibration at point P is

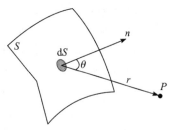

Figure 12.6.3 The Huygens-Fresnel principle

$$\boldsymbol{E} = \int_s d\boldsymbol{E}$$

$d\boldsymbol{E}$ caused by each surface element is different, and more importantly, its phase is different. Hence, in principle, the Huygens-Fresnel principle can be applied to solve the general diffraction problem, but the integral calculation is often very complicated. In the discussion of the Fraunhofer single slit diffraction, we will use the half-wave band method for clever treatment.

12.7 The Fraunhofer single slit diffraction

In 1821, J. von Fraunhofer studied single slit diffraction, as shown in Figure 12.7.1. A narrow and elongated slit is on the diffraction screen K. A monochromatic point light source S is placed at the focal point of the lens L_1, and the emitted light rays passing through the lens form a parallel beam of light that illuminates the single slit diffraction screen K. A converging lens L_2 is placed behind the diffraction screen, and the diffracted light rays passing through the slit converge on the observation screen E at the focal plane. A series of diffraction fringes parallel to the slit can be observed on the observation screen.

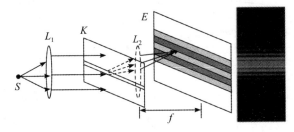

Figure 12.7.1 Experimental setup of single slit diffraction

According to the Huygens-Fresnel principle, the light vibration at any point on the receiving screen behind the single slit is the coherent superposition of the wavelets emitted by the wavelet sources located on the wavefront at the position of the single slit. As shown in Figure 12.7.2(a), when monochromatic parallel light vertically shines on the narrow slit AB with a width of a, each point on the wavefront AB at the slit can be considered as a wavelet source emitting diffracted light in various directions to the right of the slit. The angle between the direction of propagation of the diffracted light rays and the normal direction of the slit plane is called the **diffraction angle** denoted as φ. For any diffraction angle φ, the diffracted light rays emitted by the wavelets in the direction φ are a bundle of parallel rays, i.e., light 2 in Figure 12.7.2(a). Under the converging effect of the lens, these rays would converge at point P on the focal plane. With different diffraction angles, the position of point P varies. It is known that lens L does not introduce an additional optical path difference, and the incident parallel light on the AB plane has a constant phase. Therefore, the optical path difference between points A and B is maximum and can be expressed as Equation (12.7.1)

$$BC = a\sin\varphi \qquad (12.7.1)$$

Based on the Huygens-Fresnel principle, Fresnel proposed a method of dividing the wavefront into many half-wave zones with equal areas. In Figure 12.7.2(b), a series of planes parallel to plane AC are constructed, and the distance between adjacent planes is equal to half the wavelength ($\lambda/2$) of the incident light. Assuming that these planes divide the wavefront AB at the slit into several regions with equal areas (AA_1, A_1A_2, A_2B), these regions are called half-wave zones. Since the areas of half-wave zones are equal, the light amplitudes caused by them at point P are approximately equal. For any two corresponding points on adjacent half-wave zones (such as point G on the AA_1 zone and point G' on the A_2B zone), the optical path difference is always $\lambda/2$ and phase difference between the light rays emitted are always π, respectively. After passing through the lens and converging at point P, the phase difference between the rays reaching point P is still π because the lens will not introduce an extra optical path difference. As a result, any light ray emitted by adjacent half-wave zones will completely cancel out each other at point P.

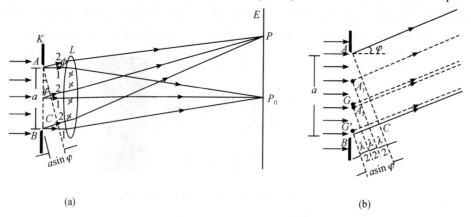

Figure 12.7.2 Analysis of single slit diffraction fringes

Thus, it can be seen that the single slit can be divided into an even number of half-wavelength zones in the corresponding diffraction angle direction, when the length of BC is equal to an even multiple of a half wavelength. The rays in these half-wavelength zones will cancel each other out in pairs, resulting in the dark fringe at point P. In contrast, when the length of BC is equal to an odd multiple of a half wavelength, the single slit can be divided into an odd number of half-wavelength zones. After the adjacent half-wavelength zones cancel each other out, one half-wavelength zone of diffracted light remains. In this case, a bright fringe will be seen at point P.

Based on above analyses, when monochromatic parallel light is incident vertically on a single slit, the relationship between the diffraction fringes and the diffraction angle is

$$a\sin\varphi = \begin{cases} 0 & \text{(central bright fringe)} \\ \pm k\lambda & (k=1, 2, 3, \cdots, \text{dark fringe}) \\ \pm(2k+1)\dfrac{\lambda}{2} & (k=1, 2, 3, \cdots, \text{bright fringe}) \end{cases} \quad (12.7.2)$$

where, k is the order of the fringe, the positive and negative signs indicate that the diffraction fringes are symmetrically distributed on both sides of the central bright fringe and φ is the diffraction angle corresponding to the center of its fringe.

For any diffraction angle φ, if the length of AB cannot be divided into an integer number of half-wavelength zones, in other words, if the length of BC is not equal to an integer multiple of half a wavelength ($\lambda/2$), the brightness of diffracted light corresponding to these diffraction angles and focused at point P through a converging lens will be between the brightest and darkest levels. Therefore, the intensity distribution in the single slit diffraction pattern is non-uniform. As shown in Figure 12.7.3, the central bright fringe is the brightest and the widest, which corresponds to the distance between the centers of the two first order dark fringes. The width is between $a\sin\varphi_0 = -\lambda$ and $a\sin\varphi_0 = \lambda$. When φ_0 is small, there is $\varphi_0 \approx \sin\varphi_0 = \pm\dfrac{\lambda}{a}$. So the angular width of the central bright fringe (the angle of the fringe to the center of the lens) equals $2\varphi_0 \approx 2\dfrac{\lambda}{a}$, and sometimes it can also be described in terms of half-angle width φ_0

$$\varphi_0 \approx \dfrac{\lambda}{a} \quad (12.7.3)$$

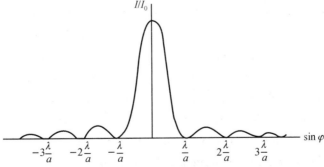

Figure 12.7.3 Intensity distribution of single-slit diffraction light

The angular width of the other bright fringes is obviously equal to half the angular width of the central bright fringe, and it is approximately given by

$$\Delta\varphi = (k+1)\frac{\lambda}{a} - k\frac{\lambda}{a} = \frac{\lambda}{a} \tag{12.7.4}$$

Assuming the focal length of the converging lens L_2 is f, the width of each order of bright fringes observed on the screen (called the line width) with small diffraction angles can be calculated by

$$\begin{cases} \Delta x_0 = 2f\tan\varphi_0 \approx 2f\sin\varphi_0 = 2f\dfrac{\lambda}{a} & \text{(central bright fringe)} \\ \Delta x = f\tan\Delta\varphi \approx f\sin\Delta\varphi = f\dfrac{\lambda}{a} & \text{(other bright fringes)} \end{cases} \tag{12.7.5}$$

It can be seen that the line width of the other bright fringes is half the line width of the central bright fringe. As the order increases, the brightness of the other bright fringes rapidly decreases. The reason is that, as the diffraction angle φ increases, the number of half-wavelength zones which the AB wavefront is divided into also increases. It makes the area of each half-wavelength zone decrease accordingly and leads to the decrease of transmitted flux of light, resulting in weaker brightness of the superimposed bright fringes formed on the screen by the remaining half-wavelength zone.

When the slit width a is constant, for the same order of diffraction fringe, the larger the wavelength λ is, the larger the diffraction angle φ is and the farther the position from the center of the central bright fringe is. Therefore, when white light is used as the light source, besides the white central bright fringe in the middle, a series of diffraction fringes ranging from violet to red will appear on both sides of the central bright fringe. It is called the **diffraction spectrum.**

From Equations (12.7.2) and (12.7.3), if the incident light is monochromatic light with a wavelength of λ, the diffraction angles φ of each order of diffraction fringes become larger with the decrease of the slit width (a cannot be smaller than λ). In other words, the diffraction phenomenon becomes more obvious. When the slit width a increases, the diffraction angles φ of each order of diffraction fringes become smaller. The diffraction fringes will be densely arranged on both sides of the central bright fringe and gradually become indistinguishable, resulting in less pronounced diffraction. When $a \gg \lambda$, the diffraction fringes of each order will merge into the central bright fringe. The image of the slit occurs after the lens, and the diffraction phenomenon disappears. At this point, light can be considered to propagate in a straight line. Therefore, the phenomenon of straight-line propagation of light occurs when the wavelength of light is much smaller than the linear dimension of the aperture or slit (or obstacle), which corresponds to the situation that the diffraction phenomenon is not significant. Only when the slit is narrow enough so that its width is comparable to the wavelength, will the diffraction phenomenon become significant.

Example 12.7.1 Monochromatic light with a wavelength of $\lambda = 500$ nm is vertically incident on a single slit with a width of $a = 0.2$ mm. The converging lens behind the slit has a focal length of $f = 1.0$ m. The observation screen is placed at the focal plane of the lens. On the observation screen, the focal point is chosen as the coordinate origin and the X axis is established in the direction perpendicular to the slit. Find

(1) the angular width and line width of the central bright fringe;

(2) the position of the first order bright fringe, and the number of half-wavelength zones;

(3) the line width of the other bright fringes.

Solution (1) The central bright fringe is the region between the upper and lower first order dark fringes. According to the single slit diffraction formula, the diffraction angle φ corresponding to the first order dark fringe should satisfy the relationship

$$\sin\varphi_0 = \frac{\lambda}{a} = \frac{500 \times 10^{-9}}{0.2 \times 10^{-3}} = 2.5 \times 10^{-3}$$

Since the value of $\sin\varphi_0$ is very small, the angular width of the central bright fringe is

$$\Delta\varphi = 2\varphi_0 \approx 2\sin\varphi_0 = 5 \times 10^{-3} \text{ rad}$$

According to the geometric relationship, the position of the first order dark fringe x_1 is

$$x_1 = f\tan\varphi_0 \approx f\sin\varphi_0 = \pm f\frac{\lambda}{a} = \pm 1.0 \times 2.5 \times 10^{-3} \text{ m} = \pm 2.5 \text{ mm}$$

Therefore, the line width of the central bright fringe is

$$\Delta x_0 = 2x_1 = 2 \times 2.5 \times 10^{-3} \text{ m} = 5.0 \text{ mm}$$

(2) The diffraction angle φ_1 corresponding to the first order bright fringe satisfies

$$\sin\varphi_1 = (2k+1)\frac{\lambda}{2a} = \frac{3 \times 500 \times 10^{-9}}{2 \times 0.2 \times 10^{-3}} = 3.75 \times 10^{-3}$$

Therefore, the coordinate of the first order bright fringe x_1 is

$$x_1 = f\tan\varphi_1 \approx f\sin\varphi_1 = \pm 1.0 \times 3.75 \times 10^{-3} \text{ m} = \pm 3.75 \text{ mm}$$

In the corresponding direction, the single slit can be divided into $(2k+1)$ half-wavelength zones, and $k = 1$. The number of half-wavelength zones is $2k + 1 = 2 + 1 = 3$.

(3) The line width Δx_k of the kth order bright fringe equals the distance between the centers of the adjacent kth and $(k+1)$th order dark fringes, which can be expressed as

$$\Delta x_k = x_{k+1} - x_k = f\sin\varphi_{k+1} - f\sin\varphi_k = f\frac{\lambda}{a} = 1.0 \times \frac{500 \times 10^{-9}}{0.2 \times 10^{-3}} \text{ m} = 2.5 \text{ mm}$$

It can be seen that the line width of the other order bright fringes is half the line width of the central bright fringe.

12.8 Diffraction grating

From the discussion in the previous section, it is possible to determine the wavelength of monochromatic light by the diffraction fringes produced by a single slit. In order to achieve an accurate value, the diffraction fringes need to be clearly separated, and the fringe should be both fine and bright. However, for single slit diffraction, it is difficult to satisfy both the requirements simultaneously. If the fringes are well separated, the width of the single slit a needs to be small, which reduces the amount of light passing through the slit, resulting in faint fringes difficult to observe. On the other hand, if the slit width a increases, the observed fringes will become brighter; however the fringe spacing becomes smaller, making it harder to distinguish them. Therefore, in practice, when measuring the wavelength of light, a diffraction grating is often used instead of a single slit, as it can meet the aforementioned measurement requirements.

12.8.1 Grating diffraction

An optical component composed of a large number of parallel narrow slits with equal widths and equal spacing is called **a diffraction grating**. The one used for transmitting light diffraction is called a transmission grating, while the one used for reflecting light diffraction is called a reflection grating, as shown in Figure 12.8.1. The commonly used transmission grating is created by etching many evenly spaced narrow parallel grooves on a glass plate. At each groove, the incident light scatters in various directions and is not easily transmitted, while the smooth parts between the grooves allow light to pass through, acting as narrow slits. If the width of the slit is a and the width of the groove is b, the distance $d=(a+b)$ is called the grating constant. Modern diffraction gratings can have thousands of grooves per centimeter, so the common grating constant is $10^{-5} \sim 10^{-6}$ m. The total number of transmitting slits in the grating is represented by N, and the grating constant and the total number of slits are two important characteristic parameters for the grating.

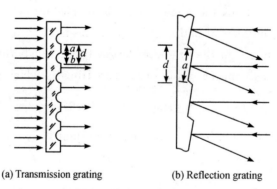

(a) Transmission grating (b) Reflection grating

Figure 12.8.1 Gratings

As shown in Figure 12.8.2, when a beam of monochromatic parallel light is vertically incident on the grating, the light rays pass through the lens L and converge on the observation screen E in the focal plane. The light passing through each slit of the grating will diffract, and the diffraction patterns of each slit completely overlap after passing through the lens. Besides, the light passing through different slits of the grating also undergoes interference. Therefore, the diffraction of a grating is the combined effect of diffraction from each individual slit and interference between different slits.

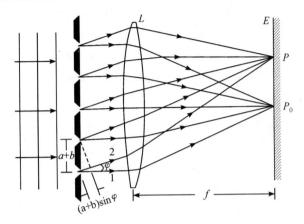

Figure 12.8.2 Grating diffraction

12.8.2 Law of grating diffraction

When monochromatic parallel light is vertically incident on a grating, each slit emits diffracted light in all directions. The light rays emanating from different slits with the same diffraction angle φ form a parallel beam that converges to a point on the observation screen, such as point P in Figure 12.8.2. These waves interfere with each other, resulting in multiple-beam interference. From the figure, any two adjacent slits with corresponding positions emit the diffracted light in the φ direction. The path difference at the final point P is given by

$$\delta = (a+b)\sin\varphi \tag{12.8.1}$$

1. Grating equation

If the path difference between diffracted light beams emitted by adjacent slits is an integer multiple of the incident wavelength λ, then these two diffracted light rays satisfy the condition of constructive interference at point P. At the same time, other diffracted light rays from any two slits in the same direction towards point P will also satisfy the constructive interference. Therefore, from the direction of φ, the diffracted light rays from all slits converge and reinforce each other, leading to a bright fringe at point P. In this case, the amplitude of light at point P is N times that of the diffracted light from a single slit, and the total intensity is N^2 times that from a single slit. Therefore, the brightness of the bright fringes formed by the multiple-beam interference of the grating is much greater than that produced by a single slit. In other words, the number of slits of the grating (N) determines the brightness of the bright fringes. The diffraction fringes that satisfy the above conditions are located at

$$(a+b)\sin\varphi = \pm k\lambda \quad (k=0, 1, 2, \cdots) \tag{12.8.2}$$

The equation is known as the **grating equation**, where k represents the order of the bright fringe. These bright fringes are very narrow, but very bright. They are commonly called the principal maximum of grating diffraction. $k = 0$ represents the zero order principal maximum, and $k = 1$ represents the first order principal maximum, and so on. The positive and negative signs indicate that the other principal maximums are symmetrically distributed on both sides of the zero order principal maximum.

2. Dark fringe condition

If the total amplitude of the light vibration at point P is zero, a dark fringe will appear. Assuming the phase difference between adjacent slits is $\Delta\varphi$, and the amplitude vectors of the vibrating components are E_1, E_2, E_3, ···, E_N. In order for the N vectors to completely cancel out each other after superimposing, they must form a closed polygon as shown in Figure 12.8.3. In this case, the relationship between the phase difference and the optical path difference is given by

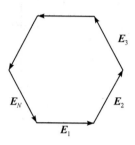

Figure 12.8.3 N-light vibration

$$\Delta\varphi = \frac{2\pi}{\lambda}\delta$$

When the N vectors form a closed polygon, they satisfy
$$N\Delta\varphi = \pm k'2\pi \quad (k'=1, 2, 3, \cdots)$$
Expressed in terms of optical path difference
$$N\delta = \pm k'\lambda$$

So
$$(a+b)\sin\varphi = \pm k'\frac{\lambda}{N} \quad (k'=1, 2, 3, \cdots) \quad (12.8.3)$$

It should be noted that the above equation includes $\Delta\varphi = \pm 2k\pi$ and $\delta = \pm k\lambda$. However, the bright fringe will be generated under the two conditions, so the condition of $k' = kN$ should be excluded. Therefore, the condition for dark fringes in grating diffraction is

$$(a+b)\sin\varphi = \pm k'\frac{\lambda}{N} \quad (k'=1, 2, 3, \cdots \text{ and } k' \neq kN) \quad (12.8.4)$$

In other words, k' does not contain the values of N, $2N$, ···, and so on, since they satisfy the condition for the bright fringes specified by Equation (12.8.2). From Equation (12.8.4), it can be seen that there are $N-1$ dark fringes between two adjacent bright fringes.

3. Sub-bright fringe

Between two adjacent principal bright fringes, there are $N-1$ dark fringes. In other words, there must be one bright fringe between two adjacent dark fringes, and there are $N-2$ bright fringes between two adjacent bright fringes. The vibration vectors in these places are not completely cancelled out, but only partially cancelled out. However, calculations show that the light intensity of these bright fringes is only about 4% that of the principle bright fringes, so they are called sub-bright fringes or sub-maximum.

In summary, due to the large number of slits (N) in the grating, there are many dark and sub-bright fringes between two adjacent principal bright fringes. In fact, there is a dark area between two adjacent principal bright fringes. The bright fringes are well separated and very fine. The light is concentrated in a small area, making bright fringes become brighter. As a result, the diffraction pattern of the grating is a series of fine bright fringes that are widely separated on a nearly dark background, as shown in Figure 12.8.4.

Figure 12.8.4　Patterns of grating diffraction

In the above discussion of multi-beam interference, the influence of slit (single slit) diffraction on the intensity distribution of fringes was not considered. In fact, because the intensity of diffracted light in different φ directions for single slit diffraction is different, the bright fringes at different positions of the receiving screen after grating diffraction are strengthened interference of diffracted light with different light intensities. That is to say, each order of bright fringe in multi-beam interference is modulated by single slit diffraction. The direction with a large single slit diffraction light intensity is also the direction with a large light intensity of the bright fringe, and the direction with a small single slit diffraction light intensity is also the direction with a small light intensity of the bright fringe. Figure 12.8.5 is a schematic diagram of the light intensity distribution of grating diffraction. The envelope line of the grating diffraction intensity at each order is similar to the intensity curve of single slit diffraction. The total light intensity distribution of grating diffraction is determined by both multi-slit interference and single slit diffraction.

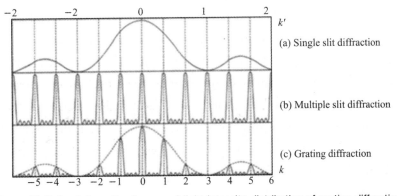

Figure 12.8.5　Schematic diagram of light intensity distribution of grating diffraction

4. Missing order phenomenon

When discussing the grating formula $(a+b)\sin\varphi=\pm k\lambda$ previously, it was only from the perspective of multi-beam interference to explain the necessary condition for producing the bright fringe with the maximum superimposing light intensity. However, when this diffraction angle φ also satisfies the condition $a\sin\varphi=\pm k'\lambda$ for dark fringes in single slit diffraction, at this position will happen "interference enhancement" with zero light intensity. Therefore, from the perspective of the grating formula, at the position where should appear a kth bright fringe is actually a dark fringe, that is, the kth bright fringe does not appear. This phenomenon is called **missing order phenomenon** of grating. Combining the above two formulas, it can be known that the missing condition is

$$k=\frac{a+b}{a}k' \quad (k'=1, 2, 3, \cdots) \tag{12.8.5}$$

It can be seen from Equation (12.8.5) that the missing order is determined by the grating constant $a+b$ and slit width a. If the grating constant $a+b$ is an integer multiple of the slit width a, the missing phenomenon will occur. For example, when $(a+b)=3a$, the missing order of grating $k=3, 6, 9, \cdots$.

12.8.3 Grating spectrum

The above discussion is about the diffraction pattern after monochromatic light passes through a grating. If white light illuminates a grating, monochromatic light with various wavelengths will diffract. According to the grating equation, for a certain grating, the diffraction angle of each order bright fringe is related to the incident wavelength. The diffraction angles of short wavelengths are small, and those of long wavelengths are large. So, the diffraction fringes of purple light are the closest to the central bright fringe while the diffraction fringes of red light are the furthest from the central bright fringe. In this way, except for the central bright fringe that is still the white light, a mixture of various colors of light, other order bright fringes on the two sides of the central bright fringe will all form symmetrical color bands arranged from purple to red. This kind of spectrum line arranged according to the wavelength produced by grating diffraction is called **grating spectrum**. As shown in Figure 12.8.6, within the same order spectrum, purple light with a shorter wavelength (represented by V in the figure) is close to the central bright fringe for the small diffraction angles of short wavelength light. Red light with a longer wavelength (represented by R in the figure) is the furthest from the central bright fringe for the large diffraction angles of light with a long wavelength. Besides, the higher order spectrum will overlap with the prior order spectrum.

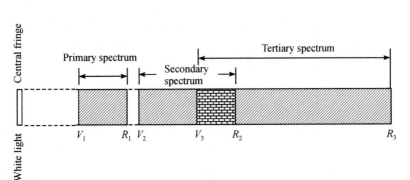

Figure 12.8.6 Schematic diagram of the grating diffraction spectrum

The grating diffraction spectrum is widely used in analysis, identification, standardization measurement, and so on. When the incident complex light only contains several discontinuous wavelength components, the grating spectrum is a discrete bright line corresponding to each wavelength, forming a linear spectrum. Since each element (or compound) has its own specific spectral lines, we can burn a certain material to be analyzed to emit light, obtain its spectral line diagram after it passes through the grating, and then compare it with known spectral lines of various elements to qualitatively analyze the elements or compounds contained in the material. By measuring the relative intensity of each spectral line, the amount of each element can be quantitatively analyzed. This method is called spectral analysis.

Example 12.8.1 Monochromatic parallel light with a wavelength $\lambda = 590$ nm is vertically incident on a grating with 500 scratches per millimeter. It is known that the width of the grating's light-transmitting slit a is 1×10^{-6} m. What is the highest order of the principle bright fringe that can be observed? How many principle bright fringes can be seen in total?

Solution According to the given grating parameters, it can be known that the grating constant of this grating is

$$a+b = \frac{1 \times 10^{-3}}{500} = 2 \times 10^{-6} \text{ m}$$

According to the grating formula Equation (12.8.2), when the diffraction angle $\varphi = \pi/2$, the maximum order of the principle bright fringe is k_m. So

$$k_m = \frac{(a+b)\sin\frac{\pi}{2}}{\lambda} = \frac{2 \times 10^{-6}}{590 \times 10^{-9}} \approx 3.4$$

Since the value of the order is an integer, take an integer for the calculated result and $k_m = 3$. The highest order of the principle bright fringe that can be observed is 3. Then according to the missing order condition of Equation (12.8.5), substitute data into calculation and

$$k = \frac{a+b}{a}k' = \frac{2 \times 10^{-6}}{1 \times 10^{-6}}k' = 2k' \quad (k'=1, 2, \cdots)$$

So the missing orders of the grating are 2, 4, 6, ⋯, and the orders of the principle bright fringe that can be seen are 0, 1, 3. A total of 5 bright fringes are symmetrically distributed on both sides of the central bright fringe.

Example 12.8.2 A diffraction grating has 200 light-transmitting slits per centimeter and the width of each light-transmitting slit is $a = 2 \times 10^{-3}$ cm. A convex lens with a focal length $f = 1$ m is placed behind the grating. If monochromatic parallel light with a wavelength $\lambda = 600$ nm is vertically incident on the grating, try to find

(1) the width of the central bright fringe of single slit diffraction with a slit width a;

(2) the number of principle bright fringes in this central bright fringe width range.

Solution (1) According to the formula of the width of the central bright fringe in single slit diffraction

$$\Delta x_0 = 2\frac{\lambda}{a}f = 2 \times \frac{600 \times 10^{-9}}{2 \times 10^{-5}} \times 1 = 0.06 \text{ m}$$

(2) Within the diffraction angle θ determined by the formula of the first order dark fringe in single slit diffraction, the maximum order of the principle bright fringe is k_m. There are

$$a\sin\theta = \lambda$$
$$(a+b)\sin\theta = k_m\lambda$$

Combine the two equations and

$$k_m = \frac{a+b}{a} = 2.5$$

Since k_m should be an integer, $k_m = 2$. The order of principle bright fringes $k = 0$, ± 1, ± 2, and there is a total of five principle bright fringes.

12.8.4 X-ray diffraction

In 1895, the German physicist W. K. Roentgen discovered a phenomenon. When high-speed electrons hit a metal plate, they would produce a ray with extremely strong penetrating power. It could make well-packaged photographic film sensitive and many substances fluoresce. This kind of ray was called X-ray. In 1901, Roentgen won the first Nobel Prize in Physics for discovering X-rays. Figure 12.8.7 shows a schematic diagram of an X-ray tube that can produce X-rays. In the figure, K is a hot cathode that emits electrons, and A is an anode. When a voltage of tens of thousands of volts is applied between the two poles, electrons emitted from the cathode are accelerated under the condition of a strong electric field. When high-speed electrons hit the anode, X-rays can be emitted from it.

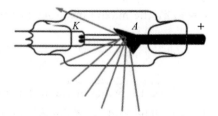

Figure 12.8.7 X-ray tube

X-rays are essentially electromagnetic waves like visible light, but their wavelengths are very short, which range between 0.01 and 10 nm. Since X-rays are electromagnetic waves, they should have the phenomena such as interference and diffraction. However, because their wavelengths are very short, the diffraction phenomenon cannot be observed with ordinary gratings, nor can be fabricated the gratings which are suitable for X-ray diffraction.

In 1912, the German physicist M. V. Laue proposed that a crystal was composed of a group of regularly arranged particles and the spacing between the particles was on the same order of magnitude as the wavelength of X-rays, so it might constitute a three dimensional spatial grating suitable for X-ray diffraction. Based on this, Laue conducted experiments and successfully obtained X-ray diffraction patterns, which confirmed that X-rays were electromagnetic waves and that atoms in crystals were evenly spaced. Laue's experimental setup is shown in Figure 12.8.8(a). In the figure, PP' is a lead plate with small holes, C is a crystal, and E is a photographic film. During the experiment, X-rays were projected onto thin crystal slices through small holes on the lead plate PP', and diffraction spots formed by diffraction were found on the photographic film. These spots are also called the Laue spots, as shown in Figure 12.8.8(b).

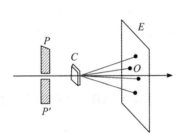
(a) Schematic diagram of the X-ray diffraction experiment

(b) X-ray diffraction pattern of a crystal

Figure 12.8.8 The Laue experiment

Laue obtained X-ray diffraction patterns through experiments, but the quantitative analysis was very complicated due to the spatial gratings. In 1931, the British physicists W. H. Bragg and his son W. E. Bragg proposed a new research method, which was to consider the X-ray diffraction pattern to be formed by coherent reflection of each lattice plane group of crystals by X-rays. The principle of and quantitative calculation by this method are relatively simple. Bragg and his son simplified the spatial lattice and imagined that the crystal was composed of a series of parallel crystal planes (i. e., atomic layers), as shown in Figure 12.8.9. Assume that the distance between crystal planes is d. When a monochromatic parallel X-ray with a wavelength λ is incident on the crystal plane at a grazing angle φ, part of it is scattered by the surface

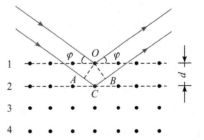
Figure 12.8.9 The Bragg reflection

crystal plane and the rest is scattered by the crystal planes inside the crystal. The intensity of the rays in the direction that conforms to the law of reflection is the greatest. From the figure, it can be seen that the optical path difference of reflected light rays between two adjacent crystal planes is

$$AC + CB = 2d\sin\varphi$$

Obviously, when φ satisfies the following condition

$$2d\sin\varphi = k\lambda \quad (k=1, 2, \cdots) \tag{12.8.6}$$

all layers of crystal planes will be mutually enhanced to form bright spots. This formula is the famous **Bragg formula**, also called the Bragg equation.

It can be seen from the formula that if the lattice constant d of a crystal is known, the wavelength of X-ray λ can be obtained by measuring φ. On the contrary, if the wavelength of X-ray λ is known, the lattice constant d can be calculated by measuring φ, which can be used to determine the crystal structure. The analysis of crystal structure by X-rays has become an important branch in applied physics, and this method has been widely used in chemistry, biology, mineralogy, and engineering technology. The famous double helix structure of deoxyribonucleic acid (DNA) was first proposed in 1953 based on the analysis of X-ray diffraction patterns of samples. Due to this finding, Wilkins, Watson, and Crick won the Nobel Prize in Physiology and Medicine in 1962.

12.9 The Fraunhofer diffraction with circular aperture

12.9.1 The Fraunhofer diffraction with circular aperture

For the Fraunhofer single slit experiment setup, if the single slit diffraction screen K in Figure 12.7.1 is replaced with a diffraction screen with a small circular aperture, as shown in Figure 12.9.1(a), the diffraction phenomenon can also be observed. When monochromatic parallel light is vertically incident on the small circular aperture K, a Fraunhofer circular aperture diffraction pattern can be observed on the screen E at the focal plane of the lens L_2, where the center is a bright circular spot and surrounded by a group of concentric circles with alternating light and dark, as Figure 12.9.1(b) shows. The central light spot surrounded by the first dark ring is called the **Airy spot**. The diameter of the Airy spot is d, and its radius corresponding to the field angle of the lens L_2 is called the Airy spot's half-angle width. Theoretical calculations can prove that the Airy spot accounts for about 84% of the total incident intensity of the light. The light intensity distribution of the diffraction pattern is shown in Figure 12.9.1(c). The half-angle width of the Airy spot is

$$\theta \approx \sin\theta = 0.610\frac{\lambda}{R} = 1.22\frac{\lambda}{D} \qquad (12.9.1)$$

where $D = 2R$ is the diameter of the circular aperture, and λ is the wavelength of the incident light. Obviously, the smaller D or the larger λ will lead to the more obvious diffraction phenomenon.

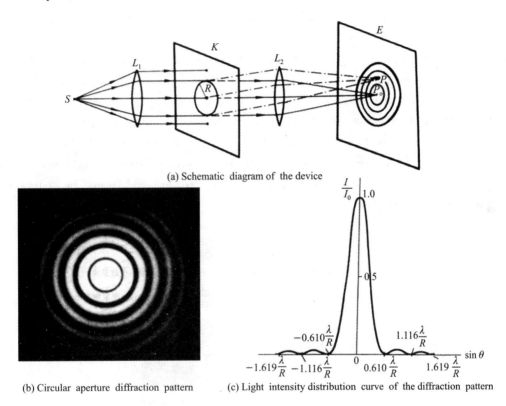

(a) Schematic diagram of the device

(b) Circular aperture diffraction pattern

(c) Light intensity distribution curve of the diffraction pattern

Figure 12.9.1　The Fraunhofer diffraction with circular aperture

12.9.2　Resolution of an optical instrument

From the perspective of geometric optics, when an object is imaged through a lens, each object point has a corresponding image point. As long as the focal length of the lens is properly selected, any tiny object can be seen in a clear image. However, from the view of wave optics, a component, such as lens, is equivalent to a small light-transmitting hole. Thus, the image we see on the screen is a diffraction pattern of a circular hole. Roughly speaking, what we see is an Airy spot with a certain size. If two object points are very close, the corresponding two Airy spots may partially overlap, which makes it difficult to distinguish them, and even they may be considered as one image point. That is to say, the diffraction phenomenon of light limits the resolution of optical instruments.

So which factors will affect the resolution of optical instruments? For simplicity, assuming that the objective lens of an optical instrument is composed of a single lens. If the two point light sources a and b are far enough away from the lens, their light shot into the lens can be regarded as parallel light. The two sets of diffraction patterns formed by them are shown in Figure 12.9.2. In Figure 12.9.2(a), image spots (the Airy spots) of light sources a and b are separated well, and there is no overlap or little overlap between them. We can distinguish the images of points a and b. Thus, we can judge that original object points are two points. If the two image spots overlap mostly, as shown in Figure 12.9.2(b), then these two light sources cannot be distinguished. To differentiate these two states, a critical position between their Airy spots produced by two point light sources can be defined. This position for two point light sources is just in the distinguishable state. The critical position was proposed by Rayleigh and accepted by other researchers, which is called the **Rayleigh criterion**. Its definition is: If the central brightest part (center of the Airy spot) of one point light source's diffraction pattern coincides with the first darkest part (edge of the Airy spot) of the other point light source's diffraction pattern, as shown in Figure 12.9.2(c), then, these two point light sources are said to just be distinguished by this optical instrument.

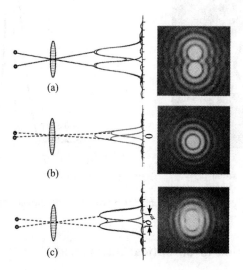

Figure 12.9.2 Resolution of optical instruments

When two point light sources are just distinguishable by an optical instrument, their field angle facing toward the lens is called **minimum resolution angle** of this instrument, represented by δ_φ. It is exactly equal to the half-angle width of the Airy spot, that is

$$\delta_\varphi = \theta \approx \sin\theta = 1.22 \frac{\lambda}{D} \qquad (12.9.2)$$

Usually, the reciprocal of the minimum resolution angle of an optical instrument is called its resolving capability, represented by R. Then, the resolving capability of a telescope is

$$R = \frac{1}{\delta_\varphi} = \frac{D}{1.22\lambda} \qquad (12.9.3)$$

The resolving capability of an optical instrument is proportional to its aperture and inversely proportional to the wavelength of incident light. Therefore, in astronomical observations, in order to distinguish several stars that are very close in distance, a telescope with a large aperture is required. For a microscope, in order to improve its resolution, short-wavelength purple light is mostly used. Modern physics experiments have confirmed that electrons also have wave properties, and their wavelengths are comparable with interatomic distances in solids (about 0.1 ~ 1 Å order of magnitude), so the resolution of an electron microscope is thousands of times higher than that of an ordinary optical microscope.

Example 12.9.1 Under normal brightness conditions, the diameter of the human eye pupil is about 3 mm. If two parallel lines are drawn on a white board with yellow-green color ($\lambda = 550$ nm) and the distance between them is 1 cm, find the distance from the person to the white board when the two parallel lines can be distinguished.

Solution Calculate the minimum resolution angle of the human eye according to Equation (12.9.2)

$$\delta_\varphi = 1.22 \frac{\lambda}{D} = 1.22 \frac{550 \times 10^{-9}}{3 \times 10^{-3}} = 2.2 \times 10^{-4} \text{ rad}$$

Assume that the distance between the person and white board is s and the distance between the parallel lines is l, the field angle to the person's eyes is $\theta \approx \frac{l}{s}$. According to the Rayleigh criterion, when the lines are just distinguishable,

$$s = \frac{l}{\theta} = \frac{l}{\delta_\varphi} = \frac{1 \times 10^{-2}}{2.2 \times 10^{-4}} = 45.5 \text{ m}$$

12.10 Polarization of light

The phenomena of interference and diffraction of light show the wave property of light, but these phenomena cannot tell us whether light is a longitudinal wave or a transverse wave. The polarization phenomenon of light clearly shows the transverse nature of light from experiments, which is consistent with the prediction of the electromagnetic theory of light. It can be said that the polarization phenomenon of light provides further evidence for the electromagnetic wave property of light. The polarization phenomenon of light is ubiquitous in nature. The reflection, refraction and birefringence of light in crystals are all related to the polarization phenomenon of light. Using this property, researchers can study the structure of crystals and can measure the stress distribution inside mechanical structures. Lasers are a type of polarized light source. In addition, saccharimeters, polarized light stereo movies, liquid crystal displays for pocket calculators and electronic watches all involve the application of polarized light.

12.10.1 Polarization of light

We already know that mechanical waves are divided into longitudinal waves and transverse waves according to the relationship between the vibration direction of the mass element and the wave propagation direction. The propagation direction of transverse waves is perpendicular to the vibration direction of mass elements, and the plane composed of the vibration direction of the mass element and the propagation direction of the wave is called the **vibration plane**. Obviously, the vibration plane has different properties from planes containing the wave propagation direction. This asymmetry between the vibration direction and the propagation direction of waves is called **polarization of waves**. Experiments have shown that only transverse waves have the polarization phenomenon. As shown in Figure 12.10.1, a slit AB is placed in the direction of wave propagation. For transverse waves, if the vibration direction is consistent with the slit direction, then the wave can pass through the slit and propagate forward, as shown in Figure 12.10.1(a). If the vibration direction is perpendicular to the slit direction, then the wave cannot pass through the slit and propagate forward, as shown in Figure 12.10.1(b). However, for longitudinal waves, no matter what the direction of the slit is, they can always pass through it and continue to propagate forward, as shown in Figures 12.10.1(c) and (d).

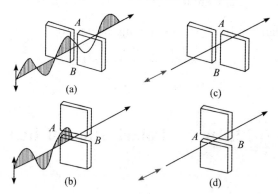

Figure 12.10.1 Difference between transverse and longitudinal waves

Light is an electromagnetic wave, and both the electric vector and magnetic vector are perpendicular to the propagation direction of light. Since human eyes can only feel the electric vector but not magnetic vector, light observed by them is transverse waves composed of vibrating electric vectors. Therefore, the light vector usually refers to the electric vector, and the **light vector plane** means the plane formed by the vibration direction of the electric vector and light propagation direction, similar to the vibration plane in mechanical waves. So light has the polarization phenomenon, which is similar to mechanical waves.

Although the light vector of a photon vibrates in a light vector plane and has the polarization property when the photon propagates in space, a beam emitted by a light source contains a large number of photons, no coherence occurs between them, and their polarization directions are random. So from the perspective of statistics, a beam does not have polarization property as a whole. After reflection or refraction on a surface or after special treatment, a light vector may have various different polarization states, called **polarization states** of light, according to which, light can be divided into five categories: natural light, linearly polarized light, partially polarized light, elliptically polarized light, and circularly polarized light. The first three types will be explained below.

1. Natural Light

The mechanism of luminescence for ordinary light sources is spontaneous radiation from numerous atoms or molecules. They are independent from each other in terms of order (phase), orientation and size (polarization and amplitude), as well as duration (length of an optical wave). So if you observe light from the plane perpendicular to the propagation direction, the fast-changing light vector vibrations with different value will occur in almost every direction. But according to the result by statistical averaging, the vibration in every direction is the same. This kind of zero-polarization state of optical transmission is called **natural light**, as shown in Figure 12.10.2(a). The optical vibration in any direction of natural light can be decomposed into vibrations in two mutually perpendicular directions. They have the same time average in each direction. Since these two components are independent and without fixed phase relation, natural light can usually be represented by two mutually independent, linearly polarized lights with equal amplitudes and mutually perpendicular vibration directions, as shown in Figure 12.10.2 (b). This is just a representation method and does not represent that natural light is composed of two perpendicular linearly polarized lights with the same intensity. Due to the symmetry, average energy of the two vibrations is equal, and each possesses half of the total energy of natural light. Figure 12.10.2(c) is the representation method of natural light, where short lines and dots are used to represent optical vibrations parallel to and perpendicular to the paper respectively. Dots and short lines are alternately and evenly drawn, indicating that the light vectors are symmetrically and evenly distributed.

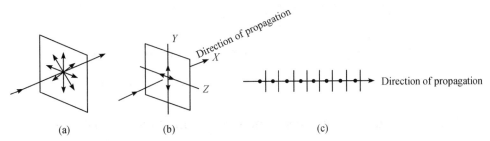

Figure 12.10.2 Schematic diagram of natural light

2. Linearly polarized light

If the direction of a beam's light vector remains unchanged and only vibrates in a fixed direction, this kind of light is called **linearly polarized light**. Since it is impossible to separate light emitted by an atom, linearly polarized light obtained in experiment is the component with parallel light vector directions among light emitted by many atoms. In experiments, linearly polarized light is usually obtained from the light emitted by an ordinary light source passing through a special device.

Figure 12.10.3 is the schematic diagram of linearly polarized light. Figure 12.10.3(a) represents linearly polarized light with the direction of light vector vibration parallel to the paper, and Figure 12.10.3(b) represents linearly polarized light with the direction of light vector vibration perpendicular to the paper.

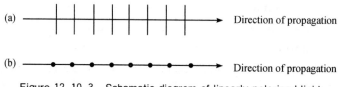

Figure 12.10.3 Schematic diagram of linearly polarized light

3. Partially polarized light

When natural light reflects or refracts on the surface of most transparent objects, its polarized state will change. In the plane perpendicular to the propagation direction of light, there are light vectors vibrating in all directions, but their amplitudes are not equal. This kind of light is called **partially polarized light**. It can be regarded as the mixed light of linearly polarized light and natural light. It is often represented as a stronger optical vibration in a certain direction and a weaker optical vibration in a direction perpendicular to the stronger vibration direction. If the contrast between these two directions of optical vibration is higher, it suggests that the light is closer to the linearly polarized light. And the lower contrast means that the light is closer to the natural light. Figure 12.10.4 is the schematic diagram of partially polarized light. Figure 12.10.4(a) represents that optical vibration in the direction perpendicular to the paper is stronger, and Figure 12.10.4(b) represents that optical vibration in the direction parallel to the paper is stronger.

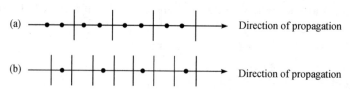

Figure 12.10.4 Schematic diagram of partially polarized light

12.10.2 Polarization generation and examination of a polarizer

Light emitted by ordinary light sources is natural light. The device used to obtain linearly polarized light from natural light is called a polarizer. Using a polarizer to obtain linearly polarized light from natural light is the most convenient method. Additionally, linearly polarized light can also be obtained through light reflection, refraction, or crystal prisms (such as the Nicol prisms and Wollaston prisms).

A polarizer is produced by depositing a layer of crystalline particles (such as iodine sulfate quinine and tourmaline) that are arranged in a fixed direction on a transparent substrate, or by brushing a certain thickness of gel containing special crystalline particles in a fixed direction. This crystal structure can selectively absorb two orthogonal polarization states of light, which strongly absorbs the component of the incident light with a certain direction of vibration and absorbs little of the component perpendicular to the direction. As a result, the polarizing film only allows the light vibrating in a special direction to pass through, which is the less absorbed direction. This direction is called the **polarization direction** of the polarizing film, or the transmission axis, denoted by the symbol ↕.

A polarizer can also be used to examine whether a light beam is linearly polarized light or not. This process is known as testing of the polarization state. The device used to analyze the polarization state of light is called an analyzer. Different polarization states of light will show distinct characteristics after passing through the analyzer. What follows is a discussion on how to distinguish natural light, linearly polarized light, and partially polarized light and how to examine their polarization states.

When a beam of natural light is vertically incident on the analyzer, the intensity of the natural light in any direction is half of the total intensity. Therefore, regardless of the orientation of the polarizing film, the intensity of the transmitted light remains unchanged. As the analyzer is rotated around the axis of light propagation, the intensity of the transmitted light remains constant and is always half of the incident light intensity.

If the incident light is linearly polarized light, the intensity of the transmitted light will be affected by the angle between the polarization direction of the polarized light and the polarization direction of the polarizing film. As the analyzer is rotated around the axis of light propagation, the intensity of the transmitted light will exhibit two maximum values and two zero values.

The examination of the polarization state of partially polarized light is similar to that of linearly polarized light. As the analyzer is rotated around the axis of light propagation, the transmitted light will also exhibit two maximum values and two minimum values of intensity, but the minimum intensity will not be zero. Figure 12.10.5 illustrates the process of polarization and polarization state testing. In the figure, A represents the polarizer, where natural light is vertically incident. And the outgoing light becomes linearly polarized light with an intensity of half that of the incident natural light. B represents the analyzer, and the linearly polarized light obtained from A is incident on B. If the polarization direction of B is parallel to the vibration direction of the linearly polarized light as shown in Figure 12.10.5(a), the light will pass through completely, and the maximum transmitted light intensity will be obtained. On the other hand, when the polarization direction of B is perpendicular to the vibration direction of the linearly polarized light as shown in Figure 12.10.5(b), the light cannot pass through. The transmitted light intensity will be zero, resulting in an extinction state.

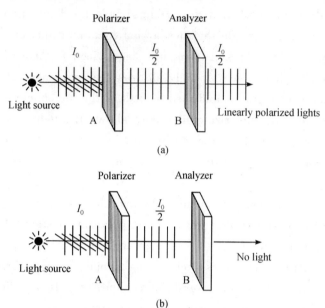

Figure 12.10.5　Schematic diagram of polarization generation and examination of a polarizer

 12.10.3　Malus's law

When linearly polarized light passes through a rotating analyzer, its intensity will change continuously; so what is the relationship between the intensity of incident linearly polarized light and the intensity of transmitted light after passing through an analyzer? In 1809, Malus derived the famous **Malus's law**: If the intensity of incident linearly polarized light is I_0, after passing through an analyzer, the intensity I of transmitted light is

$$I = I_0 \cos^2 \alpha \qquad (12.10.1)$$

where α is the angle between vibrating direction of the linearly polarized light and transmission axis direction of the analyzer. Now prove the law as follows.

As shown in Figure 12.10.6, assuming the amplitude of the light vector of the incident linearly polarized light is E_0, the polarization direction of the analyzer is the OP direction, and the angle between light vector and polarization direction is α. Since when the polarized light is incident on the analyzer, only the light component whose direction of vibration is parallel to the polarization direction can pass through, the amplitude of the light vector can be decomposed into a parallel component $E_{//}$ (the direction of the component is parallel to the polarization

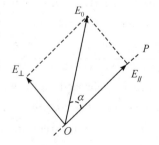

Figure 12.10.6　Malus's law

direction) and a vertical component $E_\perp$. The amplitude of the light vector of the transmitted light is $E = E_{//}$. According to the geometric relationship, it can be known that

$$E_{//} = E_0 \cos\alpha \qquad (12.10.2)$$

According to the relationship that the intensity of light is proportional to the square of the amplitude of the light vector, the ratio between the intensity of the transmitted light and that of the incident light can be obtained

$$\frac{I}{I_0} = \frac{E_{//}^2}{E_0^2} = \cos^2\alpha \qquad (12.10.3)$$

That is

$$I = I_0 \cos^2\alpha$$

It can be obtained from the above formula that, when α equals 0 or π the light vector is parallel to the polarization direction of the polarization film, the intensity is $I = I_0$, and the intensity of the transmitted light reaches the maximum. When α equals $\pi/2$ or $3\pi/2$, the light vector is perpendicular to the polarization direction of the polarization film, the intensity is $I = 0$, and the extinction phenomenon occurs.

Example 12.10.1 Two polarization films are stacked together, and the angle between their polarization directions is 60°. A beam of linearly polarized light with intensity I_0 is vertically incident on the polarization films, the angles between light vector vibration direction of the beam and polarization directions of both polarization films are both 30°. Find

(1) the intensity of the beam after passing through each polarization film;

(2) the intensity of the beam after passing through each polarization film if the original incident beam is replaced with the natural light with the same intensity.

Solution (1) According to Malus's law, the intensity I_1 of the polarized light after passing through the first polarization film can be obtained

$$I_1 = I_0 \cos^2 30° = \frac{3}{4} I_0$$

According to Malus's law, the intensity I_2 of the polarized light after passing through the second polarization film can be obtained

$$I_2 = I_1 \cos^2 60° = \frac{3}{16} I_0$$

(2) According to Malus's law, after the incident natural light passes through the first polarization film, it will become linearly polarized light with intensity $I_0/2$. And its vibration direction is parallel to the polarization direction of the first polarization film. Thus, according to Malus's law, the intensity I_2 of the polarized light after passing through the second polarization film can be obtained

$$I_2 = I_1 \cos^2 60° = \frac{1}{8} I_0$$

12.10.4 Polarization of reflection and refraction

Experiments have shown that both reflected light and refracted light will change to partially polarized light, when natural light reflects and refracts on the surface of two transparent isotropic media. As shown in Figure 12.10.7, MM' is the boundary between two media (such as air and glass), SI is the incident line of natural light, IR and IR' are the reflection line and refraction line, respectively, i is the incident angle and γ is the refraction angle. Experiments have found that vertical vibrations are

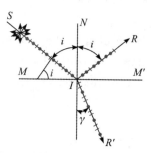

Figure 12.10.7　Polarization of reflected and refracted light

more than parallel vibrations in reflected beams while parallel vibrations are more than vertical vibrations in refracted beams. Therefore, both reflected light and refracted light are partially polarized light.

In 1815, when Brewster studied the polarization of reflected light, he found that the polarization level of reflected light was related to the incident angle. When the incident angle equals a specific value i_0, only the component perpendicular to the incident plane exists in reflected light, which is linearly polarized light. However refracted light is still partially polarized light and the component parallel with the incident plane is stronger. At the same time, reflected light and refracted light are perpendicular to each other, which means that the sum of the reflection angle and

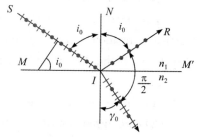

Figure 12.10.8　Brewster's law

refraction angle equals $\pi/2$. This is called **Brewster's law**. The corresponding special angle is called **Brewster's angle**, as shown in Figure 12.10.8.

Since reflected light and refracted light are perpendicular to each other, there are $i_0+\gamma=\dfrac{\pi}{2}$ and $\sin\gamma=\cos i_0$. According to the refraction law, $\dfrac{\sin i_0}{\sin \gamma}=\dfrac{n_2}{n_1}$, where n_1 and n_2 represent the refractive indexes of media containing the incident light and refracted light, respectively. So there is $\tan i_0=\dfrac{n_2}{n_1}$, and Brewster's angle can be expressed as

$$i_0=\arctan\left(\dfrac{n_2}{n_1}\right) \qquad (12.10.4)$$

When natural light reflects on the surfaces of two media, if the incident angle $i=i_0$, reflected light is linearly polarized light and refracted light is generally still partially polarized light and its polarization degree is low. This is because the intensity of refracted light is much greater than that of reflected light for most transparent media. For example, when natural light from air with a refractive index $n_1=1$ shoots towards glass with a refractive index $n_2 = 1.5$, all optical vibrations in the direction parallel to the incident plane in the incident light will be refracted if the incident angle is Brewster's angle $i = i_B = \arctan\left(\dfrac{n_2}{n_1}\right) = 56.3°$. 85% of the optical vibrations in the direction perpendicular to the incident plane are also refracted, and reflected light only accounts for about 15% in optical vibrations in the direction perpendicular to the incident plane.

The intensity of polarized light obtained by one reflection is very small and polarization degree of refracted light is low. In order to improve the intensity of reflected light and polarization degree of refracted light, many parallel glass films can be stacked together to form a glass film stack, as shown in Figure 12.10.9. When natural light is incident with Brewster's angle, it is easy to prove that reflection and refraction on each layer of the glass surface satisfy Brewster's law, which is a suitable way to improve the polarization degree of refracted light through multiple reflections and refractions. When there are enough glass films, linearly polarized light with the optical vibration direction parallel to the incident plane will be obtained in the transmission direction, which is also a wonderful method to obtain linearly polarized light.

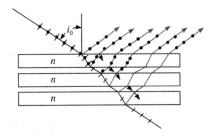

Figure 12.10.9 Stacking of glass films

Brewster's law also has many other practical usages. For example, Brewster's law can be used to measure the refractive index of a medium. If natural light is incident from air to an unknown medium, the refractive index of this medium can be calculated with the formula $\tan i_0 = n$ by measuring the polarizing angle i_0. Besides, in an external cavity laser, the seal of the laser tube is designed to be inclined and the incident angle is Brewster's angle, so that the light vector component in the direction parallel to the incident plane does not reflect and passes completely. It can reduce energy loss and improve the polarizability of laser.

Example 12.10.2 As shown in Figure 12.10.10, parallel glass plates are placed in the air, the refractive index of air is approximately 1 and the refractive index of glass is 1.50. Incident light shines to the upper surface of glass with Brewster's angle.

(1) What is the refraction angle?

(2) When refracted light reflects on the lower surface, does the reflected light belong to linearly polarized light?

Solution (1) According to Brewster's law,

$$i_0 = \arctan \dfrac{n_2}{n_1} = \arctan 1.5 = 56°18'$$

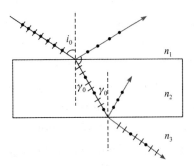

Figure 12.10.10 Figure of Example 12.10.2

Since the incident angle of light is Brewster's angle, reflected light and refracted light are perpendicular to each other. There is $i_0 + \gamma_0 = \frac{\pi}{2}$, and then

$$\gamma_0 = \frac{\pi}{2} - i_0 = 90° - 56°18' = 33°42'$$

(2) When the refracted light reflects on the lower surface, Brewster's angle is

$$i'_0 = \arctan \frac{n_3}{n_2} = \arctan \frac{1}{1.5} = 33°42'$$

It can be seen that the refracted light inside the glass plate also shines to the lower surface with Brewster's angle and the reflected light is also linearly polarized light.

12.11 Birefringence of light

12.11.1 Birefringence phenomenon of a crystal

When a beam of light enters another medium from one medium, the refracted light usually only has one beam at the boundary between the two isotropic media and it follows the law of refraction. When a beam of light passes through an anisotropic medium such as calcite ($CaCO_3$), the refracted light at the interface is often divided into two refracted light beams propagating in different directions, as shown in Figure 12.11.1. This phenomenon is called the birefringence phenomenon of a crystal.

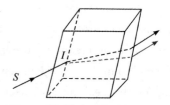

Figure 12.11.1 Birefringence of calcite

Derived from experiments, the results are as follows. When the incident angle i is changed, one of the two refracted light beams always stays within the incident plane and follows the law of refraction. This beam is called ordinary light, or o-light. The other refracted light beam is generally not within the incident plane and does not follow the law of refraction. Its propagation speed changes with the direction of the incident light. This beam is called extraordinary light, or e-light. When the incident angle is $I = 0$, the ordinary light propagates in its original direction, while the extraordinary light generally does not propagate in its original direction. When the crystal is rotated with the incident light as the axis, o-light does not move while e-light rotates around the axis. Using a polarizer, it can be found that o-light and e-light in the crystal are mutually perpendicular linearly polarized light, as shown in Figure 12.11.2.

Experiments have shown that there is a special direction in crystals such as calcite. When light propagates in this direction, o-light and e-light no longer separate and the birefringence phenomenon does not occur. This special direction is called optical axis of a crystal. It should be noted that the optical axis of a crystal is a specific direction, and any line parallel to this direction is the optical axis of the crystal.

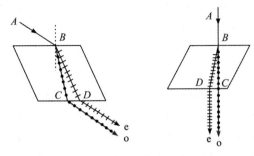

Figure 12.11.2 Ordinary light and extraordinary light

Figure 12.11.3 shows the calcite crystal with equal edge lengths and line AD is its optical axis direction. A crystal with one optical axis direction is called uniaxial crystal, such as calcite, quartz and ice. A crystal with two optical axis directions is called biaxial crystal, such as mica and sulfur. In a crystal, the plane composed of certain light and the optical axis of the crystal is called main plane corresponding to this light. The plane composed of o-light and the optical axis is called o-light main plane; the plane composed of e-light and the optical axis is called e-light main plane, as shown in Figure 12.11.4. The vibration direction of o-light light vector is perpendicular to its own main plane, and vibration direction of e-light light vector is within its own main plane.

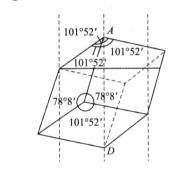

Figure 12.11.3 Optical axis of a calcite crystal

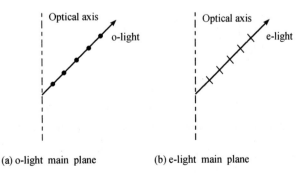

(a) o-light main plane   (b) e-light main plane

Figure 12.11.4 Main plane of light

If light shines to the surface of a crystal, the plane composed of the normal of the interface and optical axis of the crystal is called main section. When incident light is within the main section, which means that the incident plane coincides with the main section, it will be found that both refracted light is within the incident plane. At this time, it can be determined that the vibration direction of o-light is perpendicular to the main section and vibration direction of e-light is parallel to the main section by using a polarizer. The vibration directions of o-light and e-light are mutually perpendicular.

Birefringence of light is caused by the fact that the propagation rate of light in a crystal is related to the polarization state of light. When o-light propagates in a crystal, its light vector direction always remains perpendicular to the optical axis. Thus, the velocity of light in all directions is the same; the sub-wavefront at any point on the o-light wavefront is a sphere. When e-light propagates in a crystal, the angle between its light vector direction and the optical axis changes with the propagation direction, which leads to the various velocities in different directions. The subwave wavefront emitted at any point on the e-light wavefront in the crystal is the rotating ellipsoid with the optical axis as the axis. As shown in Figure 12.11.5, o-light and e-light only have the same velocities in the optical axis direction, so the wavefronts of the above wavelets are tangent to each other in the optical axis direction. In the direction perpendicular to the optical axis, the velocities of the two light differ most. Usually, we use v_o and v_e to represent the velocity of o-light propagating in a crystal and the velocity of e-light in the direction perpendicular to the optical axis, respectively. For some crystals, the sphere wavefront will surround the ellipsoid wavefront if there is the relationship $v_o > v_e$, as shown in Figure 12.11.5(a). These crystals are called the positive crystals (such as quartz). For some other crystals, the ellipsoid wavefront will surround the sphere wavefront if there is the relationship $v_o < v_e$, as shown in Figure 12.11.5(b). These crystals are called the negative crystals (such as calcite).

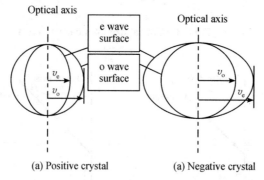

(a) Positive crystal   (a) Negative crystal

12.11.5   Wave planes of positive and negative crystals

According to the definition of refractive index, $n_o = c/v_o$ for o-light. It has nothing to do with the propagation direction of o-light and is just a constant determined by crystal material. For e-light, since the velocity varies with directions, there is no general meaning for refractive index. Usually, the ratio ($n_e = c/v_e$) of speed of light c in a vacuum to propagation velocity of e-light perpendicular to the optical axis direction is called the refractive index of e-light. n_o and n_e are both called the main refractive indices of a uniaxial crystal. If the direction of the crystal optical axis and two main refractive indices are known, the refraction directions of o-light and e-light can be determined.

12.11.2 Application of polarized light in science and technology

Polarized light has a wide range of applications in production and life, such as stereo movies, photography and liquid crystal displays.

In the stereo movie field, it is mainly shot through two cameras. For the same object, two images of the object can be separately shot by two cameras, and then these two shot images are projected onto the screen. When projecting stereo movies, two polarizers are placed in lenses of two projectors, respectively. As a result, two beams of light emitted by the two projectors are mutually perpendicular polarized light. Thus, when audiences watch stereo movies, they need to wear corresponding polarized glasses. Left and right polarizers of the glasses correspond to the polarized light of the left and right projector, respectively. In this way, the left and right eyes can only see corresponding images, which gives the audience a three-dimensional sense, as shown in Figure 12.11.6.

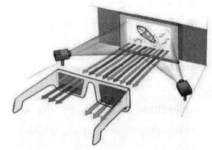

Figure 12.11.6 Schematic diagram of light path in a stereo movie

In photography, when taking photos of objects with smooth surfaces, such as glassware, water surface, display cabinets, paint surface and plastic surface, bright spots or reflection often occurs. They are caused by polarization of light. Thus, if a polarizing mirror is added during photography, the photographer can rotate it appropriately such that polarized light can be blocked to eliminate or reduce reflection and bright spots on surfaces of these smooth objects, as shown in Figure 12.11.7.

Figure 12.11.7 Comparison of photos taken before and after using a polarizing mirror

In liquid crystal displays, a liquid crystal box (optically active substance) is inserted between two polarizers with mutually perpendicular polarization directions. The upper and lower layers of the liquid crystal box are transparent electrode plates. They are engraved into shapes of numbers. External natural light becomes linearly polarized light after passing through the first polarizer. When this beam passes through the liquid crystal, the polarization direction of light will be rotated by 90° by the liquid crystal if there is no voltage between upper and lower electrodes. Thus, it can pass through the second polarizer. A reflector is below the second polarizer, and light will be reflected back. At this time, the liquid crystal box looks transparent. But if there is a certain voltage between upper and lower electrodes, the property of the liquid crystal will change and the optical rotation property will disappear, so light cannot pass through the second polarizer. The region under the upper electrode becomes dark. If the electrode is engraved into shapes of numbers, the numbers will be displayed in this way.

In scientific research, polarizing microscopes are made based on the polarization phenomenon, which are widely used in fields such as mineralogy, chemistry, biology, and botany. By adding polarizing devices into optical microscopes, different modes such as polarization, bright field and dark field can be used to observe objects by this method. Images obtained after polarization will be clearer, allowing for a detailed view of objects. In mineralogy research, using a polarizing microscope allows for the observation and measurement of crystal morphology, crystal grain size, percentage composition, cleavage, the Becke line, color and pleochroism, as shown in Figure 12.11.8. In biological samples, different fibrous protein structures exhibit distinct anisotropy. By using a polarizing microscope, detailed information about the arrangement of molecules within these fibers can be obtained, such as collagen and the spindle fibers during cell division. In the study of human and animal tissues, polarizing microscopy is commonly used to identify skeletal structures, teeth, cholesterol crystals, nerve fibers, tumor cells, striated muscles, hair, etc.

Figure 12.11.8 Mineral phase structure under a polarizing microscope

In engineering applications, stress distribution in engineering components can be detected using polarized light. When an engineering component loads, internal force at each place is generally uneven, and damage to the component always starts from the place with the maximum stress. Therefore, understanding the stress state at each point in the component and finding the location with the maximum stress are very important. For components bearing complex loads and with complex shapes, theoretical analysis and calculation are very cumbersome and even impossible to perform. Therefore, various experimental stress analysis methods are widely used, and the optical elasticity method using polarized light is one mature method. In stress detection, artificial birefringent materials can be used to fabricate component models. Then, the model is heated and external force is applied. After the model cools down the external force is removed. At this time the birefringent model made will have the stress that occurs when the external force is applied. Then the model is made into a thin slice and placed between analyzer and polarizer. When analyzing stress at different component parts, the thin slice model will produce different refractive indices at different stress parts. Since the model possesses the artificial birefringent effect, it will also produce photoelastic fringes. Engineers can identify the directions of the principal stresses and the directions of the principal stress difference by observing two types of dark fringes (isochromatic and isoclinic fringes). By separating and analyzing these fringes, it is possible to obtain the stress distribution in different parts of the engineering component, as shown in Figure 12.11.9.

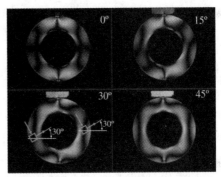

Figure 12.11.9 Isoclinic line in the photoelastic experiment of a radially compressed disk

Scientist Profile

Wang Daheng (1915 – 2011), originally from Wuxian, Jiangsu Province, was a recognized academic founder, pioneer and organizational leader in the field of optics in China. He was also a recipient of the "Two Bombs and One Satellite Merit Medal", and the academician of the Chinese Academy of Engineering.

Wang Daheng graduated from the Department of Physics at Tsinghua University in 1936. In 1938, he was admitted as a publicly-funded student to study in the UK and went to Imperial College London to study applied optics. In 1941, he transferred to the University of Sheffield to conduct research on optical glass under the guidance of the world-renowned glass scientist Professor W. E. S. Turner. In 1942, he was hired by Chance Glass

Company in Birmingham to specialize in optical glass research. He returned to China in 1948 and successively served as a professor at Dalian University, director of the Instrument Museum of the Chinese Academy of Sciences, director of Changchun Institute of Optics and Mechanics, president of Changchun Branch of the Chinese Academy of Sciences, deputy director of the Fifteenth Institute of the National Defense Science Commission, chairman of the Chinese Optical Society, and director of the Technical Science Department of the Chinese Academy of Sciences. Wang Daheng made important research achievements in laser technology, remote sensing technology, metrology science, chromaticity standards and other fields. He was a major advocate of China's "863" high-tech plan and played an active role in national science and technology decision-making. He trained a large number of experts in optical science and technology field for the country.

Starting from the 1960s, Wang Daheng led the Changchun Institute of Optics and Mechanics to optical technology and engineering research in national defense field. He made important contributions in many fields such as infrared and low-light night vision, nuclear explosion and optical measurement equipment in target field, high-altitude and space reconnaissance photography. In 1960, in order to meet the needs of national defense projects, the state proposed the task of developing large-scale precision optical tracking cinetheodolite. Under Wang Daheng's technical guidance and after five years' unremitting efforts, China's first large-scale optical measurement equipment was finally developed, which exceeded its original design specifications. This created the history that China could independently engage in research and small batch production of target field optical observation equipment. In 1965, Wang Daheng participated in the overall plan for China's first artificial earth satellite as the head of ground equipment group and deputy head of overall design group for Chinese Academy of Sciences. He put forward insightful views on the tracking system adopted by the satellite and specific technical route for the ground tracking system. Finally, his advice was adopted. Wang Daheng was a pioneer in research on Chinese aerospace camera technology. In the mid-1960s, he established the space-to-earth photography technology group at Changchun Institute of Optics and Mechanics. Later, he expanded China's first aerospace camera research team in Beijing based on this group. Under his leadership, the first aerospace camera in China was successfully developed in 1975.

In the same year, Wang Daheng hosted and completed China's first remote sensing science plan, which promoted the rapid development of China's remote sensing work. Wang Daheng also participated in the construction of Chinese remote sensing satellite ground station. He provided guidance in site selection of the ground station, computer room construction, talent training and operation services, etc. He made important contributions to making Chinese remote sensing satellite ground station system become the basic infrastructure for national space information and important technical support system for nationwide remote sensing users.

In April 1992, Wang Daheng together with five other academic committee members (academicians) jointly proposed the establishment of the Chinese Academy of Engineering to the central government, which was approved by the Communist Party of China Central Committee and State Council. It had far-reaching impact on Chinese engineering community. In June 1994, the Chinese Academy of Engineering was formally established. Wang Daheng was recommended by the Chinese Academy of Sciences and elected as one of the first academicians of the Chinese Academy of Engineering, serving as the member of the first presidium.

 Extended Reading

Controversy between "Particle Theory" and "Wave Theory" of light

The human research on light had an early origin, but the understanding of the essence of light went through a long process. Is light a particle or wave? From the 17th century to early 20th century, the two major theories ("Particle Theory" and "Wave Theory") were in fierce debate, and eventually it ended with the wave-particle duality of light. This debate promoted the development of science, and led to the major achievement in the 20th century physics—quantum mechanics.

1. First "Wave-Particle Battle"

In 1655, the mathematics professor Grimaldi in Italian Bolonia University discovered the light diffraction phenomenon while observing the shadow of a small stick placed in a light beam. Based on this, he speculated that light might be a fluid similar to water waves. Grimaldi designed an experiment: A beam of light passed through two small holes, and then shone on a screen in a dark room. At this point, he obtained an image with light and dark fringes. He believed that this phenomenon was similar to water waves and proposed that light was a fluid capable of wave-like motion. The different colors of light were attributed to varying wave frequencies. Thus, Grimaldi introduced the concept of diffraction of light, who was the earliest proponent of the wave theory of light. Shortly, the British physicist, Hooke, repeated Grimaldi's experiment and observed the colors of soap bubbles. Based on his observations, Hooke proposed the hypothesis that light was a longitudinal wave of the ether. According to this hypothesis, Hooke also believed that the color of light was determined by its frequency.

However, in 1672, Newton discussed his experiments on the dispersion of light in his paper *A New Theory of Light and Colors*. He allowed sunlight to pass through a small hole and project onto a prism inside a dark room, which resulted in a colorful spectrum on the opposite wall. He believed that the combination and separation of light were analogous to different-colored particles mixing together and then being separated again. In this paper, he used the particle theory to explain the colors of light.

The first debate between "Particle Theory" and "Wave Theory" was ignited by the

"Color of Light" fuse, and since then Hooke and Newton engaged in long and fierce debate. In 1672 the Royal Society Review Committee chaired by Hooke basically held negative attitude towards Newton's paper *A New Theory of Light and Colors*. At first Newton did not completely deny the wave theory and was not a biased supporter of the particle theory. But after the duration of debate, Newton refuted Hooke's wave theory in many papers. In 1675, Newton refuted Hooke's wave theory again in the paper *Hypothesis Explaining the Properties of Light Discussed in My Several Papers* and reiterated his particle theory. Since neither Newton nor Hooke had established complete theoretical system at the time, wave-particle debate did not fully unfold. But scientific debates, once they arise, require a clear conclusion. The old problems had not been solved, but new debates were already brewing.

In 1666, the famous Dutch astronomer, physicist and mathematician Huygens met Newton in Cambridge and exchanged views on nature of light, but at that time Huygens' views were more inclined towards the wave theory. Thus, he and Newton had differences. It was these differences that inspired Huygens' strong passion for physical optics. Huygens carefully studied Newton's optical experiments and Grimaldi's experiments, and he believed that many phenomena were inexplicable by the particle theory. So he proposed the relatively complete wave theory. Huygens believed that light was a mechanical wave, and was a longitudinal wave propagated by a substance, called the ether; each point on a wavefront is the wave source causing a medium to vibrate. Based on this theory, Huygens proved the law of reflection and refraction of light, and also relatively well explained diffraction, birefringence of light and the famous "Newton's Ring" experiment.

In 1678, Huygens submitted his optical work *Light Theory*, in which he systematically expounded light's wave theory to Paris Academy of Sciences. In the same year, Huygens gave a speech against the particle theory. Just as Huygens was actively promoting the wave theory, Newton's particle theory was gradually established as well. Newton revised and improved his optical work *Optics*. In this book, Newton put forward two reasons to refute Huygens. First, if light was a wave, it should be able to go around obstacles like sound waves without producing shadow. Second, the birefringence phenomenon of iceland spar showed that light had different properties on different sides, which could not be explained by the wave theory either. On the other hand, Newton extended his particle view to the entire natural world, and integrated it with his particle mechanics system, which provided a strong backing for the particle theory.

In order to avoid dispute with Hooke, Newton's *Optics* was officially published next year after Hooke's death (1704). At that time both Huygens and Hooke passed away, leaving no one to fight for the wave theory. Due to his enormous contributions to the scientific community, Newton became an unparalleled scientific master. As his reputation grew, people worshipped his theories, repeated his experiments, and firmly believed in the same conclusions as his. Throughout the eighteenth century, almost no one challenged the particle theory, nor did anyone conduct further research on the nature of light.

2. Second "Wave-Particle Battle"

At the end of the eighteenth century, under the influence of the German natural philosophy trend, people's thoughts gradually liberated. The famous British physicist Thomas Young began to doubt Newton's optical theory. In 1801, Thomas Young carried out the famous Young's double-slit interference experiment. The alternating black and white fringes on the white screen proved the interference phenomenon of light, which proved that light was a wave. Thomas Young believed that light was elastic vibration propagating in ether flow, and he pointed out that light propagated in form of longitudinal waves. Furthermore, he pointed out that different colors of light were similar to different frequencies of sound. Although Young's theory did not receive enough attention at that time and was even slandered, the wave theory finally made itself heard again after one hundred years of silence. At the same time, it aroused interest of the Newtonian school in optical research.

In 1808, Laplace used the particle theory to analyze the double refractive line phenomenon of light and refuted Young's wave theory. In 1809, Malus discovered the polarization phenomenon of light in experiments, and further found that light was partially polarized when it refracted. Because Huygens had proposed that light was a longitudinal wave that could not be polarized, this discovery became a favorable evidence against the wave theory. In 1811, Brewster discovered the empirical law of the polarization phenomenon of light. The discovery of the polarization phenomenon and polarization law of light put the wave theory into a dilemma at the time, and made physical optics research develop in a direction more favorable to the particle theory. Faced with this situation, Thomas Young conducted in-depth research on optics. In 1817, he abandoned Huygens' view that light was a longitudinal wave, and proposed that light was a transverse wave, which successfully explained the polarization phenomenon of light. After absorbing some opinions of the Newtonian school, he established a new wave theory again.

In 1817, the French Academy of Sciences decided to make the light diffraction theory the subject of the prize essay in 1819. The famous scientists Biot and Poisson who presided over this event were the active supporters of the particle theory. Their intention was to encourage the use of the particle theory to explain the diffraction phenomena through this prize essay in order to achieve a decisive victory for the particle theory. However, unexpectedly, an unknown scholar Fresnel used rigorous mathematical reasoning to successfully explain the polarization phenomenon of light from the perspective of transverse waves of light, and he quantitatively calculated the diffraction patterns produced by obstacles such as circular holes and circular plates using the half-wave band method. The results were consistent with experiments. Afterwards, Poisson, who presided over the prize essay activity, used Fresnel's equation to derive disk diffraction and obtained a surprising result that a bright spot would appear at the center of the disk shadow at a certain distance behind the disk on the screen. Poisson thought this was an unimaginable and absurd conclusion, and claimed to refute the wave theory of light. Later, physicists brilliantly confirmed Fresnel's theory through experiments, and indeed a bright spot appeared at the center of the disk shadow. This fact shocked the French Academy of Sciences, and Fresnel deservedly won that year's science award. Later people jokingly called this bright spot "Poisson's bright spot". As a result, the wave theory of light defeated the particle theory again and optics entered a new period—elastic ether optics period. In 1865, Maxwell established the electromagnetic field theory and predicted the existence of electromagnetic waves. He pointed out that light was also an electromagnetic wave, and established a comprehensive complete rigorous theory for light phenomena. In 1887, Hertz detected electromagnetic waves through experiments, which brilliantly proved the correctness of Maxwell's theory. Thereafter, the wave theory of light occupied the absolute dominant position.

3. Third "Wave-Particle Battle"

With the establishment of the wave theory of light, people began to search the carrier of light waves. Thus, the ether theory became active again, and some famous scientists became representatives of the ether theory. But people encountered many difficulties while searching for the ether. In 1887, the Michelson-Morley "ether drift" experiment denied the existence of the ether, which indicated the crisis for the wave theory.

At the end of the nineteenth century and beginning of the twentieth century, some new optical phenomena were discovered, such as the photoelectric effect, thermal radiation, spectrum and the Compton effect. The wave theory encountered difficulties when explaining these phenomena. But the particle theory could relatively well explain these phenomena. In this physics revolution, Planck first used the energy quantum hypothesis to explain the black body radiation experiment; then Einstein successfully explained the photoelectric effect by using the photon theory. Next, Bohr used the quantum theory of light to explain the hydrogen atom spectrum. A series of outstanding performances of the quantum theory meant that the Newton's particle theory that had been completely overturned in the past began to revive, while the wave theory of light temporarily retreated.

In fact, the quantum theory of light was not omnipotent, because it only reflected the discontinuity of light but could not explain the wave phenomenon of light. Conversely, the wave theory of light could not explain quantum of light. They both had rationality. Long fierce debate between these two major theories (particle and wave theories) effectively promoted the development of science, and eventually led to the emergence of a new idea and achievement, namely "wave-particle duality". The wave-particle duality of light was proposed by Einstein in 1905, and later confirmed by experiments. Then, under the influence of this new idea, De Broglie analogously proposed matter particles also had wave-particle duality. After pioneering work by Heisenberg, Schrödinger, Born, Dirac, et al., complete quantum mechanics theory was finally formed in 1926, which led to a series of epoch-making scientific discoveries and technological inventions. The theory made very important contributions to the progress and development of human society.

The 200 years' wave-particle controversy has given us many insights. Scientific development is a difficult and tortuous process of accumulation and struggle, and it forces us to create new concepts and theories. Scientific development cannot be separated from exploration, the environment where hundreds of schools contend, or the realistic spirit and critical spirit that people are not superstitious towards authorities and dare to challenge authorities.

Discussion Problems

12.1　If a laser beam shines on two parallel slits, an interference pattern can be observed on a distant screen. What will happen to the pattern if the distance between the two slits is slowly reduced?

12.2　At first, Young's double-slit experiment is performed in the air, and then it is repeated in water with the same device. Will the interference fringes become denser, sparser or unchanged compared to the experiment in the air?

12.3　Both wedge interference and Newton's rings belong to equal-thickness interference. What are the differences in the shape of their interference fringes and the spacing of their fringes? How do the fringes move when the thickness increases or decreases? Will the spacing change?

12.4　In the single slit diffraction experiment, what are the effects on diffraction patterns by increasing the wavelength or slit width, respectively?

12.5　Someone thinks that the dark fringe condition of grating diffraction is $(a+b)\sin\varphi = \pm(2k-1)\frac{\lambda}{2}$, $k=1, 2, \cdots$. Is it correct? If not, what is wrong?

12.6　In analyzing the distribution of bright and dark fringes in grating diffraction, if each slit is divided into several wave bands using Fresnel's half-wave band method and then all the light emitted by half-wave bands of all slits is superimposed, will the result be the same as that calculated by the grating equation? Why?

12.7　The polarization direction of a polarizer is usually not indicated. Which method can be used to determine it?

12.8　If a beam of light is incident in the direction parallel to the paper plane with a polarizing angle, will it have any reflected light?

Problems

12.9　In Young's double-slit interference experiment, monochromatic light with a wavelength of 5×10^{-7} m is vertically incident on the double slits with a distance of 0.5 mm between them. The distance from the screen to the center of the double slits is $D=1.0$ m, and the entire device is in a vacuum.

(1) Find the position of the center of the 10th bright fringe on the screen;

(2) Find the fringe width;

(3) Using a mica sheet ($n=1.58$) to cover one of the slits, the central bright fringe moves to the position that is the original 8th bright fringe. What is the thickness of the mica sheet?

12.10 Young's double-slit experiment device is in a vacuum, and the wavelength of the light source is $\lambda=6.4\times10^{-5}$ cm. The spacing between the two slits $d=0.4$ mm, and the distance from the screen to the slits is 50 cm.

(1) Find the distance between the first bright fringe and the central bright fringe on the screen;

(2) If the distance between point P and the central bright fringe is $x=0.1$ mm, what is the phase difference of the two beams at point P?

12.11 There is an oil film with a refractive index $n=1.30$ on a glass plate with a refractive index $n=1.50$. It is known that the vertical incident light produces the reflective cancellation phenomenon for the wavelengths 500 and 700 nm. There is no reflective cancellation phenomenon between these two wavelengths. What is the thickness of the oil film?

12.12 As shown in Figure T12-1, monochromatic light with a wavelength 680 nm is vertically incident on two glass plates with a length $L=0.12$ m. At one side the two glass plates touch each other, and at the other side

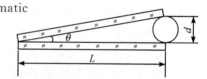

Figure T12-1 Figure of Problem 12.12

they are separated by a steel wire with a diameter $d=0.048$ mm. Find

(1) the angle between the two glass plates;

(2) the thickness difference of the air film between two adjacent bright fringes;

(3) the distance between two adjacent dark fringes;

(4) the number of bright fringes presented within this 0.12 m.

12.13 Different wavelengths ($\lambda_1=6000$ Å and $\lambda_2=4000$ Å) of light are used to observe Newton's rings, and the kth dark ring using λ_1 coincides with the $(k+1)$th dark ring using λ_2. The lens curvature radius is 190 cm.

(1) Find the radius of the kth dark ring using λ_1;

(2) If the 5th bright ring using the wavelength of 5000 Å coincides with the 6th bright ring using the unknown wavelength λ_2 in Newton's ring experiment, what is the unknown wavelength λ_2?

12.14 The Michelson interferometer can measure the wavelength of monochromatic light. When the mirror M_1 moves a distance 0.322 mm, the number of moving interference fringes is 1024. What is the wavelength of the monochromatic light?

12.15 Monochromatic parallel light with a wavelength of 600 nm is vertically incident on a single slit with a width $a=0.6$ mm. Behind the slit, there is a lens with the focal length $f=60$ cm. What is the width of the central bright fringe on the lens focal plane? What is the distance between the two third order dark fringes?

12.16 The single slit width is 0.10 mm, and the lens focal length is 50 cm. The green light $\lambda=500$ nm is vertically incident on the single slit.

(1) What are the width and half-angle width of the central bright fringe on the screen located at the lens focal plane?

(2) If this device is immersed in water ($n=1.33$), what is the half-angle width of the central bright fringe?

12.17 It is known that the angular distance between two stars relative to the telescope is 4.84×10^{-6} rad and they both emit light with a wavelength of 550 nm. What is the minimum diameter of the telescope to distinguish these two stars?

12.18 Monochromatic light with a wavelength 600 nm is vertically incident on a grating, the 2nd and 3rd order fringes appear at the places $\sin\varphi_2=0.20$ and $\sin\varphi_3=0.30$, respectively and the 4th order fringe is missing. Find

(1) the grating constant;

(2) the possible minimum width of the grating slit a;

(3) all the orders actually present on the screen according to the above selected a and b.

12.19 Monochromatic parallel light with a wavelength of 5000 Å is vertically incident on a grating with 200 lines per millimeter and the focal length of the lens behind the grating is 60 cm. Find

(1) the distance between the central bright fringe and the first order bright fringe on the screen;

(2) the displacement of the central bright fringe when light is incident on the grating with the angle that is 30° with the grating normal.

12.20 There are double slits with a distance $d=0.1$ mm, and the width of the slit is $a=0.02$ mm. Monochromatic parallel light with a wavelength $\lambda=600$ nm is vertically incident on the double slits, and a lens with a focal length $f=2.0$ m is placed behind the double slits.

(1) How many interference principal maximum fringes are within the width of the central bright fringe in single slit diffraction?

(2) If another identical single slit is in the middle of these double slits, how many interference principal maximum fringes are within the width of the central bright fringe?

12.21 Natural light is vertically incident on two overlapping polarizers. If the intensity of transmitted light is half the intensity of incident light, what is the angle between the polarization directions of these two polarizers? If the intensity of transmitted light is half of its maximum intensity, what is the angle between the polarization directions of these two polarizers?

12.22 When natural light is incident on the interface between two media with an angle of 53°, the reflected light is completely polarized. What is the refracted angle?

Challenging Problems

12.23 In Young's double-slit interference experiment, the distance between the slits is 0.2 mm. The wavelengths of blue and green line light sources are $\lambda_1 = 440$ nm and $\lambda_2 = 540$ nm, respectively. When they are vertically incident on the double slits, respectively, what are the field angles of bright fringes? When they are simultaneously incident on the double slits, the bright and dark fringes can be observed. If the white light source with a blue-green filter is used, what is the continuous light wavelength range? And what is the average wavelength? Estimate the order of the fringe, from which the fringe will become indistinguishable?

12.24 The intensity of grating diffraction is

$$I = I_0 \left(\frac{\sin u}{u}\right)^2 \left(\frac{\sin Nv}{\sin v}\right)^2$$

where N is the number of slits, $u = \pi a \sin\theta / \lambda$, and $v = \pi d \sin\theta / \lambda$. $d = a + b$ is called the grating constant where a is the width of the transparent slit, and b is the width of the opaque baffle.

(1) What is the relationship between grating diffraction fringes and number of slits when the slit width and grating constant are fixed?

(2) Explain the influence of interference on the modulation and missing order phenomenon of single slit diffraction.

(3) What is the relationship between distribution of grating diffraction fringes and slit width as well as grating constant?

Part 5 Modern Physics

By the end of the 19th century, the theory of physics had developed to a fairly complete stage, and had made great achievements. Many physicists believed that the basic problem of physical theory had been solved, and the task in the future was only to further improve the laws of physics. While physicists were rejoicing at the brilliant achievements of classical physics theories, some new experimental facts sharply contradicted them. For example, the Michelson-Morley experiment in 1887 denied the existence of an absolute frame of reference; in 1900, when Rayleigh and Jeans used classical electromagnetic theory and the equipartition theorem to explain the phenomenon of thermal radiation, what is called "ultraviolet catastrophe" occurred. These new experimental phenomena could not be properly explained by classical physics, which put it in a very difficult situation.

In order to get rid of the predicament of classical physics, some physicists who were sharp-minded but not bound by old ideas reconsidered the basic concepts of physics. Hence resulted two theoretical cornerstones—relativity and quantum theory. In this part, we will introduce the basic knowledge of modern physics, mainly including fundamentals of special relativity and quantum physics.

Chapter 13

Fundamentals of the Special Theory of Relativity

Whenever we talk about the theory of relativity, we cherish the memory of its founder, the standard bearer of the physics revolution—Einstein. Einstein was born in Ulm, Germany in 1897 and graduated from the Swiss Federal Institute of Technology in Zurich in 1900. Einstein founded the quantum theory of light, revealing for the first time that microscopic objects have wave-particle duality; he established the special and general theories of relativity, which promoted the revolution of the entire physical theory and laid a solid theoretical foundation for the use of nuclear energy; he created the radiation quantum theory and cosmology of modern science, of which the radiation quantum theory laid a theoretical foundation for the development of laser technology; his creative research on the molecular kinetic theory provided a theoretical basis for its development and the establishment of a scientific theoretical system. Einstein made epoch-making contributions in many fields of physics, won the Nobel Prize in Physics in 1921, and was recognized as the godfather scientist of physics.

In 1905, on the basis of the work of a series of scientists such as Lorenz and Poincaré, Einstein published the article *On the Electrodynamics of Moving Bodies*, established the special theory of relativity, and subverted people's understanding of time, space and matter. In 1915, Einstein further promoted the theory, created the general theory of relativity, and proposed the concept of gravitational waves. These theories have already been verified by a large number of experiments such as atomic clocks traveling around the world, Mercury's perihelion precession, and gravitational wave measurements. This chapter mainly introduces the basic concepts and basic knowledge of special relativity.

13.1　Space-time view of classical mechanics

13.1.1　Principle of mechanical relativity

In his *Dialogue Concerning the Two Chief World Systems*, Ptolemaic and Copernican published in 1632, Galileo made the following vivid description of the motion phenomena observed in a closed cabin in uniform linear motion: In a closed cabin that moves in a straight line at a uniform speed relative to the ground, you will find that the flying insects fly in any direction in exactly the same way, neither more labor-saving nor more laborious; the water droplets fall freely from the air, always fall into the bottle directly below, and never deviate; the white mist that emerges from the steam in the teacup does not deviate in any direction. Here, Galileo described the scenario that occurred in a ship's cabin moving in a straight line at a constant speed relative to the inertial frame of the earth. Thus, the following conclusions are obvious:

(1) In the reference frame in uniform linear motion relative to the inertial frame, the mechanical laws summarized will not be different due to the uniform linear motion of the entire system;

(2) Since there is no difference between the laws of mechanics in the reference frame in uniform linear motion relative to the inertial frame and those in the inertial frame, we cannot distinguish between these two reference frames, or **all reference frames in uniform linear motion relative to the inertial frame are inertial systems.**

From the above two points, we will naturally draw the following conclusion: For describing the laws of mechanics, all inertial frames are equivalent. This conclusion is **Galileo's principle of relativity**, also known as **the principle of mechanical relativity**.

13.1.2 The Galileo transformation

The motion of matter is absolute, but the description of motion is relative, and the description of motion is also different depending on the frame of reference chosen by the observer. Galileo gave the mathematical expression of the principle of mechanical relativity, that is, the Galileo transformation. As shown in Figure 13.1.1, there are two inertial reference systems S and S', and the rectangular coordinate systems $OXYZ$ and $O'X'Y'Z'$ are fixed on them respectively. The S' system makes a uniform linear motion at a speed u along the X axis relative to the S system. Under the assumption of the absoluteness of length measurement and the absoluteness of simultaneous measurement, that is, time and space are considered to be independent of each other, absolutely

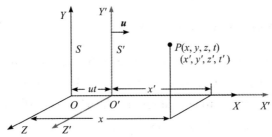

Figure 13.1.1 The Galileo coordinate transformation

invariable and independent of the motion of objects, the transformation between the S system and the S' system can be expressed as

$$\begin{cases} x' = x - ut \\ y' = y \\ z' = z \\ t' = t \end{cases} \quad \text{or} \quad \begin{cases} x = x' + ut \\ y = y' \\ z = z' \\ t = t' \end{cases} \tag{13.1.1}$$

Equation (13.1.1) is called the **Galileo coordinate transformation**.

The above-mentioned fact that the laws of mechanics are equivalent in all inertial frames means that Newton's laws of motion and other basic laws of mechanics derived from it have the same form in all inertial frames, rather than that the physical phenomena observed in different inertial frames are the same. Let's take a look at Newton's second law under the Galileo transformation. Taking the derivation of the Galileo coordinate transformation Equation (13.1.1) with respect to time, we can obtain the velocity transformation

$$\begin{cases} v'_x = v_x - u \\ v'_y = v_y \\ v'_z = v_z \end{cases} \tag{13.1.2}$$

and acceleration transformation

$$\begin{cases} a'_x = a_x \\ a'_y = a_y \\ a'_z = a_z \end{cases} \tag{13.1.3}$$

According to the Galileo acceleration transformation Equation (13.1.3), the accelerations of objects in different inertial frames are the same while classical mechanics believes that force and mass are independent of the reference frame. Newton's second law is $\boldsymbol{F} = m\boldsymbol{a}$ in the S frame, and $\boldsymbol{F}' = m\boldsymbol{a}'$ in the S' frame, that is, the mathematical form of Newton's second law remains unchanged under the Galileo transformation, known as the invariance of the Galileo transformation. It can be concluded from this that all the basic laws of classical mechanics satisfy the invariance of the Galileo transformation, or the laws of mechanics are equivalent in all inertial systems. This is **the principle of mechanical relativity**. The Galileo transformation is the mathematical expression of the principle of relativity in classical mechanics.

13.1.3 Classical space-time view

In the Galileo transformation, it is already clearly written that $t = t'$. This means that time is the same in all inertial frames, or that there is time independent of the motion state of the reference frame, that is, **time is absolute**. Since time is the same, the time interval must also be the same in all inertial frames, i.e.,

$$\Delta t = \Delta t' \tag{13.1.4}$$

The time interval is also absolute under the Galileo transformation. There is another invariant in the Galileo transformation, that is, at any time, the length of two points in space is constant for all inertial systems. At the same moment, the lengths of two points in space are expressed in the two inertial frames as

$$\Delta L = \sqrt{(x_2-x_1)^2+(y_2-y_1)^2+(z_2-z_1)^2}$$

and

$$\Delta L' = \sqrt{(x'_2-x'_1)^2+(y'_2-y'_1)^2+(z'_2-z'_1)^2}$$

It is easy to prove by the Galileo transformation

$$\Delta L = \Delta L' \tag{13.1.5}$$

Equation (13.1.5) shows that, in all inertial systems, the lengths of two points in space are the same at any given moment, or the length of space has nothing to do with the motion state of the reference frame, that is, the **length of space is absolute.**

Therefore, in the Galileo transformation, both time and space have nothing to do with the motion state of the reference frame. There is no connection between time and space and they are absolute. This is **the classical concept of space and time.** So, it can be said that the Galileo transformation is the concentrated expression of the classical space-time concept.

Newton once said, "Absolute, true, and mathematical time, of itself, and from its own nature, flows equably without relation to anything external…" "Absolute space, in its own nature, without relation to anything external, remains always similar and immovable". This is the absolute space-time view of classical mechanics. In this view, time and space are independent of each other, unrelated to each other, and independent of matter and motion. Space is like a visible and never-moving frame that holds everything in the universe, and time is like independent flowing water. By the absolute space-time theory, the time interval and space interval are absolutely unchanged, and the observation results in different inertial systems are always the same. This is the premise of the Galileo transformation, that is, the Galileo transformation can be derived from the absolute space-time theory.

13.2 Basic principles of the special theory of relativity

13.2.1 Background of the special theory of relativity

In the late 19th century, the development of electromagnetic theory led to the establishment of Maxwell's electromagnetic theory. Maxwell's equations not only fully reflected the universal law of electromagnetic motion, but also predicted the existence of electromagnetic waves and revealed the electromagnetic nature of light. This was another great achievement of classical physics after Newton's laws of mechanics.

However, for a long time, the mechanistic theory prevailed in the physics community, claiming that physics could be described by a single classical mechanical image, and its representative was the "ether hypothesis". This hypothesis held that the ether, the elastic medium that transmitted all electromagnetic waves, including light waves, filled the entire universe. The electromagnetic wave was the mechanical motion state of the ether medium, and the vibration of charged particles would cause the deformation of the ether, and the propagation of this deformation in the form of an elastic wave formed an electromagnetic wave. If the electromagnetic wave that had such a large speed and was a transverse wave really propagated through the ether, then the ether must have a very high shear modulus, and at the same time, the large and small celestial bodies in the universe traveled through the ether without any drag force from it.

At the same time, starting from Maxwell's equations, the wave equation of electromagnetic waves propagating in free space could be obtained, and in the wave equation, the speed of light in a vacuum $c = 1/\sqrt{\mu_0 \varepsilon_0}$ appeared in the form of a universal constant, that is, the propagation speed of light waves in a vacuum was related to the dielectric constant and the magnetic permeability of the vacuum, and had nothing to do with the relative motion of the reference frame and the direction of propagation. But from the perspective of the Galileo transformation, velocity was always relative to a specific reference frame, so velocity was not allowed to appear as a universal constant in the basic equations of classical mechanics. At that time it was generally believed that since the speed of light c appeared in the wave equation of electromagnetic waves, it indicated that Maxwell's equations were only true in the reference frame at rest relative to the ether, in which the propagation speed of electromagnetic waves in all directions in a vacuum was the same and equal to the constant c, but it was generally not equal to the constant c in the inertial system in motion relative to the ether.

So a situation arose: classical mechanics and classical electromagnetism in classical physics had completely different properties. The former satisfied Galileo's principle of relativity, and all inertial systems were equivalent; the latter did not satisfy Galileo's principle of relativity, and there was an optimal frame of reference at rest relative to the ether. People called this optimal reference frame the absolute reference frame, and the motion relative to the absolute reference frame the absolute motion. The earth traveled through the ether, and measuring the absolute motion of the earth relative to the ether naturally became the first concern of people at that time. The first such measurement was the famous Michelson-Morley experiment.

The device of the Michelson-Morley experiment was shown in Figure 12.5.13 in Chapter 12. The device moved relative to the ether with the earth, and the observer in the earth's reference frame would feel the oncoming ether wind. According to Galileo's speed transformation, under the influence of the ether wind, the two mutually perpendicular light paths in the device had different light speeds. Therefore, when the device was rotated 90 degrees, the optical path changed by 2δ, and the number of interference fringes that moved within the telescope's field of view was

$$\Delta N = \frac{2\delta}{\lambda} = \frac{2lv^2}{\lambda c^2} \qquad (13.2.1)$$

where, λ, c and l were all known. If the number of moving fringes ΔN could be measured, the absolute speed of the earth relative to the ether v could be calculated from the above formula, so that the ether could be used as the absolute reference system.

Michelson was the first to complete this experiment in 1881, but the expected movement of the fringes was not observed. In 1887, Michelson and Morley improved the experimental setup and extended the length of the two optical paths to 11 m. The expected number of fringe movement was 0.4, which was 40 times the smallest observable number, but no fringe movement was observed. The negative result of the Michelson-Morley experiment seemed to tell people who believed in the ether that the motion of the earth relative to the ether did not exist, and the ether as an absolute frame of reference did not exist.

13.2.2 Basic assumptions of the special theory of relativity

In 1905, Einstein took a different approach. He believed that the ether hypothesis should be completely abandoned. Like mechanical phenomena, there should not be a special optimal reference frame. The principle of relativity should be of general significance; not only the laws of classical mechanics, but also the laws of classical electromagnetism and other physical laws, should maintain the same mathematical form in all inertial systems. In this way, it was necessary to find or establish a new transformation relationship between the inertial systems to replace the Galileo transformation. We mentioned earlier that the Galileo transformation was the concentrated expression of the classical space-time concept, and establishing a new transformation relationship meant establishing a new space-time concept, which was the space-time concept of special relativity to be discussed below.

As mentioned earlier, in the classical theory of electromagnetism, namely Maxwell's equations, there was a universal constant, which was the speed of light c in a vacuum. As long as the classical electromagnetic theory was considered to satisfy a new principle of relativity, Maxwell's equations should remain in the same mathematical form under this new transformation relationship, that is to say, in all inertial systems, electromagnetic waves would propagate at the speed of light c, i.e., the invariance of the speed of light.

Einstein summarized the above discussion into two basic principles of special relativity:

(1) **The principle of relativity: In all inertial systems, all laws of physics have the same mathematical expression.** That is, all inertial systems are equivalent for the laws describing all physical phenomena. This assumption was a generalization of the principle of mechanical relativity.

(2) **The principle of constancy of the speed of light: In all inertial systems, the speed of light propagating in any direction in a vacuum is equal to the constant c, regardless of the motion state of the light source and the observer.** This hypothesis was consistent with experimental results such as the Michelson-Morley experiment.

The two basic assumptions of special relativity were incompatible with the absolute space-time theory of classical mechanics and required a new space-time coordinate transformation relationship compatible with the Galileo transformation at low speed (much smaller than the speed of light). Based on these two assumptions, Einstein derived the relative space-time coordinate transformation formula that could correctly reflect the laws of physics.

In fact, long before Einstein published his theory of relativity, Lorentz had proposed the same transformation when he was studying the electromagnetic field theory and explaining the Michelson-Morley experiment, so it was called the Lorentz-Einstein transformation, or Lorentz transformation for short.

13.2.3 The Lorentz transformation

For simplicity, we assume that the S system and the S' system are two inertial coordinate systems that move relatively in a straight line at a uniform speed, and it is stipulated that the S' system moves in the positive direction of the X axis of the S system with a velocity u relative to the S system. The X', Y' and Z' axes are parallel to the X, Y and Z axes, respectively. When the origin O of the S system coincides with the origin O' of the S' system, the clocks at the origins of both inertial coordinate systems indicate zero o'clock. We then derive a new transformation relation between these two inertial frames.

The new transformation should first satisfy two basic principles of special relativity. In addition, when the speed of motion is much less than the speed of light in a vacuum, the new transformation should transition to the Galileo transformation, because the Galileo transformation is verified to be correct in practice in this case. Finally, the new transformation should be linear, because only then can it be guaranteed that when an object is moving uniformly in a straight line in one frame of reference, it will also be observed moving in a straight line with uniform velocity in another frame of reference. Based on these requirements, we make the simplest assumption

$$x' = k(x - ut) \tag{13.2.2}$$

where k is a scaling factor independent of both x and t. According to the first basic principle of the special theory of relativity, the S system and the S' system have no

difference except for relative motion. Considering the relativity of motion, correspondingly, there should be
$$x = k(x' + ut') \tag{13.2.3}$$
The transformation of the other two coordinates is easy to write
$$\begin{cases} y' = y \\ z' = z \end{cases} \tag{13.2.4}$$

In order to obtain the transformation of the time coordinate, substitute Equation (13.2.2) in Equation (13.2.3), and we get
$$x = k^2(x - ut) + kut'$$
Solving it for t', we get
$$t' = kt + \frac{1-k^2}{ku}x \tag{13.2.5}$$

Determining k requires the use of the second fundamental principle of special relativity. According to the initial conditions we specified, when the origins of the two inertial coordinate systems coincide, there is $t = t' = 0$. If a light source at the common origin emits a light pulse at this moment, the light pulse can be observed to propagate in all directions at a rate c in both S and S' systems. So, in the S system, there is
$$x = ct \tag{13.2.6}$$
and in the S' system, we have
$$x' = ct' \tag{13.2.7}$$
Substituting Equations (13.2.6) and (13.2.7) into Equations (13.2.2) and (13.2.3) respectively, we get
$$ct' = k(c - u)t$$
and
$$ct = k(c + u)t'$$
Eliminating t and t' from the above two equations, we can get
$$k = \frac{1}{\sqrt{1 - (u/c)^2}}$$
Substitute k into Equations (13.2.2) and (13.2.5) to get a new transformation form
$$\begin{cases} x' = \dfrac{x - ut}{\sqrt{1 - (u/c)^2}} \\ y' = y \\ z' = z \\ t' = \dfrac{t - \dfrac{u}{c^2}x}{\sqrt{1 - (u/c)^2}} \end{cases} \tag{13.2.8}$$

This new transformation is called the **Lorentz transformation.**

By exchanging the primed and unprimed quantities in Equation (13.2.8), and replacing u with $-u$, we get the inverse Lorentz transformation

$$\begin{cases} x = \dfrac{x' + ut'}{\sqrt{1-(u/c)^2}} \\ y = y' \\ z = z' \\ t = \dfrac{t' + \dfrac{u}{c^2}x'}{\sqrt{1-(u/c)^2}} \end{cases} \quad (13.2.9)$$

The following are some notes about the Lorentz coordinate transformation equation:

(1) In the special theory of relativity, the Lorentz transformation occupies a central position, which embodies the space-time view of the special theory of relativity. In the Lorentz transformation, x' is a function of x and t, t' is also a function of x and t, and both are related to the relative motion velocity $\boldsymbol{u}$ of the S system and S' system, revealing the inseparable relationship between time, space and matter motion.

(2) The Lorentz transformation denies the absolute time concept of $t = t'$. In the special theory of relativity, the measurement of time and space cannot be separated from each other, and the description of physical events requires four-dimensional space-time coordinates.

(3) Both time and space coordinates are real numbers, so $u < c$, which means the speed of motion of any object in the universe cannot exceed the speed of light in a vacuum.

(4) When $u \ll c$, $\sqrt{1-\left(\dfrac{u}{c}\right)^2} \approx 1$; then $x' = \dfrac{x-ut}{\sqrt{1-(u/c)^2}} \approx x - ut$, and $t' = \dfrac{t-\dfrac{u}{c^2}x}{\sqrt{1-(u/c)^2}} \approx t$. The Lorentz transformation is transformed into the Galileo transformation, that is, the Galileo transformation is a low-speed approximation of the Lorentz transformation.

(5) According to the Lorentz transformation, the relations of time interval and space interval of two events in different inertial frames can also be obtained.

It is assumed that the space-time coordinates of two physical events 1 and 2 in the S system are respectively (x_1, y_1, z_1, t_1) and (x_2, y_2, z_2, t_2), and the coordinates in the S' system are respectively (x'_1, y'_1, z'_1, t'_1) and (x'_2, y'_2, z'_2, t'_2). In the S system, the spatial interval of two events is $\Delta x = x_2 - x_1$, $\Delta y = y_2 - y_1$ and $\Delta z = z_2 - z_1$, and the time interval is $\Delta t = t_2 - t_1$. In the S' system, the spatial interval of two events is $\Delta x' = x'_2 - x'_1$, $\Delta y' = y'_2 - y'_1$ and $\Delta z' = z'_2 - z'_1$, and the time interval is $\Delta t' = t'_2 - t'_1$. Since the relative motion velocity of the two systems $\boldsymbol{u}$ is in the X axis direction, according to the Lorentz transformation Equation (13.2.8), it is obvious that $\Delta y' = \Delta y$ and $\Delta z' = \Delta z$, that is, the spatial interval remains unchanged in the direction perpendicular to the relative motion. But in the direction in which the relative motion occurs, that is, the X axis direction, the spatial interval changes

$$\Delta x' = x'_2 - x'_1 = \frac{\Delta x - u \Delta t}{\sqrt{1 - u^2/c^2}} \tag{13.2.10}$$

The time interval relationship is

$$\Delta t' = t'_2 - t'_1 = \frac{\Delta t - \frac{u}{c^2}\Delta x}{\sqrt{1 - u^2/c^2}} \tag{13.2.11}$$

Similarly, we can also get

$$\Delta x = \frac{\Delta x' + u \Delta t'}{\sqrt{1 - u^2/c^2}} \tag{13.2.12}$$

$$\Delta t = \frac{\Delta t' + \frac{u}{c^2}\Delta x'}{\sqrt{1 - u^2/c^2}} \tag{13.2.13}$$

Equations (13.2.11) and (13.2.13) show that the spatial distance of the event will affect the measurement of the time interval by observers in different inertial systems, that is, the **spatial interval and the time interval are closely related**. This is the difference between the space-time view of special relativity and the absolute space-time view of classical mechanics.

Example 13.2.1 In the ground reference frame S, at $x = 1.0 \times 10^6$ m, a bomb explodes at time $t = 0.02$ s. If there is a high-speed train moving in the positive direction of the X axis at a rate of $u = 0.75c$, find the location (space coordinates) and time of the bomb explosion measured by an observer in the high-speed train reference frame S'.

Solution Assume that the space-time coordinates of the bomb explosion measured by the observer in the S system and in the S' system are (x, t) and (x', t'), respectively. According to Formula (13.2.8), the space-time coordinates of the bomb explosion observed in the high-speed train (S' system) are

$$x' = \frac{x - ut}{\sqrt{1 - (u/c)^2}} = \frac{1 \times 10^6 - 0.75 \times 3 \times 10^8 \times 0.02}{\sqrt{1 - (0.75)^2}} = -5.29 \times 10^6 \text{ m}$$

$$t' = \frac{t - \frac{u}{c^2}x}{\sqrt{1 - (u/c)^2}} = \frac{0.02 - \frac{0.75 \times 1 \times 10^6}{3 \times 10^8}}{\sqrt{1 - (0.75)^2}} = 0.0265 \text{ s}$$

$x' < 0$ means that the explosion location of the bomb is on the negative side of the origin O' of the X' axis in the S' system. $t' \neq t$ means that the explosion times measured in the two inertial systems are different.

Example 13.2.2 Observers A and B are at rest in two inertial reference frames S and S', respectively. S' flies at a constant speed of $u = 0.6c$ along the positive X axis relative to S. The spatial interval and the time interval between the two events measured by A are 1.0×10^9 m and 4 s, respectively. Find the spatial interval and the time interval between the two events measured by B.

Solution The time and space intervals of the two events measured by A and B are respectively

$S(A)$:
$$\Delta x = x_2 - x_1 = 1.0 \times 10^9 \text{ m}, \quad \Delta t = t_2 - t_1 = 4 \text{ s}$$

$S'(B)$:
$$\Delta x' = x'_2 - x'_1, \quad \Delta t' = t'_2 - t'_1$$

From the Lorentz transformation, we have

$$\Delta x' = \frac{\Delta x - u \Delta t}{\sqrt{1 - u^2/c^2}} = \frac{1 \times 10^9 - 0.6 \times 3 \times 10^8 \times 4}{\sqrt{1 - (0.6c/c)^2}} = 3.5 \times 10^8 \text{ m}$$

$$\Delta t' = \frac{\Delta t - \frac{u}{c^2} \Delta x}{\sqrt{1 - u^2/c^2}} = \frac{4 - \frac{0.6}{3 \times 10^8} \times 1 \times 10^9}{\sqrt{1 - (0.6c/c)^2}} = 2.5 \text{ s}$$

It can be seen that the measurement of two events in different reference systems shows different spatial intervals and different time intervals, indicating that the measurement of time and space is relative.

13.3 Space-time view of the special theory of relativity

The principle of the constant speed of light obviously does not conform to the Galilean transformation, and negates the Galilean transformation and the absolute space-time view related to it. The special theory of relativity proposes a new view of time and space for people, and many surprising conclusions that are contrary to our daily experience can be obtained by using the Lorentz transformation. For example, the distance between two points or the length of an object varies with the inertial frame in which the measurement is taken, as does the time a process takes with the inertial frame. These conclusions have been confirmed by many experiments in modern high-energy physics. Next, we will first discuss the relativity of simultaneity, which is the basis of the special theory of relativity, and then discuss the contraction effect of length and the delay effect of time.

13.3.1 Relativity of simultaneity

In the special theory of relativity, there is no identical time, and both time and time interval are related to the motion state of the observer. Let's look at the time interval between two events that occur in two inertial frames. It is assumed that the two inertial frames are still the S frame and S' frame taken in the previous section. If light pulse signals A and B are sent out at two different locations in the S system at the same time, their space-time coordinates are $A(x_1, y_1, z_1, t_1)$ and $B(x_2, y_2, z_2, t_2)$, respectively. Because they are sent out at the same time, $t_1 = t_2$. In order to ensure that the two optical pulses are sent out at the same time, an optical pulse receiving device can be placed at the midpoint M of the line connecting the two locations. If the receiving device receives the optical pulse signals at the same time, it means that the two signals are sent at the same time. However, observed in the S' system, the times when the two optical pulse signals are emitted are respectively

$$t'_1 = \frac{t - \frac{u}{c^2}x_1}{\sqrt{1-(u/c)^2}}, \quad t'_2 = \frac{t - \frac{u}{c^2}x_2}{\sqrt{1-(u/c)^2}}$$

Considering $t_1 = t_2$, the time interval is

$$\Delta t' = t'_2 - t'_1 = \frac{\frac{u}{c^2}(x_1 - x_2)}{\sqrt{1-u^2/c^2}} \neq 0 \quad (13.3.1)$$

The above expression indicates that events that occur simultaneously at two different locations in the S system do not appear to occur simultaneously in the S' system, which is the relativity of simultaneity. Because the motion is relative, the effect is reciprocal, that is, events that occur simultaneously at two different locations in the S' system do not appear to be simultaneous in the S system. It can also be seen from Expression (13.3.1) that when $x_1 = x_2$, that is, two events occur at the same place, then the events that occur at the same time are simultaneous in different inertial systems. It can be concluded from this that in the special theory of relativity, time and space are related to each other.

13.3.2 Time dilation effect

In the theory of relativity, the frame of reference that is stationary relative to the physical event to be studied is usually called the inherent frame of reference, and the physical quantity measured in the inherent reference frame is called the **intrinsic physical quantity**. According to the theory of relativity, in different inertial frames, the measurement of time intervals is related to the reference frame and is also relative. Let the inertial frame S' move relative to the S frame along the X axis at a velocity u. There is a flashing timer located at x' in the S' frame, and it sends out two flashing signals at the time t'_1 and t'_2, which are recorded as event 1 and event 2, respectively. The time interval between these two events is $\Delta t' = t'_2 - t'_1$, and the space interval is $\Delta x' = 0$. According to Equation (13.2.12), observed in the S system, the time interval between these two events becomes

$$\Delta t = \frac{\Delta t' + \frac{u}{c^2}\Delta x'}{\sqrt{1-u^2/c^2}} = \frac{\Delta t'}{\sqrt{1-u^2/c^2}}$$

where $\Delta t'$ represents the time interval between two co-located events in the S' system, called **intrinsic time**, denoted as τ_0; Δt represents the time interval between these two events (becoming remote events) in the S system, and is recorded as τ. Then

$$\tau = \frac{\tau_0}{\sqrt{1-u^2/c^2}} = \gamma\tau_0 \qquad (13.3.2)$$

The factor in the above formula $\gamma = \frac{1}{\sqrt{1-u^2/c^2}}$ is called the time delay factor. Because $\gamma > 1$, there is always $\tau_0 < \tau$, that is, the intrinsic time is always the shortest. For observers in other inertial frames moving relative to the S' frame, the clock in the S' frame slows down, which is called the **time delay effect or time dilation effect**.

The slowing down of a moving clock (the time delay effect) has nothing to do with any mechanical causes of the clock itself or atomic internal processes. It means that everything that happens to a moving object is slowed down relative to a stationary observer. It should be pointed out that the time dilation effect is relative, that is, in the eyes of the observer in the S system, the clock in the S' system has slowed down due to the movement; and conversely, the observer in the S' system will also think that the clock in the S system is slowed down.

13.3.3 Length contraction effect

In the Galileo transformation, the distance between two points or a length of an object does not change with the inertial system, that is, the length of an object is absolute. In the Lorentz transformation, the measure of the length is also related to the inertial system.

As shown in Figure 13.3.1, let the inertial frame S' move relative to the S frame along the X axis at a velocity $\boldsymbol{u}$, and a rod is placed along the X' axis and is stationary relative to the S' frame. For the S' system, it is not difficult to measure its length. It is only necessary to record the coordinates of the two ends of the rod x'_1 and x'_2, and the difference between the two coordinates $L' = x'_2 - x'_1$ is the length of the rod. We call the length of the object measured by the observer at rest relative to the object the **intrinsic length** L_0, and $L' = L_0$. When measuring the length of the same rod in the S system, the coordinates of both ends of the rod x_1 and x_2 must be measured at the same time, that is, $t_1 = t_2$. Then, the correct value of rod length $L = x_2 - x_1 = \Delta x$ can be obtained.

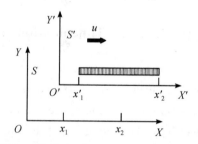

Figure 13.3.1 Length contraction effect

According to Equation (13.2.9), we can get

$$L_0 = \Delta x' = \frac{\Delta x - u\Delta t}{\sqrt{1-u^2/c^2}}$$

Because $\Delta t = 0$ and $L = \Delta x$, then

$$L_0 = \frac{L}{\sqrt{1-u^2/c^2}}$$

It can be rewritten as

$$L = L_0 \sqrt{1 - \frac{u^2}{c^2}} \tag{13.3.3}$$

It can be seen that the measured length L of the object measured in the S system is always smaller than the intrinsic length measured in the S' system, that is, when the object moves relative to the reference frame, the measured length is shortened, which is called the **length contraction effect**. The length contraction effect shows the relativity of space: Observed in any inertial frame, the length of the object moving relative to the inertial frame in its direction of motion will contract. This contraction only occurs in the direction of motion. No length shrinkage occurs in the direction perpendicular to the direction of motion. The relativistic length contraction effect is a property of space-time and has nothing to do with the specific composition, structure and interaction of matter.

The time dilation effect and length contraction effect have been confirmed by a large number of modern physical experiments.

Example 13.3.1 The lifetime of a meson at rest is $\tau_0 = 2.6 \times 10^{-8}$ s. If its velocity in the laboratory is $u = 0.9c$, find the distance that this kind of meson can fly when observed in the laboratory.

Solution The lifetime of the meson at rest is the proper time. According to the time dilation effect, the meson's lifetime observed in the laboratory is

$$\tau = \frac{\tau_0}{\sqrt{1 - \frac{u^2}{c^2}}} = \frac{2.6 \times 10^{-8}}{\sqrt{1 - \frac{(0.9c)^2}{c^2}}} \approx 5.96 \times 10^{-8} \text{ s}$$

So, the distance the meson can fly is

$$\Delta s = u\tau \approx 16 \text{ m}$$

Example 13.3.2 A meter ruler rests in the S' system at a 45° angle to the $O'X'$ axis. If the S' system moves along the OX axis at $u = \sqrt{3}c/2$ relative to the S system, what is the length of the ruler measured in the S system? What is the angle with the OX axis?

Solution The meter ruler is at rest in the S' system. The projections along the X' and Y' axes are

$$L_{0x'} = L_0 \cos 45° = \frac{\sqrt{2}}{2}, \quad L_{0y'} = L_0 \sin 45° = \frac{\sqrt{2}}{2}$$

Length contraction is only in the direction of motion, and then the projection of the meter ruler in the S system along the X and Y axes are

$$L_x = L_{0x'} \sqrt{1 - \left(\frac{u}{c}\right)^2} = \frac{\sqrt{2}}{4}, \quad L_y = L_{0y'} = \frac{\sqrt{2}}{2}$$

Observed in the S system, the length of the meter ruler is

$$L = \sqrt{L_x^2 + L_y^2} = 0.79 \text{ m}$$

The angle with the OX axis is

$$\tan\theta = \frac{L_y}{L_x} = 2, \quad \theta = 63°27'$$

13.4 Dynamic basis of the special theory of relativity

According to the relativity principle of special relativity, all physical laws have the same form in all inertial systems, which requires basic physical laws such as the law of conservation of momentum and the law of conservation of energy not only still hold and remain unchanged in the Lorentz transformation, but also return to the corresponding classical mechanic laws at low speed. So, we have to redefine some physical quantities.

13.4.1 Relativistic mass-velocity relationship

In classical mechanics, the mass of an object is a constant and has nothing to do with the speed of the object; if the object is accelerated by a constant force, as long as the force acts for a long enough time, its speed will eventually exceed the speed of light. This obviously contradicts the conclusion in the theory of relativity that there is a limit speed c for object movement. Therefore, the special theory of relativity believes that the mass of the object should not be constant. It is related to the selection of the inertial frame, that is, related to the speed of the object.

Considering that the law of conservation of momentum is a universal law, it should also hold true in the theory of relativity. If the momentum of a system is conserved in one inertial system, then after the Lorentz transformation, it is also conserved in the other inertial system. Based on the law of conservation of momentum and the relationship of relativistic velocity transformation, the relationship between the mass m of a moving object and its velocity v can be derived as

$$m = \frac{m_0}{\sqrt{1-v^2/c^2}} = \gamma m_0 \tag{13.4.1}$$

Among them m_0 is the mass measured when the object is stationary relative to the inertial system, which is called **static mass**; m is the mass measured when the object moves at a speed of v, which is called **relativistic mass**. As shown in Figure 13.4.1, when the object is accelerated to a speed close to the speed of light, the speed no longer increases linearly and cannot exceed the speed of light.

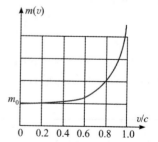

Figure 13.4.1 Relativistic mass-velocity relationship

The electron linear accelerator put into operation at Stanford University in 1966 has a total length of 3×10^3 m and an accelerating potential difference of 7×10^6 V · m^{-1}. It can accelerate electrons to $0.999,999,999,7c$, which is close to the speed of light but cannot exceed the speed of light. It strongly proves the correctness of the relativistic mass-velocity relationship.

13.4.2 Basic equation of relativistic dynamics

Relativistic momentum is the product of relativistic mass and velocity, namely

$$\boldsymbol{p} = m\boldsymbol{v} = \frac{m_0}{\sqrt{1-v^2/c^2}} \boldsymbol{v} \tag{13.4.2}$$

Therefore, the classical Newton's second law can still be expressed as

$$\boldsymbol{F} = \frac{\mathrm{d}(m\boldsymbol{v})}{\mathrm{d}t} = \frac{\mathrm{d}}{\mathrm{d}t}\left(\frac{m_0 \boldsymbol{v}}{\sqrt{1-v^2/c^2}}\right) \tag{13.4.3}$$

This is the basic equation of relativistic particle dynamics. Obviously, when the moving speed of an object is much less than the speed of light, i.e., $v \ll c$, the relativistic mass in the above formula is approximately equal to the static mass, and the equation approximately becomes the original Newton's equation, or the Newton's equation is a low-speed approximation of the relativistic dynamic equation.

13.4.3 Mass-energy relationship

Let a particle start to move from rest under the action of force $\boldsymbol{F}$. Since Equation (13.4.3) is formally consistent with Newton's second law of classical mechanics, it can be considered that the kinetic energy theorem derived from it is still valid, that is, the work done by the force $\boldsymbol{F}$ is equal to the increment of the kinetic energy of the particle

$$\mathrm{d}E_k = \boldsymbol{F} \cdot \mathrm{d}\boldsymbol{r}$$

Substituting $\boldsymbol{F} = \dfrac{\mathrm{d}(m\boldsymbol{v})}{\mathrm{d}t}$ into the above equation, we can get

$$\mathrm{d}E_k = \mathrm{d}(m\boldsymbol{v}) \cdot \frac{\mathrm{d}\boldsymbol{r}}{\mathrm{d}t} = \mathrm{d}(m\boldsymbol{v}) \cdot \boldsymbol{v} = (\mathrm{d}m)\boldsymbol{v} \cdot \boldsymbol{v} + m\mathrm{d}\boldsymbol{v} \cdot \boldsymbol{v} = v^2 \mathrm{d}m + mv\mathrm{d}v \tag{13.4.4}$$

The mass-velocity relation Equation (13.4.1) can be rearranged as $m^2 v^2 = m^2 c^2 - m_0^2 c^2$. Differentiating both sides of the equation, we can get

$$v^2 \mathrm{d}m + mv\mathrm{d}v = c^2 \mathrm{d}m \tag{13.4.5}$$

Substituting Equation (13.4.5) into Equation (13.4.4), we can get

$$\mathrm{d}E_k = c^2 \mathrm{d}m$$

Integrating both sides at the same time, we can get the **relativistic kinetic energy**.

$$E_k = \int_{m_0}^{m} c^2 \mathrm{d}m = mc^2 - m_0 c^2 \tag{13.4.6}$$

Obviously, the relativistic kinetic energy is completely different from the classical kinetic energy, but when the speed of the moving object is much less than the speed of light, the mass-velocity relationship can be expanded as

$$m = \frac{m_0}{\sqrt{1-v^2/c^2}} = m_0 \left[1 + \frac{1}{2}\left(\frac{v}{c}\right)^2 + \cdots\right] \approx m_0 \left[1 + \frac{1}{2}\left(\frac{v}{c}\right)^2\right]$$

At this time, the kinetic energy is approximated to $E_k \approx \frac{1}{2}m_0 v^2$ and returns to the classical kinetic energy form.

The term $m_0 c^2$ in Equation (13.4.6) is the **static energy** of the object, that is

$$E_0 = m_0 c^2 \qquad (13.4.7)$$

The term mc^2 is defined as the **total energy of the object**, i.e.

$$E = mc^2 = \frac{m_0 c^2}{\sqrt{1-v^2/c^2}} \qquad (13.4.8)$$

This formula is the **mass-energy relationship**.

Equation (13.4.6) can also be written as

$$E = E_0 + E_k \qquad (13.4.9)$$

That is, **the total energy of an object consists of two parts: static energy and kinetic energy.**

The mass-energy relationship shows that the energy of an object is proportional to its mass, and even if the object is stationary, it still has static energy. In the theory of relativity, mass and energy are inseparable, and the change of mass must lead to the change of energy, namely

$$\Delta E = \Delta m c^2 \qquad (13.4.10)$$

When the mass of an object decreases, it means that it releases a huge amount of energy, which is the theoretical basis for the utilization of atomic energy (nuclear energy). The atomic bomb and hydrogen bomb technology are both applications of the mass-energy relationship of the special theory of relativity, and their success has also become a strong proof of the correctness of the special theory of relativity.

Example 13.4.1 The energy released when an isolated nucleus forms an atomic nucleus is the binding energy of the nucleus. The static masses of protons and ions are known to be $m_p = 1.672,62 \times 10^{-27}$ kg and $m_n = 1.674,93 \times 10^{-27}$ kg, respectively, and the static mass of the deuteron composed of them is $m_D = 3.343,59 \times 10^{-27}$ kg. Find the binding energy of the deuteron.

Solution The mass loss in the process of combining protons and neutrons into a deuteron nucleus is

$$\Delta m = m_p + m_n - m_D = (1.672,62 + 1.674,93 - 3.343,59) \times 10^{-27} \text{ kg} = 3.96 \times 10^{-30} \text{ kg}$$

The static energy corresponding to the mass loss is

$$\Delta E_k = \Delta m_0 c^2 = 3.96 \times 10^{-30} \times (2.998 \times 10^8)^2 = 2.22 \text{ Mev}$$

13.4.4 Momentum-energy relationship

The mass-velocity relationship $m = \dfrac{m_0}{\sqrt{1-v^2/c^2}}$ can be rewritten as $m^2 v^2 c^2 = m^2 c^4 - m_0^2 c^4$. According to the definitions of relativistic momentum, energy and static energy, the above formula can be written as

Chapter 13 Fundamentals of the Special Theory of Relativity   235

$$p^2c^2 = E^2 - E_0^2 \quad \text{or} \quad E^2 = E_0^2 + p^2c^2 \qquad (13.4.11)$$

This is the relationship between momentum and energy in special relativity.

This relationship can be represented graphically by the three sides of a right triangle, as shown in Figure 13.4.2.

Some particles with zero static mass, such as photons, must have $E = pc = mc^2$, that is, the particles' momentum $p = mc$ and the particles' velocity $v = c$. This suggests that particles with zero static mass must move at the speed of light.

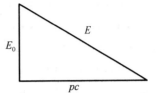

Figure 13.4.2 Relativistic momentum-energy relationship

For objects moving at low speeds, that is, the speed of motion $v \ll c$, we get

Momentum: $p \approx m_0 v$

Kinetic energy: $E_k \approx \dfrac{1}{2} m_0 v^2 = \dfrac{p^2}{2m_0}$

Total energy: $E = E_0 + \dfrac{p^2}{2m_0}$

Example 13.4.2 A stationary atom with mass m_0 is hit by a photon with energy E, the atom absorbs all the energy of the photon, and the energy and momentum are conserved during the collision process. Find the velocity and static mass of the combined system.

Solution Let the velocity of the merging system be v, the mass be M, and the static mass be M_0.

The momentum of the atom before the collision is 0, the momentum of the photon is E/c, and the momentum of the system after the collision is Mv.

From the conservation of momentum before and after the collision, we get

$$p = E/c = Mv$$

From the conservation of energy before and after the collision, we get

$$m_0 c^2 + E = Mc^2$$

From the above two equations, we get

$$v = \dfrac{Ec}{m_0 c^2 + E}, \quad M = \dfrac{m_0 c^2 + E}{c^2}$$

According to $M = \dfrac{M_0}{\sqrt{1-(v/c)^2}}$, we get

$$M_0 = \dfrac{m_0 c^2 + E}{c^2} \sqrt{1 - \left(\dfrac{E}{m_0 c^2 + E}\right)^2} = \dfrac{1}{c^2} \sqrt{(m_0 c^2 + E)^2 - E^2} = m_0 \sqrt{1 + \dfrac{2E}{m_0 c^2}}$$

 13.4.5 Application of relativistic mass-energy relationship in nuclear energy

The nucleus of an atom is composed of protons and neutrons, which are collectively called nucleons. Experimental data have found that the mass of any atomic nucleus is always less than the sum of the masses of all its nucleons, that is, nucleons have a mass deficit in the process of composing a nucleus, and the deficit is equal to the reduction in mass Δm when they are combined into a nucleus. According to Einstein's mass-energy relationship Equation (13.4.10), when several nucleons are combined into an atomic nucleus, energy is released, and this energy $\Delta E = \Delta mc^2$ is called the binding energy of the nucleus. The average binding energy of all nucleons in a nucleus is called the average binding energy (also called the specific binding energy). The greater the average binding energy, the stronger the binding force of nucleons in the nucleus and the more stable the nucleus is. Figure 13.4.3 shows the distribution curve of the average binding energy of different nuclei versus mass number. It can be seen that the average binding energy of medium-mass nuclei is larger, while the average binding energy of light and heavy nuclei is smaller. This means that when a heavy nucleus splits into two medium-mass nuclei or when two very light nuclei fuse into a heavier nucleus, there will be energy released, which is atomic energy (also known as nuclear energy). In 1939, the German female physicist Meitner predicted that the fission energy of a uranium nucleus was 200 Mev based on the relationship between mass and energy. Shortly thereafter, the Austrian physicist Frisch observed this energy in experiments. This energy is much larger than the general reaction energy. In complete fission of one gram of uranium, the energy released is 8×10^{10} J, which is equivalent to the combustion heat of 2.5 tons of coal. When a deuterium nucleus and a tritium nucleus are aggregated into a helium nucleus, an energy of 17.6 Mev is released, which is 4 times the fission energy of uranium of the same mass. It can be seen that there is a large amount of available energy in the nucleus. The fission of heavy nuclei and the fusion of light nuclei are the two main ways to obtain atomic energy.

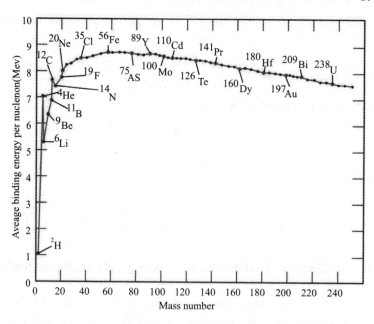

Figure 13.4.3 Average binding energies of different nuclei

To use fission energy peacefully on a large scale, heavy nuclear fission must form a chain reaction that must be controllable. A device that realizes a controllable chain reaction is called a reactor. In 1942, the University of Chicago successfully started the world's first nuclear reactor, which marked the beginning of a new era of the use of atomic energy. Atomic energy was first used in the military in World War II. In 1945, the United States dropped two atomic bombs on Hiroshima and Nagasaki, Japan, with an energy equivalent to 2,000 tons of explosives, killing more than 240,000 people. In 1954, the Soviet Union built the world's first nuclear power plant, the Oblinsk nuclear power plant, which opened the prelude to the peaceful use of atomic energy by mankind. By 1960, 5 countries had built 20 nuclear power plants with an installed capacity of 1279 megawatts. Due to the development of nuclear enrichment technology, by 1966, the cost of atomic power generation was lower than that of thermal power generation, and atomic power generation really entered the practical stage. At present, 33 countries and regions in the world have nuclear power plants to generate electricity, accounting for 17% of the world's total electricity generation. Among them, nuclear electricity generation accounts for more than 25% of total electricity generation in more than a dozen countries, and even exceeds 79% in France. As of now, there are 52 nuclear power units in operation in mainland China, with a rated installed capacity of 53,485.95 MW. In 2021, the power generation of nuclear power plants accounted for 4.99% of the country's cumulative power generation. China's nuclear power technology is at the world's leading level with the third-generation nuclear power technology of "Hualong No. 1" shown in Figure 13.4.4, which represents the highest level of nuclear power research and development capabilities in the world.

Figure 13.4.4 China's "Hualong No. 1" nuclear power set

After human beings mastered the principle of nuclear fusion, they successfully tested the world's first hydrogen bomb as early as 1952. However, the explosion of a hydrogen bomb is an uncontrollable nuclear fusion reaction and cannot be used as a means of providing energy. Since then, humans have been committed to artificially controlling nuclear fusion reactions on the earth, that is, controlled nuclear fusion research, hoping to use the principle of solar heating to provide a steady stream of energy. Therefore, the experimental device of controlled nuclear fusion is also called "artificial sun". The condition for a controlled fusion reaction is to heat a certain density of plasma to a high enough temperature and hold it long enough for the fusion reaction to proceed. Due to the extremely high temperature of nuclear fusion plasma (up to hundreds of millions of degrees), any physical container cannot withstand such a high temperature, so special methods must be used to confine the high-temperature plasma. The sun and other stars use huge gravity to bind plasma at 10 to 15 million degrees Celsius to maintain fusion reactions. However, the earth does not have such a large gravity and can only heat low-density plasma to a higher temperature (above 100 million degrees) to cause fusion reactions, namely inertial constraints and magnetic constraints. In 1954, the "Tokamak" approach invented by Soviet scientists showed unique advantages and became the mainstream approach for fusion energy research. The Tokamak device, also known as the circulator, is a "magnetic cage" composed of an annular closed magnetic field. The plasma is confined in this "magnetic cage", and a large ring current is induced in the plasma ring. In December 2020, a new generation of Tokamak device developed by the Southwest Institute of Physics of China's nuclear industry, the China Circulator No. 2 M device, was completed in Chengdu and achieved its first discharge. In December 2021, the Tokamak device (EAST) developed by the Hefei Institute of Physical Science, Chinese Academy of Sciences, realized the long-pulse high-parameter plasma operation of 1056 seconds, which was the longest time for the high-temperature plasma operation of the Tokamak device in the world, and the highest temperature was as high as 160 million degrees Celsius, as shown in Figure 13.4.5.

Figure 13.4.5 Chinese Tokamak device (EAST)

Scientist Profile

Albert Einstein(1879 – 1955), born in Ulm, Baden-Württemberg, Germany, was the pioneer, synthesizer and one of the founders of modern physics.

On March 14, 1879, Einstein was born in a Jewish family in Ulm, Germany. Influenced by his uncle, an engineer, he was enlightened by natural science and philosophy since childhood. In 1896, he entered the Teachers Department of the Swiss Federal Institute of Technology (ETH) Zurich, Switzerland, to study physics. In 1901, he obtained Swiss nationality. In the following year, he was hired as a technician by the Swiss Patent Office in Bern, engaged in the technical appraisal of invention patent applications. In 1909, Einstein left the Swiss Patent Office and became an associate professor of theoretical physics at the University of Zurich. In 1912, he became a professor at the ETH Zurich. In 1914, he returned to Germany as director of the Kaiser Wilhelm Institute for Physics and professor at the University of Berlin. Persecuted by fascists in 1933, Einstein was forced to leave Germany and emigrated to the United States, where he served as a professor at the Institute for Advanced Study in Princeton until his retirement in 1945.

During his work at the Swiss Patent Office in Bern, Einstein used his spare time to conduct scientific research, and in 1905, he made a historic achievement. His epoch-making achievements are in the following five aspects.

1. Quantum theory of light

In March 1905, Einstein published the paper *On a Heuristic Viewpoint Concerning the Production and Transformation of Light* in *Journal of Physics* in Leipzig, Germany. He extended the quantum concept proposed by Professor Planck in 1900 to the emission and absorption of optical radiation, and established the quantum theory of light. For the first time in the history of science, this theory linked the particle nature and wave nature of light through mathematical expressions, successfully explaining the "photoelectric effect" that had puzzled people for a long time and could not be explained by classical physics, and Einstein won the Nobel Prize in Physics in 1921.

2. Theoretical research on the Brownian motion

In April 1905, Einstein successively completed two articles on molecular physics research. One was a doctoral thesis *A New Determination of Molecular Dimensions* submitted to the University of Zurich, and the other one was the paper *On The Motion of Small Particles in Liquids at Rest Required by the Molecular Kinetic Theory of Heat*, published in *Journal of Physics* in Leipzig. These two works successfully explained the Brownian motion and gave the atomic theory the final victory. This theory made people realize that heat is a kind of energy, which is caused by the random motion of molecules, and that molecules exist objectively, not being an assumption made for the convenience of research. Einstein's research achieved the final victory of the atomic theory, so that extreme anti-atomists like Ostwald and Mach had to turn to the atomic theory from then on. Three years later, Perrin made experimental measurements based on this, confirming Einstein's theoretical predictions.

3. Special relativity

In June 1905, Einstein published a more-than-30-page paper *On the Electrodynamics of Moving Objects* in *Journal of Physics*, fully proposing the special theory of relativity. And he put forward a new concept of four-dimensional space-time (three-dimensional space and one-dimensional time), which caused a change in the theoretical basis of physics and had the significance of creating a new era of science. As a corollary of the theory of relativity, he also proposed the mass-energy relationship, which theoretically opened up the way for the application of atomic energy.

In 1905, the scientific achievements of the 26-year-old Einstein established his status as a scientific master of the 20th century, and this year was therefore called "Einstein's miracle year". After that, Einstein also made great achievements in the general relativity and unified field theory.

4. General relativity

After the establishment of the special theory of relativity, Einstein was not satisfied, and tried to extend the scope of application of the principle of relativity to non-inertial frames. In 1907, Einstein proposed that it be necessary to generalize the theory of relativity from constant velocity motion to accelerated motion based on the equivalence of inertial mass and gravitational mass. Beginning in 1912, he collaborated with Grossmann to find mathematical tools for establishing the general theory of relativity in the Riemannian geometry and tensor analysis. Then, he proposed the *Outline of the General Theory of Relativity and Theory of Gravity* in 1913. Finally, in 1915, he published *On General Relativity* and *Field Equations of Gravitation*, which proposed the complete form of the gravitational field equations of general relativity, and established the general theory of relativity. The general theory of relativity made up for the two flaws of the special theory of relativity and established a new gravity theory. General relativity unified space-time, matter and motion, perfectly explained the abnormal precession of Mercury's perihelion, proposed that the propagation speed of gravity is the speed of light, and predicted the existence of gravitational waves. General relativity predicted that light will bend when passing through the gravitational field of the sun and the gravitational red shift effect, which has been confirmed experimentally. General relativity also established modern cosmology, opened up new ways to study the origin, evolution and structure of the universe, and predicted the existence of black holes. As a result, various previously unknown new celestial objects and new astronomical phenomena were discovered, which greatly deepened our understanding of the structure of the universe.

5. Unified field theory

After the general theory of relativity was built, Einstein was still not satisfied. He believed that the world was unified, so he wanted to generalize the general theory of relativity, unify the gravitational field and the electromagnetic field, and unify the theory of relativity and quantum theory. He believed that this was the third stage of the development of the theory of relativity, namely the unified field theory. In 1923, Einstein published the paper *Can Field Theory be Used to Solve Quantum Problems*? and began to embark on the road of studying the unified field theory. In the following 30 years, Einstein almost devoted all his scientific creation energy to the exploration of the unified field theory. Unfortunately, he didn't succeed. Since then, scientists have continued Einstein's work, and the idea of the unified field theory has shown its vitality in a new form and has made great progress.

Einstein was one of the greatest scientists of all time and was included by many historians as one of the hundred most influential people of the past millennium, and his name became synonymous with "genius". Einstein's achievements were closely related to his scientific spirit and thought. Influenced by philosophers such as Hume and Mach, Einstein had a strong spirit of independent thinking, doubt and criticism.

 Extended Reading

General relativity and gravitational waves

1. General relativity

In 1687, Newton proposed the law of universal gravitation in his masterpiece *Mathematical Principles of Natural Philosophy*, one of the first natural laws discovered by human beings. In 1905, Einstein established the special theory of relativity based on the principle of relativity and the principle of constant speed of light. In the special theory of relativity, the inertial frame has a superior position. Einstein believed that all laws of physics should be invariant in all frames of reference. Considering the universality of the law of gravity and the Mach principle, Einstein discovered the connection between the non-inertial frame and gravity, and in 1915 he obtained the Einstein gravitational field equation

$$R_{\mu\nu} - \frac{1}{2} g_{\mu\nu} R = -\kappa T_{\mu\nu} \quad (\mu, \nu = 0, 1, 2, 3)$$

The left side of this gravitational field equation represents the curvature of space-time, which is a geometric quantity. In the above equation, $g_{\mu\nu}$ is the metric tensor, which can be understood as the gravitational potential in the law of universal gravitation; $R_{\mu\nu}$ and R are the Ricci tensor and the curvature scalar, respectively, which are nonlinear functions composed of the metric and its first and second derivatives. The right side of the equation is the matter term, where $T_{\mu\nu}$ is composed of energy, momentum, energy flow, and momentum flow and κ is a constant

$$\kappa = \frac{8\pi G}{c^4}$$

where G is the gravitational constant and c is the speed of light in a vacuum. Einstein's gravitational field equation shows how the energy and motion of matter determine the curvature of space-time. As long as the metric tensor $g_{\mu\nu}$ is known, the curvature of space-time can be calculated, so as to understand how space-time is curved.

According to the general theory of relativity, there is no gravity in the local inertial frame, and one-dimensional time and three-dimensional space form a four-dimensional flat Euclidean space. In any reference frame, there is gravity, and gravity causes space-time curvature, so space-time is a four-dimensional curved Riemann space. The curved structure of time and space depends on the distribution of material energy density and momentum density in time and space, and the curved structure of time and space in turn determines the motion trajectory of objects. When the gravitational force is weak and the curvature of time and space is small, the predictions of validity of general relativity tend to be consistent with the predictions of Newton's law of universal gravitation and Newton's law of motion. The general theory of relativity predicted the anomalous precession of Mercury's perihelion, the gravitational redshift of light frequency, the gravitational deflection of light, and the delay of radar echoes, all of which have been confirmed by astronomical observations or experiments, and the general theory of relativity has been widely recognized. In 1859, the astronomer Le Verrier discovered the observed value of Mercury's perihelion precession, which was 38 arcseconds per hundred years faster than the theoretical value calculated by Newton's law of universal gravitation. He conjectured that there might be an asteroid within Mercury. The asteroid's gravitational pull on Mercury caused the deviation. However, after years of searching, the asteroid was never found. In 1882, S. Newcomb recalculated the excess precession of Mercury at perihelion of 43 arcseconds per hundred years, and he began to doubt whether gravity obeyed the inverse square law. In 1915, Einstein regarded the motion of a planet around the sun as its motion in the gravitational field of the sun according to the general theory of relativity. Due to the curvature of the surrounding space caused by the mass of the sun, he calculated that the perihelion precession of Mercury was 43 arcseconds per hundred years, which was consistent with Newcomb's result. It solved a case that had been unsolved for many years in Newton's theory of validity of gravity and became the most powerful evidence of the validity of general relativity at that time.

2. Gravitational waves

In Einstein's theory of general relativity, gravity is thought to be an effect of the curvature of space-time, which is caused by the presence of mass. In general, the greater the mass contained within a given volume, the greater the resulting space-time curvature at the boundaries of that volume. Under certain circumstances, accelerating objects can change this curvature and make wave travel at the speed of light. This propagation phenomenon is called gravitational waves.

Einstein predicted the existence of gravitational waves in the 2nd year (1916) after his general theory of relativity. He predicted that gravitational waves are similar to the waves produced by throwing rocks on a calm water surface, and are the "ripple" caused by the accelerated motion of matter in the gravitational field. In 1918, he wrote the paper *On Gravitational Waves*, proposing that the accelerated mass can produce gravitational waves and predicting how gravitational waves propagate. Later in 1937, he cooperated with N. Rosen to further clarify the generation and propagation of gravitational waves and other issues, showing that the gravitational field, like the electromagnetic field, can radiate and propagate in the form of waves, and the speed of the gravitational wave is the speed of light c.

Through research, it has also been found that there are three types of gravitational wave radiation in nature. The first is a pulse-type, high-intensity, short-lived gravitational wave emitted by sudden events such as a supernova explosion at the end of a star's life, the gravitational collapse of a high-density object and the collision of two black holes with a speed of light. The second is the frequency-determined gravitational waves emitted by binary star systems, neutron stars and white dwarfs, and other rotating objects. The third type is gravitational waves with irregular backgrounds, such as gravitational waves emitted in the very early physical process of the universe. These cosmic waves carry information about their origins as well as clues about the nature of gravity itself, so scientists can use gravitational waves to go back even further into the past and gain a lot of information about the Big Bang.

Generally, the earth is very far away from the source of these gravitational waves, so the gravitational waves leave a weak and short trail as they pass by the earth, and objects on the earth are slightly "squeezed" resulting in a positional shift about 10^{-21} m (equivalent to one-millionth of the size of a proton). Gravitational waves are so weak that Einstein himself thought they might be impossible to observe, but it wasn't until the mid-1950s that physicists actually proved that gravitational waves carry energy and are a detectable objective reality. In terms of experiments, the first great attempt to directly detect gravitational waves was made by Joseph Weber from the United States. In the early 1960s, he pioneered the gravitational wave detector, as shown in Figure ER 1. He used a cylindrical solid aluminum rod with a length of 2 m and a diameter of 0.5 m as the antenna. The middle of the rod was suspended on the isolation stack with a thin wire, and the side of the rod pointed to the direction of the gravitational wave. When the gravitational wave arrived, the two ends of the aluminum rod would be squeezed and stretched alternately. When the frequency of the gravitational wave was consistent with the design frequency of the aluminum rod, the aluminum rod would resonate, and the chip attached to the surface of the aluminum rod would generate a corresponding voltage signal. The resonance rod detector has obvious limitations. Its resonance frequency is determined. Although we can adjust the resonance frequency by changing the length of the resonance rod, the same detector can only detect the gravitational wave signal of its corresponding frequency. If the frequencies of the gravitational wave signals don't match, the detector can't do anything. Although Weber's resonant rod detector finally failed to find gravitational waves, Weber pioneered the experimental exploration of gravitational waves. After him, more than ten groups from the United States, Germany, China, etc. has continued this work. High quality factor rod-shaped material with large mass is used as an antenna and is placed in a low-temperature isolation environment that excludes acoustic, electrical and mechanical interference. The sensitivity of the detector at ultra-low temperature can reach 10^{-21} m. The gravitational wave antenna of the Department of Gravitational Physics of Sun Yat-sen University in our country uses the same principle, but the installation method has been improved. The sensitivity of the detector ranks among the top of similar antennas in the world. But so far, no direct evidence of gravitational waves has been obtained.

Figure ER 1 Weber's gravitational wave detector

At the same time that Weber designed and built the resonant rod, Rainer Weiss and Robert Forward from the United States proposed the use of laser interference to detect gravitational waves. The advantages of the laser interferometer compared to the resonant rod are obvious. It can detect gravitational wave signals in a certain frequency range, and the arm length of the laser interferometer can be designed to be very long. For example, the arm length of the ground gravitational wave interferometer is generally in the range of kilometers, which far exceeds the length of the resonance rod. In February 2016, the US Laser Interferometer Gravitational Wave Observatory (LIGO) announced the first direct detection of the gravitational wave event (GW150914) generated by the merger of stellar-mass binary black holes. It was emitted when two black holes 36 times and 29 times the mass of the sun respectively merged and formed a spinning black hole with a mass of 21 times that of the sun, converting material roughly the same mass as that of the sun into energy. Subsequently, LIGO-Virgo jointly discovered a series of double black hole merger events and double neutron star merger events (GW170817). These new discoveries opened a new era in gravitational wave astronomy and cosmology. Rainer Weiss, Kip Stephen Thorne and Barry Clark Barish were awarded the 2017 Nobel Prize in Physics for their outstanding contributions to LIGO.

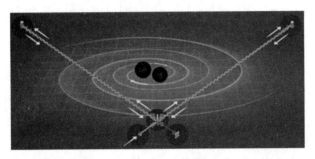

Figure ER 2 Schematic diagram of the merger of two mutually attractive black holes

Discussion Problems

13.1 What is the fundamental difference between space-time view of the Newtonian mechanics and relativistic space-time view? What is the connection between them?

13.2 What are the two basic assumptions of special relativity?

13.3 A friend argues with you that the theory of relativity is absurd, "Clocks in motion obviously don't go slower, and objects in motion can't be shorter than what they are when they are at rest. "How should you respond?

13.4 A 100m runner starts from the starting line (Event 1) and reaches the finish line at a constant speed (Event 2). In which frame of reference is the time measured by the observer is the proper time? In which frame of reference the length of the runway measured from the start line to the finish is the proper length?

13.5 Does the mass of the phone battery change when you answer the phone? If yes, will it increase or decrease?

Problems

13.6 A spaceship moves away from the earth at a speed of $0.13c$ and sends a radio signal to the earth.

(1) According to the Galileo transformation, what is the speed of the signal relative to the earth?

(2) Using the Einstein hypothesis, what is the speed of the signal relative to the earth?

13.7 An airplane flew for 8 hours, and the average speed relative to the earth during the flight was 220 m·s^{-1}. What is the difference between atomic clocks on the plane and atomic clocks on the ground (Assuming they were synchronized before the flight and ignoring the complication of general relativity introduced by gravity and aircraft acceleration)?

13.8 A spaceship flies towards the earth at a speed of $0.97c$, and the bodies of the passengers on board are all parallel to the motion direction of the spaceship. Observers on the earth measure these passengers to be approximately 0.50 m high and 0.50 m wide. What are the heights of these passengers measured in the spacecraft's frame of reference? What are the widths?

13.9 The inertial system S' moves in a straight line at a uniform speed along the X axis relative to another inertial system S. Take the time when the two coordinate origins coincide as the starting point for timing. The space-time coordinates of the two events measured in the S system are $x_1 = 6 \times 10^{-4}$ m, $t_1 = 2 \times 10^{-4}$ s, and $x_2 = 12 \times 10^{-4}$ m, $t_2 = 1 \times 10^{-4}$ s, respectively. Both events are known to occur simultaneously in the S' system. Please find

(1) the velocity of the S' system relative to the S system;

(2) the spatial interval between the two events measured in the S' system.

13.10 There is a stationary square in the S system with an area of 100 m^2. The observer S' moves along the diagonal of the square at a speed of $0.8c$. What is the area of the square measured in the system S'?

13.11 The length of a spacecraft that is stationary relative to the earth is 35.2 m. When the earth travels to other planets, the ground observers measures its length to be 30.5 m. The observers on the ground also notices that an astronaut on the spacecraft exercises for 22.2 min. So how long does the astronaut think he has exercised?

13.12 Electron positron colliders can accelerate electrons to a kinetic energy of $E_k = 2.8 \times 10^9$ eV. How much less than the speed of light is the speed of this electron? What is the momentum of such an electron (The energy corresponding to the static mass of the electron is $E_0 = 0.511 \times 10^8$ eV)?

13.13 In the inertial system, there are two particles A and B whose static mass is both m_0. They move towards each other at the same speed of v. After collision, they are combined into one particle. Find the static mass of this particle.

13.14 How much work must be done on an electron to increase its speed from $v_1 = 0.4c$ to $v_2 = 0.8c$ (c is the speed of light in a vacuum and the electron static mass is $m_e = 9.11 \times 10^{-31}$ kg)?

Challenging Problems

13.15 (1) According to the Lorentz coordinate transformation, derive the speed transformation formula in the X direction. Spaceship A is moving at the speed of $0.5c$ on the ground and spaceship B is moving at a speed of $0.8c$ in the same direction on the ground. What is the speed of spaceship B relative to spaceship A? What if spaceship B moves towards spaceship A at the same speed of $0.8c$ on the ground? What are the characteristics of the speed transformation surface in the X direction?

(2) According to the Lorentz coordinate transformation, derive the speed transformation formula in the Y direction or the Z direction and the transformation formula of the total speed. Observed in the solar reference system, a beam of starlight is directed perpendicular to the ground at a speed of c, while the earth is moving perpendicular to the light at a speed of u. What are the magnitude and direction of the speed of this beam of starlight measured on the ground? What are the characteristics of the speed transformation surface in the Y direction? What are the characteristics of the transformation of the magnitude and direction of the total speed?

Chapter 14

Fundamentals of Quantum Physics

Quantum mechanics is a theory that reflects the laws of motion of microscopic particles (molecules, atoms, nuclei, elementary particles, etc.). Quantum mechanics and relativity are considered to be the two theoretical pillars of modern physics. Many basic concepts, laws, and methods of quantum mechanics, different from those of classical physics, caused profound changes in physics and revolutionized physics in the 20th century. Although its philosophical significance and interpretation are still debated among scientists, its application in modern science and technology has achieved great success, such as in material science, life science (DNA helical structure and organic biomolecules), doping of semiconductors, nanotechnology, microscopic physical properties, chemistry (quantum chemistry) and electronic technology (quantum dot, quantum wire and quantum switch). Quantum mechanics has become one of the foundations of modern physics.

In the late 19th century, the discovery of X-rays (1895), radioactivity (1896) and electrons (1897) opened the prelude to the development of modern physics. By the beginning of the 20th century, many new phenomena had been discovered from a large number of precise experiments. These new phenomena, including thermal radiation, photoelectric effect, and linear spectroscopic phenomena of atoms, could not be explained by classical physics theory. In 1900, Planck proposed the quantization hypothesis of radiation energy, which perfectly overcame the difficulties encountered by classical physics in thermal radiation. In 1913, Bohr applied the concept of quantization to atomic orbits and successfully explained the linear spectrum of hydrogen atoms; in 1925, the exclusion principle proposed by Pauli, combined with the electron spin hypothesis of Uhlenbeck and Goudsmit successfully explained the element periodicity. Although a series of experimental results could be satisfactorily explained then, the theories of quantum mechanics were not completely free from the constraints of classical physics theory. A mixture of classical physics and modern physics, they were called old quantum theory.

In 1923, de Broglie proposed the concept that matter has wave-particle duality; in 1925, Heisenberg founded matrix mechanics; in 1926, Schrödinger established wave mechanics; and later, Schrödinger and Dirac proved the equivalence of matrix mechanics and wave mechanics. From 1926 to 1930, Dirac made a comprehensive summary of quantum mechanics and developed it into relativistic quantum mechanics. Thus, it was not until the early 1930s that quantum mechanics was established and developed.

During the creation of quantum mechanics, dozens of physicists won the Nobel Prize in Physics, including Planck (1918), Bohr (1922), Compton (1927), de Broglie (1929), Heisenberg (1932), Dirac and Schrödinger (1933), Fermi (1934), Pauli (1945) and Bonn (1954). In addition to their major contributions, many other scientists devoted their lives to the development and refinement of quantum theory.

Up to now, quantum theory has penetrated different fields of different disciplines. Many branches of science have been developed and established on its basis, for example, particle physics, atomic physics, quantum chemistry and condensed matter physics. In the meantime, some epoch-making new technologies have also been formed and produced, such as modern quantum communication and quantum computing technology. The scientific and technological achievements brought by quantum theory have had a huge impact on human life and social and economic development.

14.1 Planck's energy quantum hypothesis

14.1.1 Thermal radiation

The concept of quantum was first proposed by M. Planck when he studied black body radiation. Any object is emitting electromagnetic waves of various wavelengths at any temperature. We call this phenomenon of emitting electromagnetic waves due to the thermal excitation of molecules and atoms in the object **thermal radiation.**

The thermal radiation of an object has a continuous spectrum, but the radiated energy and its distribution by the wavelength vary with temperature. The higher the temperature, the greater the radiation energy and the more short-wave components contained in the radiation energy spectrum. For example, an iron block being heated in a furnace, when the initial temperature is low, radiates less energy, the color is dark red, and the radiated electromagnetic wave has a longer wavelength. With the increase in temperature, the energy radiated from the iron block is larger and larger, the short-wave components in the radiation energy spectrum increase gradually, and the color changes from dark red to orange and then to blue-white. This shows that the amount of radiated energy of an object in a certain period of time and the distribution of radiated energy by the wavelength are related to temperature.

To describe the change of electromagnetic wave energy radiated by an object with the wavelength, the concept of monochromatic radiance is introduced. When the temperature of the object is T, the wavelength of the electromagnetic wave it radiates is within the interval of $\lambda \sim \lambda + d\lambda$, the radiance over this interval is $dM(T)$ and the **monochromatic radiance** is $M_\lambda(T)$, defined as

$$M_\lambda(T) = \frac{dM(T)}{d\lambda} \tag{14.1.1}$$

$M_\lambda(T)$ is a function of the temperature of the radiating object and the wavelength of the radiation, reflecting the distribution of the radiation energy of the object according to the wavelength at different temperatures, and the unit is $W \cdot m^{-3}$.

At a certain temperature T, the total radiant energy of various wavelengths emitted from a unit area of an object surface in a unit time is called the radiant emittance of an object, or radiance for short, which is a function of the thermodynamic temperature T of the object, expressed by $M(T)$. Obviously, the radiance can be obtained by Equation (14.1.1)

$$M(T) = \int dM(T) = \int M_\lambda(T) d\lambda \tag{14.1.2}$$

In addition to radiating electromagnetic waves, objects can also absorb electromagnetic waves from the environment. When electromagnetic waves are incident on the surface of an object, a part of the electromagnetic waves is reflected, a part is transmitted from the object, and the remaining part is absorbed by the object. Research has found that the ability of objects to absorb electromagnetic waves is proportional to that to radiate electromagnetic waves, that is, the strong radiation ability of one object corresponds to the strong absorption.

Different objects have different abilities to absorb electromagnetic radiation. An object is called a **black body if it absorbs all electromagnetic radiation of any wavelength without reflection at any temperature.** A black body is an idealized model since a true black body does not exist. However, some substances are very close to an absolute black body; For example, bituminous coal can absorb 99% of the incident light energy, which is close to a black body. In the daytime, the windows of a building far away always look black, which is also an approximation of a black body. In the field of physics, a cavity with a small hole on its wall is usually used to study the black body. As shown in Figure 14.1.1, after the electromagnetic wave enters the small hole of the

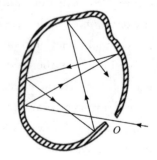

Figure 14.1.1 Black body model

cavity, the wave is continuously reflected and absorbed in the cavity, and almost no longer emerges from the small hole, so the small hole can be seen as a black body.

14.1.2 Experimental law of black body radiation

For a black body, its monochromatic radiance is only related to the wavelength λ and temperature T, and has nothing to do with its material, size, shape, and surface condition. Therefore, the black body has become an important model for the study of thermal radiation theory. Figure 14.1.2 shows the experimental curve of the experimentally measured monochromatic radiance $M_\lambda(T)$ of a black body as a function of wavelength λ at different temperatures T. According to the experimental curve, the following two black body radiation laws are summarized.

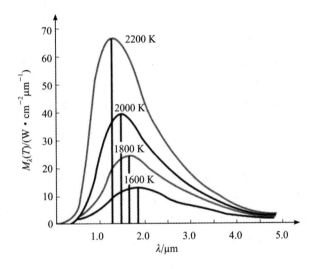

Figure14.1.2 Experimental curves of black body monochromatic radiance

1. The Stefan-Boltzmann law

In 1879, after comparing a large number of experimental results, J. Stefan first discovered that the area under a certain curve, i.e. the radiance $M(T)$, is proportional to the fourth power of the temperature. This was rigorously proved five years later by L. E. Boltzmann using thermodynamic theory. The Stefan-Boltzmann law is as follows:

$$M(T) = \sigma T^4 \qquad (14.1.3)$$

where the constant $\sigma = 5.670 \times 10^{-8}$ W $\cdot$ m^{-2} $\cdot$ K^{-4}. This equation becomes the Stefan-Boltzmann equation.

2. Wien's displacement law

It can be seen from the experimental curve that each thermal radiation curve of a specific temperature in the black body radiation has a maximum value, and the wavelength at the maximum is called λ_m. The higher the temperature, the smaller λ_m. In 1893, W. Wien obtained Wien's displacement law from the theory of electromagnetism and thermodynamics

$$\lambda_m T = b \qquad (14.1.4)$$

in which $b = 2.898 \times 10^{-3}$ m $\cdot$ K is Wien's constant.

The Stefan-Boltzmann law and Wien displacement law are the basic laws of black body radiation. They have a wide range of applications in modern science and technology. For instance, they are the physical basis for measuring high temperature and technologies such as remote sensing and infrared tracking. The effective temperature of a star is often measured in this way.

14.1.3 Planck's quantum hypothesis

At the end of the 19th century, one of the most striking topics in physics was how to theoretically derive the mathematical expression of the black body monochromatic radiance $M_\lambda(T)$, so that it could be consistent with the experimental curve. In 1896, Wien, starting from thermodynamic theory and analysis of experimental data, assumed that the frequency distribution of harmonic oscillator energy was similar to Maxwell's speed distribution, and derived the following semi-empirical equation according to classical statistical physics

$$M_\lambda(T) = C_1 \lambda^{-5} e^{-\frac{C_2}{\lambda T}} \tag{14.1.5}$$

This equation is called Wien's equation, where C_1 and C_2 are two empirical parameters that need to be determined by experimental data. The Wien equation is only consistent with the experimental curve in the short-wave band, and there is a significant deviation from the experimental curve in the long-wave band, as shown in Figure 14.1.3.

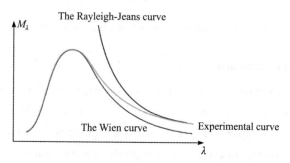

Figure 14.1.3 Experimental and theoretical curves of black body monochromatic radiance

In 1900, I. Rayleigh derived an equation for black body radiation based on the theory of classical electrodynamics and statistical physics. In 1905, J. H. Jeans corrected a numerical factor to give the current Rayleigh-Jeans equation

$$M_\lambda(T) = \frac{2\pi c k T}{\lambda^4} \tag{14.1.6}$$

where k is the Boltzmann constant and c is the speed of light in a vacuum. This equation is only applicable to the long-wave band, and obviously does not match the real curve in the ultraviolet region. Its short-wave limit $M_\lambda(T) \to \infty$, as shown in Figure 14.1.3, and was historically called the "ultraviolet catastrophe".

In 1900, on the basis of in-depth research on the results of predecessors, Planck combined the Wien equation representing the short-wave band with the experimental results representing the long-wave band, and came up with a new distribution equation

$$M_\lambda(T) = C_1 \lambda^{-5} \frac{1}{e^{C_2/\lambda T} - 1}$$

Planck was not satisfied with the equation he had come up with and tried to derive it theoretically. After two months of intense work, Planck finally made a great discovery with far-reaching historical significance by introducing a new concept—energy quantum to physics and proposing the epoch-making Planck's quantum hypothesis. Its basic points are as follows:

(1) The radiator is composed of many charged linear harmonic oscillators (For example, the vibration of molecules and atoms can be regarded as linear harmonic oscillators), which can radiate or absorb electromagnetic waves and exchange energy with the surrounding electromagnetic fields.

(2) The energy states of these linear harmonic oscillators are not continuous, and each harmonic oscillator can only be in some special and discrete states; in these states, the corresponding energy can only take an integer multiple of a certain minimum energy ε (called the energy quantum) that is, ε, 2ε, 3ε, ···, nε. When radiating or absorbing energy, the harmonic oscillator can only transition from one of these states to another state, and the radiated or absorbed energy can only be an integer multiple of ε too.

(3) The energy quantum ε is proportional to the frequency of the linear harmonic oscillator ν, namely

$$\varepsilon = h\nu$$

where h is called Planck's constant, and its magnitude is $h = 6.626 \times 10^{-34}$ J·s.

According to the energy quantization hypothesis and the Boltzmann distribution law, Planck theoretically deduced a black body radiation equation that was completely consistent with the experimental curve, and it was called Planck's equation

$$M_\lambda(T) = \frac{2\pi hc^2}{\lambda^5} \frac{1}{e^{hc/k\lambda T} - 1} \tag{14.1.7}$$

The Planck equation was in good agreement with the experimental results. When $hc/\lambda \ll kT$, this equation boiled down to the Rayleigh-Jeans equation; when $hc/\lambda \gg kT$, this equation boiled down to the Wien equation.

Planck's quantum hypothesis not only successfully explained the law of black body thermal radiation, but more importantly, put forward the concept of quantum for the first time in the history of physics. This revolutionary new concept created a new field of human understanding of the microscopic world and caused a revolution in physics. On December 14, 1900, Planck read out his paper at the German Physical Society, which was considered as the birthday of quantum theory. Planck was awarded the 1918 Nobel Prize in Physics for his pioneering contributions to quantum theory.

However, this new concept of energy quantization was far from the way of thinking that physicists had long been accustomed to. In the early stage of the quantum theory, people only accepted the equation derived by Planck (which was considered to be the empirical equation of the radiation theory), but could not accept his theory. Planck himself also felt uneasy for a long time, thinking that he had taken a "desperate action" and tried many times to incorporate his theory into the framework of classical physics, all of which ended in failure. Not until 1905 when Einstein proposed the quantum theory of light with the help of the energy quantum hypothesis and successfully explained the photoelectric effect, did the thought of quantum begin to be gradually accepted.

14.2 Photoelectric effect

In 1887, H. R. Hertz made an unexpected discovery during an experiment to confirm the existence of electromagnetic waves that light can make metal discharge and create an electrical spark. Thomson, who discovered the electron in 1899, found that the charge-to-mass ratio of the particles emitted by the photoinduced metal is the same as that of the electron, so he stated that this phenomenon is due to the release of electrons from the metal surface after being illuminated by light. Later, people called this phenomenon that electrons escape from the metal surface when light is illuminating the metal surface the **photoelectric effect**, and the released electrons were called **photoelectrons**. In 1905, Einstein proposed the concept of light quantum, and successfully explained the experimental law of the photoelectric effect in theory. The photoelectric effect is of great significance for understanding the nature of light, and is widely used in production, scientific research and national defense.

14.2.1 Experimental law of the photoelectric effect

The photoelectric effect experimental device is shown in Figure 14.2.1.

Two electrodes are installed in a vacuum glass tube S, the anode is A, the cathode is K, and its material is metal. There is a quartz window at the mouth of the tube, and the incident light passes through the window and illuminates the surface of the cathode, so that the cathode emits photoelectrons. When a voltage is applied across the electrodes, the photoelectrons fly to the anode under the action of the accelerating electric field, forming a photocurrent in the loop. The electric field is

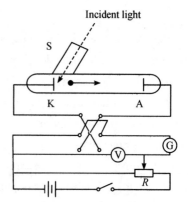

Figure14.2.1 Schematic diagram of the photoelectric effect

adjusted in magnitude by the sliding rheostat in the circuit and is changed in direction with a bidirectional switch. Voltage and current can be obtained from the ammeter G and voltmeter V. The experimental results of the photoelectric effect are summarized as follows.

(1) **The number of photoelectrons emitted by the cathode per unit time is proportional to the intensity of the incident light.**

As shown in Figure 14.2.2, with a certain intensity of monochromatic light, the higher the accelerating voltage, the greater the photocurrent. When the voltage increases to a certain value, the photocurrent does not increase anymore. The photocurrent value at this time is called the **saturation photocurrent intensity.** When the intensity of incident monochromatic light increases, its saturated photocurrent also increases, and the saturated photocurrent is proportional to the incident light intensity, that is, the number of photoelectrons escaping from the cathode per unit time is proportional to the incident light intensity.

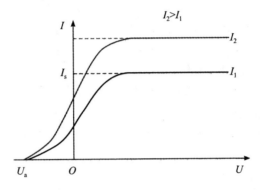

Figure 14.2.2 Volt-ampère characteristic curve of the photoelectric effect

(2) **Each metal has a cutoff frequency.**

Experiments show that when light illuminates a certain metal, if the frequency of the light ν is less than a certain frequency ν_0, no matter what the intensity of the light is, the photoelectric effect will not occur. This frequency is **called the cutoff frequency or the red limit.** Different substances have different cutoff frequencies, corresponding to a certain red limit wavelength λ_0, as shown in Table 14.2.1.

Table 14.2.1 **Red limit frequencies and work functions of several metals**

Metal	Red limit frequency ν_0/Hz	Red limit frequency λ_0/nm	Work function A/eV
Na	4.39×10^{14}	684	1.82
Cs	4.60×10^{14}	652	1.90
K	5.45×10^{14}	550	2.30
Ca	6.53×10^{14}	459	2.71
Be	9.40×10^{14}	319	3.90
Ti	9.9×10^{14}	303	4.10
W	1.08×10^{15}	278	4.50
Hg	1.09×10^{15}	275	4.50
Au	1.16×10^{15}	258	4.8
Pd	1.21×10^{15}	248	5.0

(3) **The maximum initial kinetic energy of photoelectrons has nothing to do with the incident light intensity, but increases linearly with the increase of the incident light frequency.** As can be seen from Figure 14.2.2, when $U=0$, the photocurrent is not zero. Only when a reverse voltage U_a is applied between the two poles, the photocurrent is zero. The voltage when the photocurrent is zero is called the **stopping voltage** U_a. This shows that the photoelectrons escaping from the cathode have initial kinetic energy. When the kinetic energy has all been consumed to overcome the reverse electric field force to do work, the electrons just cannot reach the anode. Hence, there is the equation

$$eU_a = \frac{1}{2}mv_m^2 \tag{14.2.1}$$

where m is the mass of the electron and v_m is the maximum initial velocity of the photoelectron.

The experimental results also show that the stopping voltage U_a increases linearly with the increase of the incident light frequency ν, independent of the incident light intensity, as shown in Figure 14.2.3. The relationship between U_a and ν can be expressed by the following equation

$$U_a = K(\nu - \nu_0) \tag{14.2.2}$$

where K is a constant irrelevant to the properties of the metal material, and ν_0 is the intercept of the curve on the abscissa, which is equal to the cutoff frequency of the metal.

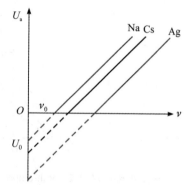

Figure 14.2.3　Relationship between stopping voltage and incident light frequency

Combining Equations (14.2.1) and (14.2.2), we can get

$$\frac{1}{2}mv_m^2 = eK(\nu - \nu_0) \tag{14.2.3}$$

It can be seen that the initial kinetic energy of photoelectrons increases linearly with the increase of the incident light frequency, regardless of the incident light intensity.

(4) **The photoelectric effect occurs instantaneously.**

Experiments find as long as the incident light frequency $\nu > \nu_0$, no matter how weak the intensity of the light is, the time from the light hitting the cathode to the photoelectron escaping does not exceed 10^{-9} s.

14.2.2　Difficulties of classical electromagnetic theory

Using the electromagnetic wave theory of classical physics to explain the photoelectric effect has encountered insurmountable difficulties. First of all, the existence of the cutoff frequency cannot be explained. According to the wave theory of light, no matter what frequency of incident light, as long as its intensity is large enough, electrons can obtain enough energy to escape from the metal surface, so that the photoelectric effect should be produced by light with any frequency, and there should be no cutoff frequency. In addition, it cannot explain the experimental fact that the stopping voltage has nothing to do with the intensity of the incident light. According to the wave theory of light, the initial kinetic energy of the photoelectrons should be determined by the intensity of the incident light, that is, determined by the amplitude of the light and not by the frequency of the light. Classical theory cannot explain the instantaneous nature of the photoelectric effect either. According to classical theory, the energy required for electrons to escape from the metal plate is gradually accumulated. The weaker the light intensity, the longer it takes for electrons to escape, but actually the effect is always instantaneous. These all show that there is an unresolved contradiction between the classical theory and the experimental law of the photoelectric effect.

14.2.3 Einstein's photon theory

In order to explain the experimental phenomenon of the photoelectric effect, in 1905, Einstein, inspired by Planck, put forward the light quantum hypothesis about the nature of light. He believed that light, not only as Planck had pointed out, has particle nature at emission or absorption, but also has particle nature when it propagates in space, that is, a beam of light is a particle stream moving at the speed of light c. The light particle was called light quantum. In 1926, an American chemist (G. H. Lewis) called it a **photon**. For monochromatic light of frequency ν, the energy of the photon is $\varepsilon = h\nu$, where h is Planck's constant.

According to the photon hypothesis, the photoelectric effect is a phenomenon in which electrons in the metal absorb the energy of the photons and escape from the surface by overcoming the constraints of the metal. Assuming that the frequency of the incident light is ν and after one electron in the metal absorbs one photon's energy, part of it overcomes the work function A, and if there is no other energy loss, the other part is converted into the initial kinetic energy of the photoelectron. Hence, according to the conservation of energy, there is

$$h\nu = A + \frac{1}{2}mv_m^2 \qquad (14.2.4)$$

This equation only expresses the energy conversion process of the photoelectron with the largest initial kinetic energy, which is called **Einstein's photoelectric effect equation.**

Due to the existence of work function A, if the energy of the incident photon is low, electrons cannot be excited out of the metal. The energy of the photon is related to the frequency, so there must be a cutoff frequency ν_0 that satisfies

$$A = h\nu_0 \qquad (14.2.5)$$

that is, the minimum energy that the electron needs to obtain. Different metals have different work functions and different natural cutoff frequencies. Since the electron absorbs the entire photon instantaneously, and no accumulation time is required, the emission time is also very short. The light beam is composed of photons, so we can regard the light wave as the directional movement of photons. The number of photons reaching the metal surface per unit time depends on the light intensity. The stronger the light intensity, the more photoelectrons are excited and the larger the saturation current.

Photoelectrons escape from the metal with different kinetic energy depending on the initial state of the electron. But there will always be some electrons with the highest kinetic energy, and if these electrons cannot reach the anode, the current is zero. According to Equations (14.2.1), (14.2.4) and (14.2.5), we can get

$$U_a = \frac{h}{e}(\nu - \nu_0) \tag{14.2.6}$$

It shows that the stopping voltage and frequency are linear and the slope $K = h/e$ is constant.

Einstein's photon hypothesis successfully explained the experimental law of the photoelectric effect that had been pending for many years. But at that time, almost all the older generation of physicists were opposed to the quantum theory of light, and even Planck, who proposed the quantum hypothesis, could not agree with it. The American physicist RA Millikan did not believe in the quantum theory of light at first. He spent ten years testing Einstein's photoelectric effect equation, hoping to prove that the equation was wrong. But the experimental results forced him to assert the correctness of this theory, and the value of h was accurately determined from the equation by him. Millikan was awarded the Nobel Prize for confirming this equation, and Einstein was awarded the Nobel Prize for his contributions to theoretical physics, especially the discovery of the photoelectric effect equation.

14.2.4 Wave-particle duality of light

When light propagates, phenomena such as interference and diffraction will occur, which indicates that light has wave properties, and the photoelectric effect indicates that light has particle properties. Therefore, the correct theory about the nature of light is that light has wave-particle duality. The wave nature and particle nature seem to be contradictory, but they coexist in the microscopic field. If one party is dominant, the other party is subordinate, and the nature of the performance is determined by the dominant party. For example, in the process of light propagation, the wave nature of light dominates, thus showing phenomena such as interference and diffraction while in the process of light radiation, absorption, and light-matter interaction, the particle nature of light becomes the main aspect, hence results the photoelectric effect and other phenomena.

We use the physical quantities of frequency ν, wavelength λ and period T to describe the wave nature; The particle nature of light, like physical particles, can be described by physical quantities such as energy, mass and momentum. Because a photon is a particle moving at the speed of light, the theory of relativity must be used to discuss its energy, mass, and momentum.

According to relativity and Einstein's photon hypothesis, the energy of a photon is
$$E = m_\varphi c^2 = h\nu \tag{14.2.7}$$
m_φ is the mass of the photon, expressed as
$$m_\varphi = \frac{h\nu}{c^2} = \frac{h}{c\lambda} \tag{14.2.8}$$
The momentum of the photon is
$$p = m_\varphi c = \frac{h\nu}{c} = \frac{h}{\lambda} \tag{14.2.9}$$

It can be seen that there is a one-to-one correspondence between the wavelength and momentum, frequency and energy of light, which is called the **wave-particle duality of light**. The photoelectric effect experiment and Einstein's photoelectric effect equation not only further proved that the quantum hypothesis of Planck is correct, but also revealed that matter has wave-particle duality, which deepens human's understanding of the microscopic world.

14.2.5 Application of the photoelectric effect

There are many applications of the photoelectric effect. For example, the "optical cat" connected after the optical fiber enters the home, converts the optical signal into an electrical signal, and then sends it to the terminal device by wired or wireless means. The conversion from the optical signal into electrical signal can also realize the control of mechanical devices and the automation of production and monitoring. In addition, photovoltaic power generation is a clean energy, and its basic principle is the photoelectric effect of semiconductors.

Example 14.2.1 It is known that the red limit wavelength corresponding to a certain metal is λ_0. When monochromatic light of $\lambda < \lambda_0$ illuminates the metal, find the maximum momentum of the escaping electron. Let the mass of the electron be m and the speed of light be c.

Solution According to Einstein's photoelectric effect equation
$$\frac{1}{2}mv_m^2 = h\nu - A$$
The work function of the metal is $A = h\nu_0 = h\dfrac{c}{\lambda_0}$ and $h\nu = h\dfrac{c}{\lambda}$, so
$$v_m = \sqrt{\frac{2hc(\lambda_0 - \lambda)}{m\lambda_0\lambda}}$$
The maximum momentum is
$$p = mv_m = \sqrt{\frac{2mhc(\lambda_0 - \lambda)}{\lambda_0\lambda}}$$

It should be noted that Einstein's photoelectric effect equation is the embodiment of energy conservation, there is no collision in this process, and it has nothing to do with momentum conservation.

14.3 The Compton effect

When visible light travels in matter, due to the existence of other particles in the medium or due to the unevenness of the medium, some part of the light deviates from the original direction. This phenomenon is called **scattering**. X-rays illuminate matter and produce scattered light in all directions. In 1923, the American physicist A. H. Compton and later the Chinese scientist Wu Youxun, through a series of X-ray scattering experiments, found that when monochromatic X-rays are scattered by matter, X-rays with two wavelengths appear in the scattered light; one of them has a wavelength equal to the wavelength of the incident X-ray, while the other is longer than the incident ray in wavelength, and the amount of change in its wavelength has nothing to do with the scattering material and increases with the increase of the scattering angle. This kind of scattering with changing wavelengths is called the **Compton effect**. This scattering effect can be satisfactorily explained by the photon theory, for which Compton won the Nobel Prize.

14.3.1 Experimental law of the Compton effect

Figure 14.3.1 is a schematic diagram of the Compton effect experiment. The X-rays with a wavelength of λ_0 illuminate the graphite through the aperture slot. After the X-rays are scattered, scattered rays are emitted in all directions, the angle between the scattering direction and the incident direction is called the scattering angle φ, and the wavelength and intensity of the scattered light can be measured with a spectrometer.

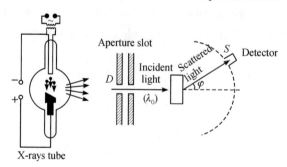

Figure 14.3.1　Schematic diagram of the Compton effect experiment

Figure 14.3.2 shows Compton's and Wu Youxun's experimental results. The results are as follows:

(1) In addition to the rays of the same wavelength λ_0 as the original wavelength, there are also rays of wavelengths $\lambda > \lambda_0$ in the scattered rays.

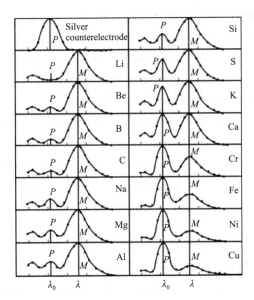

Figure 14.3.2 Compton's and Wu Youxun's experimental results

(2) The amount of change in wavelength $\Delta\lambda=\lambda-\lambda_0$ increases with the increase of the scattering angle φ.

(3) For scattering matter of different elements, the amount of change $\Delta\lambda$ is the same at the same scattering angle, and the intensity of scattered light of wavelength λ decreases with the increase of the atomic number of the scattering matter.

 14.3.2 Theoretical explanation

According to the classical electromagnetic wave theory, X-rays are a kind of electromagnetic wave. When electromagnetic waves pass through matter, they cause forced vibrations of electrons in the matter, and each vibrating electron radiates electromagnetic waves around it. Since the frequency of the forced vibration of electrons is equal to the frequency of the incident X-ray, and the frequency of the electromagnetic wave radiated outward is also the same as the frequency of the incident X-ray, the classical electromagnetic theory cannot explain the Compton effect. Applying Einstein's quantum theory of light and considering the Compton effect as the process of elastic collision between X-ray photons and electrons in matter, the experimental law of the Compton effect can be fully explained.

When X-rays illuminate a crystal, photons collide with electrons in the crystal. The electrons in the crystal can be divided into two categories according to their states: electrons located deep in the atomic shell and strongly bound; the other electrons located in the outer layer of atomic shell, weakly bound by the nucleus and that can be regarded as free electrons.

The inner electrons in an atom are tightly bound, and when a photon collides with these electrons, the photon collides with the entire atom. Since the mass of the atom is much larger than that of the photon, after the collision, the photon changes the direction of motion, but almost does not lose energy, so the frequency or wavelength of the scattered photon is almost unchanged. This is the reason why the scattered light contains the wavelength the same as that of the incident X-rays.

When a photon and a free electron collide, the energy of the thermal motion of the electron is much smaller than the energy of the X-ray (a difference of 2 to 3 orders of magnitude) so the kinetic energy of the electron can be ignored and the collision can be regarded as a completely elastic collision process between a photon and a static electron. As shown in Figure 14.3.3, the energies of the incident photon and the scattered photon are $h\nu_0$ and $h\nu$ respectively; the corresponding momentums are $\dfrac{h\nu_0}{c}$ and $\dfrac{h\nu}{c}$, respectively; the energies of the electron before and after the collision are m_0c^2 and mc^2 respectively; the corresponding momentums are 0 and mv, respectively. According to the laws of conservation of momentum and energy, we have

$$\frac{h\nu_0}{c} = \frac{h\nu}{c}\cos\varphi + mv\cos\theta \tag{14.3.1}$$

$$0 = \frac{h\nu}{c}\sin\varphi - mv\sin\theta \tag{14.3.2}$$

$$h\nu_0 + m_0c^2 = h\nu + mc^2 \tag{14.3.3}$$

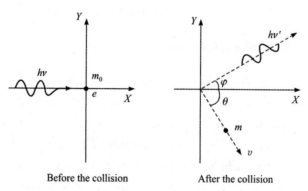

Before the collision   After the collision

Figure 14.3.3 Collision of a photon and a free electron

After rearranging the above three equations, we can get

$$\Delta\lambda = \lambda - \lambda_0 = \frac{c}{\nu} - \frac{c}{\nu_0} = \frac{h}{m_0c}(1 - \cos\varphi) = \frac{2h}{m_0c}\sin^2\frac{\varphi}{2} \tag{14.3.4}$$

This is the Compton scattering equation. Let $\lambda_c = \dfrac{h}{m_0c} \approx 0.0024$ nm, called the **Compton wavelength**, and then the above equation can be written as

$$\Delta\lambda = 2\lambda_c\sin^2\frac{\varphi}{2}$$

In the Compton effect, the interaction of a photon with a free electron or a weakly bound electron in the scattering material can be regarded as an elastic collision between a photon and an electron. Since the electron acquires a part of the energy of the photon after the collision, the energy of the photon is reduced, so the frequency becomes smaller (the wavelength becomes longer). From Equation (14.3.4), it can be seen that the amount of change in wavelength $\Delta\lambda$ increases with the scattering angle φ and is independent of the scattering material. It should be noted that when $\lambda_0 \gg \lambda_c$, $(\lambda-\lambda_0)/\lambda_0 \to 0$, that is, no Compton effect can be observed. This is why we cannot use visible light but X-rays to do the Compton scattering experiment. The theoretical analysis of the Compton effect is highly consistent with the experimental result, which not only strongly confirms the photon theory, but also shows that photons do have the same mass, energy and momentum as physical particles. In particular, the interactions between individual photons and individual electrons also obey the laws of conservation of energy and momentum. That is to say, in the microscopic field, the interaction between individual microscopic particles also strictly obeys the law of conservation of energy and the law of conservation of momentum.

Example 14.3.1 X-rays with a wavelength of $\lambda_0 = 0.020$ nm collide with free electrons, and the scattered rays are observed in a direction that forms an angle of 90° with the incident angle. Find

(1) the wavelength of the scattered rays;

(2) the kinetic energy of the recoil electron;

(3) the momentum of the recoil electron.

(Planck's constant $h = 6.63 \times 10^{-34}$ J·S and the electron rest mass $m_e = 9.11 \times 10^{-31}$ kg).

Solution (1) Because of the scattering angle 90°, the wavelength change of the Compton scattered photon is

$$\Delta\lambda = \frac{h}{m_0 c}(1-\cos\theta) = 0.0024 \text{ nm}$$

Therefore, the wavelength of the scattered rays is

$$\lambda = \lambda_0 + \Delta\lambda = 0.0224 \text{ nm}$$

(2) According to the energy conservation relationship in the collision process, the kinetic energy of the recoil electron is obtained

$$E_k = mc^2 - m_0 c^2 = h\nu_0 - h\nu = \frac{hc}{\lambda_0} - \frac{hc}{\lambda} = \frac{hc}{\lambda_0 \lambda}\Delta\lambda = 1.08 \times 10^{-15} \text{ J}$$

(3) According to the conservation of momentum in the collision process, draw a vector diagram shown in Figure 14.3.4, and the momentum of the recoil electron is

$$p = \sqrt{\left(\frac{h}{\lambda_0}\right)^2 + \left(\frac{h}{\lambda}\right)^2} = 4.5 \times 10^{-23} \text{ kg·m·s}^{-1}$$

Let the angle between the momentum of the recoil electron and

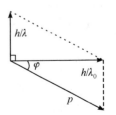

Figure 14.3.4 Figure of Example 14.3.1

the incident ray be φ, and

$$\tan\varphi = \frac{\dfrac{h}{\lambda}}{\dfrac{h}{\lambda_0}} = \frac{10}{11}$$

$$\varphi = \arctan\frac{10}{11} = 42°16'$$

14.4 Bohr's theory of the hydrogen atom

In the 1880s, great progress was made in spectroscopy and a large amount of experimental data about spectra was accumulated. In 1885, Balmer summarized the seemingly irregular hydrogen atom spectrum into a regular equation, which prompted people to realize that the essence of the spectrum law lies in the internal mechanism of the atom.

In 1913, Bohr established a semi-classical quantum theory of the structure of the hydrogen atom on the basis of the Rutherford nuclei-type structure, and satisfactorily explained the regularity of the spectrum of the hydrogen atom. However, Bohr's theory of hydrogen atoms had certain flaws. Ten years later, the quantum theory established on the basis of wave-particle duality completely solved the difficulties encountered by Bohr's theory with more correct concepts and theories. Even so, Bohr's theory played an important leading role in the development of quantum mechanics. At the same time, Bohr's concept of steady state and assumption of spectral line frequencies are still very important concepts in modern theories of atomic and molecular structures.

14.4.1 Experimental law of the hydrogen atom spectrum

The spectrum emitted by solid heating is continuous, but the atomic spectrum is composed of discrete line spectra. Different atoms emit different spectra. That is to say, the line spectrum reflects the important information on the internal structure of the atom, and the study of the spectral law has become an important means to explore the structure of matter. Among all the elements, the spectrum of the hydrogen atom is the simplest, so the study of the law of the hydrogen atom spectrum has become a breakthrough in the study of atomic spectrum. By 1885, 14 hydrogen spectral lines had been observed in the spectra of some stars. A Swiss middle school mathematics teacher (J. J. Balmer) found that the spectral lines in the visible part of these spectral lines can be summarized as the following equation

$$\lambda = B\frac{n^2}{n^2-4} \quad (n=3, 4, 5)$$

where $B = 365.46$ nm. The above equation is called Balmer's equation, and the value obtained by the equation is in good agreement with the experimental value. The set of spectral lines it expresses is called the Balmer series. If the reciprocal of the wavelength $\tilde{\nu} = \dfrac{1}{\lambda}$ is used to represent the number of wavelengths contained in an interval of a unit length, called the wavenumber, the Balmer equation can be written as

$$\tilde{\nu} = \frac{1}{\lambda} = R_H \left(\frac{1}{2^2} - \frac{1}{n^2} \right) \quad (n = 3, 4, 5) \tag{14.4.1}$$

where $R_H = \dfrac{4}{B}$, called the Rydberg constant of the hydrogen spectrum, and its measured value is $R_H = 1.096,775,8 \times 10^7 \, \text{m}^{-1}$.

Other spectral lines of the hydrogen atomic spectrum have also been discovered successively, one in the ultraviolet region, discovered by Lyman, and three in the infrared region, respectively by Paschen, Brackett, and Pfund. These line systems, like the Balmer system, can be expressed by a simple equation. In 1889, the Swiss physicist J. R. Rydberg proposed a universal equation, written as

$$\tilde{\nu} = R_H \left(\frac{1}{k^2} - \frac{1}{n^2} \right) \tag{14.4.2}$$

where $k = 1, 2, 3, \cdots$; $n = k+1, k+2, k+3, \cdots$, and it was called the Rydberg equation. For $k = 1$ and $n = 2, 3, 4, \cdots$, Lyman series, ultraviolet region; $k = 2$ and $n = 3, 4, 5, \cdots$, Balmer series, visible region; $k = 3$ and $n = 4, 5, 6, \cdots$, Paschen series, infrared region; $k = 4$ and $n = 5, 6, 7, \cdots$, Brackett series, infrared region; $k = 5$ and $n = 6, 7, 8, \cdots$, Pfund series, infrared region. These spectrum lines are shown in Figure 14.4.1.

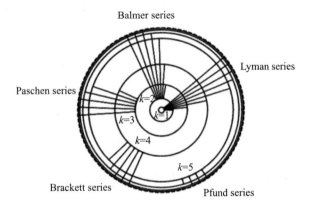

Figure 14.4.1 Different spectrum series of the hydrogen atom

14.4.2 Bohr's theory of the hydrogen atom

In 1911, the British physicist E. Rutherford proposed the nuclear structure of an atom based on the alpha particle scattering experiment, that is, an atom is composed of a positively charged nucleus and electrons orbiting outside the nucleus. According to the classical electromagnetic wave theory, the rotation of electrons around the nucleus must have acceleration, and the accelerated electrons will emit electromagnetic waves whose frequency should be equal to the frequency of electrons revolving around the nucleus. The energy of electrons will decrease continuously due to radiation, the frequency will gradually change, and the spectrum of radiation should be continuous. At the same time, due to the continuous reduction of energy, the electron movement radius will gradually decrease, and finally the electron will fall into the nucleus, so the structure of the atom is unstable. However, in fact, an atom is a stable system and the spectrum it emits is a line spectrum. It can be seen that the classical electromagnetic theory cannot explain the connection between the nuclear structure model and the experimental law.

In 1913, the Danish physicist N. H. D. Bohr extended Planck's and Einstein's quantum theory of light to the atomic system based on the nuclear structure, and proposed three basic assumptions as the starting point of his hydrogen atom theory such that the hydrogen atom spectrum was well explained.

Bohr's three basic assumptions are:

(1) **Steady state assumption**: There are only some discontinuous energy states in the hydrogen atom system, electrons can only make circular motions in orbits corresponding to these energies, and electrons do not radiate electromagnetic waves to the outside; such a state is called a stable state (steady state).

(2) **Orbital quantization assumption**: The angular momentum L of an electron in these steady state orbits satisfies the quantization, namely

$$L = n\hbar \tag{14.4.3}$$

where $n = 1, 2, 3, \cdots$ and is an integer, called the angular momentum quantum number, and $\hbar = \dfrac{h}{2\pi}$ is the reduced Planck constant.

(3) **Transition assumption**: When the electron in the hydrogen atom transitions from a E_n steady state of energy to another E_k steady state of energy, it will emit or absorb a photon with a frequency of ν_{nk}

$$\nu_{nk} = \frac{|E_n - E_k|}{h} \tag{14.4.4}$$

$E_n > E_k$ represents the emission of a photon, while $E_n < E_k$ represents the absorption of a photon.

Among these three assumptions, the first one is empirical, but it is Bohr's major contribution to the theory of atomic structure because he made a huge modification to the classical theory, thus solving the problem of atomic stability. The quantization of angular momentum expressed in the second item is artificially set, and it was later known that it can be derived naturally from the de Broglie hypothesis. The third item is derived from the Planck quantum hypothesis, which can explain the line spectrum formation.

According to the above three assumptions, combined with classical mechanics, the energy level equation of the hydrogen atom can be deduced, and the law of the hydrogen atom spectrum can be explained.

When the electron moves around the nucleus, the Coulomb force of the nucleus on the electron is equal to the centripetal force of the circular motion around the nucleus, namely

$$\frac{1}{4\pi\varepsilon_0}\frac{e^2}{r^2}=m\frac{v^2}{r} \qquad (14.4.5)$$

where m is the mass of the electron, v is the velocity of the electron, r is the radius of the orbit, and e is the electric quantity of the electron. The mechanical energy of the electron in the circular motion is

$$E=\frac{1}{2}mv^2-\frac{e^2}{4\pi\varepsilon_0 r}=-\frac{e^2}{8\pi\varepsilon_0 r} \qquad (14.4.6)$$

And in the circular motion the angular momentum satisfies the quantization condition

$$L=mvr=n\frac{h}{2\pi} \qquad (14.4.7)$$

Using the above three equations, we can get

$$r_n=n^2\left(\frac{\varepsilon_0 h^2}{\pi m e^2}\right) \quad (n=1, 2, 3, \cdots) \qquad (14.4.8)$$

This is the radius of the nth stable orbit in the atom, and the radius corresponding to $n=1$ is $r_1=\dfrac{\varepsilon_0 h^2}{\pi m e^2}=0.0529$ nm, called the **Bohr radius**. It can also be seen from this result that the orbital radius of the hydrogen atom is proportional to the square of the integer n and cannot be continuously changed. The quantum number n is usually used to mark different steady states; combined with Equation (14.4.8), the energy at the nth steady state is

$$E_n=-\frac{1}{8\pi\varepsilon_0}\frac{e^2}{r_n}=-\frac{1}{n^2}\frac{me^4}{8\varepsilon_0^2 h^2} \qquad (14.4.9)$$

It shows that the energy of a hydrogen atom can only take some discrete values, which is called energy quantization. The value corresponding to the energy is called the energy level. When $n=1$, the corresponding energy level $E_1=-\dfrac{me^4}{8\varepsilon_0^2 h^2}=-13.6$ eV is the lowest, and the corresponding state is called **the ground state**. States corresponding to other energy levels are called **excited states** with energy of $E_n=\dfrac{E_1}{n^2}$.

Equation (14.4.3) can be obtained by using the steady state assumption of Bohr's hydrogen atom theory. When the electron transitions from a high energy level E_n to a low energy level E_k, the frequency of the emitted photon is

$$\nu_{nk}=\frac{E_n-E_k}{h}=\left(\frac{1}{n^2}-\frac{1}{k^2}\right)E_1 \qquad (14.4.10)$$

The wavenumber is

$$\tilde{\nu} = \frac{1}{\lambda_{nk}} = \frac{\nu_{nk}}{c} = \frac{E_1}{hc}\left(\frac{1}{n^2}-\frac{1}{k^2}\right) = R_H\left(\frac{1}{k^2}-\frac{1}{n^2}\right) \tag{14.4.11}$$

where $R_H = -\dfrac{E_1}{hc}$. The theoretical calculation result is $R_H = 1.097,373,1 \times 10^7$ m^{-1}, very close to the experimental value.

Example 14.4.1 In the Balmer line series of the hydrogen atom spectrum, the wavelength of a spectral line is $\lambda = 434$ nm.

(1) How many electron volts is the photon energy corresponding to this spectral line?

(2) If this spectral line is produced by the transition of the hydrogen atom from energy level E_n to energy level E_k, what are the values of n and k?

(3) How many spectral lines in total can a large number of hydrogen atoms with the highest energy level E_n emit? Which spectral line has the shortest wavelength?

Solution (1) $\lambda = 434$ nm corresponds to the energy of the photon

$$E = \frac{hc}{\lambda} = \frac{6.63 \times 10^{-34} \times 3.0 \times 10^8}{434 \times 10^{-9}} = 4.58 \times 10^{-19} \text{ J} = 2.86 \text{ eV}$$

(2) $k = 2$ in the Balmer series. According to the transition hypothesis and the energy level relationship $E_n = \dfrac{E_1}{n^2}$, we can get

$$h\nu = E_n - E_2 = \left(\frac{1}{n^2} - \frac{1}{2^2}\right)E_1$$

in which $E_1 = -13.6$ eV, so the solution is $n = 5$ and the spectral line is generated by the transition of the hydrogen atom from energy level E_5 to energy level E_2.

(3) The excited state of $n = 5$ can transition to the steady state of $k = 4, 3, 2, 1$, and there are 4 spectral lines, which are Brackett, Paschen, Balmer and Lyman series respectively. Similarly, there are respectively 3, 2, and 1 spectral lines from the excited state transitions of $n = 4, 3, 2$, that is, there are $4+3+2+1 = 10$ spectral lines in total.

Among them, the wavelength is the shortest during the transition from $n = 5$ to $k = 1$. According to the transition assumption $h\nu = E_5 - E_1$ and energy level $E_5 = \dfrac{E_1}{5^2} = -0.544$ eV, the shortest wavelength is

$$\lambda = \frac{hc}{E_5 - E_1} = \frac{6.63 \times 10^{-34} \times 3 \times 10^8}{1.6 \times 10^{-19} \times (13.6 - 0.544)} = 9.52 \times 10^{-8} \text{ m}$$

14.5 Wave-particle duality of a physical particle

Interference and diffraction of light illustrate the wave nature of light, while black body radiation, photoelectric effect and the Compton scattering fully demonstrate the particle nature of light, so light has wave-particle duality. De Broglie proposed the concept of matter waves inspired by the wave-particle duality of light, which was soon confirmed by experiments, opening up a new way for the development of quantum theory.

14.5.1 The de Broglie hypothesis

Inspired by the wave-particle duality of light, in 1924, the French physicist de Broglie extended this feature to all physical particles. He believed that all physical particles such as electrons, protons, and neutrons have the wave-particle duality.

Analogous to the case of photons, assuming that the physical particle energy is E, momentum is p, its corresponding matter wave frequency is ν, and wavelength is λ, then its momentum and wavelength, energy and frequency have the following correspondence

$$p = mv = \frac{h}{\lambda} \tag{14.5.1}$$

$$E = mc^2 = h\nu \tag{14.5.2}$$

Considering the relativistic effect, wavelength and frequency can be expressed as follows:

$$\lambda = \frac{h}{p} = \frac{h}{mv} = \frac{h}{m_0 v}\sqrt{1 - v^2/c^2} \tag{14.5.3}$$

$$\nu = \frac{E}{h} = \frac{mc^2}{h} = \frac{m_0 c^2}{h\sqrt{1 - v^2/c^2}} \tag{14.5.4}$$

The waves corresponding to physical particles are called the **de Broglie waves** or **matter waves**. The wave-particle duality is an inherent property of matter. Electron microscopes manufactured by using the wave nature of electrons have been widely used in metals, semiconductors, biology, chemistry, medicine and new material field.

Example 14.5.1 Calculate the wavelength of the matter wave of a bullet with mass of $m = 0.01$ kg when flying at a velocity of $v = 300$ m·s^{-1}.

Solution According to the de Broglie relation Equation (14.5.3)

$$\lambda = \frac{h}{mv} \approx 2.2 \times 10^{-34} \text{ m}$$

Since this wavelength is very small relative to the size of the bullet, it has little effect on the motion of the bullet, so in the macroscopic world, the quantum effect of objects can be ignored.

14.5.2 Experimental verification of a matter wave

The correctness of de Broglie's hypothesis of matter waves must be verified by experiments. In 1927, the electron diffraction experiment made by the American physicists C. J. Davisson and L. H. Germer confirmed the wave nature of electrons. The experimental setup is shown in Figure 14.5.1(a). The electron beam emitted by the electron gun is accelerated by the electric field, projected onto the specially selected crystal face of the nickel crystal and then scattered by the crystal face into the electron detector, and the intensity of the scattered electron beam can be measured. During the experiment, keep the glancing angle of the electron beam θ unchanged, change the accelerating voltage U, and observe the current intensity I of the scattered electron current. The results of the experiment show that I does not increase with the increase of U, but occur some maximum and minimum peaks, as shown in Figure 14.5.1(b). The above results cannot be explained by particle motion, but can be analyzed by X-ray diffraction of crystals. Treat electron beams exactly like X-rays, and the whole experiment is completely similar to the X-ray diffraction on the crystal lattice structure, satisfying the Bragg equation

$$a\sin\theta = k\lambda$$

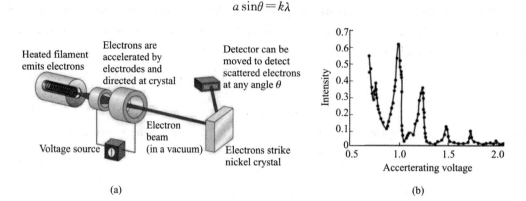

Figure 14.5.1 The Davisson-Germer experiment

Substitute the de Broglie wavelength equation (14.5.3) into Bragg's equation, and the relationship between the electron current intensity and the accelerating voltage can be obtained

$$a\sin\theta = k \frac{h}{\sqrt{2m_0 e}} \frac{1}{\sqrt{U}}$$

The U value corresponding to each maximum point could be calculated from the above equation, and the result was in complete agreement with the experiment. This not only proved that electrons do have the nature of wave, but also proved the correctness of de Broglie's equation.

Later in the same year, the physicist G. P. Thomson conducted another electron diffraction experiment, where he hit the electron beam accelerated by the electric field on the gold foil and found the electron diffraction pattern on the photographic plate behind the gold foil, as shown in Figure 14.5.2. According to the radii of these diffraction rings, the wavelength of the electron wave could be calculated, which further confirmed the de Broglie wavelength equation and the wave-particle duality of electrons. After the 1930s, diffraction phenomena of protons, neutrons, helium atoms, and hydrogen molecules all were confirmed. Especially, the neutron diffraction technique has become one of the most effective methods to study the microstructure of solids. Therefore, the wave property is the inherent property of the particle itself, and the de Broglie equation is the basic equation reflecting the wave-particle duality of the physical particle. De Broglie won the 1929 Nobel

Prize in Physics for his theory of matter waves.

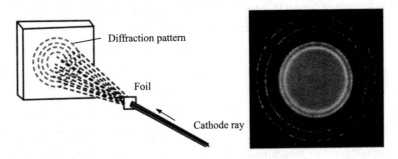

Figure 14.5.2 Diffraction of electrons through a metal foil

Example 14.5.2 Calculate the de Broglie wavelength of an electron after it is accelerated by accelerating voltages $U_1=100$ V, and $U_2=10,000$ V, respectively.

Solution After being accelerated, the kinetic energy of the electron is

$$\frac{1}{2}mv^2 = eU$$

Hence

$$v = \sqrt{\frac{2eU}{m}}$$

According to the de Broglie equation (14.5.3)

$$\lambda = \frac{h}{mv} = \frac{h}{\sqrt{2eUm}}$$

When the accelerating voltage is $U_1 = 100$ V,

$$\lambda_1 = \frac{h}{\sqrt{2emU_1}} = 0.123 \text{ nm}$$

When the accelerating voltage is $U_2 = 10,000$ V,

$$\lambda_1 = \frac{h}{\sqrt{2emU_1}} = 0.0123 \text{ nm}$$

14.5.3 Statistical interpretation of a matter wave

In classical physics, particles and waves are two distinct concepts. Particles are discrete, have definite orbits, and have definite positions and velocities at any time; waves are continuous, can be superimposed, and can produce the interference and diffraction phenomena. How can two opposing concepts be unified? How are the wave nature and particle nature related? In 1926, M. Born made a convincing explanation for the wave nature of physical particles, and unified the wave nature and particle nature of physical particles.

For the problem of light intensity in the diffraction phenomenon of light, based on the photon theory and statistics, Einstein proposed that the light intensity is proportional to the number of photons falling on the screen per unit time. Where the light is strong, the probability of photons arriving is high; Where the light is weak, the probability of photons arriving is small. For the electron diffraction phenomenon, Born used the same point of view that the probability of electrons appearing at the peak of the electron flow (or the bright fringe in the diffraction pattern) is high, and the probability of electrons appearing at the non-peak (or the dark fringes in the diffraction pattern) is small. In the double-slit diffraction experiment of electrons, if the electron flow is so weak that the electrons pass through the double slits almost one by one, the bright spots appearing on the phosphor screen are irregular at the beginning. As the number of incident electrons increases, a certain regularity gradually shows. When the number of incident electrons is quite large, regularity is clearly displayed. Figure 14.5.3 is the experimental photo of electron double-slit diffraction, which is the same as the result of Young's double-slit interference experiment, and both bright and dark fringes appear. This shows that the behavior of individual particles is uncertain, while a large number of particles follow a certain statistical law. The de Broglie matter wave is essentially a probability wave in a statistical sense, which is essentially different from the wave in classical physics.

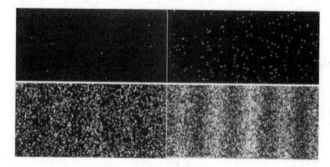

Figure 14.5.3   Photographs of electron diffraction through double slits

14.6   Uncertainty relation

In classical mechanics, the position and momentum of an object can be determined simultaneously. If the position and momentum of the object at a certain moment and its force are known, by solving the equation of motion, the position, momentum, and path of the object can be accurately determined at any time. For a microscopic particle, due to its particle nature, it is possible to talk about its momentum and position; but because of its wave nature, the particle does not have a definite position at any time. Therefore, due to the duality of wave and particle, the position and momentum of microscopic particles at any time have an uncertain quantity, that is, the motion of microscopic particles cannot be accurately described by position and momentum at the same time.

In 1927, the German physicist W. K. Heisenberg quantitatively expressed this uncertainty relationship after analyzing several ideal experiments, which is the famous **uncertainty principle**, also known as the **uncertainty relation**. Heisenberg was awarded the Nobel Prize in Physics in 1932. Now take the electron single slit diffraction experiment as an example.

As shown in Figure 14.6.1, a beam of electrons moving in the Y direction (horizontal direction) passes through a single slit with a width of $a = \Delta x$ in the X direction. Due to the wave nature of the electrons, they fall on the screen to form the Fraunhofer single slit diffraction similar to that produced by a beam of light, that is, a series of bright and dark fringes. For an electron, it can pass through any position in the single slit, so its position in the X direction has an indeterminate amount: the width of the slit Δx, called **uncertainty of position**. Bright fringes of electron diffraction appear in many places on the screen, indicating the momentum of the electron in the X direction $p_x \neq 0$. Due to the diffraction effect, p_x is only determined by its diffraction angle φ. If we only consider the electron appears in the central bright fringe $\sin\varphi = \dfrac{\lambda}{a}$, then the uncertainty range of the momentum in the X direction p_x when it passes through the slit is

$$\Delta p_x = p\sin\varphi = p\,\frac{\lambda}{a} = p\,\frac{\lambda}{\Delta x} \tag{14.6.1}$$

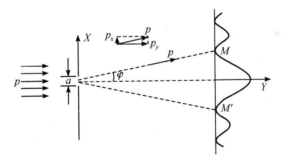

Figure 14.6.1 Schematic diagram of electron single slit diffraction

Combining the de Broglie equation (14.5.1), we get

$$\Delta p_x = p\sin\varphi = \frac{p\lambda}{\Delta x} = \frac{h}{\Delta x}$$

Therefore

$$\Delta x \Delta p_x = h \tag{14.6.2}$$

Considering that electrons may still fall in areas other than the central bright fringe, Δp_x should be larger than $p\sin\varphi$

$$\Delta x \Delta p_x \geqslant h \tag{14.6.3}$$

The above equation shows that the product of the electron's position uncertainty in the X direction and the momentum uncertainty in the same direction is greater than or equal to

Planck's constant h. Of course, the position uncertainty and momentum uncertainty of the electron in other directions also have a similar relationship.

The above is only a rough estimate based on the special case of electron diffraction. The strict derivation from the quantum theory points out that the uncertainty relation between the coordinate and momentum of a microscopic particle is

$$\Delta x \Delta p_x \geqslant \frac{\hbar}{2} \tag{14.6.4}$$

Equation(14.6.4) is called the **uncertainty relation between position and momentum** or the **Heisenberg uncertainty relation**, where $\hbar = \dfrac{h}{2\pi}$ is the reduced Planck constant. The uncertainty relation is not only applicable to electrons, but also to other microscopic particles. This relation shows that a microscopic particle can't have a definite position and momentum at the same time. Specifically, the more concentrated the particles are in the spatial distribution, that is, the smaller the spatial uncertainty, the larger the distribution of momentum, that is, the larger the momentum variation interval, and vice versa. This property of microscopic particles derives directly from their wave-particle duality, which represents the limit of accuracy with which a particle's position and momentum can be measured simultaneously. The uncertainty relation is the objective law of matter, not a problem of measurement technology or subjective ability.

The uncertainty relation is a universal principle that widely exists in the microscopic world and has many forms of expression. There is also such relationship between time and energy, angular position and angular momentum, namely

$$\Delta E \Delta t \geqslant \hbar \tag{14.6.5}$$

$$\Delta \theta \Delta L \geqslant \hbar \tag{14.6.6}$$

The uncertainty associated with energy and time exists in the energy level of an atom. In fact, the energy level is not a single value, but has a certain width ΔE. That is to say, when an electron is at a certain energy level, its actual energy has an indeterminate range of ΔE. In a large number of atoms of the same type, some electrons staying at the same energy level have a long residence time, and some have a short residence time, which can be represented by an average lifetime Δt. The width of the energy level can be measured by the width of the spectral line, so that the average lifetime of the energy level can be inferred. Since the ground state energy level of an atom is stable, there are $\Delta E = 0$ and $\Delta t \to \infty$ for the ground state. This principle applies not only to the energy levels of electrons outside the nucleus in atoms, but also to nuclei and elementary particles.

Example 14.6.1 A bullet with a mass of 0.01 kg is ejected from a muzzle with a diameter of 5 mm. Try to find the uncertainty of velocity of the bullet at the muzzle by the uncertainty relation.

Solution The diameter of the muzzle is the uncertainty amount of the position Δx, and due to the uncertainty relation and $\Delta p_x = m \Delta v_x$, the uncertainty of velocity of the

bullet is

$$\Delta v_x = \frac{\hbar}{m\Delta x} = \frac{6.63 \times 10^{-34}}{2 \times 3.14 \times 0.01 \times 5 \times 10^{-3}} = 2.1 \times 10^{-30} \text{ m} \cdot \text{s}^{-1}$$

It can be seen that the uncertainty of the position and velocity of macroscopic particles like bullets is negligible. So, for a macroscopic object like a bullet, the uncertainty relation doesn't work, and its wave nature doesn't have any real effect on its "classic" movement.

Example 14.6.2 Find the uncertainty of the velocity of electrons in an atom.

Solution The uncertainty of the position of the electron in an atom is the linearity of the atom 10^{-10} m. According to the uncertainty relationship $\Delta x \Delta p_x \geqslant \frac{\hbar}{2}$ and $\Delta p_x = m\Delta v_x$ we can get

$$\Delta v_x \geqslant \frac{\hbar}{2m\Delta x} = \frac{6.63 \times 10^{-34}}{2 \times 2 \times 3.14 \times 9.1 \times 10^{-31} \times 10^{-10}} \approx 5.8 \times 10^5 \text{ m} \cdot \text{s}^{-1}$$

The magnitude of this velocity uncertainty is comparable to the magnitude of the orbital velocity (10^6 m $\cdot$ s^{-1}) of the electrons in the hydrogen atom calculated according to Newtonian mechanics. Thus, for electrons in an atom, talking about their speed is of little practical significance. The wave nature of electrons is so prominently displayed that the concept of orbits must be completely abandoned for the description of electron motion, which can only be described by the probability distribution of electrons in space—**the electron cloud image.**

14.7　Wave function and the Schrödinger equation

All matter has the duality of wave and particle. For an macroscopic object, because of its insignificant wave nature, classical physics can be used to describe its motion law. Newton's equation of motion is a general equation to describe the motion of macroscopic objects. But for a microscopic particle, its wave nature cannot be ignored, and its motion cannot be described by the method of classical mechanics. Then, how to describe the state of motion of the microscopic particle? And what is its equation of motion? Shortly after de Broglie proposed the matter wave hypothesis the Austrian physicist Schrödinger proposed using the wave function to describe the motion state of microscopic particles, in 1925, and established the equation followed by the wave function, namely the Schrödinger equation. This section mainly introduces some basic concepts of non-relativistic quantum mechanics and the Schrödinger equation.

14.7.1　Wave function

Schrödinger believed that the motion of electrons, neutrons, protons and other microscopic particles with the wave-particle duality could be described by wave functions like mechanical waves or light waves, but frequency and wavelength that are the physical quantities describing the wave nature should follow the de Broglie relationship.

For simplicity, we first consider the wave function of a free particle. What is called a free particle refers to a particle that is not affected by any external force, and its motion is a uniform linear motion, so it has constant energy E and momentum p. Take the trajectory of a particle as the X axis. According to the wave-particle duality, the wavelength of the matter wave corresponding to the particle is $\lambda = \dfrac{h}{p}$, and the frequency $\nu = \dfrac{h}{E}$, and they remain unchanged throughout the motion process. In classical physics, a wave with a constant wavelength and frequency is a plane simple harmonic wave and its wave function is

$$y(x, t) = A\cos\left[2\pi\left(\nu t - \frac{x}{\lambda}\right)\right] \tag{14.7.1}$$

Rewrite the above equation in the form of the complex number

$$y(x, t) = A e^{-i2\pi(\nu t - \frac{x}{\lambda})} \tag{14.7.2}$$

If only the real part is taken, this equation corresponds exactly to the classical wave function. Use the wave-particle duality equation, replace frequency and wavelength with energy and momentum respectively, and use the reduced Planck constant $\hbar = \dfrac{h}{2\pi}$ and the plane matter wave equation of a free particle is

$$\Psi(x, t) = \Psi_0 e^{-\frac{i}{\hbar}(Et - px)} \tag{14.7.3}$$

In order to distinguish it from the classical plane wave function, Ψ and Ψ_0 are used to represent A and y in Equation (14.7.1) respectively. This is the matter wave function corresponding to the one-dimensional free moving particle.

Extend this wave function to the general case, and any single particle wave function can be expressed as $\Psi(r, t)$, and its wave function can be expressed as

$$\Psi(r, t) = \Psi_0 e^{-\frac{i}{\hbar}(Et - p \cdot r)}$$

or

$$\Psi(x, y, z, t) = \Psi_0 e^{-\frac{i}{\hbar}[Et - (p_x x + p_y y + p_z z)]}$$

According to Born's statistical interpretation of matter waves, the de Broglie waves are probability waves. Where there are more particles, the intensity of the de Broglie waves of particles is stronger, and the number of particles in the spatial distribution is proportional to the probability of particles appearing there. According to the proportional relationship between the intensity of the wave and the square of its amplitude, it can be known that at a certain moment, the probability of appearing in the volume element dV near the space r is proportional to the product of the square of the amplitude of the wave function in the volume element and the volume element $\Psi_0^2 \, \mathrm{d}V$. Because the wave function $\Psi(\boldsymbol{r}, t)$ is a complex exponential function, the algorithm of the complex exponential function can be used to obtain

$$\Psi_0^2 = |\Psi(\boldsymbol{r}, t)|^2 = \Psi(\boldsymbol{r}, t)\Psi^*(\boldsymbol{r}, t)$$

So, the probability that the wave function is in the volume element $\mathrm{d}V$ near the space $\boldsymbol{r}$ is

$$\mathrm{d}W = |\Psi(\boldsymbol{r}, t)|^2 \mathrm{d}V \tag{14.7.4}$$

This is the physical meaning of the wave function.

It must be noted that matter waves are fundamentally different from classical mechanical and electromagnetic waves. Mechanical waves are the propagation of mechanical vibrations in space, and electromagnetic waves are the propagation of electromagnetic fields in space. Their wave functions represent their law of motion, having exact physical meanings. In contrast, matter waves are probabilistic waves. The specific physical meaning can only be reflected by the square of the absolute value of the wave function.

Since the probability of particles appearing at any point in space should be **unique and finite**, the wave function must satisfy **singularity, finiteness** and **continuity**. In addition, the wave function should also satisfy **the normalization condition**

$$\int |\Psi(\boldsymbol{r}, t)|^2 \mathrm{d}V = 1 \tag{14.7.5}$$

14.7.2 The Schrödinger equation

In order to obtain the differential equation satisfied by the wave function of the matter wave, in 1926, through analysis and analogy, Schrödinger established the wave equation satisfied by microscopic particles moving at low speed in a potential field, which is now called the **Schrödinger equation**. The way he obtained this equation is not completely based on classical physics, and the equation cannot be strictly deduced. The correctness of the equation is only verified by experiments. In the following part, we introduce the main idea in establishing the Schrödinger equation.

For the sake of simplicity, we still take one-dimensional free particles as an example. As mentioned earlier, moving along the X axis, a free particle has momentum p and energy E, and its wave function is Equation (14.7.3).

Taking the first-order partial derivative of the equation with respect to t, we get

$$\frac{\partial \Psi}{\partial t} = -\frac{\mathrm{i}}{\hbar} E \Psi \tag{14.7.6}$$

Taking the second-order partial derivative with respect to x, we get

$$\frac{\partial^2 \Psi}{\partial x^2} = -\frac{p^2}{\hbar} \Psi \tag{14.7.7}$$

Considering that the energy of the free particle is kinetic energy, and when the speed of the free particle is much less than the speed of light, in the non-relativistic range, the relationship between the kinetic energy and the momentum of the free particle is

$$E = E_k = \frac{p^2}{2m}$$

By Equations (14.7.6) and (14.7.7), we get

$$-\frac{\hbar^2}{2m}\frac{\partial^2 \Psi}{\partial x^2} = i\hbar \frac{\partial \Psi}{\partial t} \qquad (14.7.8)$$

This is the law followed by the wave function of one-dimensional free particles, which is called the time-dependent Schrödinger equation of one-dimensional free particles.

Free particles are only a special case. Generally speaking, microscopic particles are usually affected by a force field. If the potential energy of a particle in a potential field is U, its energy is

$$E = E_k + U = \frac{p^2}{2m} + U$$

Replacing E in Equation (14.7.6) with the above equation, and making use of Equation (14.7.8), we can get

$$-\frac{\hbar^2}{2m}\frac{\partial^2 \Psi}{\partial x^2} + U\Psi = i\hbar \frac{\partial \Psi}{\partial t} \qquad (14.7.9)$$

This is the time-dependent Schrödinger equation of a particle moving in one dimension in a potential field, which describes the law of the state of a particle of mass m changing with time in a potential field of potential energy U.

If the potential field U in which the particle is located is independent of time and only related to the spatial coordinates, that is, $U = U(x)$, the Schrödinger equation can be solved by the method of separation of variables. Let

$$\Psi(x, t) = \psi(x) f(t) = \psi(x) e^{-\frac{i}{\hbar} px} \qquad (14.7.10)$$

in which

$$\psi(x) = \psi_0 e^{-\frac{i}{\hbar} px}$$

Substituting Equation (14.7.10) into Equation (14.7.9), we can get

$$\frac{\hbar^2}{2m}\frac{\partial^2 \psi}{\partial x^2} + (E - U)\psi(x) = 0 \qquad (14.7.11)$$

The energy of the system in the above equation is also a definite value independent of time, and the state in which the energy does not change with time is called the steady state. Therefore, the above equation is called the **steady-state Schrödinger equation** of one-dimensional moving particles, and $\psi(x)$ is a one-dimensional steady-state wave function. Since $\psi(x)$ is only a function of coordinates, its probability density $\psi(x)\psi^*(x)$ is also a function of coordinates and has nothing to do with time, so the probability distribution of steady-state particles in space will not change with time.

Similarly, the steady-state Schrödinger equation in three dimensions can be expressed as

Chapter 14 Fundamentals of Quantum Physics

$$\frac{\hbar^2}{2m}\left(\frac{\partial^2 \psi}{\partial x^2}+\frac{\partial^2 \psi}{\partial y^2}+\frac{\partial^2 \psi}{\partial z^2}\right)+[E-U(x,y,z)]\psi=0 \qquad (14.7.12)$$

14.7.3 Application of the Schrödinger equation

Solving the Schrödinger equation is far beyond this book, but it has huge application value in our life; for example, semiconductor chips, quantum dot light emission and other problems can be explained by Equation (14.7.12). There are many steady-state problems in physics, such as hydrogen atom and one-dimensional infinitely deep potential well, whose results obtained by using the steady-state Schrödinger equation have important applications in many related fields. In the following, the steady-state Schrödinger equation for the one-dimensional infinitely deep potential well is solved to discuss its physical meanings.

Under the action of a force field, a particle of mass m is restricted to move within a certain range. The simplest case is that the motion of the particle in the external force field is one-dimensional, such as the motion in a one-dimensional infinitely deep potential well as shown in Figure 14.7.1. In the well, since the potential energy is constant, the particle moves freely without any force. At the boundaries $x=0$ and $x=a$, the potential energy

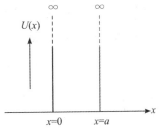

Figure 14.7.1 One-dimensional infinitely deep potential well

suddenly increases to infinity. Therefore, the position of the particle is restricted in the well, and this motion state of the particle is called the bound state.

The potential energy function for the one-dimensional infinitely deep potential well shown in Figure 14.7.1 can be expressed as

$$U(x)=\begin{cases} 0 & (0<x<a) \\ \infty & (x\leqslant 0, x\geqslant a) \end{cases} \qquad (14.7.13)$$

Thus, it can be seen that $U(x)$ is only a function of space coordinates and has nothing to do with time, and the wave function can be solved by the steady-state Schrödinger equation.

In the regions of $x\leqslant 0$ and $x\geqslant a$, since the potential function is infinite, only the wave function of 0 can satisfy the Schrödinger equation, and the steady-state wave function is $\psi(x)=0$.

There is a steady-state Schrödinger equation when $0<x<a$

$$-\frac{\hbar^2}{2m}\frac{d^2\psi(x)}{dx^2}=E\psi(x) \qquad (14.7.14)$$

Let $k=\sqrt{\dfrac{2mE}{\hbar^2}}$, and then

$$\frac{d^2\psi}{dx^2}+k^2\psi=0$$

When $k>0$, its general solution is

$$\psi(x)=A\sin kx+B\cos kx \tag{14.7.15}$$

where A and B are undetermined constants.

The wave function is continuous at $x=0$ and $x=a$, hence, there should be $\psi(0)=\psi(a)=0$.

$\psi(0)=0$ makes $B=0$, and the steady-state wave function Equation (14.7.15) becomes

$$\psi(x)=A\sin kx \tag{14.7.16}$$

From $\psi(a)=0$, that is $A\sin ka=0$, we can get $ka=n\pi$, that is

$$k=\frac{n\pi}{a} \quad (n=1,2,\cdots) \tag{14.7.17}$$

Substitute $k=\sqrt{\frac{2mE}{\hbar^2}}$ into Equation (14.7.17), and the quantized energy of the particle in the one-dimensional infinitely deep potential well is obtained as

$$E_n=\frac{\pi^2\hbar^2}{2ma^2}n^2 \quad (n=1,2,\cdots) \tag{14.7.18}$$

From Equation (14.7.18), it can be known that the energy can only take some discrete values, that is, the energy is quantized. Normalize the wave function $\psi_n(x)=A\sin\frac{n\pi}{a}x$ of the one-dimensional infinitely deep potential well, and there is

$$\int_{-\infty}^{0}0\,dx+\int_{0}^{a}\psi_n^2\,dx+\int_{a}^{\infty}0\,dx=1$$

Then the constant A of the wave function in Equation (14.7.16) is obtained as

$$A=\sqrt{\frac{2}{a}} \tag{14.7.19}$$

So far, the wave function for a one-dimensional infinitely deep potential well can be written as

$$\psi_n(x)=\begin{cases}\sqrt{\dfrac{2}{a}}\sin\dfrac{n\pi x}{a} & (0<x<a)\\ 0 & (x\leqslant 0,\ x\geqslant a)\end{cases} \tag{14.7.20}$$

From the above results we get the conclusions that follow.

(1) When $n=1$, $E_1=\dfrac{\pi^2\hbar^2}{2ma^2}$, called the ground state energy, and there is a simple relationship between the energy on other energy levels and the ground state energy

$$E_n=n^2E_1$$

It can be seen that the energy level distribution is not uniform. The higher the energy level, the greater the energy level density. When n is large, the energy level can be regarded as continuous.

(2) The steady-state wave function corresponding to E_n is

$$\psi_n(x) = \sqrt{\frac{2}{a}} \sin\left(\frac{n\pi}{a}\right) x$$

The probability density is

$$|\psi_n(x)|^2 = \frac{2}{a} \sin^2\left(\frac{n\pi x}{a}\right)$$

Their distribution is shown in Figure 14.7.2.

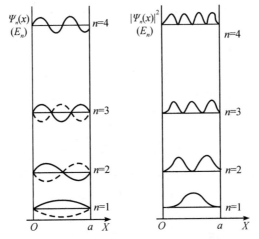

Figure14.7.2 Wave functions and probability density diagrams in a one-dimensional infinitely deep potential well

As the quantum number n increases, the number of peaks in the probability density distribution curve also increases and is equal to the quantum number, and the distance between two adjacent peaks decreases with the increase of n. When n is large, the distance between adjacent peaks will be very small, and they will be very close to each other, which is very close to the situation in classical physics that the probability of particles in the potential well is the same everywhere.

14.8 Application of quantum mechanics to the hydrogen atom

Bohr's theory of hydrogen atoms has great limitations because it cannot satisfactorily explain the structure of the hydrogen atom and the law of electron movement in it, while quantum mechanics can solve the hydrogen atom problem more satisfactorily. Due to the mathematical difficulties in solving the Schrödinger equation for the hydrogen atom, we do not provide strict calculations here but only focus on some important results.

14.8.1 The Schrödinger equation for the hydrogen atom

In a hydrogen atom, since the mass of the nucleus is thousands of times that of the electron, it is assumed that the nucleus is at rest at the origin of coordinates, and the electron moves in the Coulomb electric field of the proton, so the potential energy function of the electron is

$$U(r) = \frac{e^2}{4\pi\varepsilon_0 r} \tag{14.8.1}$$

$r = \sqrt{x^2 + y^2 + z^2}$ is the distance of the electron from the nucleus. Since the potential function is only a function of spatial position, the steady-state Schrödinger equation is

$$\frac{\hbar^2}{2m}\left(\frac{\partial^2 \psi}{\partial x^2} + \frac{\partial^2 \psi}{\partial y^2} + \frac{\partial^2 \psi}{\partial z^2}\right) + \left(E + \frac{e^2}{4\pi\varepsilon_0 r}\right)\psi = 0 \tag{14.8.2}$$

Considering that the potential energy U is a function of r and has spherical symmetry, the spherical coordinate (r, θ, φ) is more convenient; the origin of the coordinates is taken on the nucleus, and then Equation (14.8.2) is transformed into

$$\frac{1}{r^2}\frac{\partial}{\partial r}\left(r^2 \frac{\partial \psi}{\partial r}\right) + \frac{1}{r^2 \sin\theta}\frac{\partial}{\partial \theta}\left(\sin\theta \frac{\partial \psi}{\partial \theta}\right) + \frac{1}{r^2 \sin^2\theta}\frac{\partial^2 \psi}{\partial \varphi^2} + \frac{2m}{\hbar^2}\left(E + \frac{e^2}{4\pi\varepsilon_0 r}\right)\psi = 0 \tag{14.8.3}$$

The solution to this equation can be expressed as the product of three functions, namely

$$\psi = R(r)\Theta(\theta)\Psi(\varphi)$$

Because R is only related to the radial variable, it is also called the radial wave function; $Y = \Theta(\theta)\Psi(\varphi)$ is related to the square of the angular momentum, so Y is the wave function of the square of the angular momentum; Ψ is related to the projection of the angular momentum in the Z direction, so Ψ is called the angular momentum projection wave function. Solving Equation (14.8.3) requires more and deeper mathematical knowledge. Here, we mainly introduce the important conclusions from the equation.

1. Quantization of energy

The energy of the electron can be obtained by solving the equation by quantizing the energy

$$E_n = -\frac{1}{n^2}\left(\frac{me^4}{8\varepsilon_0^2 h^2}\right) \tag{14.8.4}$$

$n = 1, 2, 3, \cdots$, called the principal quantum number, and this result is the same as the energy level equation obtained by Bohr's theory. On the one hand, it confirms the correctness of the Schrödinger equation; on the other hand, it is the natural result of the Schrödinger equation without any artificial assumptions. $n = 1, 2, 3, 4, 5, \cdots$, can be represented by K, L, M, N, O, $\cdots$ respectively.

2. Quantization of angular momentum

The electron moving around the nucleus has angular momentum. In Bohr's theory, its quantization is just an assumption; but now, by solving the Schrödinger equation, the magnitude of the angular momentum can be obtained

$$L = \sqrt{l(l+1)}\hbar \quad (l = 0, 1, 2, 3, \cdots, n-1) \tag{14.8.5}$$

where L is called angular quantum number or sub-quantum number, so the angular momentum of the electron moving around the nucleus is also quantized, but its magnitude is different from that in Bohr's theory. For example, when $n=3$, the angular momentum in Bohr's theory is $L=3\hbar$, while the Schrödinger equation gives the value of the angular momentum $L=0, \sqrt{2}\hbar, \sqrt{6}\hbar$, consistent with the experiment result.

3. Space quantization of angular momentum

In addition to the quantized orbital angular momentum, the projected value of the angular momentum in the spatial direction (taking the Z axis direction as an example) is also quantized.

$$L_z = m\hbar \quad (m=0, \pm 1, \pm 2, \cdots, \pm l) \quad (14.8.6)$$

in which m is called the magnetic quantum number. For example, for the angular quantum number $l=1$, the magnetic quantum number $m=0, \pm 1$ and its projection value $L_z = 0$, $\pm\hbar$, that is, when the angular momentum is $L=\sqrt{2}\hbar$, its projection in the Z axis direction can only take three discontinuous values, as shown in Figure 14.8.1(a); the orientation of angular momentum in space is quantized, which is called spatial quantization. The situation is similar when the magnitude of angular momentum takes other values, as shown in Figures 14.8.1(b) and 14.8.1(c) which correspond to $l=2$ and $l=3$, respectively. When the atom is placed in a magnetic field, the effect of the magnetic quantum number will appear, and the corresponding phenomenon is called the **Zeeman effect**.

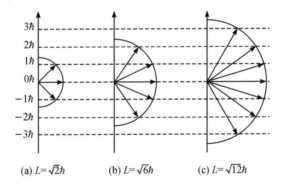

(a) $L=\sqrt{2}\hbar$ (b) $L=\sqrt{6}\hbar$ (c) $L=\sqrt{12}\hbar$

Figure 14.8.1 Schematic diagram of spatial quantization of angular momentum

14.8.2 Electron spin

As shown in Figure 14.8.2, an atomic emission source O emits a beam of atomic rays. After passing through the magnetic field, the atoms are deposited on the base plate P. According to theoretical analysis, due to the different magnetic quantum numbers of electrons in different atoms, the effect of the magnetic field will also be different, and hence, the deflection angle under the action of the magnetic field should be different. The greater the magnetic quantum number, the greater the deflection angle. The magnetic quantum number is represented by m. The atoms with $m=0$ could not be affected by the magnetic field and hence should deposit in the center of the base plate. The atoms with $m=1$ are deflected upward by the upward force, while the atoms with $m=-1$ are deflected downward. Since the possible values of m are always odd, an odd number of deposition lines should be formed on the base plate. However, in 1921, the German physicists Stern and Gerlach experimented with silver atoms, but only two deposition lines appeared on the bottom plate, as shown in Figure 14.8.2. The same phenomenon was observed when other types of atoms were used, indicating that electrons have an additional property.

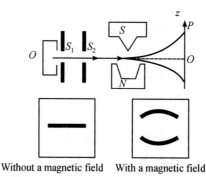

Figure 14.8.2 Schematic diagram of the Stern and Gerlach experimental setup

In 1925, Uhlenbeck and Goudsmit proposed the electron spin hypothesis, which satisfactorily explained the above phenomenon. They believed that the angular momentum corresponding to the electron's spin can be analogous to the orbital angular momentum

$$S=\sqrt{s(s+1)}\hbar \tag{14.8.7}$$

where s is the spin quantum number and can only take one value, namely

$$s=\frac{1}{2} \tag{14.8.8}$$

The spin angular momentum on the Z axis is also similar to the spatial quantization of angular momentum, namely

$$S_z=m_s\hbar, \quad m_s=\pm\frac{1}{2} \tag{14.8.9}$$

where m_s is called the spin magnetic quantum number and can only take two values of $\pm\frac{1}{2}$.

After considering the electron spin, the state of the electron in the hydrogen atom can be completely determined by four quantum numbers:

(1) the principal quantum number $n=1, 2, 3, \cdots$, which determines the energy of the electron;

(2) the angular quantum number $l=0, 1, 2, \cdots, n-1$, which determines the angular momentum of the electron moving around the nucleus;

(3) the magnetic quantum number $m=0, \pm1, \pm2, \cdots, \pm l$, which determines the spatial orientation of the angular momentum of the electron moving around the nucleus;

(4) the spin magnetic quantum number $m_s = \pm \dfrac{1}{2}$, which determines the spatial orientation of the spin angular momentum.

14.9 Electron shell structure of an atom

Except for hydrogen atoms or hydrogen-like ions, other elements have two or more electrons outside the nucleus. At this time, the interaction between the electrons will also affect their motion states. The mathematical solution of the Schrödinger equation for multi-electron atoms is quite difficult. However, the approximate calculation method in quantum mechanics can prove that the state of each extranuclear electron is still determined by four quantum numbers.

There are only two points different from the hydrogen atom:

(1) The energy of the electron is jointly determined by the principal quantum number n and the angular quantum number l. Electrons with the same principal quantum number but different angular quantum numbers have slightly different energies, so for a multi-electron system, the principal quantum number n only roughly determines the energy of the electrons.

(2) Different electrons have a certain distribution outside the nucleus. In 1916, W. Kossel proposed the hypothesis that electrons outside a multi-electron nucleus are distributed in certain shells, which is called the shell structure model of electrons. He believed that electrons with different principal quantum numbers are distributed in different shells. Such kind of shell is called the main shell labelled as $n = 1, 2, 3, \cdots$, which can also be represented by K, L, M, $\cdots$. Electrons with the same principal quantum number but different angular quantum numbers are distributed in different subshells, and the subshells corresponding to $l = 0, 1, 2, 3, \cdots$ are called s, p, d, f, $\cdots$ respectively. Generally speaking, the smaller the main quantum number n of the shell, the lower the energy level. In the same main shell, the smaller the angular quantum number l, the lower the energy level. The specific distribution of extranuclear electrons in each shell should also follow the following two principles.

1. The Pauli exclusion principle

In 1925, Pauli concluded the following rule based on spectral experiments: In the same atom, two or more electrons can't be in exactly the same quantum state. That is, in an atom, it is impossible for any two electrons to have a same set of quantum numbers (n, l, m, m_s), which is called **the Pauli exclusion principle**. According to this principle, the maximum number of electrons in one main shell n can be deduced. For the sub-shell l, the allowable magnetic quantum number is $m = 0, \pm 1, \pm 2, \cdots, \pm l$, that is, $2l+1$ values in total, and each magnetic quantum number m can correspond to two spin states, so in the sub-shell $4l+2$ electrons at most can be arranged. The main shell contains n sub-shells in total, so the maximum number of electrons it can hold is

$$Z_n = \sum_{l=0}^{n-1}(4l+2) = 2n^2 \qquad (14.9.1)$$

2. The principle of minimum energy

When an atom is in a normal state, electrons tend to occupy the lowest possible energy level. Therefore, the shell with lower energy level, that is, closer to the nucleus, is filled first with electrons, and the remaining electrons fill the lowest energy level that has not been occupied in turn, until all electrons fill the lowest possible energy levels. The energy level of an atom is determined by the primary quantum number n and the secondary quantum number l. Generally speaking, the smaller the primary quantum number, the lower the energy level; with the same primary quantum number, the lower secondary quantum number correspond to the lower energy. For example, the electron arrangement of a lithium atom can be recorded as $1s^2 2s^1$ and the symbol in the upper right corner represents the number of electrons in the sub-shell. The energy arrangement of different sub-shells in the atom is shown in Figure 14.9.1, in which the main quantum number increases gradually from bottom to top, and the angular quantum number gradually increases from left to right. The energy of the sub-shell is arranged from low to high in the direction of the dashed arrow, that is $1s^2 2s^2 2p^6 3s^2 \cdots$. It should be noted that the sub-shell

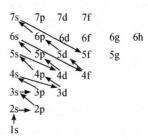

Figure 14.9.1 Atomic shell structure

with a large principal quantum number is not necessarily higher in energy than the sub-shell with a small principal quantum number. For example, the energy of the 4s sub-shell is lower than that of the 3d sub-shell. There are other special cases among which the outermost electron arrangement of a potassium atom is $1s^2 2s^2 2p^6 3s^2 3p^6 4s^1$, skipping the 3d sub-shell. In such cases, Chinese scientists have summed up a method of using $n+0.7l$ to determine energy. By using the atomic shell structure theory, scientific problems such as the source of element periodicity, the spectrum of atomic luminescence, chemical reactions, and material properties can be well explained.

Scientist Profile

Niels Henrik David Bohr(1885 - 1962), a Danish physicist, won the Nobel Prize in Physics in 1922 and as the founder of the Copenhagen School, he had a profound influence on the development of physics in the 20th century.

Bohr began his scientific career in 1905 and conducted scientific research for 57 years. His research work began in the unknown atomic structure and ended in the era when atomic physics was widely used. Bohr proposed the Bohr model to explain the spectrum of hydrogen atoms, and proposed the complementary principle and the Copenhagen interpretation to explain quantum mechanics. His contributions to atomic science made him one of the greatest physicists alongside Einstein in the first half of the 20th century.

Bohr founded the theory of atomic structure in his long paper, *On Atomic Structure and Molecular Structure*, which opened the way for atomic physics in the 20th century. In 1921, at the initiative of Bohr, the Institute of Theoretical Physics at the University of Copenhagen was established, which Bohr led for 40 years. From this institute emerged a large number of outstanding physicists, which became the most important and active academic center in the world during the rise of quantum mechanics, and still has a high international status. In 1927, Bohr first put forward the "complementary principle", which laid the foundation of the Copenhagen interpretation of quantum mechanics, and began a years-long debate with Einstein about the significance of quantum mechanics. Einstein proposed one thought experiment after another, trying to prove the contradictions and errors of the new theory, but Bohr cleverly refuted Einstein's opposition. This long debate promoted the perfection of Bohr's view in many aspects, enabling him to relate not only physics but also other disciplines to his later study of the complementary principle.

Extended Reading

Application of quantum effects

1. Introduction to the scanning tunneling microscope

When a particle with lower energy moves towards a potential barrier with higher energy, the particle should be reflected from the view of classical physics. However, in quantum physics, this particle has a certain ability to pass through the potential barrier due to its wave nature, which is called quantum tunneling. An important application of the quantum tunneling effect is the scanning tunneling microscope (STM). A STM is a device based on the tunneling effect and allows one to make detailed maps of surfaces, revealing features on the atomic scale with a resolution much greater than can be obtained with an optical or electron microscope.

The working principle of the STM is shown in Figure ER 1, where a very fine probe (atomic size) and the surface of the material under study are used as two electrodes. When the distance between the sample and the tip of the probe is very short (usually less than 1 nm), under the action of an external electric field, electrons will pass through the potential barrier between the two electrodes to form a tunneling current. The tunneling current is very sensitive to the distance between the tip and the sample. If the tunneling current is constant, the height change of the probe in the direction perpendicular to the sample can reflect the fluctuation of the sample surface. Using the STM, the steps, platforms and atomic arrays of atoms on the surface of the sample can be distinguished, so that the three-dimensional image of the surface can be directly drawn.

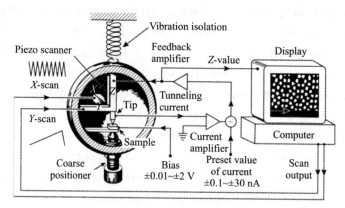

Figure ER 1　Schematic diagram of a scanning tunneling microscope

The tunneling current is the result of the overlapping of the electron wave functions of the probe tip and the sample, and is related to the distance d between them and the effective local work function U. The tunneling current I can be written as

$$I \propto V_0 \exp(-2ks)$$

in which V_0 is the bias voltage applied between the tip and the sample; $k = 2\pi\sqrt{2mU}/h$, where m is the electron mass, h is Planck's constant, and the effective local work function U is approximately equal to the potential barrier height. Under typical conditions, $k = 10$ nm^{-1}, and then when d changes by 0.1 nm, I will change by an order of magnitude. If the feedback circuit is used to keep I constant, then when the probe scans along the sample surface, d needs to remain constant, so that the probe will move up and down along the undulations of the sample surface. Then the data of the probe moving up and down and the coordinate position of the probe along the horizontal plane, that is, the three-dimensional coordinate data of each point on the surface of the sample is obtained, and the three-dimensional topography of the material surface can be drawn.

The STM provides scientists with a tiny laboratory and can be used to study some novel phenomena and some new effects in the nano world, which cannot be achieved by other technologies.

In 1981, the scanning tunneling microscope invented by the physicists G. Binning and H. Rohrer of the Zurich Institute of IBM Corporation observed the clear atomic structure on the surface of Si (111), thus enabling human beings to enter the atomic world for the first time and directly observe single atoms on the surface of matter. In recognition of the outstanding contributions of G. Binning and H. Rohrer, they were awarded the Nobel Prize in Physics in 1986.

The STM can not only be used as an "eye" to observe the fine structure of the material surface, but also can be used as a "hand" to manipulate individual atoms. It can pick up an isolated atom with its probe tip and place the atom in another location. This is the first step for human beings to build ideal materials with "bricks" such as single atoms. Figure ER 2 is a computer photo of the "quantum fence" elaborated by IBM scientists. They planted 48 iron atoms on a refined copper surface one by one with the probe tip of a STM at a temperature of 4 K, forming a circle potential well, and the atoms moving on the copper surface were restricted by this circle potential well. The circular ripple in the iron circle in the figure is the wave picture of these electrons, and its size and pattern are in good agreement with the predictions of quantum mechanics. The lateral resolution of the STM has reached 0.1 nm, and the vertical resolution has reached 0.01 nm. The emergence of the STM enables us to observe the arrangement state of single atoms on the surface of matter and the properties related to surface electron behavior for the first time. In 1990, scientists such as Eiger used a STM in the laboratory of IBM Corporation in California and successfully wrote the world's smallest company name "IBM", 3 letters within the range of no more than one virus (about 100 nm) in length and width, as Figure ER 3 shows. It was the first realization of the direct arbitrary manipulation of atoms which was predicted by Feynman.

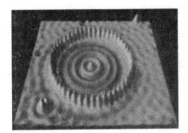

Figure ER 2 48 Fe atoms form a "quantum fence", and the electrons in the fence form a standing wave

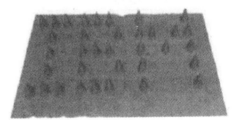

Figure ER 3 35 Selenium atoms arranged into three letters "IBM" on the clean nickel (110) surface using a STM

2. Overview of quantum communication technology

Quantum communication is a new communication method that uses quantum teleportation (transmission) to transmit information based on the effect of quantum entanglement shown in Figure ER 4. Quantum communication is a new type of interdisciplinary subject developed in the past two decades, and it is a new research field combining quantum theory and information theory.

The main principle of quantum communication is the application of quantum entanglement: Entanglement occurs if two similar quanta are close enough, and the states of the two quanta interact with each other no matter how far apart they are. Let's make an analogy with a pair of twins. When one is happy, the other will laugh. When one cries, the other must be sad. It's one of the strangest phenomena in science, and Einstein called this "induction" ghostly action at a distance.

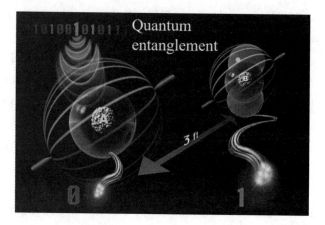

Figure ER 4 Quantum entanglement

The quantum teleportation in the process of quantum communication can be simply described in Figure ER 5.

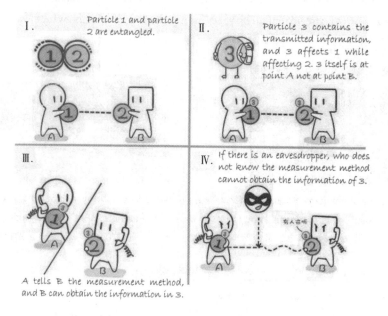

Figure ER 5 Quantum teleportation

The first step is to prepare an entangled particle pair, and particle 1 is sent to point A, and particle 2 is sent to point B; in the second step, at point A, another particle 3 carries a message to be transmitted, so particle 1 at point A and particle 2 at point B will form a total state with particle 3 together. At point A, measure particle 1 and particle 3 at the same time to obtain a measurement result. This measurement will collapse the entangled state of particle 1 and particle 2, but at the same time particle 1 and particle 3 are entangled together; in the third step, people at point A use the classical communication method to inform people at point B of the measurement result; in the fourth step, after receiving the measurement result, people at point B know the state of particle 2 at point B. Through a simple operation on particle 2, it becomes the state of particle 3 before the measurement. That is, the information carried by particle 3 is transmitted from point A to point B without loss, while particle 3 itself only stays at point A. Quantum communication is characterized by absolute security that traditional communication methods do not have. It has huge application value and prospects in important fields such as national security, finance and other information security. On August 16, 2016, China's self-developed world's first quantum science experimental satellite "Mozi" shown in Figure ER 6 was successfully launched. On September 12, 2017, China's first commercial quantum secure communication network was officially launched in Jinan.

Figure ER 6 "Mozi" quantum science experiment satellite

Discussion Problems

14.1 What is the difference between an absolute black body and an ordinary black object?

14.2 Use the photon model to explain why ultraviolet radiation is bad for your skin, but visible light is not.

14.3 In both the Compton scattering and the photoelectric effect, electrons gain energy from incident photons. What is the essential difference between the two processes?

14.4 What is the physical meaning of the wave function? How does it describe the state of motion of microscopic particles? How is it different from the classical wave function?

14.5 What does the Stern-Gerlach experiment show about the quantization of space? How do you prove electron spin?

Problems

14.6 The photoelectric effect occurs when two monochromatic light of frequencies ν_1 and ν_2 respectively are used to illuminate the same metal. The red limit frequency of the metal is ν_0. The relationship between the stopping voltage corresponding to two exposures is $U_{a1} = 3U_{a2}$. Try to calculate the relationship between ν_1 and ν_2.

14.7 The red limit wavelength of the photoelectric effect of lithium is 50 nm. Find

(1) the lithium electron escape work;

(2) the stopping voltage when ultraviolet light with a wavelength of 330 nm is incident.

14.8 Try to find the photon energy, momentum and mass of (1) red light ($\lambda = 700$ nm); (2) X-ray ($\lambda = 0.025$ nm); (3) γ-ray ($\lambda = 1.24 \times 10^{-3}$ nm).

14.9 The paraffin wax is illuminated with X-rays of wavelength $\lambda = 0.0708$ nm. Calculate the X-ray wavelength with the scattering angle in the $\frac{\pi}{2}$ direction.

14.10 The energy of an X-ray photon is 0.60 MeV, and the wavelength changes by 20% after the Compton scattering. Find the energy of the recoil electron.

14.11 In the Balmer series, from which energy level transition is visible light with a wavelength of 486.1 nm produced?

14.12 When the de Broglie wavelength of the electron is equal to the Compton wavelength, find

(1) the momentum of the electron;

(2) the ratio of the electron speed to the speed of light.

14.13 In a uniform magnetic field with magnetic induction of B, what is the de Broglie wavelength of a particle with mass m and electric quantity q moving in a circle of radius R?

14.14 In the fifth generation communication technology (5G), the electromagnetic wave frequency of the high frequency band used is more than 20 GHz. Try to calculate the wavelength and energy corresponding to the electromagnetic wave at 28 GHz.

14.15 Light with a wavelength of 300 nm travels along the X axis, and the uncertainty of its wavelength is $\Delta \lambda = 1$ nm. What is the uncertainty of its position?

14.16 Let the wave function of a one-dimensional moving particle be $\psi(x) = \begin{cases} A e^{-ax} & (x \geq 0) \\ 0 & (x < 0) \end{cases}$, where a is a constant greater than zero. Find the normalization constant A.

14.17 When the principal quantum number $n = 4$, calculate

(1) the energy value of the hydrogen atom;

(2) the value of the angular momentum that the electron may have;

(3) the angular momentum component L_z that the electron may have;

(4) the number of possible states of the electron.

Challenging Problems

14.18 The longitude distribution function of the Schrödinger equation for a hydrogen atom is

$$\Phi_m(\varphi) = \frac{1}{\sqrt{2\pi}} e^{im\varphi}$$

where, $1/\sqrt{2\pi}$ is the normalization constant. The latitude distribution function is

$$\Theta_{lm}(\theta) = N_{lm} P_l^{|m|}(\cos\theta)$$

where $P_l^m(x)$ is the associated Legendre polynomial and N_{lm} is the normalization constant. Find the angular probability density and the characteristics of the curve of the angular probability density and the three-dimensional surface.

14.19 According to the solution of the Schrödinger equation for hydrogen atoms,

(1) what are the characteristics of the electron cloud map that expresses the probability density with the density of points?

(2) what are the characteristics of probability density surfaces in hydrogen atoms? The projection of a surface represents a color electron cloud image. What are the characteristics of this electron cloud image?

(3) For each probability density, take 1% of the maximum probability density to form a surface. How does this three-dimensional equal probability density surface show the distribution of the probability density?